SOLID STATE NMR SPECTROSCOPY FOR BIOPOLYMERS

Solid State NMR Spectroscopy for Biopolymers

Principles and Applications

Hazime Saitô
Himeji Institute of Technology, Himeji, Japan
and Hiroshima University, Hiroshima, Japan

Isao Ando
Tokyo Institute of Technology, Tokyo, Japan

Akira Naito
Yokohama National University, Yokohama, Japan

Springer

A C.I.P. Catalogue record for this book is available from the Library of Congress.

ISBN-13 978-90-481-7100-2
ISBN-10 1-4020-4303-1 (e-book)
ISBN-13 978-1-4020-4303-1 (e-book)

Published by Springer,
P.O. Box 17, 3300 AA Dordrecht, The Netherlands.

www.springer.com

Printer on acid-free paper

Printed in the Netherlands.

TABLE OF CONTENTS

PREFACE

''Biopolymers'' are polymeric materials of biological origin, including globular, membrane, and fibrous proteins, polypeptides, nucleic acids, polysaccharides, lipids, etc. and their assembly, although preference to respective subjects may be different among readers who are more interested in their biological significance or industrial and/or medical applications. Nevertheless, characterizing or revealing their secondary structure and dynamics may be an equally very important and useful issue for both kinds of readers. Special interest in revealing the 3D structure of globular proteins, nucleic acids, and peptides was aroused in relation to the currently active Structural Biology. X-ray crystallography and multidimensional solution NMR spectroscopy have proved to be the standard and indispensable means for this purpose. There remain, however, several limitations to this end, if one intends to expand its scope further. This is because these approaches are not always straightforward to characterize fibrous or membrane proteins owing to extreme difficulty in crystallization in the former, and insufficient spectral resolution due to sparing solubility or increased effective molecular mass in the presence of surrounding lipid bilayers in the latter.

It is a natural consequence to expect solid state NMR as an alternative means for this purpose, because major developments in the past 30 years witnessed remarkable progress of this technique. The expected NMR line widths available from achieved high-resolution solid state NMR can be manipulated experimentally and are not any more influenced by motional fluctuation of proteins under consideration as a whole, in contrast to solution NMR. Accordingly, detailed structural information such as mutual orientation of molecules, interatomic distances, torsion angles, etc. is available from NMR parameters contained in solid state NMR, although most of them are lost due to time-averaging in solution NMR.

This book is intended to give a comprehensive account as to how polymorphic, secondary, and dynamic structures of a variety of the above-mentioned biopolymers are revealed by (high-resolution) solid state NMR.

In particular, a special emphasis is made toward the following two aspects: historical or chronological consequences of a variety of applications and dynamic aspect of the biopolymers. It is unfortunate that the latter aspect seems to be sacrificed in recent years in exchange for better structural information available from spectra taken either at lower temperature or hydration level far from biological significance. Indeed, revealing dynamics aspect of biopolymers is really an excellent and unrivaled means as revealed by solid state NMR study, as compared with other spectroscopic or diffraction methods. In this connection, it is emphasized that special attention to recording spectra by the most simple DD-MAS (one-pulse excitation with high-power decoupling) NMR is the unrivaled and only means to be able to record signals from such a flexible but structured portion of a variety of biopolymer systems, even if a number of complicated pulse techniques have been developed to obtain more detailed information. Readers may realize that very valuable structural information is still available from such a simple NMR technique. Indeed, we must bear in our mind *Simplex Sigillum Veri* (the truth is reflected in simplicity), as known since old times.

This book consists of the two parts: principles and applications. In the first part, a brief account of basic principles, NMR parameters, experimental strategies, and dynamics available from solid state NMR spectroscopy is made. In the second part, on the other hand, illustrative examples of these techniques to various subjects such as hydrogen-bonded systems, fibrous proteins, polysaccharides, polypeptides as new materials, globular and membrane proteins, membrane-associated peptides with biological functions, and amyloid and related biomolecules, are given. The organization of this book is thus like textile fabrics woven from the warps of biopolymers and the weft of solid state NMR.

The references in this book are of course by no means to intend to provide a complete compilation of related subjects. For this purpose, please consult references cited in a variety of review articles or books. Many of the data from our contributions are collaborative works with our colleagues and students at National Cancer Center Research Institute, Himeji Institute of Technology (currently University of Hyogo), Tokyo Institute of Technology, Yokohama National University, Gunma University, Tokyo University of Agriculture and Technology, especially with Professor A. Shoji, Professor T. Asakura, and Professor S. Tuzi, to whom the authors are deeply indebted. We are also grateful to Dr. S. Kuroki, Dr. K. Nishimura, Dr. S. Yamaguchi, Dr. M. Tanio, and Mr. I. Kawamura for their help in preparation of this manuscript.

August 2005

Hazime Saitô
Isao Ando
Akira Naito

ACKNOWLEDGEMENTS

We are indebted to the following authors and publishers for permission to reproduce figures.

A. Abragam: Fig. 2.11
T. Asakura: Fig. 9.6
A. Bax: Fig. 6.28
M. F. Brown: Fig. 14.6
E. E. Burnell: Fig. 4.8
T. A. Cross: Figs. 6.3, 6.4, 6.8, 14.2, 14.9
J. H. Davis: Fig. 4.16
T. Erata: Fig. 10.7
L. Emsley: Fig. 5.9
R. R. Ernst: Figs. 5.7
B. C. Gerstein: Fig. 2.13
A. M. Gil: Fig. 9.3
R. G. Griffin: Fig. 12.2
R. E. W. Hancock: Fig. 15.6
J. Herzfeld: Fig. 14.7
M. Hong: Fig. 5.1
L. W. Jelinski: Fig. 9.8
K. K. Kumashiro: Fig. 9.2
M. H. Levitt: Figs. 6.25, 6.26
J. R. Lyerla, Jr.: Fig. 3.4
G. E. Maciel: Fig. 3.7
R. H. Marchessault: Fig. 10.3
B. H. Meier: Figs. 5.8, 5.12, 6.21, 9.7, 12.1
E. Oldfield: Figs. 4.10, 4.13, 7.3
S. J. Opella: Figs. 4.4, 4.15, 6.1, 6.5, 6.6, 6.7, 14.1, 14.3, 14.4
H. Oschkinat: Fig. 5.11
M. Pruski: Fig. 2.15

H. A. Scheraga: Fig. 4.14
J. Seelig: Fig. 4.11
H. Shindo: Figs. 4.1, 4.2, 4.3
B. Schnabel: Fig. 2.14
I. C. P. Smith: Figs. 2.16, 4.5, 4.6, 4.7, 4.12
S. O. Smith: Fig. 14.8
H. W. Spiess: Fig. 9.9
D. A. Torchia: Figs. 7.4, 7.7, 7.8
R. Tycko: Fig. 16.1
D. L. VanderHart: Fig. 10.6
R. R. Vold: Fig. 6.11
A. Watts: Figs. 6.9, 6.10, 14.5, 14.10, 14.11, 14.12
C. S. Yannoni: Figs. 3.10
G. Zaccai: Fig. 7.2

American Association for the Advancement of Science
Science: Fig. 6.4
American Chemical Society
Anal. Chem.: Fig. 3.7
Biochemistry: Figs 4.1, 4.2, 4.3, 4.4, 4.5, 4.11, 4.12, 6:3, 7.4, 14.6, 14.9, 16.1
Chem. Rev.: Fig. 14.1
J. Amer. Chem. Soc.: Figs. 4.14, 5.9, 5.13, 5.14, 6.9, 6.10, 6.15, 6.16, 6.21, 6.23, 6.24, 6.25, 6.28, 8.3, 8.4, 9.6, 9.8, 9.10, 12.1, 12.2, 14.7
J. Phys. Chem.: Figs. 2.12, 6.15, 6.16, 6.22, 6.23
Macromolecules: Figs. 3.1, 6.27, 9.4, 9.5, 9.7, 9.10, 9.11, 9.12, 10.1, 10.6, 10.7, 11.2, 11.3, 11.4, 11.5, 11.6, 11.7
American Society for Biochemistry and Molecular Biology
J. Biol. Chem.: Figs. 4.10, 4.13, 7.3, 7.7
Biophysical Society
Biophys. J.: Figs. 2.7, 5.4, 5.5, 6.11, 6.12, 6.13, 9.2, 15.1, 15.3, 15.5, 15.7
Chemical Society of Japan
Bull. Chem. Soc. Jpn., Figs. 3.5, 10.4
Cold Springer Harbor Laboratory Press
Protein Sci. Fig. 14.2
Elsevier
Annu. Rep. NMR Spectrosc.: Figs. 10.2, 10.5, 13.1, 13.2
Biochim. Biophys. Acta: Figs. 3.9, 4.8, 4.16, 13.5, 13.9
Carbohydr. Res.: Fig. 10.3
Chem. Phys. Lett.: Figs. 4.21, 5.7, 6.21, 8.5, 8.8, 8.9
Chem. Phys. Lipids: Fig. 13.8
FEBS Lett.: Figs. 13.7, 14.12
J. Magn. Reson.: Figs. 6.5, 6.7, 6.8

J. Mol. Biol.: Fig. 14.4
J. Mol. Struct.: Figs. 3.2, 3.3, 3.8, 4.17, 4.18, 4.19, 4.20, 6.19, 7.5, 7.6, 8.1, 8.2, 8.6, 8.7, 8.10, 11.1
Methods in Enzymol.: Figs. 4.7, 4.15, 6.1
Peptides: Fig. 15.6
Polymer: Figs. 11.9, 11.10
Physica: Fig. 2.14
Industrial Polysaccharides: Genetic Engineering, Structure/Property Relations and Applications: Fig. 10.8
Phosphorous-31 NMR. Principles and Applications: Figs. 4.5, 4.6, 4.7
Transient Techniques in NMR of Solid: Fig. 2.13
IBM
IBM J. Res. Develop.: Fig. 3.10
IOS Press
Spectroscopy: Fig. 13.6
Japanese Biochemical Society
J. Biochem. (Tokyo): Fig. 2.8, 5.2, 5.3, 7.9, 9.1, 13.3,
Marcel Dekker
Polysaccharides: *Structural Diversity and Functional Versatility:* Fig. 10.12
National Academy of Sciences
Proc. Natl. Acad. Sci. USA: Figs. 6.6, 7.2, 7.8, 14.3, 14.8, 14.11
Nature Publishing Group
Nature: Fig. 14.5
Oxford University Press
Principles of Nuclear Magnetism: Fig. 2.11
Royal Society of Chemistry
Magnetic Resonance in Food Scienc: Fig. 10.11
Springer
Eur. Biophys. J.: Fig. 13.4, 13.6, 14.10
J. Biomol. NMR: Figs. 5.1, 5.8, 5.12
The Multinuclear Approach to NMR Spectroscopy: Fig. 2.15
Wiley
Biopolymers: Figs. 9.3, 10.10, 11.8,
Chembiochem.: Fig. 5.11
Encyclopedia of Nucl. Magn. Reson.: Figs. 2.14, 5.15, 6.18
Magn. Reson. Chem.: Figs. 5.6, 16.3
High Resolution NMR of Synthetic Polymes in Bulk: Fig. 3.4

Chapter 1

INTRODUCTION

Over five decades since its discovery, high-resolution NMR spectroscopy has been widely accepted as an indispensable tool to analyze the structure and dynamics of isolated molecules in pure and applied sciences and a variety of complex molecular systems from rocks to biological tissues.[1] Solution NMR approach has been extensively explored to this end, because complicated spectral feature of solid state NMR[2–7] arising from inherent dipolar interactions with neighboring nuclei and chemical shift anisotropy (CSA) is substantially simplified in solution NMR due to time-averaging caused by rapid motional fluctuations. Alternatively, high-resolution solid state NMR can be achieved if such interactions were successfully removed by specific manipulations of spin system, although a wealth of very important structural data such as interatomic distance, orientation to the applied magnetic field, etc. are naturally lost by this procedure. The most widely used cross polarization-magic angle spinning (CP-MAS) is a means to utilize magnetization transfer from abundant spin such as ^{1}H to rare spin of ½ nuclei such as ^{13}C or ^{15}N. It is, therefore, difficult to record NMR signals from portions undergoing fluctuation motions with frequency $>10^8$ Hz in the solid sample by CP-MAS NMR. Indeed, NMR signals of N- or C-terminal residues of membrane proteins protruding from the membrane surface are inevitably suppressed by time-averaged dipolar interactions and are accordingly very often ignored by solid state NMR spectroscopists. In such a case, single pulse excitation and dipolar decoupled-magic angle spinning (DD-MAS) NMR[8,11] is a suitable means to record such NMR signals.

High-resolution solid state NMR spectroscopy based on CP- and DD-MAS techniques has been recognized as a very important tool for the characterization of various types of polymers[7–12] because *in situ* characterization is especially important for naturally occurring or synthetic organic and inorganic polymers used as materials and also for polymers having sparing solubility in ordinary solvents, as in the case of network polymers. For this purpose, ^{13}C and ^{15}N nuclei are naturally utilized as the most suitable structural probes for carbon and nitrogen backbones, owing to the enhanced spectral resolution because of wide chemical shift range over 200–1000 ppm as compared with 20 ppm of ^{1}H nuclei. Several other nuclei such as ^{31}P, ^{19}F, ^{17}O and ^{2}H, etc. are also used depending upon the type of problems.

In general, ^{13}C NMR spectra of biopolymers consisting of limited numbers of repeating units such as structural proteins (collagen, elastin, silk fibroin, etc.) and polysaccharides (cellulose, amylose, etc.) can be recorded without any specific isotope labeling (or enrichment), because individual peaks from samples of natural abundance are well resolved and rather intense, owing to their limited numbers of variety of residues. This approach is especially useful for biopolymers present usually in a crystalline state, taking one of the polymorphic structures depending upon history of sample treatments,[13] in contrast to the cases of most amorphous synthetic polymers. Accordingly, conformational characterization of respective polymorph is simply feasible by examination of their ^{13}C chemical shifts with reference to the database of the conformation-dependent displacement of ^{13}C NMR peaks,[14–16] as an indispensable constraint for structural determination as well as history of sample treatments.

It should be anticipated that biopolymers consisting of nonrepeating units, such as biologically active peptides, globular proteins, membrane proteins, and polynucleotides, yield heavily crowded ^{13}C NMR signals arising from superposed peaks of constituting residues. To avoid such complexity, it is strongly advised to prepare selectively ^{13}C- or ^{15}N-labeled samples by feeding specific ^{13}C-labeled amino acid residues either by biosynthesis from cell culture or incorporation of selectively labeled amino acids for peptide synthesis. This is also important to enhance selectivity of certain signals from particular sites. Such resolved ^{13}C NMR signals could be then assigned to carbons of respective residues by site-directed manner, based upon comparison of peaks between wild-type and site-directed mutants in which the amino acid residue of interest is replaced by others, deleted peaks by enzymatic cleavage, etc.[15–17] Subsequently, local conformational analysis of particular sites is feasible based on the above-mentioned conformation-dependent displacements of ^{13}C chemical shifts.

It is cautioned that ^{13}C NMR signals of such biopolymers are not always fully visible at ambient temperature; ^{13}C NMR signals are in many instances suppressed from certain side-chains of amino acid residues even in crystalline peptides[18] or fibrous proteins,[19] and also from backbone carbons of fully hydrated membrane proteins.[17] This happens very frequently, when attempted peak-narrowing process by both CP-MAS and DD-MAS NMR fails due to interference of incoherent fluctuation frequency of internal motions with coherent frequencies of either proton decoupling (10^5 Hz) or MAS (10^4 Hz).[20,21] Indeed, this situation occurs very frequently for fully hydrated membrane proteins in lipid bilayers present as a monomer rather than a two-dimensional (2D) crystal, because they are not always rigid as conceived at ambient temperature in spite of a picture as obtained from crystalline preparation at low temperature, but undergo fluctuation motions with a variety of motional frequencies from 10^2 to 10^8 Hz. Especially, low-frequency local motions (10^4–10^5 Hz) are present for fully hydrated membrane proteins, especially when their 2D crystalline lattices were distorted or disorganized due to modified lipid–protein interactions, as in a mutant protein of bacteriorhodopsin (bR), or reconstituted proteins into lipid bilayers.[22,23] Such portion can be readily located when the suppressed ^{13}C NMR signals could be assigned to certain amino acid residues with reference to those of fully visible [3-^{13}C]Ala- or [1-^{13}C]Val-bR from 2D crystal. ^{13}C NMR signals of uniformly or densely ^{13}C-labeled membrane proteins could be outrageously broadened by shortened spin–spin relaxation times coupled with the low-frequency motions.[24] This comes from an increased number of relaxation pathways through a number of ^{13}C–^{13}C homonuclear dipolar interactions and scalar *J*-couplings in the CP-MAS experiments, etc.[24,25] Nevertheless, it should be emphasized that revealing dynamic picture by analyzing such low-frequency motions by solid state NMR is a very unique means, and cannot be obtained by other diffraction or spectroscopic techniques.

Recently, considerable progress has been achieved in multidimensional solid state NMR toward sequential assignment of ^{13}C signals from fully ^{13}C, ^{15}N-labeled peptides and proteins, utilizing intra-residue and inter-residue ^{13}C–^{15}N correlation with proton-driven ^{13}C–^{13}C spin-diffusion studies[26], ^{13}C–^{13}C dipolar recoupling, 2D NCACX (CX = any carbon atom), NCOCX experiments,[27] and total through-bond correlation spectroscopy (TOBSY) using the *J*-coupling as a transfer mechanism.[28] Secondary structures of peptides and small proteins[29–31] were also constructed by means of constraints such as distances[32–34] and torsion (or dihedral) angles,[35] as well as the conformation-dependent displacement of ^{13}C peak mentioned above. Nevertheless, the presence of strong *intermolecular* interactions in the crystalline lattice may result in additional constraints, which are not present in solution NMR. In fact, it turned out that molecular dynamics simulation, taking into account all the molecules in a unit cell, are consistent with the experimental data for small peptides in an environment of strong intermolecular interactions.[36] This is because numbers of experimentally available constraints are limited, as compared with constraints due to intermolecular interactions. Currently, solid state NMR approach is still under development to be used as a standard methodology to reveal three-dimensional (3D) structure of a variety of biopolymers, especially for intact membrane proteins.

The above-mentioned brief account of high-resolution solid state NMR approach has proved that this technique is very useful and promising for revealing conformation and dynamics of biopolymers, although its capacity to reveal 3D structures of globular and membrane proteins is still premature. However, their 3D structures, if any, could be very easily obscured when high- and low-frequency motional fluctuations are persistent at ambient temperature of physiological conditions, as encountered for bR as a typical membrane protein. Under such circumstances, dynamic aspect of membrane proteins rather than 3D structure is readily available from careful evaluation of site-directed solid state NMR techniques. In this connection, readers should realize that membrane proteins are not anymore rigid solids in spite of the picture revealed by X-ray diffraction, but are soft matter as represented by fully hydrated pellet samples.

This book is intended as a practical guide, rather than a compilation of new developments, for readers who are interested in unraveling the conformation and dynamics of a variety of biopolymers by solid state NMR. The authors advise readers to pay more attention to the potential application of DD-MAS NMR, as a complementary means to X-ray diffraction which is unable to obtain data from very flexible portions of a variety of biopolymer systems such as fully hydrated membrane-associated peptides, surface of membrane proteins, polysaccharide gels, etc. This is because no signals could be detected by CP-MAS NMR from such a region at ambient temperature of biological significance. We emphasize to pay special attention on the unrivaled feasibility of solid state NMR, as compared with crystallographic and other spectroscopic techniques, to reveal dynamic aspect of biopolymer systems. Such knowledge about dynamics revealed by solid state NMR is very important, as a complementary means to 3D structure revealed by diffraction studies so far examined.

References

[1] D. M. Grant and R. K. Harris, Eds., 1996, *Encyclopedia of Nuclear Magnetic Resonance*, John Wiley, Chichester.

[2] M. Mehring, 1983, *Principles of High Resolution NMR in Solids*, Second Edition, Springer-Verlag, Berlin.

[3] C. P. Slichter, *Principles of Magnetic Resonance*, Harper & Row, New York, 1963; Third Enlarged and Updated Edition, Springer-Verlag, 1989.

[4] C. A. Fyfe, 1983, *Solid State NMR for Chemists*, CFC Press, Guelph.

[5] B. C. Gerstein and C. R. Dybowski, 1985, *Transient Techniques in NMR of Solids: An Introduction to Theory and Practice*, Academic Press, New York.

[6] U. Haeberlen, 1976, *High Resolution NMR in Solids: Selective Averaging*, Academic Press, New York.

[7] K. Schmidt-Rohr and H. W. Spiess, 1994, *Multidimensional Solid-state NMR and Polymers*, Academic Press, New York.
[8] R. A. Komoroski (Ed.), 1986, *High Resolution NMR Spectroscopy of Synthetic Polymer in Bulk*, VCH Publishers, Deerfield Beach.
[9] V. J. McBriety and K. J. Packer, 1993, *Nuclear Magnetic Resonance in Solid Polymers*, Cambridge University Press, Cambridge.
[10] G. A. Webb and I. Ando, Eds., 1997, *Annu. Rep. NMR Spectrosc.*, vol. 34 (Special Issue: NMR in Polymer Science).
[11] I. Ando and T. Asakura, Eds., 1998, *Solid State NMR of Polymers*, Elsevier, Amsterdam.
[12] H. Saitô and I. Ando, 1989, *Annu. Rep. NMR Spectrosc.*, 21, 209–290.
[13] H. Saitô, S. Tuzi, and A. Naito, 1998, in *Solid State NMR for Polymers*, I. Ando and T. Asakura, Eds., Elsevier, Amsterdam, pp. 891–921.
[14] H. Saitô, 1986, *Magn. Reson. Chem.*, 24, 835–852.
[15] H. Saitô, S. Tuzi, and A. Naito, 1998, *Annu. Rep. NMR Spectrosc.*, 36, 79–121.
[16] H. Saitô, S. Tuzi, M. Tanio, and A. Naito, 2002, *Annu. Rep. NMR Spectrosc.*, 47, 39–108.
[17] H. Saitô, S. Tuzi, S. Yamaguchi, M. Tanio, and A. Naito, 2000, *Biochim. Biophys. Acta*, 1460, 39–48.
[18] M. Kamihira, A. Naito, K. Nishimura, S. Tuzi, and H. Saitô, 1998, *J. Phys. Chem. B*, 102, 2826–2834.
[19] H. Saitô, R. Tabeta, A. Shoji, T. Ozaki, I. Ando, and T. Miyata, 1984, *Biopolymers*, 23, 2279–2297.
[20] D. W. Suwelack, W. P. Rothwell, and J. S. Waugh, 1980, *J. Chem. Phys.*, 73, 2559–2569.
[21] W. P. Rothwell and J. S. Waugh, 1981, *J. Chem. Phys.*, 75, 2721–2732.
[22] H. Saitô, T. Tsuchida, K. Ogawa, T. Arakawa, S. Yamaguchi, and S. Tuzi, 2002, *Biochim. Biophys. Acta*, 1565, 97–106.
[23] H. Saitô, K.Yamamoto S. Tuzi, and S. Yamaguchi, 2003, *Biochim. Biophys. Acta*, 1616, 127–136.
[24] S. Yamaguchi, S. Tuzi, K. Yonebayashi, A. Naito, R. Needleman, J. K. Lanyi, and H. Saitô, 2001, *J. Biochem. (Tokyo)*, 129, 373–382.
[25] A. Naito, A. Fukutani, M. Uitdehaag, S. Tuzi, and H. Saitô, 1998, *J. Mol. Struct.*, 441, 231–241.
[26] S. K. Straus, T. Bremi, and R. R. Ernst, 1998, *J. Biomol. NMR*, 12, 39–50.
[27] J. Pauli, M. Baldus, B. van Rossum, H. de Groot, and H. Oschkinat, 2001, *Chembiochem*, 2, 272–281.
[28] A. Detken, E. H. Hardy, M. Ernst, M. Kainosho, T. Kawakami, S. Aimoto, and B. H. Meier, 2001, *J. Biomol. NMR*, 20, 203–221.
[29] A. Naito, K. Nishimura, S. Kimura, S. Tuzi, M. Aida, N. Yasuoka, and H. Saitô, 1996, *J. Phys. Chem.*, 100, 14995–15004.
[30] C. M. Rienstra, L. Tucker-Kellogg, C. P. Jaroniec, M. Hohwy, B. Reif, M. T. McMahon, B. Tidor, T. Lozano-Perez, and R. G. Griffin, 2002, *Proc. Natl. Acad. Sci. USA*, 99, 10260–10265.
[31] F. Castellani, B. van Rossum, A. Diehl, M. Schubert, K. Rehbein, and H. Oschkinat, 2002, *Nature*, 420, 98–102.
[32] A. Naito and H. Saitô, 2002, *Encyclopedia of Nuclear Magnetic Resonance*, vol. 9, pp. 283–291.
[33] V. Ladizhansky and R. G. Griffin, 2004, *J. Am. Chem. Soc.*, 126, 948–958.

[34] R. Ramachandran, V. Ladizhansky, V. S. Bajaj, and R. G. Griffin, 2003, *J. Am. Chem. Soc.*, 125, 15623–15629.
[35] C. M. Rienstra, M. Hohwy, L. J. Mueller, C. P. Jaroniec, B. Reif, and R. G. Griffin, 2002, *J. Am. Chem. Soc.*, 124, 11908–11922.
[36] M. Aida, A. Naito, and H. Saitô, 1996, *J. Mol. Struct. (Theochem.)*, 388, 187–200.

Chapter 2

SOLID STATE NMR APPROACH

2.1. CP-MAS and DD-MAS NMR

For spin-1/2 nuclei such as ^{13}C or ^{15}N, solid state NMR spectra of biopolymers, including fibrous, globular, and membrane proteins, biomembranes, polysaccharides, etc. yield enormously broadened signals with line widths in the order of 20 kHz. Such broadened signals arise from nuclear interactions such as dipolar interactions and chemical shift anisotropy (CSA),[1–3] which are in the order of 10^2–10^4 Hz and 10^3–10^4 Hz, respectively. Such broadened NMR signals, therefore, are usually not easy to interpret in terms of chemical or biological significance, unless otherwise an attempt is made to narrow line widths to give rise to high-resolution solid state NMR signals, as encountered in solution NMR. Nevertheless, a wealth of structural data in relation to interatomic distances and nuclear orientation to the applied magnetic field is contained in such broadened NMR signals in the crystalline or noncrystalline state, and should be carefully analyzed in a site-specific manner using site-specific isotope labeling.

2.1.1. Dipolar Decoupling

From a classical viewpoint, the component of local magnetic field at spin I, caused by magnetic moment $\boldsymbol{\mu}$ from neighboring nucleus S (in many cases, proton), parallel to the applied static magnetic field $\boldsymbol{B}_0$ is

$$\boldsymbol{B}_L = \pm \boldsymbol{\mu} r^{-3}\left(3\cos^2\theta - 1\right), \tag{2.1}$$

where r is the distance between two nuclei and θ is the angle between $\boldsymbol{B}_0$ and $\boldsymbol{r}$ (Figure 2.1), leading to splitting of peak to doublet. This equation can be rewritten using quantum mechanical formalism as

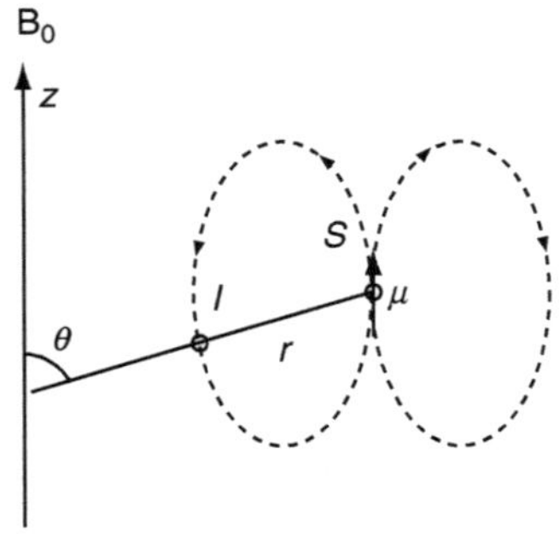

Figure 2.1. Dipolar field at the spin I caused by the neighboring spin S.

$$\mathcal{H}_D = -\gamma_I \gamma_S \hbar^2 r^{-3}(3\cos^2\theta - 1)\boldsymbol{I}_z \boldsymbol{S}_z, \tag{2.2}$$

where $\boldsymbol{I}$ and $\boldsymbol{S}$ are nuclear spin numbers for nuclei I and S and γ_I and γ_S are gyromagnetic ratios for I and S spin, respectively, and $\hbar$ is Planck's constant (h) divided by 2π. NMR signals of rare spins of spin number 1/2 such as ^{13}C or ^{15}N nuclei are mainly broadened by dipolar interactions with neighboring proton nuclei, because homonuclear ^{13}C–^{13}C or ^{15}N–^{15}N dipolar interactions are very weak under the condition of natural abundance, and hence negligible in most cases. Therefore, such dipolar couplings can be effectively removed by irradiation of strong radio-frequency (rf) at proton resonance frequency (dipolar decoupling) to result in time-averaged magnetic dipolar field to zero ($\langle\mu\rangle = 0$), as in a similar manner to proton decoupling in solution NMR to remove scalar C–H spin couplings which range from 0 to 200 Hz. In solids, high-power, dipolar decoupling is required to use decoupling fields of 40 kHz or more to reduce the stronger dipolar couplings of 20 kHz.

The peak narrowing by the dipolar decoupling, however, is not always sufficient to achieve high-resolution signals as found in solution NMR, because a unique spectral pattern arising from CSA that spreads 10^3–10^4 Hz (for magnetic field of 5–10 T) is persistent. In general, the resonance frequency of a particular nucleus in a molecule depends on the electronic environment produced by surrounding electrons that shield the nucleus from the external magnetic field. This means that the nucleus experiences a different shielding, and hence has a different chemical shift, depending upon the direction of the applied magnetic field to the molecule, as demonstrated for the carbonyl ^{13}C chemical shifts of crystalline benzophenone[4] as an example (Figure 2.2). The chemical shielding constant in the solid state, σ, is not anymore a scalar quantity like in solution NMR, but should be expressed as a second-rank tensor in the chemical shift interaction:

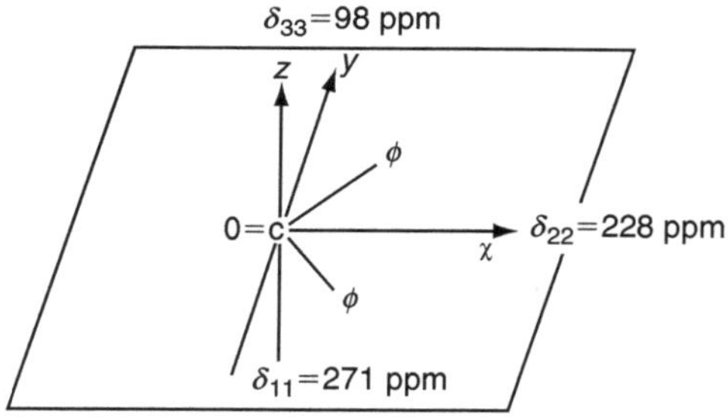

Figure 2.2. Carbonyl ^{13}C chemical shift anisotropy (CSA) of benzophenone.[4]

$$\mathcal{H}_{\mathrm{CSA}} = \gamma\hbar \boldsymbol{B}_0 \boldsymbol{\sigma} \boldsymbol{I}$$
$$= \gamma\hbar \left(B_{0x}, B_{0y}, B_{0z}\right) \cdot \begin{pmatrix} \sigma_{xx} & \sigma_{xy} & \sigma_{xz} \\ \sigma_{yx} & \sigma_{yy} & \sigma_{yz} \\ \sigma_{zx} & \sigma_{zy} & \sigma_{zz} \end{pmatrix} \cdot \begin{pmatrix} I_x \\ I_y \\ I_z \end{pmatrix} \tag{2.3}$$

Diagonalization of the chemical shielding tensor in Eq. (2.3) to the principal axis system (PAS) leads to the chemical shielding tensor with three principal values,

$$\boldsymbol{\sigma} = \begin{pmatrix} \sigma_{11} & 0 & 0 \\ 0 & \sigma_{22} & 0 \\ 0 & 0 & \sigma_{33} \end{pmatrix}. \tag{2.4}$$

The experimentally available chemical shift tensor $\boldsymbol{\delta}$ is defined by difference of the shielding constant σ_{ii} with respect to that of reference compound σ_{R} such as TMS,

$$\boldsymbol{\delta} = \boldsymbol{\sigma}_{\mathbf{R}} \mathbf{1} - \boldsymbol{\sigma}, \tag{2.5}$$

where **1** is a unit tensor. The chemical shift δ_{zz} in a particular orientation of the molecule in the external magnetic field (B_0 parallel to z axis, $B_0 \| z$ axis) is thus expressed as follows:

$$\delta_{zz} = \lambda_1^2 \delta_{11} + \lambda_2^2 \delta_{22} + \lambda_3^2 \delta_{33} \tag{2.6}$$

$$\begin{aligned} \lambda_1 &= \cos\alpha \sin\beta \\ \lambda_2 &= \sin\alpha \sin\beta \\ \lambda_3 &= \cos\beta, \end{aligned} \tag{2.7}$$

where α and β are Euler angles which relate PAS (X, Y, and Z) in a molecule to laboratory frames (x, y, and z in which z is taken along the direction of applied magnetic field B_0) (Figure 2.3(a)). Principal values of the chemical shift tensor and the angles to relate the relative orientation to the molecular coordinates can be obtained by examination of ^{13}C NMR of single crystals or oriented samples. Microcrystalline or powder samples yield ^{13}C or ^{15}N signals arising from sum of all possible contributions of the Euler angles (powder pattern) described later, for either axially symmetric (left) ($\delta_{11} = \delta_{22} \neq \delta_{33}$) or axially asymmetric (right) ($\delta_{11} \neq \delta_{22} \neq \delta_{33}$), as illustrated in Figure 2.4. The principal values are very important parameters to characterize solid state NMR spectral feature, electronic state, molecular orientation, and dynamics, and can be obtained directly from the powder anisotropy pattern. Namely, δ_{11}, δ_{22}, and δ_{33} can be obtained from the peak at δ_{zz} of the powder pattern arising from the divergence of the lineshape function and two edges at δ_{11} and δ_{33} of the shoulders, respectively. In many systems of interest for the chemist, overlapping anisotropies thus obtained by dipolar decoupling are still not interpretable. Such CSA dispersions achieved by dipolar decoupling during the course of cross polarization (CP)[5,6] are averaged to give their isotropic values,[7] $\delta_{iso} = (\delta_{11} + \delta_{22} + \delta_{33})/3$, [7–9] by means of MAS at 2–10 kHz.

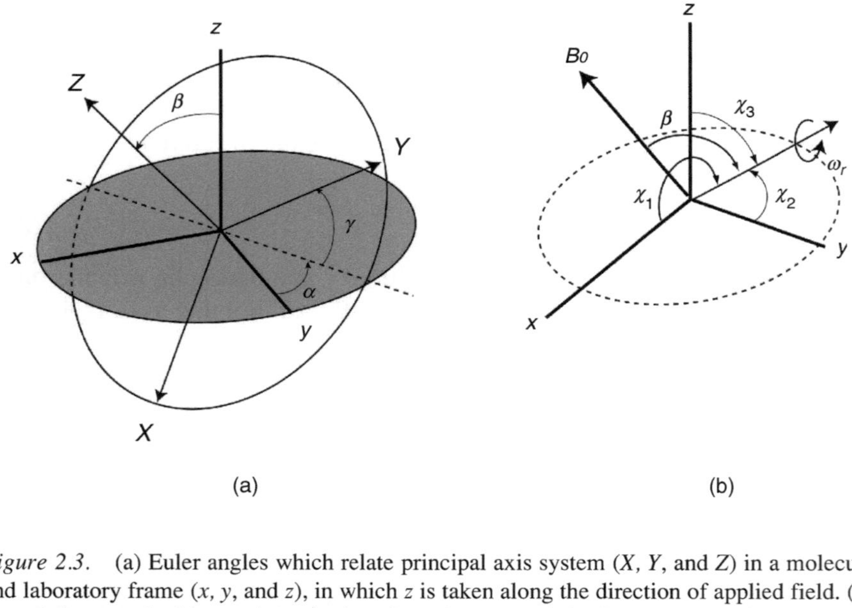

Figure 2.3. (a) Euler angles which relate principal axis system (*X*, *Y*, and *Z*) in a molecule and laboratory frame (*x*, *y*, and *z*), in which *z* is taken along the direction of applied field. (b) Sample is rotated with angular velocity of ω_r about an axis inclined at an angle of β to the applied field B_0 at angles χ_1, χ_2, and χ_3 to the principal axes of σ. Magic angle for $\beta = 54°44'$.

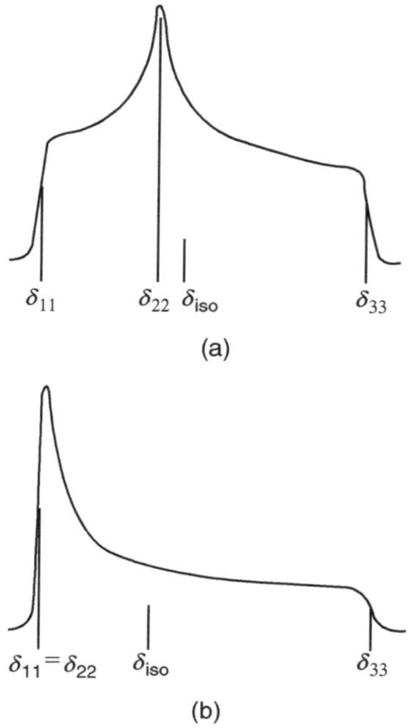

Figure 2.4. Powder pattern spectra for spin-1/2 nuclei: (a) axially asymmetric ($\delta_{11} \neq \delta_{22} \neq \delta_{33}$), (b) axially symmetric ($\delta_{11} = \delta_{22} \neq \delta_{33}$).$\delta_{iso} = (1/3)(\delta_{11} + \delta_{22} + \delta_{33})$.

2.1.2. Magic Angle Spinning

Taking into account of the isotropic averages of each λ_i^2 being reduced to 1/3, Eq. (2.5) by isotropic tumbling motions in liquid state is:

$$\delta_{zz} = (1/3)\,\mathrm{Tr}\,(\boldsymbol{\delta}) = (1/3)(\delta_{11} + \delta_{22} + \delta_{33}). \tag{2.8}$$

When the rigid array of nuclei in a solid is rotated with angular velocity of ω_r about an axis inclined at an angle of β (Figure 2.3(b)) to the applied field B_0, at angles χ_1, χ_2, χ_3 to the principal axes of δ, we have

$$\lambda_p = \cos\beta\cos\chi_p + \sin\beta\sin\chi_p\cos(\omega_r t + \varphi_P), \tag{2.9}$$

where φ_P is the azimuth angle of the pth principal axis of $\boldsymbol{\delta}$ at $t = 0$. Substituting Eq. (2.9) to Eq. (2.6) and taking the time-average we find

$$\delta_{zz} = (1/2)\sin^2\beta\,\mathrm{Tr}(\boldsymbol{\delta}) + (1/2)\left(3cos^2\beta - 1\right)\Sigma\delta_{pp}\cos^2\chi_p. \tag{2.10}$$

The second term vanishes when β is the magic angle $\sec^{-1}\sqrt{3}$, namely $54°44'$, the time-averaged value is reduced to

$$\delta_{zz} = (1/3)\,\mathrm{Tr}(\boldsymbol{\delta}), \tag{2.11}$$

which is the same as for a liquid. In this treatment, spinning rate ω_r should be satisfied as

$$\omega_r \gg |\delta_{11} - \delta_{33}|, \tag{2.12}$$

where $|\delta_{11} - \delta_{33}|$ is the width of the dispersion of the CSA up to approximately 150 ppm for ^{13}C and ^{15}N nuclei.[10,11] This means that rotation at 3–10 kHz is sufficient for a variety of aliphatic carbons of smaller CSA, but several spinning sidebands remain for quaternary, carbonyl, or aromatic carbons with larger CSA. Of course, this requirement is more stringent for high-frequency NMR spectrometer utilizing a magnetic field of 20 T. In such cases, it is possible to remove such side bands by a pulse sequence total suppression of spinning sidebands[12] (TOSS).

There remains a problem of prolonged spin–lattice relaxation times, T_1 s, for rare spins such as ^{13}C or ^{15}N nuclei in rigid solids in the order of tens of seconds or minutes. This is really troublesome for recording spectra with such longer T_1 in order to improve signal-to-noise (S/N) ratio by time-averaging, because repetition time for signal acquisition should be ideally taken longer than at least $5T_1$ for 90° pulse or shorter when the optimum pulse rotation angle decreases,[13] to achieve the maximum S/N ratio. Instead, enhanced NMR signals from the rare spin can be achieved through a technique called CP rather than the direct observation, because repetition times for spectral accumulation for such case can be taken based on shorter spin–lattice relaxation times of abundant spin. This is because spin–lattice relaxation times of abundant spin such as ^{1}H are naturally very short (ms) because of large amounts of spins present.

As demonstrated already, MAS plays an essential role to achieve high-resolution NMR signals in solid NMR comparable to those in solution NMR, once dipolar broadenings are removed by the dipolar decoupling. CP technique described below is naturally a very useful means to record signals from rigid portion of samples, although DD-MAS (single pulse or Bloch decay technique) is also a very effective means to detect ^{13}C NMR signals from rather flexible areas of fully hydrated biopolymers such as membrane proteins and swollen polypeptides.

2.1.3. Cross Polarization

This technique relies on polarization transfer from the abundant I spin such as ^{1}H to the rare spin S such as ^{13}C through matched H_1 fields for both the ^{1}H and ^{13}C, known as Hartmann–Hahn condition:

$$\gamma_I B_{1I} = \omega_I = \omega_S = \gamma_S B_{1S}, \tag{2.13}$$

where γ_I and γ_S are the gyromagnetic ratios, and ω_I and ω_S are radio frequencies, B_{1I} and B_{1S} are the frequency strength for abundant I and rare S nuclei, respectively. This procedure starts when a 90° rf pulse B_{1I} in the x-direction brings the magnetization along the y-direction. This magnetization $\boldsymbol{M_I}$ is then ''spin-locked'' along with B_{1I} when the phase of the rf field is shifted by 90° from the x to the y directions, as far as it is greater than any local dipolar fields and there is no static magnetic field in the z-direction as viewed from the rotating frame,[14] which is rotating at Larmor frequency $\omega_0 = \gamma_I B_0$, as illustrated in Figure 2.5 and Figures 2.6(a) and(b). The μ_{zI}, the z-component of $\boldsymbol{M_I}$, oscillates as

$$\mu_{zI} \propto \cos \omega_I t \tag{2.14}$$

and its time-averaged value

$$\langle \mu_{zI} \rangle = 0 \tag{2.15}$$

consistent with the explanation already described. In the presence of B_{1I}, the rare spins S (^{13}C or ^{15}N) are naturally dipolar decoupled and their magnetization $\boldsymbol{M_S}$ is exactly aligned with their applied rf field B_{1S} in the y-axis (Figure 2.6(d)), while individual S spins precess about the rf field B_{1S} in the y-axis, as viewed from the Cartesian coordinates in the rotating frame, instead of the z-axis in the laboratory frame. The μ_{zS}, the z-component of $\boldsymbol{M_S}$, also oscillates as

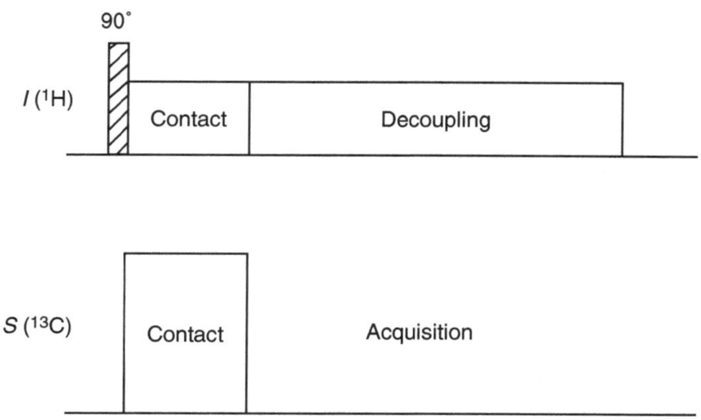

Figure 2.5. Pulse sequence for cross polarization. After 90°rf pulse, B_{1I} pulse is applied in the x direction, the I magnetization at the y-axis is spin-locked along with B_{1I} when the phase of the rf field is shifted by 90° from the x to the y axis. The I spins transfer polarization to S spins during the contact time. The S signal is then acquired.

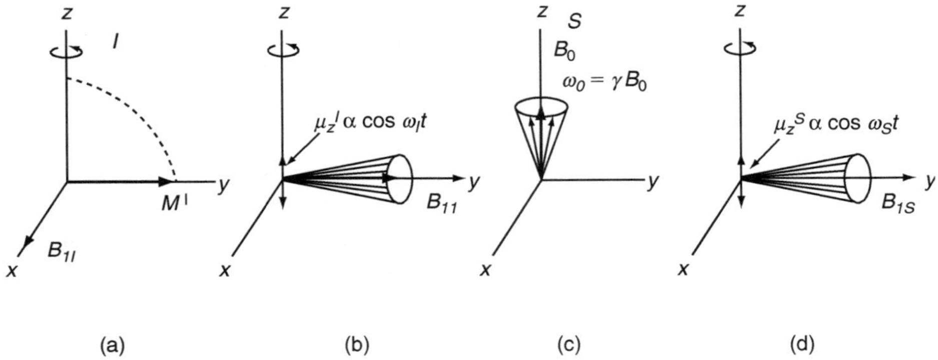

Figure 2.6. Spin-lock in the rotating frame and cross polarization.

$$\mu_{zS} \propto \cos \omega_S t \tag{2.16}$$

The z-axis components of the I and S spins, μ_{zI} and μ_{zS}, have the same time-dependence in Eqs. (2.14) and (2.16), when the Hartmann–Hahn condition is satisfied as demonstrated in Eq.(2.13). Therefore, in a doubly rotating coordinate system, efficient CP transfer occurs from the abundant spin I to rare spin S through cross relaxation by dipolar coupling between I and S spins, in view of the same time dependence ($\omega_I = \omega_S$) in μ_{zI} and μ_{zS} in Eqs. (2.14) and (2.16). In the laboratory reference frame, this interaction can be expressed as[15,16]

$$\mathcal{H}_{IS} = \gamma_I \gamma_S \hbar^2 \Sigma\, \Sigma r_{im}^{-3} (3\cos^2\theta_{im} - 1) I_{iZ} I_{mZ} \tag{2.17}$$

where $r_{\rm im}$ is the distance between $i(I)$ and $m(S)$ spins and θ_{im} is the angle between the vector $\boldsymbol{r_{im}}$ and $\boldsymbol{B_0}$. In the doubly rotating reference frame, the interaction B_{IS} transforms approximately to

$$\begin{aligned} \mathcal{H}_P &\sim \sin\theta\, \Sigma r_{im}^{-3}\, (3\cos^2\theta_{im} - 1)\, I_{ix} S_{mx} \\ \theta &= \tan^{-1}(B_{1I}/B_L). \end{aligned} \tag{2.18}$$

The spin operator part of this interaction involves $I_x S_x$ and so $I_{\pm} S_{\mp}$. Therefore, the spin-lock CP transfer results from a mutual spin flip between S spins (carbons or nitrogens) and I spins (protons) by the combination of spin-raising and spin-lowering operators in Eq. (2.18). The requirement of a CP transfer for a strong static interaction is contained in r_{im}^{-3} (3 $\cos^2\theta_{im} - 1$) and this term averages to 0 in the presence of rapid, isotropic molecular motions. In such case, ^{13}C or ^{15}N signals could be suppressed by CP-MAS NMR, and *recording these NMR signals by DD-MAS or Bloch decay method is essential as an alternative means.* In the absence of such motions, therefore, the optimum magnetization achieved by CP-MAS NMR, following a single spin-lock CP transfer, is B_{1S}/B_{1I} (or $\gamma_I/\gamma_S = 4$ for carbons) greater than the magnetization that would be observed in a fully relaxed Fourier transform (FT) experiment. This is the major advantage to record CP-MAS NMR spectra, as viewed from gaining sensitivity as well as feasibility of spectral accumulation with shorter repetition times for abundant spins such as proton, as described above.

As an illustrative example, we show the ^{13}C DD-MAS and CP-MAS NMR spectra of fully hydrated [3-^{13}C]Ala-labeled preparation of bacteriorhodopsin (bR) from purple membrane (centrifuged pellet) adopting 2D crystal, illustrated in Figures 2.7(a) and (b), respectively.[17] This kind of site-directed ^{13}C-labeling at Ala C_β as circled carbons is essential to give rise to well-resolved ^{13}C NMR signals of membrane proteins among possible ^{13}C NMR signals from a variety of amino acid residues involved in this protein (Figure 2.8).[18] The following acquisition parameters are usually used: spectral width, 40 kHz; acquisition time, 50 ms; and repetition time, 4 s.[17] Up to 12 ^{13}C NMR signals, including the five single carbon signals among 29 Ala residues in bR, are well-resolved in the two types of spectra under the condition of excess hydration. The three intense ^{13}C NMR signals marked by gray in the DD-MAS NMR spectrum (Figure 2.7(a)) are suppressed in the corresponding CP-MAS NMR spectrum (Figure 2.7(b)), although the spectral features of the rest are unchanged. This is obviously caused by the presence of rapid motional fluctuations with correlation times shorter than 10^{-8} s in the N- or C-terminal region protruding from the membrane surface. The intensity gain in recording ^{13}C NMR signals for rigid portions such as the transmembrane α-helices by CP-MAS is approximately four times over the DD-MAS as mentioned above. Only three peaks were resolved when a lyophilized preparation was used instead of the pelleted samples for the ^{13}C NMR measurement.[18a] This is because individual line widths were substantially broadened in the presence of distorted conformers arising from lyophilization. For this reason, keeping intact proteins under physiological conditions is essential to record well-resolved ^{13}C NMR spectra.

2.1.4. Powder Pattern NMR Spectra of Spin-1/2 Nuclei

Instead of isotropic NMR signals achieved by the attempted peak-narrowing as described above, careful analysis of powder pattern consisting of sum of all resonance lines resulting from all possible orientations of the nuclear site (Figure 2.4) gives rise to a more detailed picture of the dynamics and orientation of certain moiety in a given residue. Principal values of the

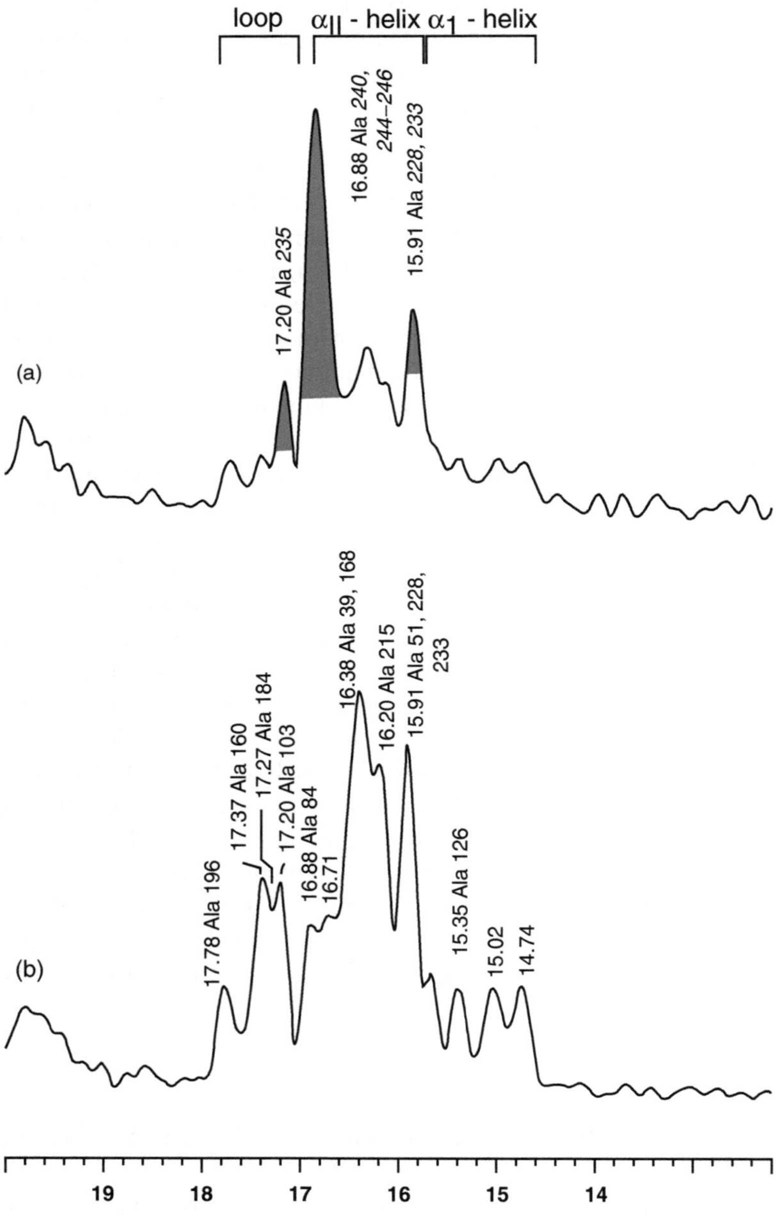

Figure 2.7. ^{13}C DD-MAS (a) and CP-MAS (b) NMR spectra of [3-^{13}C]Ala-labeled bacteriorhodopsin from purple membrane.[17]

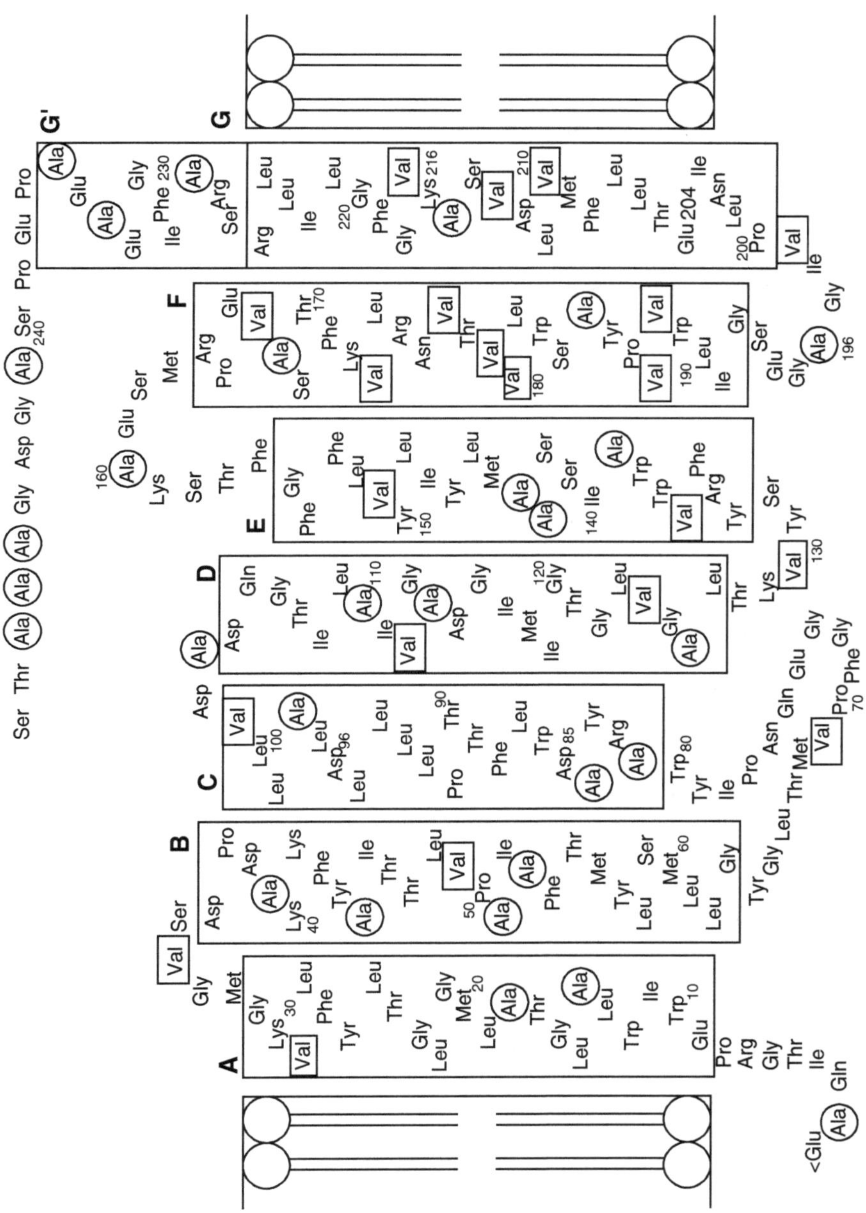

Figure 2.8. Amino acid sequence of bR by taking into account the secondary structure. [3-^{13}C]Ala and [1-^{13}C]residues are indicated by the circles and boxes, respectively.[18]

chemical shift tensor are readily available from the powder pattern based on the following treatment.

Following Eq. (2.6), the observed frequency ω_{zz} of a spectral line can be expressed by the Euler angles (α, β) that relates the principal axes of the tensor (ω_{11}, ω_{22}, ω_{33}) to the laboratory frame as[1]

$$\omega_{zz} = \omega_{11}\cos^2\alpha\sin^2\beta + \omega_{22}\sin^2\alpha\sin^2\beta + \omega_{33}\cos^2\beta. \tag{2.19a}$$

For axially symmetric tensor, we write

$$\omega_{zz} = (\omega_{\parallel} - \omega_{\perp})\cos^2\beta + \omega_{\perp}, \tag{2.19b}$$

where β is the angle between the 3-axes and the direction of the magnetic field B_0, and where

$$\omega_{\parallel} = \gamma_S\delta_{33}B_0, \quad \omega_{\perp} = \gamma_S\delta_{11}B_0 = \gamma_S\delta_{22}B_0$$

leading to the total anisotropy

$$\Delta\omega = \omega_{\parallel} - \omega_{\perp}.$$

If $I(\omega)$ is the intensity of the NMR signal at the frequency ω and $p(\Omega)\mathrm{d}\Omega$ is the probability of finding the tensor orientation in the range between the solid angle Ω and $\Omega + \mathrm{d}\Omega$, we may write

$$P(\Omega)\mathrm{d}\Omega = I(\omega)\mathrm{d}\omega \tag{2.20}$$

In a powder, all angles Ω are equally probable i.e., $p(\Omega) = 1/4\pi$ leading to

$$I(\omega) = (1/4\pi)/|\mathrm{d}\omega/\mathrm{d}\Omega|, \tag{2.21}$$

based on ''differential conservation of the integral''.[19] In the case of axial symmetry $\mathrm{d}\Omega = \sin\beta\mathrm{d}\beta$ we arrive at the line-shape function, using Eqs. (2.19) and (2.20) for an axially symmetric shift tensor ($\delta_{11} = \delta_{22}$)

$$I(\omega) = (1/2)\{(\omega_{\parallel} - \omega_{\perp})(\omega - \omega_{\perp})\}^{-1/2} \tag{2.22}$$

where
$\omega_{\parallel} \leq \omega \leq \omega_{\perp}$.

The powder pattern has a divergence at $\omega = \omega_{\perp}$ and a shoulder at $\omega = \omega_{\parallel}$. The general case with $\omega_{33} > \omega_{22} > \omega_{11}$ has been calculated by Bloembergen and Rowland[20] as

$$I(\omega) = \pi^{-1}(\omega - \omega_{11})^{-1/2}(\omega_{33} - \omega_{22})^{-1/2}K(m) \tag{2.23}$$

with

$$m = (\omega_{22} - \omega_{11})(\omega_{33} - \omega)/(\omega_{33} - \omega_{22})(\omega - \omega_{11}) \text{ for } \omega_{33} \geq \omega > \omega_{11}$$

and

$$I(\omega) = \pi^{-1}(\omega_{33} - \omega)^{-1/2}(\omega_{22} - \omega_{11})^{-1/2}K(m) \tag{2.24}$$

with

$$m = (\omega - \omega_{11})(\omega_{33} - \omega_{22})/(\omega_{33} - \omega)(\omega_{22} - \omega_{11}) \quad \text{for}\, \omega_{22} > \omega > \omega_{11}$$

$I(\omega) = 0$ in case $\omega > \omega_{33}$ and $\omega < \omega_{11}$.

$K(m)$ is the complete elliptic integral of the first kind

$$K(m) = \int_0^{\pi/2} \mathrm{d}\varphi[1 - m^2 \sin^2 \varphi]^{-1/2}$$
$$K(0) = \pi/2; \quad K(1) = \infty \tag{2.25}$$

In this case, a divergence occurs at $\omega = \omega_{22}$ and shoulders at $\omega = \omega_{11}$ and $\omega = \omega_{33}$, as shown in Figure 2.4. Usually, a residual line broadening should be taken into account by convoluting $I(\omega)$ with a Lorentzian or Gaussian broadening function.

2.2. Quadrupolar Nuclei

The nuclei of spin quantum number $I = 1/2$ such as ^1H, ^{13}C, ^{15}N, ^{31}P, etc. have a spherical distribution of positive electric charge. In contrast, the nuclei with spin quantum number $I > 1/2$ such as ^2H, ^{14}N, ^{17}O, etc. have an asymmetric distribution of nonspherical positive electric charge. The asymmetric charge distribution in the nucleus is described by the nuclear electric quadrupole moment (eQ), which is a measure of the departure of the nuclear charge distribution from spherical symmetry.

2.2.1. Quadrupolar Interaction

The quadrupole moment eQ changes its sign, either positive or negative, depending upon the prolate or oblate nuclei, respectively, as demonstrated in Figure 2.9. Accordingly, the quadrupolar nucleus interacts with the surrounding electric field gradient (EFG) in an asymmetric environment of electrons around the nucleus to result in much larger splitting of peaks in the solid and broadened spectral lines in solution, because of shorter relaxation times. The Hamiltonian for a quadrupolar nuclei can be written as[21]

$$\begin{aligned} \mathcal{H} = {} & -\gamma\hbar B_0 I_z + [e^2qQ/4I(2I-1)] \\ & [3I_z^2 - I(I+1) + \tfrac{1}{2}\eta(I_+^2 + I_-^2)], \end{aligned} \tag{2.26}$$

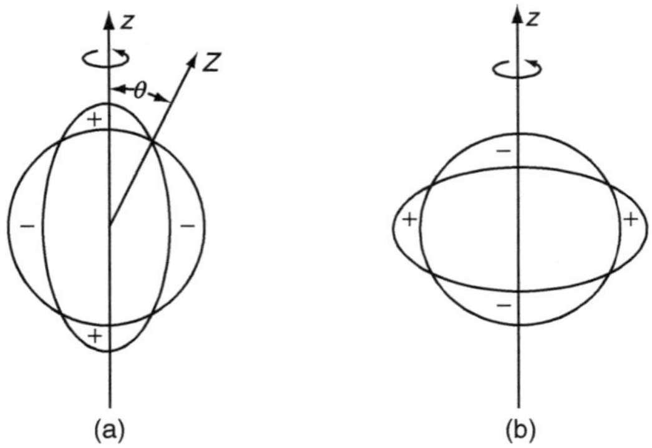

Figure 2.9. Two types of quadrupolar nuclei: (a) prolate nucleus with $eQ > 0$, (b) oblate nucleus with $eQ < 0$.

where B_0 is the applied magnetic field in the z-axis and I_z is z-component of the spin angular momentum of the nucleus. The axes X, Y, and Z are the principal axes of the field gradient tensor V_{ij} and

$$\begin{aligned}&|V_{ZZ}|\geq|V_{XX}|\geq|V_{YY}|\\&eq = V_{ZZ}\\&\eta = (V_{XX} - V_{YY})/V_{ZZ}, \quad 0\leq\eta\leq 1\end{aligned} \tag{2.27}$$

Taking the z-axis along the applied field B_0, the energy level calculated by the first-order perturbation theory is,

$$\begin{aligned}E_m^{(1)} &= -\gamma\hbar B_0 m + [3e^2qQ/8I(2I-1)][3\cos^2\theta - 1\\&\quad + \eta\sin^2\theta\cos 2\phi] \times [m^2 - I(I+1)/3],\end{aligned} \tag{2.28}$$

in which θ and ϕ are the polar and azimuthal angles in the principal axes of the field gradient tensor (frequently the C–^{2}H bond vector) (Figure 2.10), respectively, and m is the quantum number for the z-component of the angular momentum I. This is because the Zeeman interaction in the first term in Eq. (2.28) dominates the quadrupolar interaction in the second term in high field strengths. The resonance frequency $\nu(m-1 \rightarrow m)$ is written as

$$\begin{aligned}\nu(m-1 \rightarrow m) &= \left(E_{m-1}^{(1)} - E_m^{(1)}\right)/h\\&= \gamma B_0/2\pi - [3e^2qQ/4I(2I-1)][3\cos^2\theta - 1\\&\quad + \eta\sin^2\theta\cos 2\phi)(m - 1/2)]\end{aligned} \tag{2.29}$$

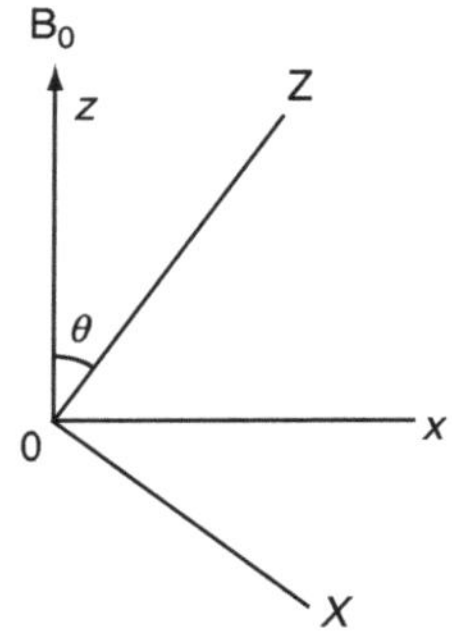

Figure 2.10. The orientation of the principal axes of the EGF tensor with respect to the polar (θ) and azimuthal (φ) angles with respect to the applied field B_0.

NMR spectral feature of the quadrupolar nuclei is different to some extent between integer (^2H or ^{14}N for $I = 1$) and half-integer spins (^{23}Na, ^{27}Al, ^{17}O for spin numbers 3/2, 5/2, 5/2, respectively).

For nucleus of spin number $I = 1$, two transitions $\nu(0 \rightarrow 1)$ and $\nu(-1 \rightarrow 0)$ of equal intensity are expected. Therefore, the difference between the two transitions is

$$\Delta\nu_Q = [3e^2qQ/4I(2I-1)h](3\cos^2\theta - 1 + \eta\sin^2\theta\cos 2\phi) \qquad (2.30)$$

In many instances, EFG tensor has axial symmetry, so that the asymmetry parameter η vanishes and Eq. (2.30) reduces to

$$\Delta\nu_Q = (3/4)[e^2qQ/I(2I-1)h](3\cos^2\theta - 1) \qquad (2.31)$$

2.2.2. ^{2}H NMR

^2H is a quadrupolar nucleus with exceptionally smaller quadrupole moments ($eQ = 0.286$ in unit of 10^{-26} cm^2) as compared with other nuclei such as ^{14}N (2.01), ^{17}O (−2.56) and ^{23}Na (10.06). Because of instrumental limitation, current NMR technology permits one to record whole NMR spectra for nuclei with such smaller quadrupole moment as ^2H alone. Besides, ^2H turns out to be a very convenient nucleus that has the ability to probe the manner of partially ordered systems and molecular dynamics in solid and semi-solids,[22–25] although specific incorporation of this nuclei either by biosynthesis or chemical synthesis is necessary for this purpose due to its low sensitivity (0.0097 relative to ^{1}H) and natural abundance (0.015%). In case of ^{2}H, Eq. (2.31) is reduced to

$$\Delta\nu_Q = (3/4)(e^2qQ/h)(3\cos^2\theta - 1). \tag{2.32}$$

This orientation-dependent splitting of peaks yields a clue on how a C^{2}H vector of a molecule under consideration is oriented to the applied magnetic field when single crystalline, mechanically, or magnetically oriented samples are utilized.

In many instances, however, most of samples are not always homogeneously oriented with respect to B_0, but are randomly distributed as in microcrystalline or powder samples. Following the similar treatment of the powder pattern NMR spectra of spin-1/2 nuclei discussed above, ^{2}H NMR powder pattern can be evaluated as follows. The observed ^{2}H NMR frequency for $-1 \leftrightarrow 0$ and $0 \leftrightarrow 1$ transitions can be written by

$$\nu_{zz} = \tfrac{1}{2}\delta(3\cos^2\theta - 1) \quad \text{for } -1 \leftrightarrow 0 \text{ transition} \tag{2.33a}$$

$$\nu_{zz} = -\tfrac{1}{2}\delta(3\cos^2\theta - 1) \quad \text{for } 0 \leftrightarrow 1 \text{ transition,} \tag{2.33b}$$

where $\delta = (3/4)(e^2qQ/h)$, following Eq. (2.28), analogous to the treatment yielding the powder pattern from the CSA, as discussed in the previous section. The spectral intensity for $-1 \leftrightarrow 0$ transition, as defined in Eq. (2.22)

$$I(\omega) = (1/4\pi)/|d\omega/d\theta| = (1/4\pi)/|3\delta\sin\theta\cos\theta|, \tag{2.34}$$

is obtained based on Eq. (2.33).

Therefore,

$$I(\omega) = (1/4\pi)(1/3\delta|\cos\theta|),\ |\cos\theta| = (\sqrt{2}/\sqrt{3})\sqrt{\omega/\delta + 1/2}$$

$$I(\omega) = (1/4\pi)(1/\sqrt{6\delta})(1/\sqrt{(\omega + \delta/2)}) \quad -\delta/2 \leq \omega \leq \delta \tag{2.35}$$

There is a (integrable) square root singularity at $\omega = -1/2\delta$, that is, at one edge of the spectrum (Figure 2.11). At $\omega = \delta$ (i.e., $\sin\theta = 0$), there is also a $1/(\sin\theta = 0)$ singularity of the angular dependence $1/(\mathrm{d}\omega/\mathrm{d}\theta)$ in Eq.(2.35), but it is canceled by the vanishingly small contribution of the north pole, $\theta = 0$, surface element, $P(0°) = \sin(0°) = 0$, to yield the finite value of $1/(3\delta)$. Combined with the peak intensity of another $0 \leftrightarrow 1$ transition, the ‘‘powder pattern,’’ so-called ‘‘Pake doublet’’ with two ‘‘horns,’’ is formed as a result of inhomogeneously broadened peaks, as shown in Figure 2.11. The more complicated case of the powder spectrum for $\eta \neq 0$ has been treated by Bloembergen and Rowland.[20] The separation of the ‘‘horns’’ in the powder sample is $\Delta\delta$ namely

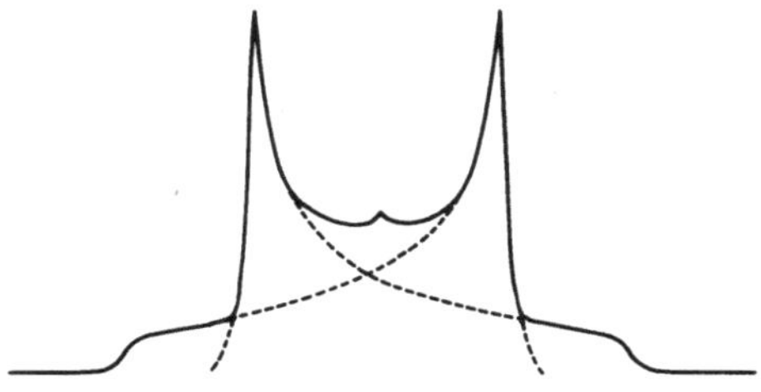

Figure 2.11. ^{2}H powder pattern spectrum arising from two transitions in ^{2}H NMR lines.

$$\Delta\nu_{\text{powder}} = (3/4)e^2qQ/h \tag{2.36}$$

The deuterium quadrupole coupling constant, e^2qQ/h, for C–^{2}H fragments ranges from160 to 210 kHz, depending upon the type of hybridization.

2.2.3. NMR Spectra of Half-Integer Spins

The first-order shift $\nu_m^{(1)} = \nu(m-1 \leftrightarrow m)$ given in Eq. (2.26) vanishes for $m = 1/2$. Accordingly, for half-integer spins such as ^{17}O and ^{23}Na, the frequency of the central transition $-1/2 \leftrightarrow 1/2$ is not shifted in the first order by the quadrupolar interaction, as illustrated in Figure 2.12.[21] In fact, for $I = 3/2$

$$\begin{aligned}
&\nu_{3/2}^{(1)}(3/2 \leftrightarrow 1/2) = -(1/4)e^2qQ(3\cos^2\theta - 1)/h \\
&\nu_{1/2}^{(1)}(1/2 \leftrightarrow -1/2) = 0 \\
&\nu_{-1/2}^{(1)}(-1/2 \leftrightarrow -3/2) = (1/4)e^2qQ(3\cos^2\theta - 1)/h
\end{aligned} \tag{2.37}$$

are easily evaluated. In such cases, the quadrupole coupling constant is readily evaluated from the separation of the satellite signals which being $(1/2)e^2qQ/h$. The frequencies of the other lines, however, are shifted, and satellite lines corresponding to transitions $m \leftrightarrow (m-1)$ with $m \neq 1/2$ appear on each side of the central line. The quadrupolar coupling constants for most of the quadrupolar nuclei vary from zero to several hundred MHz, depending upon the values of eQ and EFG of particular nuclei. Therefore, their NMR spectra are spread over a frequency range that for most nuclei exceeds the range of chemical shifts and the bandwidth of spectrometer under pulse excitation. The second-order shift $\nu_m^{(2)} = (E_{m-1}^{(2)} - E_m^{(2)})/h$ is readily available from analysis of the second-order perturbation of the

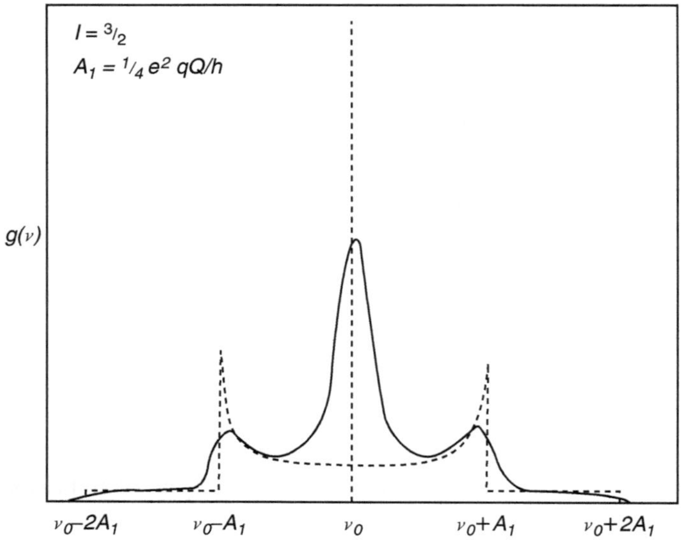

Figure 2.12. First-order quadrupole perturbation. Line shape for $I = 3/2$ nuclei (from Abragam[21])

Hamiltonian, as shown in Eq. (2.26). In particular, for the central line $-1/2 \leftrightarrow 1/2$, it is given by[21]

$$\nu_{1/2}^{(2)} = (-\nu_Q^2/16\nu_L)(a - 3/4)(1 - \mu^2)(9\mu^2 - 1), \tag{2.38}$$

where $\nu_Q = 3e^2qQ/h2I(2I - 1)$, $a = I(I + 1)$, $\mu = \cos\theta$, $\nu_L = \gamma H/2\pi$. The detection of broad quadrupole line shapes can rarely be accomplished with single pulse excitation, because of the loss of magnetization within the dead time of the receiver. In such cases, the acquisition of spectra can be accomplished by scanning the frequency or the magnetic field B_0, and via point-by-point acquisition of the echo intensity.[26] Instead, it is practical to record spectra of the central $-1/2 \leftrightarrow 1/2$ transition alone by a single pulse excitation. The theoretical line shape for powder pattern of the central line $-1/2 \leftrightarrow 1/2$ for a spin with $I = 3/2$ and axial symmetry is illustrated in Figure 2.13.[27] It is noted that the extent of the split central peaks caused by the second-order perturbation is decreased by a factor of ν_Q^2/ν_L. Interestingly, MAS eliminates the first-order broadening completely and reduces the quadrupolar contribution to the central $-1/2 \leftrightarrow 1/2$ transition line width by a factor of approximately 3. The second-order quadrupolar interaction, however, is not averaged to zero by MAS alone. A general formalism for calculating the MAS-NMR spectra of half-integer quadrupole nuclei in powdered solids has been presented by Behrens and Schnabel[28] and

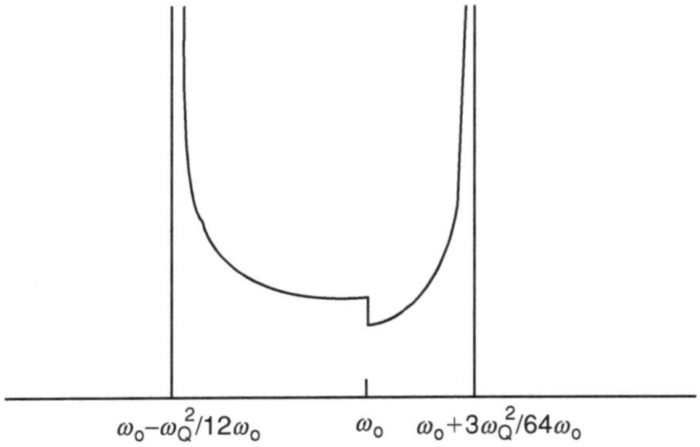

Figure 2.13. The powder spectrum for the central transition for a spin with $I = 3/2$ and axial symmetry (from Gerstein and Dybowski [27]).

Lippmaa and coworkers,[29,30] as demonstrated in Figure 2.14. The resulting spectral pattern depends strongly on the asymmetrical factor ν_Q^2/ν_L ratio and η. The most serious point, however, is that the center of gravity of the powder pattern of the central $-1/2 \leftrightarrow 1/2$ line in MAS-NMR spectrum does not coincide with the frequency ν_L containing the isotropic shift, as shown by

$$\nu_{CG} - \nu_L = (-\nu_Q^2/30\nu_L)\{I(I+1) - 3/4\}\{1 + \eta^2/3\} \tag{2.39}$$

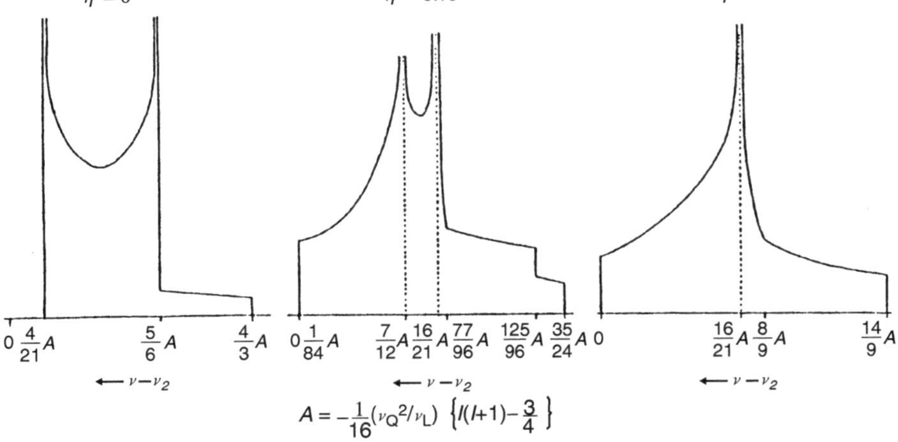

Figure 2.14. Calculated peak-narrowing by MAS for the central transition of quadrupolar nuclei (from Behrens and Schnabel[28]).

This is because the quadrupole interaction consists of spherical harmonics of the second and fourth order, and the latter effect cannot be eliminated by rotation of single axis such as MAS.[31] To eliminate the quadrupolar interactions completely, double rotation (DOR) experiment was proposed: the sample is spun in a small inner rotor, which is rotating inside an outer rotor.[31] The rotation axis of the outer rotor is set at the magic angle, while the rotation of axis of the inner rotor moves continuously on a cone of an aperture of 61.12°. In contrast, dynamic angle spinning (DAS) uses two discrete rotational reorientations (commonly tilted 37.38° and 79.19° away from the applied magnetic field and correlates the corresponding resonance frequencies in a 2D experiment.[32] In any case, these two methods seem to be far from routine applications for the study of biopolymers because of inherent technical problems.

Alternatively, Frydman and coworkers[33,34] showed that the narrowing of the central transition can be obtained without changing the orientation of the spinning axis, as long as the motion of this axis in space is replaced by changing the coherent state of the observed spins. This technique is referred to as multiple quantum magic angle spinning (MQMAS). MQMAS is a 2D method that correlates the phase evolution of the MQ and single quantum (1Q) coherence, and allows for observation of a purely isotropic echo. The MQMAS precludes most of the shortcomings of DOR (presence of closely spaced spinning sidebands) and DAS (loss of magnetization during the reorientation of the spinner axis) and is technically straightforward. The resolution enhancement that this technique can provide is demonstrated in Figure 2.15, which compares static, MAS, and MQMAS-NMR spectra of ^{87}Rb in crystalline $LiRbSO_4$.[35,36] The observed line width in $LiRbSO_4$ changed from approximately 400 ppm in a static spectrum to 150 ppm under MAS and 1.5 ppm in MQMAS.

The most simple and effective remedy, however, to circumvent these problems, is to use high-frequency NMR spectrometer, because the second-order effect, ν_Q^2/ν_L, is diminished in proportion to the applied magnetic field. In fact, Gan *et al.*[37] showed that MAS spectrum at very high fields (25 and 40 T) is nearly free from the second-order quadrupolar broadening, and can be interpreted quantitatively in a very simple manner.

2.2.4. Quadrupole Echo

Often, the resulting powder patterns for ^2H nuclei are very broad (up to 100–200 kHz), although they are one order of magnitude smaller than those of other quadrupolar nuclei such as ^{14}N, implying that much of the information is

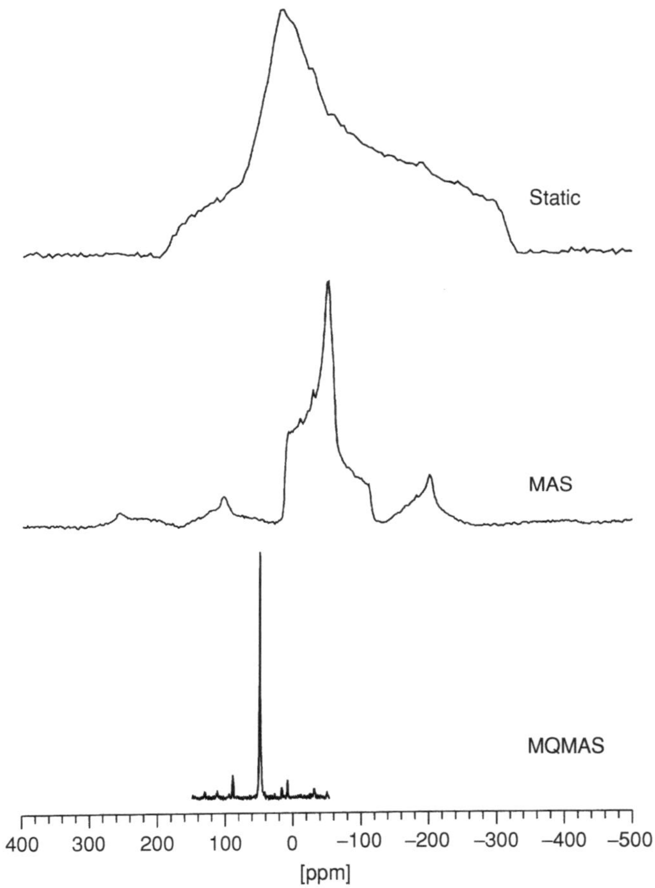

Figure 2.15. Static, MAS, and MQMAS (isotropic projection) spectra of ^{87}Rb in polycrystalline $LiRbSO_4$ taken at 130.88 MHz (from Amoureux and Pruski[35]).

contained in the very early part of the FID. Since the dead-time of most spectrometers operating in this frequency range is several tens of microseconds, much of the information about these broad components is lost and cannot be faithfully reproduced. The effect of receiver dead time on an ^2H powder pattern is represented in Figure 2.16(a). In order to circumvent this problem, Davis *et al.*[38] proposed a pulse sequence which involves a 90° pulse followed after time τ by a second 90° pulse, phase shifted by $\pi/2$ relative to the first pulse. A quadrupole echo is formed at $t = 2\tau$ due to the refocusing of the nuclear magnetization. Starting the Fourier transform at the echo maximum gives an undistorted spectrum, as shown in Figure 2.16(b).

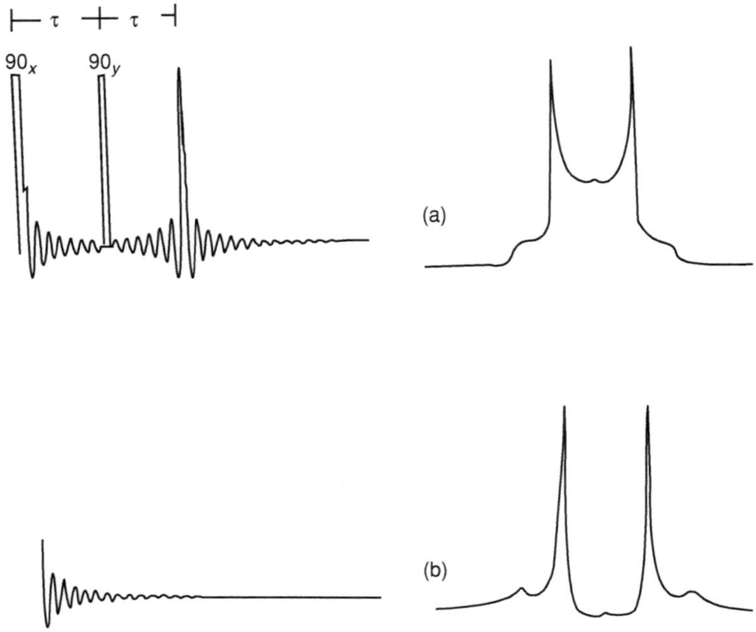

Figure 2.16. (a) Simulated FID after a 90°pulse and 30 μs delay corresponding to the receiver dead time and the Fourier transform of the FID. (b) Simulated quadrupole echo and its Fourier transform starting at the echo maximum (from Jarrell and Smith[24]).

References

[1] M. Mehring, 1983, *Principles of High Resolution NMR in Solids*, Second Edition, Springer-Verlag, Berlin.
[2] C. P. Slichter, *Principles of Magnetic Resonance*, Harper & Row, New York, 1963; Third Enlarged and Updated Edition, Springer-Verlag, 1989.
[3] C. A. Fyfe, 1983, *Solid State NMR for Chemists*, CFC Press, Guelph.
[4] J. Kempf, H. W. Spiess, V. Haeberlen, and H. Zimmerman, 1972, *Chem. Phys. Lett.*, 17, 39–42.
[5] S. R. Hartmann and E. L. Hahn, 1962, *Phys. Rev.*, 128, 2042–2053.
[6] A. Pines, M. G. Gibby, and J. S. Waugh, 1973, *J. Chem. Phys.*, 59, 569–590.
[7] J. Schaefer and E. O. Stejeskal, 1976, *J. Am. Chem. Soc.*, 98, 1031–1032.
[8] E. R. Andrew, A. Bradbury, and R. G. Eades, 1958, *Nature*, 182, 1659.
[9] E. R. Andrew, 1971, *Prog. Nucl. Magn. Reson. Spectrosc.*, 8, 1–39.
[10] W. S. Veeman, 1984, *Prog. Nucl. Magn. Reson. Spectrosc.*, 16, 193–235.
[11] A. Shoji, S. Ando, S. Kuroki, I. Ando, and G. A. Webb, 1993, *Annu. Rep. NMR Spectrosc.*, 26, 55–98.
[12] W. T. Dixon, 1981, *J. Magn. Reson.*, 44, 220–223.
[13] R. R. Ernst, G. Bodenhausen, and A. Wokaun, 1987, *Principles of Nuclear Magnetic Resonance in One and Two Dimensions*, Clarendon Press, Oxford.

[14] E. D. Becker, 1980, *High Resolution NMR, Theory and Chemical Applications*, Second Edition, Academic Press, New York.
[15] D. E. Demco, H. Tegenfeldt, and J. S. Waugh, 1975, *Phys. Rev.*, B11, 4133–4155.
[16] J. Schaefer, 1979, in *Topics in Carbon-13 NMR Spectroscopy*, vol. 3, pp.283–324.
[17] S. Tuzi, S. Yamaguchi, M. Tanio, H. Konishi, S. Inoue, A. Naito, R. Needleman, J. K. Lanyi, and H. Saitô, 1999, *Biophys. J.*, 76, 1523–1531.
[18] S. Yamaguchi, S. Tuzi, M. Tanio, A. Naito, J. K. Lanyi, R. Needleman, and H. Saitô, 2000, *J. Biochem. (Tokyo)*, 127, 861–869.
[18a] S. Tuzi, A. Naito, and H. Saitô, 1993, Eur. J. Biochem., 218. 837–844.
[19] K. Schmidt-Rohr and H. W. Spiess, 1994, *Multidimensional Solid-state NMR and Polymers*, Academic Press, London.
[20] N. Bloembergen and J. A. Rowland, 1953, *Acta Metall.*, 1, 731–746.
[21] A. Abragam, 1961, *The Principles of Nuclear Magnetism*, Claredon Press, Oxford.
[22] H. H. Mantsch, H. Saitô and I. C. P. Smith, 1977, *Prog. Nucl. Magn. Reson. Spectrosc.*, 11, 211–271.
[23] J. Seelig, 1977, *Q. Rev. Biophys.*, 10, 353–418.
[24] H. C. Jarrell and I. C. P. Smith, 1982, in *The Multinuclear Approach to NMR Spectroscopy*, J. B. Lambert and F. G. Riddell, Eds., D. Reidel, Dordrecht, The Netherlands, pp. 151–168.
[25] E. Oldfield, R. A. Kinsey, and A, Kintanar, 1982, *Methods Enzymol.*, 88, 310–325.
[26] T. J. Bastow, 1994, *Solid State Nucl. Magn. Reson.*, 3, 17–22.
[27] B. C. Gerstein and C. R. Dybowski, 1985, *Transient Techniques in NMR of Solids, An Introduction to Theory and Practice*, Academic Press, Orlando.
[28] H.-J. Behrens and B. Schnabel, 1982, *Physica*, 114B, 185–190.
[29] E. Kundla, A. Samoson, and E. Lippmaa, 1981, *Chem. Phys. Lett.*, 83, 229–232.
[30] A. Samoson, E. Kundla, and E. Lippmaa, 1982, *J. Magn. Reson.*, 49, 350–357.
[31] F. Chmelka, K. T. Mueller, A. Pines, J. Stebbins, Y. Wu, and J. W. Zwanziger, 1989, *Nature*, 339, 42–43.
[32] K. T. Mueller, B. Q. Sun, G. C. Chingas, J. W. Zwanziger, T. Terao, and A. Pines, 1990, *J. Magn. Reson.*, 86, 470–487.
[33] L. Frydman and J. S. Harwood, 1995, *J. Am. Chem. Soc.*, 117, 5367–5368.
[34] A. Medek, J. S. Harwood, and L. Frydman, 1995, *J. Am. Chem. Soc.*, 117, 12779–12787.
[35] J.-P. Amoureux and M. Pruski, 2002, *Encl. Nucl. Magn. Reson.*, 9, 226–252.
[36] M. Pruski, J. W. Wiench, and J.-P. Amoureux, 2000, *J. Magn. Reson.*, 147, 286–295.
[37] Z. Gan, P. Gor'kov, T. A. Cross, A. Samoson, and D. Massiot, 2002, *J. Am. Chem. Soc.*, 124, 5634–5635.
[38] J. H. Davis, K. R. Jeffrey, M. Bloom, M. I. Valic, and T. P. Higgs, 1976, *Chem. Phys. Lett.*, 42, 390–394.

Chapter 3

BRIEF OUTLINE OF NMR PARAMETERS

3.1. Chemical Shift

The chemical shift is one of the most important NMR parameters used for structural elucidation. The chemical shifts provide detailed information on the structure and electronic structures of biopolymers and polymers in solution, noncrystalline, and crystalline states.[1–12] In solution state, the chemical shifts of a polypeptide consisting of an enormous number of chemical bonds are often the averaged values for all possible conformations, owing to rapid interconversion by rotation about chemical bonds. In solids, however, chemical shifts are characteristic of specific conformations because of strongly restricted rotation about the bonds. The chemical shift is affected by a change of the electronic structure through the structural change. Solid-state chemical shifts, therefore, give useful information about the electronic structure of a polypeptide with a fixed structure. Further, the chemical-shift tensor components can often be experimentally determined, and provide information about a local symmetry of the electron cloud around the nucleus, leading to much more detailed knowledge of the electronic structure of the polymer compared with the average chemical shifts in solution. In order to clarify a relationship between the chemical shift and the electronic structure of biopolymers, a sophisticated theoretical method that can take into account the characteristics of biopolymers is needed.

Some methodologies for obtaining structures and electronic structures of polypeptides in the solution and solid states use a combination of the observation and calculation of NMR chemical shifts, and these methods have been applied to various polypeptide systems. Theoretical calculations of chemical shifts for a polypeptide have been done mainly by two approaches. One uses model molecules such as dimer, trimer, etc. for local structures of biopolymers for calculations that combine quantum chemistry with statistical mechanics. In particular, this approach has been initially applied to polymer and biopolymer systems in the solution state.[13–21] However, it should be recognized that quantum chemical calculations on model molecules are not readily applicable to polypeptides in the crystalline state, because of the existence of long-range, intrachain, and interchain interactions. Electrons are constrained to a finite region of space in small molecules whereas this is not necessarily the case for biopolymers, and thus some additional approaches are required. Another approach is to employ the tight-binding molecular orbital (TBMO) theory,[22–24] which is well known in the field of solid state physics, to describe the electronic structures of biopolymers with periodic structure within the framework of a linear combination of atomic orbitals (LCAO) approximation for the electronic eigenfunctions.[25–37] This method can take into account of long-range, intra- and interchain interactions in the chemical shift calculation.

3.1.1. Chemical Shifts and Electronic States: Approach Using Model Molecules

3.1.1.1. The origin of chemical shift

Chemical shift of atom in a molecule arises from the nuclear shielding effect of applied magnetic field, caused by induced magnetic field owing to circulation of surrounding electrons. The magnitude of such induced magnetic field is proportional to the applied magnetic field $\boldsymbol{B}_0$, so that the effective field $\boldsymbol{B}_{\text{eff}}$ at the nucleus will be

$$\boldsymbol{B}_{\text{eff}} = \boldsymbol{B}_0(\mathbf{1} - \boldsymbol{\sigma}), \tag{3.1}$$

where $\boldsymbol{\sigma}$ is the second-rank tensor of nuclear shielding and $\mathbf{1}$ is the unit matrix. The chemical shift δ is given by the difference in resonance frequencies between the nucleus of interest and a reference nucleus, usually taken from the resonance frequency of TMS, ν_{TMS}:

$$\delta = 10^6(\nu - \nu_{\text{TMS}})/\nu_{\text{TMS}} \approx 10^6(\sigma_{\text{TMS}} - \sigma) \tag{3.2}$$

The nuclear shielding constant corresponding to the chemical shifts for atom A can be precisely estimated as a sum of the following terms according to the theory of chemical shift.[3,38–43]

$$\sigma_{\text{A}} = \sigma^{\text{d}} + \sigma^{\text{p}} + \sigma', \tag{3.3}$$

where σ^{d} is the diamagnetic term, σ^{p} the paramagnetic term, and σ' the other term which comes from the magnetic anisotropy effects, polar effects, and ring current effects. For nuclei with 2p electrons such as ^{13}C, ^{15}N, etc. the relative chemical shift is predominantly governed by the paramagnetic term, and for the ^{1}H nucleus by the first and third terms.

The paramagnetic term is expressed by a function of excitation energy, bond order, and electron density according to the sum-over-state (SOS) method in the simple form as follows:

$$\sigma^{\mathrm{p}} = -\left(\mu_0 \hbar^2 e^2 / 4\pi m_{\mathrm{e}}^2\right) \Sigma \left\langle r^{-3} \right\rangle_{2\mathrm{p}} (E_m - E_n)^{-1} Q, \tag{3.4}$$

where $E_m - E_n$ is the singlet–singlet excitation energy of the nth occupied and the mth unoccupied orbitals, and Q is a factor including the bond order and electron density. The quantity $\langle r^{-3} \rangle_{2\mathrm{p}}$ is the spatial dimension for a 2p electron, μ_0 is the permeability of free space, $\hbar$ is the Planck constant (h) divided by 2π, e is the electric charge of electron, and m_{e} is the mass of electron. The paramagnetic term is calculated by semiempirical MO or *ab initio* MO methods. This term governs the chemical shift behavior associated with the structure and/or the electronic structure. On the other hand, the diamagnetic term is estimated from the calculated electron density. Using these procedures, shielding constants σ_i of the model molecule with any specified conformation have been calculated.

3.1.1.2. Medium effects on chemical shifts of mobile phase

Most MO calculations of nuclear shielding relate the case of a molecule or molecules in a vacuum except for the tight-binding MO calculations on polypeptide systems. For nuclei forming the molecular skeleton such as ^{13}C and nuclei with small shielding ranges such as ^{1}H, this may not be an unreasonable approximation. This is true if comparisons of the theoretical results are made with experimental data taken on a molecule dissolved in an inert solvent.

Atoms with lone pair electrons, such as ^{14}N, ^{15}N, ^{17}O, and ^{19}F, are very likely to have their nuclear shieldings influenced by interactions with solvent molecules. Such interactions may be specific, e.g., hydrogen bonding, or nonspecific, e.g., polarisability/polarity, or perhaps involve a combination of specific and nonspecific solute–solvent interactions. A theoretical procedure has been developed for quantitatively unravelling the contributions made to the shielding of solute nuclei by specific and nonspecific interactions. Nonspecific solute–solvent effects on nuclear shielding and specific solute–solvent (medium) interactions, such as hydrogen bonding or protonation, may be included in the calculation of the shielding of solute nuclei by a supermolecule approach. The appropriate structure of the solute–solvent supermolecule may be obtained by the use of molecular mechanics simulations. At the semiempirical MO level, this approach has been successfully used to describe the effects of hydrogen bonding on the nuclear shielding of small molecules.

Ando *et al.*[44–46] have developed the theoretical formalism for taking into account the nonspecific medium effect on ^{13}C chemical shifts of polypeptides by incorporating polypeptide chain to statistical mechanics where the Born-type reaction field model for nonspecific solute–medium interaction in the sophisticated chemical shift theory is expressed as a function of $(\varepsilon - 1)/\varepsilon$ where ε is the dielectric constant of medium. This is the so-called ''solvaton/chemical shift theory''. It has been successfully used to elucidate the ^{13}C spectral behavior of some molecular systems in solution as a function of the dielectric constant of the medium.

3.1.1.3. Chemical shifts of polypeptides in solution

A polypeptide chain can assume an enormous number of conformations because of the various possibilities of rotation around the chain bonds, which are undergoing molecular motion.[47] Thus, the factors governing the appearance of the NMR spectra include the structures, the relative energies of the rotational isomers, the chemical shifts, and the spin couplings. If molecular motion in the polymer chain is extremely slow on the NMR timescale, spectrum consists of superposed contributions from various conformations. However, if the rotation around the chain bonds is very fast on the NMR timescale, the experimentally observable chemical shift for nucleus A is given by[2, 13–18]

$$\langle \sigma_A \rangle = \sum_{i=1}^{n} P_i \sigma_i, \tag{3.5}$$

where the numerical indices refer to the preferred conformations, and P_i and σ_i are the probability of occurrence and the chemical shift of the preferred conformation i, respectively. This indicates that the chemical shift of a given nucleus can be obtained from a combination of a quantum chemical method and a statistical mechanical method as described elsewhere.[13–18]

3.1.1.4. Conformation-dependent ^{13}C chemical shifts for polypeptides

In the crystalline state, the structural information obtained from the chemical shift corresponds to the fixed conformation of polymer chains. As a result, the observed ^{13}C chemical shifts of C_α, C_β, and C=O carbons for a variety of polypeptides exhibit significant conformation-dependent changes up to 8 ppm among right-handed α-helix, β-sheet, or other conformations[48–50] (to be described more detail in Chapter 6, Tables 6.1 and 6.2). As an illustrative example, ^{13}C chemical shifts of Ala residue in polypeptide and proteins were calculated for a dipeptide fragment (*N*-acetyl-*N*′-methyl-L-alanine amide) [Ac-L-Ala-NHMe] as a model using the finite perturbation theory (FPT) INDO method,[51,52] in order to understand and predict the conformation-dependent ^{13}C chemical shift behavior, associated with the secondary structure elements such as an α-helix, β-sheet, etc. and to determine secondary structure through the observation of the ^{13}C chemical shift. Such sizeable displacements of the ^{13}C chemical shifts can be characterized by variations of the electronic structures of the local conformation as defined by the torsion angles (dihedral angles) (ϕ, ψ). We can estimate the ^{13}C chemical shift for any specified conformation from the calculated contour map of the shielding constant for the Ala C_β carbon as shown in Figure 3.1.[53] This is a very useful representation of the chemical shift map as a function of the torsion angles as in a Ramachandran map. It is possible to predict the ^{13}C chemical shifts of Ala residues in polypeptides and proteins taking particular conformations,[53–55] on the basis of comparative inspection of the experimental and calculated ^{13}C chemical shifts. For example, ^{13}C chemical

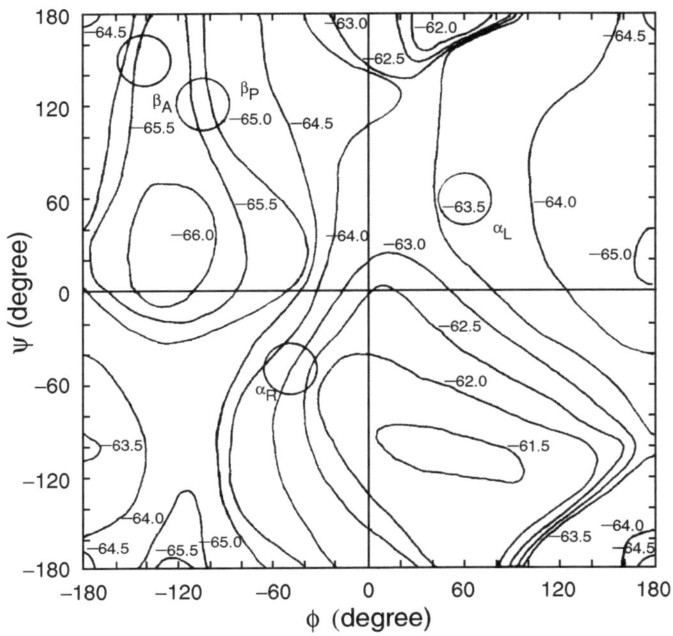

Figure 3.1. The calculated ^{13}C chemical shift (shielding constant) map of the C_β carbon of *N*-acetyl-*N*′-methyl-L-alanine amide by using the FPT INDO method. The chemical shielding constants were calculated at 15° intervals for the torsion angles (ϕ, ψ).[53]

shift of the right-handed α-helix form appears downfield by 2.5 ppm from that of the β-sheet form, consistent with the experimental results. Similar approach is feasible for ^{15}N chemical shift calculation of biopolymers.

Ab initio calculations for the NMR chemical shifts are available for medium-size molecules as a consequence of the remarkable advances in performance of workstations, personal computers, and supercomputers.[10,56–60] This leads to a quantitative discussion on the chemical shift behavior. For example, *ab initio* MO calculation with the 4–31G basis set using the gauge-independent atomic orbital-coupled Hartree–Fock (GIAO-CHF) method on *N*-acetyl-*N'*-methyl-L-alanine amide,[57,58] which is the same model molecule used as in the case of the above FPT INDO calculation, is introduced. All the geometrical parameters are energy-optimized. The isotropic ^{13}C chemical shift map of the C_β carbon as functions of the torsion angles can be calculated as shown in Figure 3.2[57] where the positive sign indicates shielding. The overall trend of this map is similar to that obtained by the FPT INDO method. The calculated isotropic shielding constants (σ) for the C_β carbon are 186.4 ppm for the torsion angles (ϕ, ψ), corresponding to the antiparallel $\beta(\beta_A)$-sheet conformation, 189.4 ppm for the right-handed $\alpha(\alpha_R)$-helix, 189.6 ppm for the left-handed $\alpha(\alpha_L)$-helix; on the other hand, the observed isotropic chemical shifts(δ) are 21.0 ppm for the β_A-sheet, 15.5 ppm for the α_R-helix, and 15.9 ppm for the α_L-helix. Such experimental chemical shift behavior is well explained by the calculated data. It is found that the change of the torsion angles dominates the isotropic chemical shift behavior of the C_β carbon for L-alanine residue.

The orientations of the PAS of the CSAs of Ala C_β carbons are calculated as shown in Figure 3.3[58], whose Ala moieties have different main-chain torsion angles, (ϕ, ψ) = (− 57.4°, − 47.5°) [α_R-helix], (−138.8°, 134.7°) [β_A-sheet], (−66.3°,−24.1°) [3^R_{10}-helix], and (−84.3°, 159.0°) [3_1-helix]. The σ_{33} component nearly lies along the $C_\alpha - C_\beta$ bond for all the peptides considered here, and also the σ_{11} is nearly perpendicular to the plane defined by the C_β, C_α, and N nuclei in Ala residue; on the other hand, σ_{22} is parallel to the plane. These results agree with the experimentally determined direction for σ_{33} of the C_β carbon in L-Ala amino acid by Naito *et al.*[78] The σ_{11} component for the torsion angles corresponding to the β_A-sheet conformation is 37.06 ppm. This shows a downfield shift of about 9 ppm with respect to that for the α_R-helix conformation. Further, the σ_{11} dominates the downfield shift on the isotropic chemical shift of the C_β carbon for the β_A-sheet conformation. Since the σ_{11} does not orient along with a specified chemical bond, it is not easy to comprehend intuitively the chemical shift tensor behavior of the C_β carbon. However, it is obvious that the through-space interaction between the C_β methyl group and its surrounding might be important for understanding the σ_{11} behavior.

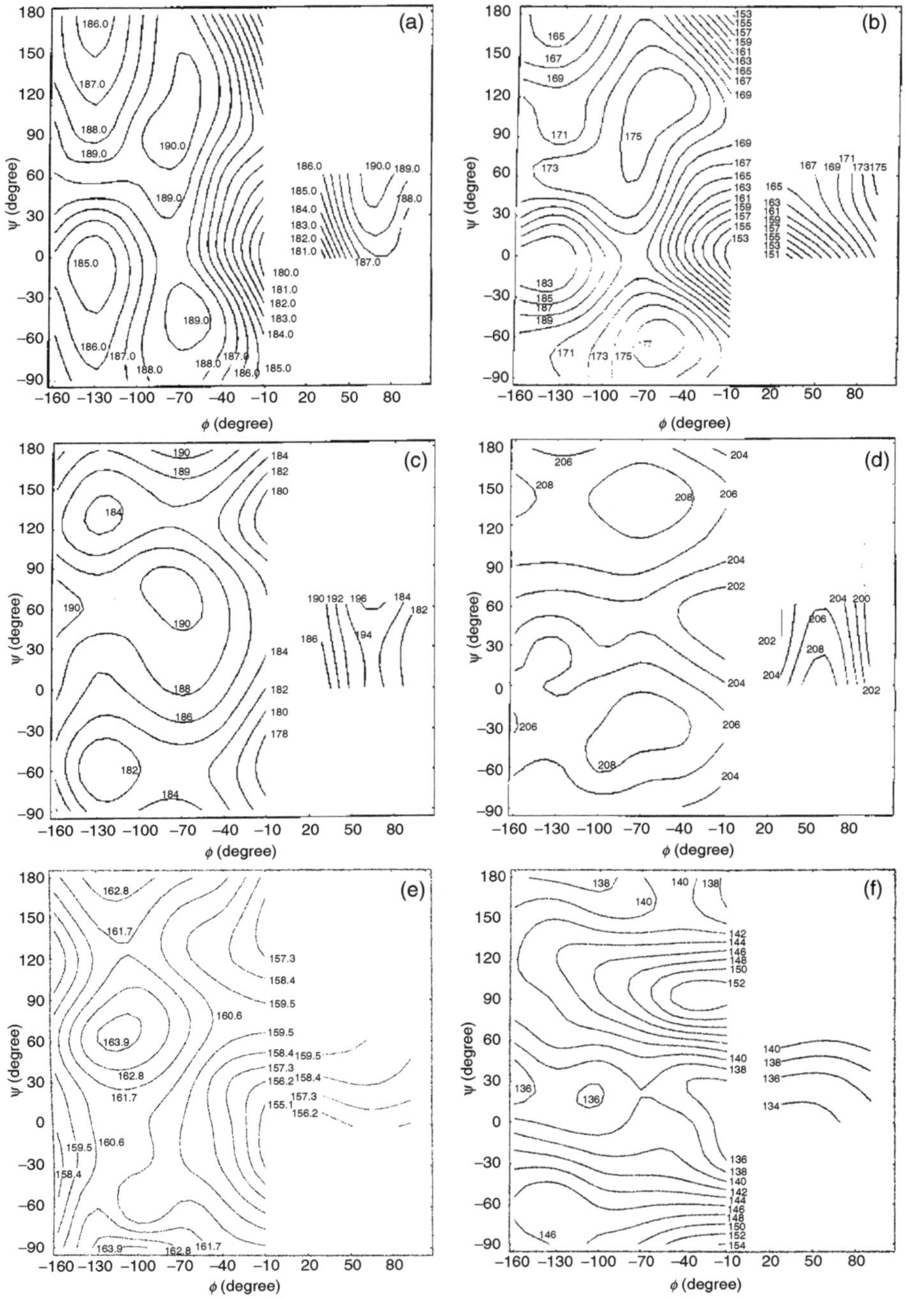

Figure 3.2. The calculated ^{13}C chemical shift (shielding constant) map of the C_β (a–d) and C_α (e–h) carbons of *N*-acetyl-*N*′-methyl-L-alanine amide by using the GIAO-CHF method with 4-31G *ab initio* MO basis sets. The 4–31G optimized geometries for the peptide were employed. (a) isotropic; (b) σ_{11}; (c) σ_{22}; and (d) σ_{33} for the C_β carbon (in ppm), and (e) isotropic; (f) σ_{11};

(Continues)

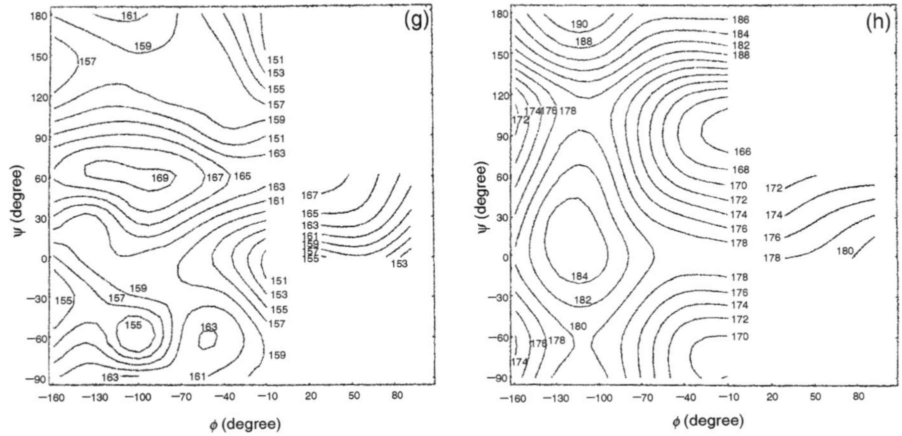

Figure 3.2. (*Continued*) (g) σ_{22}; and (h) σ_{33} for the C_α carbon (in ppm).[57]

In contrast, the calculated orientation of the principal axis system for the C_α carbon is quite different from sample to sample. For all the torsion angles used for the calculations, the σ_{33} component of the chemical shift tensor for Ala C_α carbons always lies along the C_α–C′ bond, the σ_{11} component lies in a slightly deviated direction from the $C_\alpha-C_\beta$ bond: and for $(\phi,\psi) = (-138.8°,134.7°)$ [β_A-sheet], the σ_{11} component is along this direction. The tensor component which is nearly along the $C_\alpha-C_\beta$ bond is 47.53 ppm for the β_A-sheet form, 61.93 ppm for the α_R-helix form, 64.74 ppm for the 3^R_{10}-helix, and 65.79 ppm for $(\phi,\psi) = (-84.3°,159.0°)$ [3_1-helix]. The change of the torsion angles causes the large deviation of the chemical shift tensor component that is along the $C_\alpha-C_\beta$ bond. Moreover, since σ_{33} depends on changes from one torsion angle to another, it is obvious that there exists the explicit torsion angle dependency on σ_{33}. It is thought that if the carbonyl group in the Ala residue forms a hydrogen bond, σ_{33} will be probably affected.

de Dios *et al.*[59] have studied ^{13}C chemical shift behavior of polypeptides by *ab initio* MO calculation of ^{13}C chemical shifts of several kinds of oligopeptides and by employing the representation of FPT INDO ^{13}C chemical shift map as function of the torsion angles as mentioned above. Sternberg *et al.*[12] have published a work on *ab initio* MO calculations of peptides.Undoubtedly, inclusion of electron correlation effect is essential for accurate chemical shift calculation by *ab initio* MO, even though most of the calculations on large molecules have been carried out without taking electron correlation into account. GIAO-MP2[60] is frequently utilized for the *ab initio* chemical shift calculation based on the second-order perturbation expansion approaches to

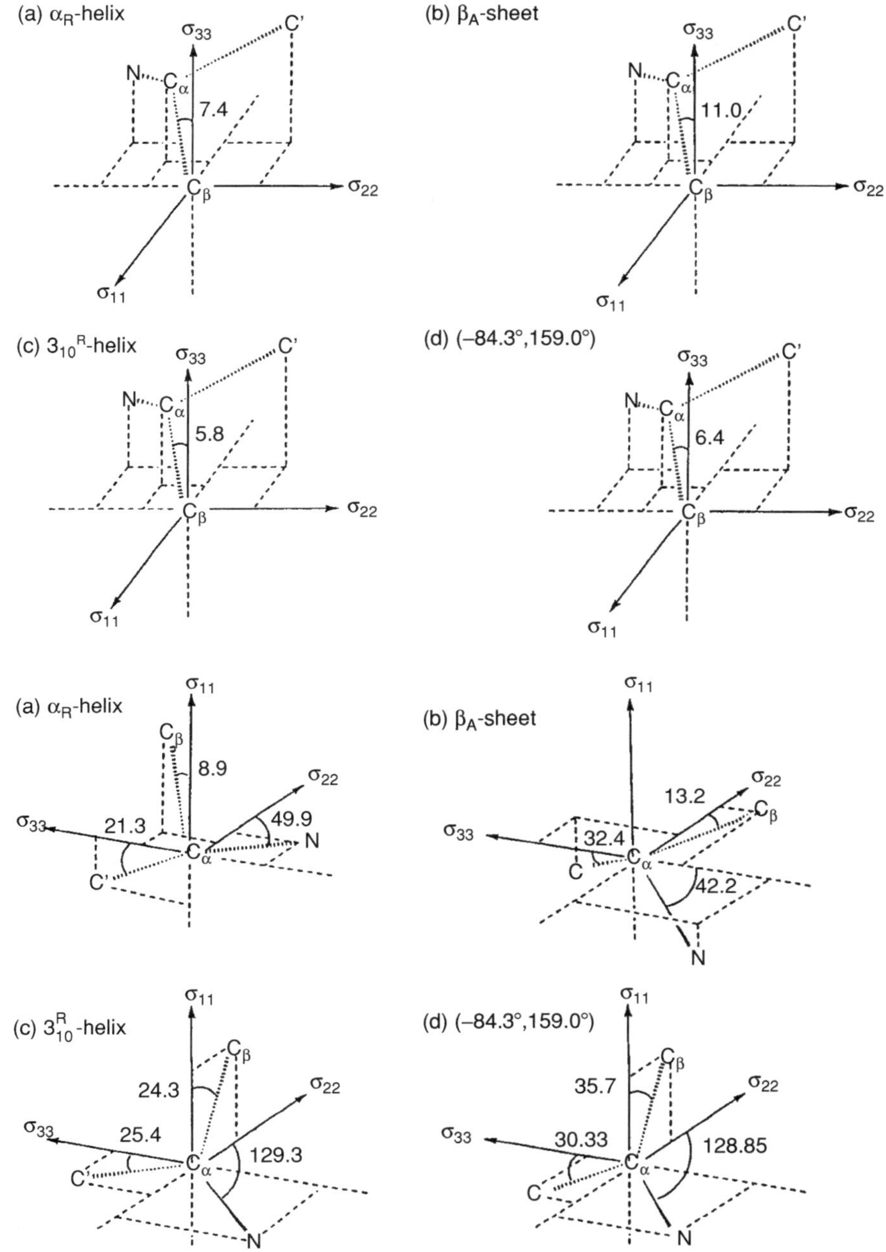

Figure 3.3. Orientation of the principal axes of the calculated ^{13}C chemical shift tensor components of the L-alanine residue C_β-carbon in *N*-acetyl-*N'*-methyl-L-alanine amide: (a) $(\phi, \psi) = (-57.4°, 47.5°)$ (α_R-helix); (b) $(-66.3°, 134.7°)$ (β_A-sheet); (c) $(-66.3°, -24.1°)$ (3^R_{10}-helix); and (d) $(-84.3°, 159.0°)$.[58]

electron correlation by Møller–Plesset. Highly precise calculations by *ab initio* MO, however, are inevitably restricted to small molecular systems because of computational costs. Instead, density functional theory[61] (DFT) method, based on Beck's[62] and Lee–Yang–Parr's[63] (BLYP) gradient-corrected exchange correlation functionals in combination with the 3–21G or 6–31G basis set using the GIAO method, is frequently utilized as an alternative means useful for shift calculation of medium or larger molecular systems taking into account the modest computational costs. Various program packages capable of calculating NMR shielding constants implemented with the above-mentioned methods and others are available.[64] For instance, ^{13}C chemical shifts for retinylidene chromophore and chlorophylls were compared with the experimental data treated in this book (see Chapter 14).[65,66]

3.1.2. Approach Using Infinite Polymer Chains

In order to understand the relationship between the NMR chemical shift and the electronic structure of polymers, it is necessary to use a quantum chemical approach.[25–37] The electronic structure of polypeptides with a periodic regular structure has been studied using the TBMO theory within the band theory that is known in the field of solid state physics.[22–24] The formulae for calculating the ^{13}C NMR chemical shift and its tensor for a single infinite biopolymer or polymer chain and for multi-infinite polypeptide chains have been presented using the TB theory based on some semi-empirical MO methods such as extended Hückel, CNDO/2, and INDO/S methods incorporated with the SOS method of the NMR chemical shift theory.[25–35,68,69] Such approaches are useful for obtaining appropriate knowledge of the relationship between the electronic structure and the ^{13}C NMR chemical shifts of solid polypeptides. From this, we can obtain useful information about the intramolecular short-range interactions, but cannot obtain exact information about the intermolecular long-range interactions in the crystalline state. As polypeptide crystals have a three-dimensional periodicity, one needs to employ infinite polypeptide chains with three-dimensional periodicity for obtaining the electronic structure.[36,37]

Sometimes, the estimation of the electronic structures of polypeptide chains necessitates the inclusion of long-range and intermolecular interactions in the chemical shift calculations. To do this, it is necessary to use a sophisticated theoretical method which can take into account the characteristics of polypeptides. In the TB approximation, the wave function $\psi(\mathbf{k})$ for an electron at position $\mathbf{r}$, which belongs to the nth crystal orbital (CO) is expressed with Bloch's theory by

$$\psi_n(\mathbf{k}) = N^{-1/2} \sum_{\nu} \sum_{\mathbf{R}} C_{\nu n}(\mathbf{k}) \phi_{\nu}(\mathbf{r} - \mathbf{R}) \exp(i\mathbf{k} \cdot \mathbf{R}), \tag{3.6}$$

where v is an index of atomic orbital, N is the total number of unit cells, **k** is the wave vector expressed by $k_x + k_y + k_z$, and **R** is the lattice vector. The symbol i denotes the imaginary number, and $C_{\nu n}(\mathbf{k})$ is the expansion coefficient for atomic orbital $\phi_\nu(\mathbf{r{-}R})$. Using the expansion coefficients obtained $C_{\nu n}(\mathbf{k})$, we can estimate the ^{13}C NMR chemical shift, which is related to the nuclear shielding constant $\sigma(\mathbf{k})$. In general, the nuclear shielding constant $\sigma_A(\mathbf{k})$ for atom A can be written like Eq. (3.7) as[25–36]

$$\sigma_A(\mathbf{k}) = \sigma_A^d(\mathbf{k}) + \sigma_A^p(\mathbf{k}) + \sigma'(\mathbf{k}) \tag{3.7}$$

where σ_A^d and σ_A^p are the diamagnetic and paramagnetic contributions, respectively, and σ' is a contribution from neighboring atoms such as the ring current effect, magnetic anisotropy, etc. For the carbon atom σ' is much smaller than 1 ppm, and so can be negligible compared with σ_A^d and σ_A^p. Thus, σ can be estimated by the sum of σ_A^d and σ_A^p. The formalism for calculating the NMR chemical shift in the framework of the LCAO approximation with the neglect of integrals involving more than two centers is derived. In this work, although the *ab initio* crystal orbital (CO) method[67] was used to calculate COs, the same concept was applied to derive the formalism of the NMR chemical shift as the first step to combining the three-dimensional CO method and the SOS method for calculating ^{13}C NMR chemical shifts.[36,37] Inclusion of the multicenter integrals in the calculation of ^{13}C NMR chemical shifts with *ab initio* COs is important.

In order to compare the total shielding constant $\sigma_A(\mathbf{k})$ with the experimental data, one needs to integrate over the first Brillouin zone (BZ) as expressed by

$$\sigma_A = \frac{\Omega}{8\pi^3} \int_{BZ} \{\sigma_A^d(\mathbf{k}) + \sigma_A^p(\mathbf{k})\} d\mathbf{k}, \tag{3.8}$$

where Ω is the volume of the primitive cell. This method has been successfully applied to the ^{13}C chemical shift calculations of a single infinite chain of various polymers[25,26,30–35,68,69] and polypeptides[27–29,71,72] to explain the observed ^{13}C chemical shift behavior.

Uchida *et al.*[36,37] have derived formulae for calculating the NMR chemical shift of a 3D polymer crystal by a combination of *ab initio* TBMO theory and the SOS method of the chemical shift theory.[44] This formalism was applied to the calculation of the ^{13}C NMR chemical shifts of 3D polyethylene chains in the orthorhombic and the monoclinic crystallographic forms by using the STO-3G minimal basis set. The inter- and intrachain interactions on the ^{13}C NMR chemical shift and the band structure were evaluated by changing the *a*, *c*, and *b* axis lattice constants. Further, the calculated results on *cis*- and *trans*-polyacetylene crystals explain reasonably the experimental data. This means that 3D *ab initio* TBMO theory can be applied to biopolymer crystals.

3.2. Relaxation Parameters

As in solution NMR, spin relaxation times are also very important parameters in the solid, not only to record spectra under the optimum conditions but also to gain insight into dynamic feature of biopolymers in the solid, gels, or membrane environment. In particular, several kinds of relaxation parameters including those sensitive to low-frequency motions are available from CP-MAS and DD-MAS NMR measurements for spin-1/2 nuclei, and they can be utilized as convenient probes to determine frequencies of local molecular fluctuations at particular sites for a variety of biopolymers. They include the ^{13}C and ^{1}H spin–lattice relaxation times in the laboratory frame (T_1^{C} and T_1^{H}, respectively), the carbon and proton spin–lattice relaxation times in the rotating frame ($T_{1\rho}^{\mathrm{C}}$ and $T_{1\rho}^{\mathrm{H}}$, respectively), spin–spin relaxation time (T_2), and the C−H cross relaxation times (T_{CH}).[73–75] In general, fully hydrated biopolymers such as membrane proteins, polysaccharide gels, etc. are not always dynamically homogeneous, but heterogeneous arising from contributions of rigid portions of solid-like, solution-like, or flexible portions undergoing a variety of motions like in solution, in spite of their solid-like appearance. To identify the latter portion, recording ^{13}C NMR spectra by both DD-MAS and CP-MAS NMR is strongly recommended, together with measurements of the ^{13}C spin–lattice relaxation time (T_1^{C}) as determined by the conventional inversion-recovery method (180°–τ–90°-acquire). This is because no ^{13}C NMR signals are available from such flexible portions by CP-MAS NMR technique alone.

3.2.1. ^{13}C Spin–Lattice Relaxation Times in the Laboratory Frame (T_1^{C})

The ^{13}C spin–lattice relaxation times in the laboratory frame, T_1^{C} s, from the solid-like portions are conveniently measured by CP-MAS method with an inversion of carbon spin temperature, to remove signals established by shorter spin–lattice relaxation from such mobile components rather than the cross-polarization and transient signals generated by magnetoacoustic ringing of the receiver coil.[76] This pulse sequence consists of two similar but not identical pulse sequences (Figure 3.4, top, right). The first sequence begins with the application of a resonant field B_{1H}, in the rotating frame of the protons which rotates the proton polarization into the *xy* plane (Figure 3.4, middle). After the desired ^{13}C polarization along B_{1C}, contact is broken by turning off $B_{1H.}$ Simultaneously, B_{1C} is phase shifted 90° and rotates the carbon magnetization from the *x*-axis to the *z*-axis, at which time B_{1C} is turned off. In the absence of an B_{1C} field, the proton-enhanced longitudinal magnetization, $M_{\mathrm{CP}}(t)$, changes exponentially from its initial value, $M_{\mathrm{CP}}(0)$, to its equilibrium value, M_0, with the time constant equal to T_1^{C}. Carbon magnetization, $M_{\mathrm{A}}(t)$, which does not arise from CP may also be present as demonstrated above. The total longitudinal magnetization $M_{z1}(t)$ is thus given by

$$M_{z1}(t) = [M_{\mathrm{CP}}(0) - M_0] \exp\left(-t/T_1^{\mathrm{C}}\right) + M_0 + [M_{\mathrm{A}}(t)]_z, \tag{3.9}$$

where $[M_{\mathrm{A}}(t)]_z$ is the *z*-component of the carbon magnetization without cross polarization (Figure 3.4, bottom). In the second sequence, the total magnetization $M_{z2}(t)$ is given by

$$M_{z2}(t) = [-M_{\mathrm{CP}}(0) - M_0] \exp\left(-t/T_1^{\mathrm{C}}\right) + M_0 + [M_{\mathrm{A}}(t)]_z \tag{3.10}$$

in which the initial B_1^{C} pulse is phase shifted by 180° relative to the initial $B_{1\mathrm{C}}$ pulse in the first sequence. The net signal after *N* scans is proportional to

$$\begin{aligned} M_{\mathrm{net}}(t) &= N[M_{z1}(t) - M_{z2}(t)] \\ &= 2NM_{\mathrm{CP}}(0) \exp\left(-t/T_1^{\mathrm{C}}\right), \end{aligned} \tag{3.11}$$

if the signal derived from the second sequence (Eq. (3.10)) is subtracted from that derived from the first (Eq. (3.9)) after *N* scans. A semilog plot of $M_{\mathrm{net}}(t)$ vs *t* yields the T_1^{C} value from its gradient $-1/2.303T_1^{\mathrm{C}}$.

As an illustrative example, Figure 3.5 exhibits such a stacked plot of $M_{\mathrm{net}}(t)$ of (1 → 3)-β-D-glucan (curdlan)[77] (Figure 3.6) lyophilized from DMSO solution against the delay time *t*. T_1^{C} values of lyophilized curdlan obtained from a plot of log $M_{\mathrm{net}}(0)$ vs *t* are summarized in Table 3.1, together with those of annealed sample at 150°C taking the triple helix conformation. The shortest T_1^{C} values (1.2–1.6 s) are observed from the C-6 hydroxylmethyl carbon undergoing rotational reorientation, although the longest ones (10–18 s) are from the C-1 to C-3, and the C-4 and C-5 carbons exhibit intermediate values (8–10 s). It appears that the presence of methyl or hydroxymethyl groups, undergoing rapid reorientatonal motions with correlation times in the order of 10^{-8} s, strongly affects the spin–lattice relaxation times of nearby carbons due to mutual dipolar couplings, instead

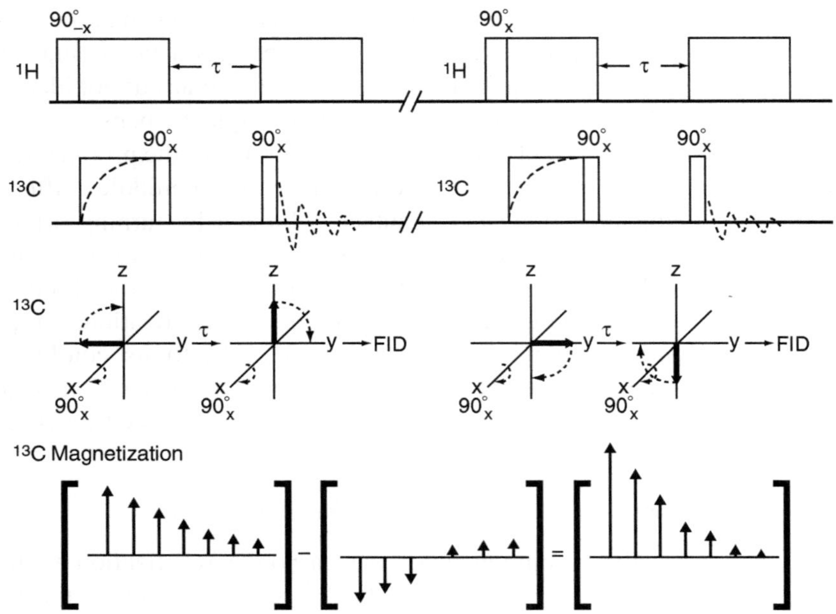

Figure 3.4. Pulse sequences used to measure the ^{13}C spin–lattice relaxation times in the Zeeman frame (T_1^C).[76]

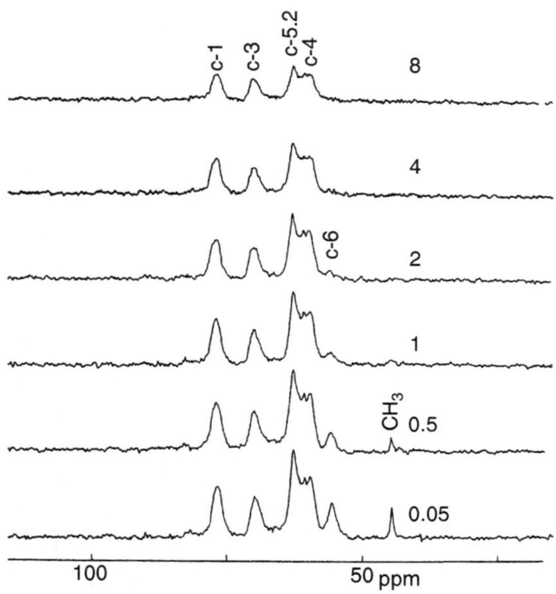

Figure 3.5. Stacked plot of ^{13}C NMR peak-intensities of curdlan lyophilized from DMSO, obtained by CP for the T_1^C measurements. Delay times (s) are given in the right-hand side.[77]

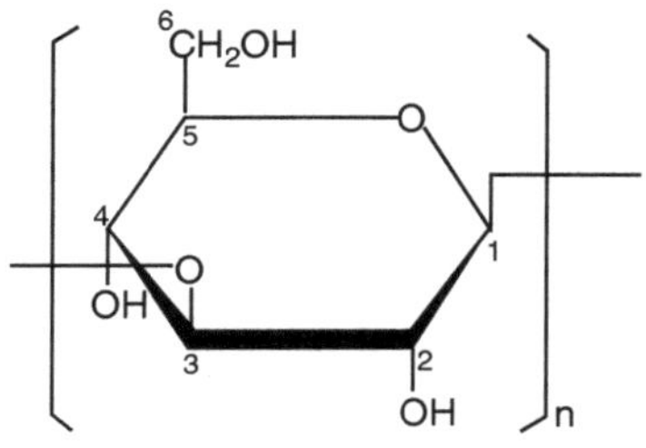

Figure 3.6. Chemical structure of (1 → 3) -β-D-glucan (curdlan).

of slow overall motions, if any, ineffective to the T_1 process of the laboratory frame.[78,79] In fact, it was shown that very long T_1^C values (50–800 s) were reported for highly crystalline cellulose samples,[80,81] where there exists no rapid reorientational motion in the C-6 hydroxymethyl group. Thus, the T_1^C values of the C-4 and C-5 carbons are appreciably shortened by protons undergoing rapid internal rotation, since the extent of the dipolar coupling is proportional to r^{-6}, where r is the interatomic distance. The ^{13}C magnetization decays, in principle, nonexponentially in the CP-MAS experiment due to the transient Overhauser effect if fast relaxing moiety such as methyl groups are present in a system and longer delay times are taken. In such case, Naito *et al.*[78] showed that the proton irradiation between the 90°_y and 90°_{-y} pulses in the pulse sequence in Figure 3.4 is effective to remove such effect and to yield the exponential decays of the magnetization as demonstrated in Eq. (3.11). In any case, motional information is unavailable from the ^{13}C T_1^C values in the laboratory frame for such a rigid system.

In contrast, it is emphasized that this method is best suited for T_1^C measurements in two-phase systems very often encountered for a variety of

Table 3.1. ^{13}C spin–lattice relaxation times in the laboratory frame (T_1^C) for (1→3)-β-D-glucans taking single chain and triple helix (s).[79]

	Single chain[a]	Triple helix[b]
C-1	15	16
C-2	11	18
C-3	16	16
C-4	10	8.2
C-5	9.7	9.3
C-6	1.6	1.2

[a]Lyophilized from DMSO solution.
[b]Annealed curdlan at temperature 150°C.

biopolymers under the condition of fully hydrated state, because the presence of intense ^{13}C NMR signals, if any, from soluble or mobile components can be very easily eliminated in the CP-MAS experiment. As an illustrative example for a typical membrane protein, Tuzi *et al.*[82] measured the T_1^C values of cytochrome *c* oxidase (M_r = 200 kDa) in 3D crystal in the presence of BL8SY as a detergent, compared to those of crystalline lysozyme as a typical globular protein, and which are summarized in Table 3.2. The backbone carbons (C=O and C_α) of cytochrome *c* oxidase exhibit longer T_1^C values than the side-chain carbons because of lower mobility in the rigid lattice of the crystal. Decreased T_1^C values in some aromatic and aliphatic side-chain carbons, such as tyrosine or phenylalanine, and methyl carbons of leucine, valine, alanine, isoleucine, etc. (19.9, 16.0, and 12.0 ppm), are explained in terms of internal motions, such as continuous diffusion about the $C_\beta-C_\gamma$ axis or rotation about the C_3 in a timescale of 10 ns, as observed for fibrous proteins. Nevertheless, the T_1^C values of backbone carbons (C=O and C_α) of cytochrome *c* oxidase are almost one-half of those observed for lysozyme

Table 3.2. T_1^C, T_{CH} and $T_{1\rho}^H$ of lysozyme and cytochrome *c* oxidase in the presence of BL8SY in the crystalline state.[82]

	^{13}C NMR peaks	ppm	T_1^C (s) (in liquid)	T_{CH} (ms)	$T_{1\rho}^H$ (ms)
Lysozyme	C=O	174.4	32.9	0.39	2.5
	C_α	53.3	16.3	0.19	2.7
	Side-chains aliphatic	40.5	8.2 (52)[a] 0.4 (48)[a]	0.15	2.8
		24.8	6.2 (38)[a] 0.6 (62)[a]	0.19	2.7
		20.2	0.51	0.23	2.8
		16.2	0.38	0.24	2.8
Cytochrome-c oxidase	C=O	175.5	15.0	0.27	5.1
	C_α	56.2	6.4	0.073	13
	Side-chains: aromatic	128.4	0.55	n.d.	n.d.
	Aliphatic	40.2	4.8 (28)[a] 0.39 (72)[a]	0.10	25.7
		24.8	4.2 (82)[a] 0.48 (18)[a]	n.d.	n.d.
		20.6	0.59	0.16	4.5
		16.0	0.96	0.18	4.9
		12.0	0.99	0.4	2
BL8SY	CH_3	14.1	n.d. (2.1)[b]	n.d.	n.d.
	CH_2	30.0	0.71 (0.87)[b]	1.01	57.5
	OCH_2CH_2	70.1	0.46 (0.61)[b]	3.83	542

[a] Relative intensity for the component.
[b] ^{13}C T_1^C values from liquid components determined by DD-MAS experiment.

crystal, in spite of the similarity in T_1^C of the side-chain carbons. Therefore, it is pointed in view of the T_1^C data that the backbone carbons of cytochrome *c* oxidase fluctuate much more in the crystal containing detergents than in crystalline lysozyme. Further, the T_1^C data of BL8SY as a detergent are very similar, as expected by both CP-MAS and DD-MAS NMR.

3.2.2. ^{13}C Resolved, ^{1}H Spin–Lattice Relaxation Time T_1^H and Carbon and Proton Spin-Lattice Relaxation Times ($T_{1\rho}^C$ and $T_{1\rho}^H$) in the Rotating Frame

The ^{13}C resolved, ^{1}H spin–lattice relaxation time T_1^H is more conveniently measured via detection of ^{13}C resonances,[83] instead of direct measurements of substantially broadened signals, as illustrated in the pulse sequence demonstrated in Figure 3.7. In the proton channel, a 180°-τ–90° pulse sequence routinely used in solution NMR experiments is inserted prior to the spin-lock in the cross polarization (Figure 3.7, top). Behavior of spins during the sequence was also shown in Figure 3.7, bottom. Here, one indirectly obtains a measure of ^{1}H relaxation during the τ period, by observing the well-resolved ^{13}C magnetization generated during the contact time as a function of τ. It is expected, however, that in most pure organic solids rapid spin diffusion maintains a uniform spin temperature among all of the protons in the sample. Thus, they exhibit a single T_1^H relaxation time as manifested from the data of single chain (0.30–0.34 s) and triple helix forms[77] (0.48–0.62 s) of (1→3)-β-D-glucan, in spite of the significantly different corresponding carbon relaxation times T_1^C (Table 3.1).

The lifetime of carbon and proton magnetization for spin-locked along the carbon B_1^C and proton B_1^H is designated as the ^{13}C and ^{1}H spin–lattice relaxation times in the rotating frame, $T_{1\rho}^C$ and $T_{1\rho}^H$, respectively. In contrast to the spin–lattice relaxation times in the laboratory frame which are sensitive to the motions with frequencies $>10^8$ Hz as discussed above, these parameters are expected to be sensitive to lower frequency of molecular motions in the regions of tens of kHz. Unlike the corresponding ^{1}H $T_{1\rho}^H$ experiment, ^{13}C $T_{1\rho}^C$ should not be confused by spin diffusion between carbons, because their low natural abundance ensures a physical separation within the solid. In this connection, Schaefer *et al.* [84] examined ^{13}C $T_{1\rho}^C$ data of glassy polymers in terms of distinct main- and side-chain motions in the 10–50 kHz regime. Nevertheless, these values are mainly determined by spin–spin process rather than motions, unless otherwise the carbon B_{1C} is

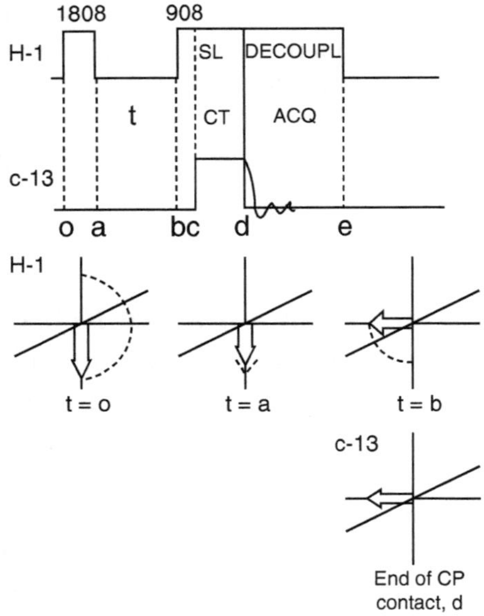

Figure 3.7. Top, Pulse sequence of T_1^{H} determination via ^{13}C CP-MAS method. DECOUPL, decoupling; SL, spin lock. Bottom, Behavior of spins during the sequence (from Sullivan and Maciel[83]).

less than the local dipolar field. In addition, there is no simple NMR theory to interpret the ^{13}C $T_{1\rho}^{\mathrm{C}}$ values.[85] In contrast, ^{13}C *resolved,* ^{1}H $T_{1\rho}^{\mathrm{H}}$ values can be conveniently utilized as a probe to detect such above-mentioned slow motions because of experimental simplicity, although there exists no distinction about the relaxation times among protons because of rapid spin diffusion process. This parameter is easily evaluated by a nonlinear least-squares fit of the ^{13}C peak intensities $I(t)$ against the contact time t, from a stacked spectral plot of ordinary spectra as a function of the contact time, together with cross polarization time (T_{CH}).

$$I(t) = [I(0)/T_{\mathrm{CH}}] \times \left[\exp\left(-t/T_{1\rho}^{\mathrm{H}}\right) - \exp(-t/T_{\mathrm{CH}})\right]\left[1/T_{\mathrm{CH}} - 1/T_{1\rho}^{\mathrm{H}}\right]^{-1}, \quad (3.12)$$

where $I(0)$ denotes the initial peak intensity.[86]

In Table 3.2, the $T_{1\rho}^{\mathrm{H}}$ and T_{CH} values of crystalline lysozyme and cytochrome-*c* oxidase are summarized. Interestingly, the $T_{1\rho}^{\mathrm{H}}$ values for lysozyme crystal are almost identical among protons at different sites (2.5–2.8 ms), because they are averaged by dominant spin diffusion process.[82] In fact, the range of the proton spin diffusion on $T_{1\rho}^{\mathrm{H}}$ is effective over 2 nm in the case of a rigid solid.[87] By contrast, the $T_{1\rho}^{\mathrm{H}}$ values of cytochrome-*c* oxidase crystal in the presence of detergent BL8SY ($CH_3(CH_2)_{11}(OCH_2$-$CH_2)_8OH$) are remarkably different (2–542 ms) among protons at different sites of the protons and also detergent molecules. This is because spin diffusion is no longer effective in the presence of tumbling motions within a time scale of at least 100 kHz especially for detergent molecules and side-chains located at the surface. This means that membrane proteins surrounded by detergents or lipids are not always rigid as conceived but are very flexible, in contrast to the case of crystalline lysozyme. In particular, information as to the presence of restricted molecular motion in the detergent molecules at the protein–detergent interface is available from the *cross polarization time* (T_{CH}). T_{CH} of the polar head region of the detergent is longer than that of aliphatic region (Table 3.2). T_{CH} is expressed by[75]

$$1/T_{CH} = C_{IS}M_2^{IS}/(M_2^{II})^{1/2}, \tag{3.13}$$

where M_2^{IS} and M_2^{II} are the ^{13}C and ^{1}H second moments, respectively, and C_{IS} is a constant depending on geometrical factors. Since T_{CH} is a function of the second moment, and T_{CH} is sensitive to the static component and reflects the existence of slow motion. Therefore, the higher T_{CH} of the polar head region, compared to that of the aliphatic region, implies that the aliphatic region of the detergent is strongly bound to the hydrophobic surface of the protein molecules.

3.2.3. Carbon Spin–Spin Relaxation Times T_2^C Under CP-MAS Condition

Carbon spin–spin relaxation times T_2^C under CP-MAS condition can provide motional information with frequencies of 10^4–10^5 Hz about the individual carbon site of interest, in contrast to the case of the proton spin–lattice relaxation time in the rotating frame $T_{1\rho}^H$, in which information on individual sites would be masked by the presence of a rapid spin–spin process. The T_2^C values under the condition of magic angle spinning and ^1H decoupling can be measured[88] using a CPMG (Carr–Purcell—Meiboom–Gill) spin echo pulse[89,90] sequence by adjusting the interval between 180° and the starting point of acquisition to be multiples of the rotor cycle, as illustrated in Figure 3.8. In general, the T_2^C values strongly depend on coherent frequencies of the proton decoupling or MAS which are interfered with incoherent frequency of molecular fluctuation motions. Therefore, the overall relaxation rate, T_2^C can be given by[88,91,92]

$$1/T_2^C = (1/T_2^C)^S + (1/T_2^C)_{DD}^M + (1/T_2^C)_{CS}^M, \tag{3.14}$$

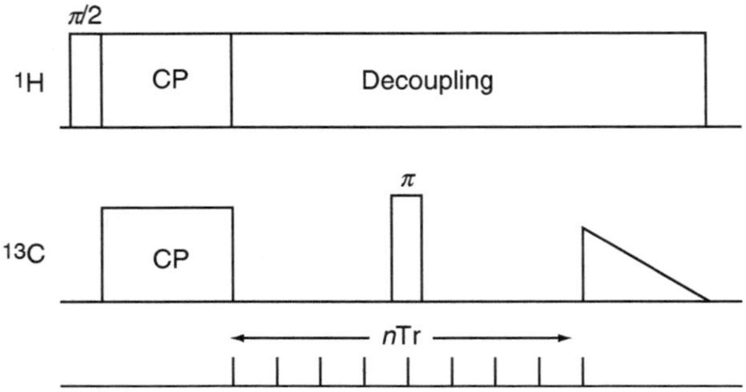

Figure 3.8. Pulse sequence for determination of spin–spin relaxation time (T_2^{C}) using a Carr–Purcell–Meiboom–Gill sequence.[88]

where $(1/T_2^{\mathrm{C}})^S$ is the transverse component due to static C−H dipolar interactions, and $(1/T_2^{\mathrm{C}})^M_{DD}$ and $(1/T_2^{\mathrm{C}})^M_{\mathrm{CS}}$ are the transverse components due to the fluctuation of the dipolar and chemical shift interactions respectively. The latter two terms are given as a function of correlation time, τ_{c}, by

$$(1/T_2^{\mathrm{C}})^M_{\mathrm{DD}} = \frac{4\gamma_I^2\gamma_S^2\hbar^2}{15r^6} I(I+1)\frac{\tau_{\mathrm{c}}}{1+\omega_I^2\tau_{\mathrm{c}}^2} \tag{3.15}$$

$$(1/T_2^{\mathrm{C}})^M_{\mathrm{CS}} = \frac{\omega_0(\Delta\delta)^2\eta^2}{45}\left(\frac{\tau_{\mathrm{c}}}{1+4\omega_{\mathrm{r}}^2\tau_{\mathrm{c}}^2} + \frac{2\tau_{\mathrm{c}}}{1+\omega_{\mathrm{r}}^2\tau_{\mathrm{c}}^2}\right), \tag{3.16}$$

where γ_I and γ_S are the gyromagnetic ratios of the I and S nuclei respectively, r is the internuclear distance between spins I and S, ω_0 and ω_{r} are the carbon resonance frequency and the amplitude of the proton decoupling rf field respectively, ω_r is the rate of spinner rotation, $\Delta\delta$ is the CSA, and η is the asymmetric parameter of the chemical shift tensor. Obviously, the transverse relaxation rate is dominated by modulation of either dipolar interactions or CSAs, if the presence of internal fluctuations cannot be ignored as in membrane proteins. In general, it is expected that a decoupling field of 50 kHz is sufficient to reduce the static component, and the second $(1/T_2^{\mathrm{C}})^M_{\mathrm{CS}}$ term will be dominant in the overall $1/T_2^{\mathrm{C}}$ in Eq. (3.14), as far as carbonyl groups with larger chemical shift anisotropies are concerned. In addition, it is expected the C_α carbon signal could be affected by both the $(1/T_2^{\mathrm{C}})^M_{\mathrm{DD}}$ and $(1/T_2^{\mathrm{C}})^M_{\mathrm{CS}}$ terms, depending upon the frequency range of either 50 kHz (ω_I) or 4 kHz (ω_{r}) respectively. As an illustrative example, typical T_2^{C} values for simple crystalline peptides, Gly-Gly and Val-Gly-Gly, are summarized in Table 3.3, respectively, together with their T_1^{C} values.[88] It turned out that the T_2^{C} values of the CO carbons are shorter than those of the

Table 3.3. ^{13}C T_2^C (ms) and T_1^C (s) values of Gly-Gly[a] and Val-Gly-Gly[a] in the crystalline state.[88]

	Val					Gly		Gly	
	CO	C_α	C_β	C_γ		CO	C_α	COO^-	C_α
Gly-Gly									
T_2^C						16.9	7.8	32.0	3.4
T_1^C						212	249	186	69
Val-Gly-Gly									
T_2^C	21.3	8.7	10.5	36.0	41.1	15.5	7.0	40.8	7.0
T_1^C	*43.4*	7.6	8.0	0.7	0.35	103.6	71.7	28.9	71.7

[a]The first and second Gly residues correspond to Gly[1] and Gly[2], and Gly[2] and Gly[3] for Gly-Gly and Val-Gly-Gly, respectively.

COO^- for both Gly-Gly and Val-Gly-Gly, indicating that the CO carbons are more mobile than those of the COO^- carbons. The T_2^C values of the CO carbons in Leu- and Met-enkephalins (3.1–5.9 s) (not shown) are much shorter than those of the other two peptides, indicated in Table 3.3. The difference suggests that the peptide planes of enkephalins are more flexible than those of Gly-Gly and Val-Gly-Gly. It is interesting to note that the methyl carbons of the Val residue are very long because the reorientation of the methyl groups is so rapid that the T_2^C values cannot be affected by the fluctuation of the C−H dipolar interactions and the CSA in the proton decoupling of 50 kHz. This is consistent with the fact that the quite long T_1^C are obtained for Gly-Gly as compared with those of Val-Gly-Gly which undergoes local motions with a frequency of 10^8 Hz.

3.3. Dynamics-Dependent Suppression of Peaks

The spectral line width $\nu_{1/2}$ is related to *the spin–spin relaxation time* T_2^C *by*

$$\nu_{1/2} = 1/\pi T_2^C \tag{3.17}$$

for a Lorentzian line. In the solids, however, it should be anticipated that the spectral line width is also strongly affected by the manner of sample preparation: narrow spectral lines in crystalline preparation arise from the presence of a unique conformation, while broader lines in amorphous preparations are from superimposed peaks arising from a number of different conformations in which individual peaks are appreciably displaced due to the conformation-dependent displacement of peaks to be described later.

In addition, it is noted that ^{13}C NMR signals of fully hydrated biopolymers are not always fully visible at ambient temperature. If there are isotropic or large-amplitude motions, whose correlation times are shorter than 10^{-8} s, as in the case of the N- and C-terminal residues in bR, several peaks from such areas could be selectively suppressed in CP-MAS NMR, although they are not in DD-MAS (see Figure 2.7(a)) as schematically demonstrated in the blanked area **a** in Figure 3.9. Therefore, presence of this kind of motion can be very easily detected by observation of the prolonged spin–lattice relaxation times by DD-MAS experiments, as illustrated by the horizontal bar in Figure 3.9.[91–94] This type of motion was readily recognized at ambient temperature, because ^{13}C CP-MAS NMR signals from the N- or C-terminal residues were partially suppressed as compared with the ^{13}C DD-MAS NMR spectra.[95] The latter approach is suitable for detecting signals from the whole area of [3-^{13}C]Ala-labeled proteins in view of the relatively shorter spin–lattice relaxation time of the Ala C_{β} carbons in the order of 0.5 s,[96] if there is no additional *incoherent*

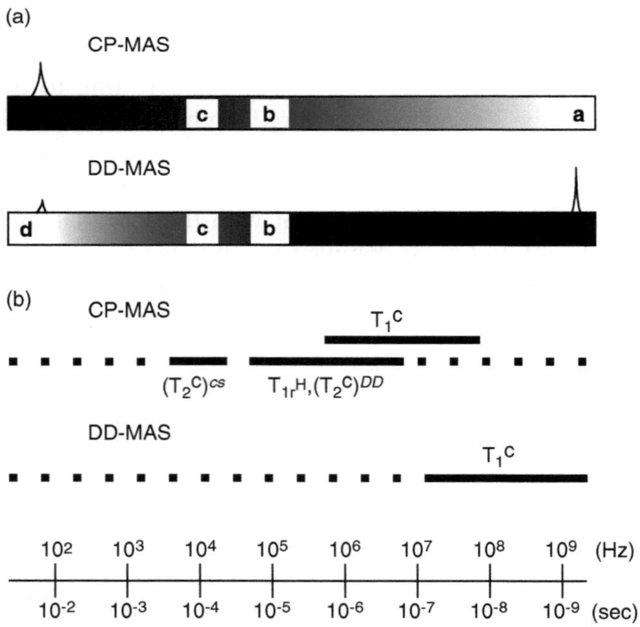

Figure 3.9. Detection of several types of motions either by observation of suppressed peaks (a) or measurements of relaxation parameters as a function of respective motional frequency (Hz) or correlation times (b). NMR peaks were suppressed by: fast isotropic motions (**a**), interferenece with proton decoupling frequency (b), and MAS frequency (c), and longer spin–lattice relaxation times as compared with repetition times (d).[93]

random motions which result in interference with the peak-narrowing process by *coherent* proton decoupling and/or the MAS process in CP-MAS or DD-MAS spectra.[91,92]

It is demonstrated that selective suppression of peaks is noticed as a result of failure of attempted peak-narrowing process by the proton decoupling or MAS, when either correlation time of incoherent molecular motion, if any, τ_c, is equal to the modulation period of the decoupling ($1/\omega_I$) or MAS ($1/\omega_r$), respectively, as well documented for ^{13}C CP-MAS NMR spectra for a number of synthetic polymers (see $(T_2^C)^M_{DD}$ or $(T_2^C)^M_{CS}$ terms in Eq. (3.14)). Figure 3.10 illustrates a typical example of such phenomenon arising from temperature-dependent peak-suppression of methyl group for ^{13}C CP-MAS NMR spectra of polypropylene (PP).[97] At temperatures around 105 K, the methyl signal resonating at the highermost peak position completely disappears as a result of the interference of fluctuation frequency with frequency of the proton decoupling, as discussed above. This peak, however, turns out to be completely recovered at higher temperatures (300 K) in the "short correlation time" limit ($\omega_1\tau_c \ll 1$) (high temperature) which allows the rapid motional averaging of methyl group about the C_3 axis free from such interference. At temperatures below 77 K, on the other hand, this peak seems to be partially recovered because of taking the "long correlation

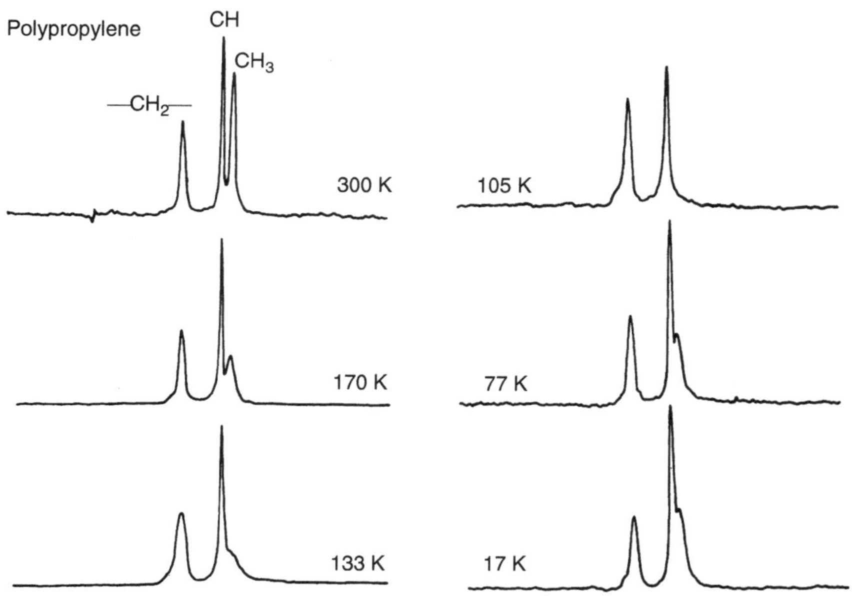

Figure 3.10. ^{13}C CP-MAS NMR spectra of polypropylene (PP) as a function of temperature.[97]

time'' regime ($\omega_1\tau_c \gg 1$) of slow motion, to result in situation free from such interference. The maximum broadening can be predicted to occur at $\tau_c = 2.8$ μs for a value of $\omega_2/2\pi = 57\,kHz$ based on Eq. (3.15). Using the proton $T_{1\rho}^{H}$ relaxation time data on isotactic PP to assess the temperature at which this value of τ_c obtains[98] the maximum broadening is predicted to occur at 105 K, which is in excellent agreement with the observations.

It should be realized that any biopolymer, especially membrane proteins, is able to undergo a variety of flexible backbone or side-chain motions, depending upon the functional groups and sites where amino acid residues under consideration are located, in the fully hydrated state corresponding to physiological condition. In fact, molecular motions of such intermediate or slow frequency can be very easily detected when certain ^{13}C NMR signals from both the DD-MAS and the CP-MAS NMR spectra are simultaneously suppressed (blanked area **b**), as illustrated in Figure 3.9. Such motion was first recognized when the ^{13}C NMR signals of the C-terminus were almost completely suppressed both in the CP-MAS and DD-MAS NMR spectra, when the temperature was lowered to between −40° and −110°C.[99] Yamanobe *et al.*[100] showed that ^{13}C NMR signals of the backbone C_α and carbonyl (amide) carbons of poly(γ-*n*-octadecyl-L-glutamate) disappeared at a temperature above 40°C accompanied by undergoing reorientational motion of the α-helical main-chain with frequency of ca. 60 kHz, as a result of the melting of side-chain crystallites (Section 11.1.1). It should be noted that this kind of peak suppression occurs very often at ambient temperature for the whole range of transmembrane α-helices or a part of the loop for a variety of ^{13}C-labeled membrane proteins, including bR from PM[101] and blue membrane,[102] its mutants[103,104] and peptide fragments,[105] bacterioopsin,[106] phoborhodopsin,[107] its transducer,[108] *Escherichia coli* diacylglycerol kinase,[109] etc. depending upon the type of amino acid residues and also 2D crystal or monomer.

References

[1] K. A. K. Ebraheem and G. A.Webb, 1977, *Prog. NMR Spectrosc.*, 11, 149–181.

[2] I. Ando and T. Asakura, 1979, *Annu. Rep. NMR Spectrosc.*, 10A, 81–132.

[3] I. Ando and G. A. Webb, *Theory of NMR Parameters*, Academic Press, London, 1983.

[4] H. Saitô, R. Tabeta, A. Shoji, I. Ando and T. Asakura, in G. Govil, C. L. Khetrapal and A. Saran (Eds.), 1985, *Magnetic Resonance in Biology and Medicine*, pp.195–215, Tata McGraw-Hill, New Dehli.

[5] H. Saitô and I. Ando, 1989, *Annu. Rep. NMR Spectrosc.*, 21, 210–290.

[6] I. Ando, T. Yamanobe and T. Asakura, 1990, *Prog. NMR Spectrosc.*, 22, 349–400.

[7] A. Shoji, S. Ando, S. Kuroki, I. Ando and G. A. Webb, 1993, *Annu. Rep.. NMR Spectrosc.*, 26, 55–98.

[8] R. Born and H. W. Spiess, 1997, *Ab initio Calculations of Conformational Effects on ^{13}C NMR Spectra of Amorphous Polymers*: *NMR*, Vol. 35, Springer, Berlin.
[9] I. Ando and T. Asakura (ed.), 1998, *Solid State NMR of Polymers*, Elsevier Science, Amsterdam.
[10] J. C. Facelli and A. C. de Dios (ed.), 1999, *Modeling NMR Chemical Shifts: Gaining Insight into Structure and Environment, ACS Symp. Ser. 732,* American Chemical Society, Washington, DC.
[11] I. Ando, S. Kuroki, H. Kurosu and T. Yamanobe, 2001, *Prog. NMR Spectrosc.*, 39, 79–133
[12] U. Sternberg, R. Witter and A. S. Ulrich, 2004, *Annu. Rep. NMR Spectrosc.*, 52, 53–104.
[13] I. Ando, A. Nishioka and T. Asakura, 1975, *Makromol. Chem.*, 176, 411–437.
[14] I. Ando and A. Nishioka, 1975, *Makromol. Chem.*, 176, 3089–3101
[15] T. Asakura, I. Ando and A. Nishioka, 1975, *Makromol. Chem.*, 176, 1151–1161.
[16] I. Ando, Y. Kato and A. Nishioka, 1976, *Makromol. Chem.*, 177, 2759–2771.
[17] T. Asakura, I. Ando and A. Nishioka, 1977, *Makromol. Chem.*, 177, 1493–1500.
[18] I. Ando,Y. Kato, M. Kondo and A. Nishioka, 1977, *Makromol. Chem.*, 178, 803–816.
[19] I. Ando, 1996, *Encyclopedia of NMR*, eds. D. M. Grant and R. K. Harris, pp.176–180, John Wiley & Sons, New York.
[20] I. Ando, T. Yamanobe, H. Kurosu and G. A. Webb, 1990, *Annu. Rep. NMR Spectrosc.*, 22, 205–248.
[21] I. Ando, S. Kuroki, H. Kurosu, M. Uchida and T. Yamanobe, (J. C. Facelli and A.C. de Dios (ed.)), 1999, *ACS Symp. Ser.*, 732, 24–39.
[22] J.-M. Andre and J. Ladik, 1974, *Electronic Structure of Polymer and Molecular Crystals*, Plenum Press, New York.
[23] J.-M. Andre, J. Delhalle and J. Ladik, 1978, *Quantum Theory of Polymers*, Reidel, Dordrecht.
[24] J. Ladik, 1988, *Quantum Theory of Polymers as Solids*, Plenum Press, New York.
[25] T. Yamanobe, R. Chujo and I. Ando, 1983, *Mol. Phys.*, 50, 1231–1249.
[26] T. Yamanobe and I. Ando, 1985, *J. Chem. Phys.*, 83, 3154–3160.
[27] T. Yamanobe, I. Ando, H. Saitô, R. Tabeta, A. Shoji and T. Ozaki., 1985, *Chem. Phys.*, 99, 259–264.
[28] T. Yamanobe, I. Ando, H. Saitô, R. Tabeta, A. Shoji and T. Ozaki, 1985, *Bull. Chem. Soc. Jpn.*, 58, 23–29.
[29] T. Yamanobe, T. Sorita, T. Komoto, I. Ando and H. Saitô, 1987, *J. Mol. Struct.*, 151, 191–201.
[30] H. Kurosu, T. Yamanobe, T. Komoto and I. Ando, 1987, *Chem. Phys.*, 116, 391–398.
[31] T. Ishii, H. Kurosu, T. Yamanobe and I. Ando, 1988, *J. Chem. Phys.*, 89, 7315–7319.
[32] H. Kurosu, T. Yamanobe and I. Ando, 1988, *J. Chem. Phys.*, 89, 5261–5223.
[33] H. Kurosu and I. Ando, 1991, *J. Mol. Struct. (Theochem)*, 231, 231–242.
[34] M. Kikuchi, H. Kurosu and I. Ando, 1992, *J. Mol. Struct.*, 269, 183–195.
[35] T. Yamanobe and H. Kurosu, 1996, *Encyclopedia of NMR*, eds. D.M.Grant and R. K. Harris, pp. 4468–4474, John Wiley & Sons, New York.
[36] M. Uchida, Y. Toida, H. Kurosu and I. Ando, 1999, *J. Mol. Struct.*, 508, 181–191.
[37] K. Fujii, M. Uchida, H. Kurosu, S. Kuroki and I. Ando, 2002, *J. Mol. Struct.*, 602/603, 3–8.
[38] A. Saika and C. P. Slichter, 1954, *J. Chem. Phys.*, 22, 26–28.
[39] J. A. Pople, 1957, *Proc. R. Soc. London*, A239, 541–549.
[40] J. A. Pople, 1957, *Proc. R. Soc. London*, A239, 550–556.

[41] J. A. Pople, 1962, *J. Chem. Phys.*, 37, 53–59.
[42] J. A. Pople, 1962, *J. Chem. Phys.*, 37, 60–66.
[43] J. A. Pople, 1964, *Mol. Phys.*, 7, 301–306.
[44] I. Ando, A. Nishioka and M. Kondo, 1974, *Chem. Phys. Lett.* 25, 212–214.
[45] I. Ando, A. Nishioka and M. Kondo, 1976, *J. Mag. Reson.*, 21, 429–436.
[46] I. Ando and G. A. Webb, 1981, *Org. Magn. Reson.*, 15, 111–130.
[47] P. J. Flory, 1969, *Statistical Mechanics of Chain Molecules*, Interscience, NewYork.
[48] H. Saitô, R. Tabeta, A. Shoji, T. Ozaki and I. Ando, 1983, *Macromolecules*, 16, 1050–1057.
[49] T. Asakura, K. Taoka, M. Demura and M. P. Williamson, 1995, *J. Biomol. NMR*, 6, 227–236.
[50] H. Saitô, 1986, *Magn. Reson. Chem.*, 24, 835–852
[51] P. D. Ellis, G. E. Maciel and J. W. McIver, Jr., 1972, *J. Am. Chem. Soc.*, 94, 4069–4076.
[52] M. Kondo, I. Ando, R. Chujo and A. Nishioka, 1976, *J. Mag. Reson.*, 24, 315–326.
[53] I. Ando, H. Saitô, R. Tabeta, A. Shoji and T. Ozaki, 1984, *Macromolecules,*17, 457–461.
[54] H. Saitô, R. Tabeta, A. Shoji, T. Ozaki, I. Ando and T. Miyata, 1984, *Biopolymers*, 23, 2279–2297.
[55] H. Saitô, R.Tabeta, T. Asakura, Y. Iwanaga, A.Shoji, T.Ozaki and I.Ando, 1984, *Macromolecules*, 17, 1405–1412.
[56] D. B. Chesnet, 1989, *Annu. Rep. NMR Spectrosc.*, 21, 51.
[57] N. Asakawa, H. Kurosu and I. Ando, 1994, *J. Mol. Struct.*, 323, 279–285.
[58] N. Asakawa, H. Kurosu, I. Ando, S. Shoji and T. Ozaki, 1994, *J. Mol. Struct.*, 317, 119–129.
[59] A. C. de Dios, D. J. L. Roach and A. E. Walling, (J.C.Facelli and A.C. de Dios (ed.)), 1999, *ACS Symp. Ser.*, 732, 220–239.
[60] J. Gauss, 1992, *Chem. Phys. Lett.*, 191, 614–620.
[61] R. G. Parr and W. Yang, 1989, *Density-Functional Theory of Atoms and Molecules*, Oxford University Press, New York.
[62] A. D. Beck, 1986, *J. Chem. Phys.*, 84, 4524–4529.
[63] C. T. Lee, W. T. Yang, and R. G. Parr, 1988, *Phys. Rev.*, 37, 785–789.
[64] P. B. Karadakov, 2006, in *Modern Magnetic Resonance*, Springer, in press.
[65] F. Buda, P. Giannozzi and F. Mauri, 2000, *J. Phys. Chem. B*, 104, 9048–9053.
[66] A. J. van Gammeren, F. Buda, F. B. Hulsbergen, S. Kihne, J. G. Hollander, T. A. Egorova-Zachernyuk, N. J. Fraser, R. J. Cogdelli, and H. J. M. de Groot, 2005, *J. Amer. Chem. Soc.,* 127, 3213–3219.
[67] R. Dovesi, C. Pisani, C. Roetti, M. Causa and V. R. Saunders, Quantum Chemistry Program Exchange, Department of Chemistry, Indiana University, Indiana 47405, Program No. 577.
[68] T. Yamanobe, I. Ando and G. A. Webb, 1987, *J. Mol. Struc.,* 151, 191–201.
[69] H. Kurosu, I. Ando and T. Yamanobe, 1989, *J. Mol. Struc.,* 201, 239–247.
[70] G. N. Ramachandran and V. Sasisekharan, 1968, *Adv. Protein Chemistry*, 23, 283.
[71] M. Sone, H. Yoshimizu, H. Kurosu and I. Ando, 1994, *J.Mol.Struc.* 317, 111–118.
[72] M. Sone, H.Yoshimizu, H. Kurosu and I. Ando, 1993, *J.Mol.Struct.*, 301, 227–230.
[73] C. A. Fyfe, 1983, *Solid State NMR for Chemist*, C. F. C. Press, Guelph.

[74] R. A. Komoroski (ed.) 1986, *High Resolution NMR Spectroscopy of Synthetic Polymer in Bulk,* VCH Publishers, Deerfield Beach
[75] J. Schaefer and E. O. Stejskal, in *Topics in Carbon-13 NMR Spectroscopy*, vol. 3, pp.283–324.
[76] A. Torchia, 1978, *J. Magn. Reson.*, 30, 613–616
[77] H. Saitô and M. Yokoi, 1989, *Bull. Chem. Soc. Jpn*, 62, 392–398.
[78] A. Naito, S. Ganapathy, K. Akasaka and C. A. McDowell, 1983, *J. Magn. Reson.*, 226–235
[79] H. Saitô, R. Tabeta, M. Yokoi, and T. Erata, 1987, *Bull. Chem. Soc. Jpn.*, 60, 4259–4266.
[80] F. Horii, A. Hirai, and R. Kitamaru, *J. Carbohydr. Chem.*, 1984, 3, 641–662.
[81] D. L. VanderHart, 1987, *J. Magn. Reson.* 72, 13–42.
[82] S. Tuzi, K. Shinzawa-Itoh, T. Erata, A. Naito, S. Yoshikawa, and H. Saitô, 1992, *Eur. J. Biochem.*, 208, 713–720.
[83] M. J. Sullivan and G. E. Maciel, 1982, *Anal. Chem.*, 54, 1615–1623.
[84] J. Schaefer, E. O. Stejskal and R. Buchdahl, 1977, *Macromolecules*, 10, 384–405.
[85] D. A. McArther, E. L. Hahn, and R. E. Walstadt, 1969, *Phys. Rev.*, 188, 609–638.
[86] M. Mehring, *Principles of High Resolution NMR in Solids*, Second Edition, Springer-Verlag, Berlin
[87] M. Linder, P. M. Henrichs, J. M. Hewitt, and D. J. Massa, 1985, *J. Chem. Phys.* 82, 1585–1598.
[88] A. Naito, A. Fukutani, M. Uitdehaag, S. Tuzi, and H. Saitô, 1998, *J. Mol. Struct.*, 441, 231–241.
[89] H. Y. Carr and E. M. Purcell, 1954, *Phys. Rev.*, 94, 630–638.
[90] S. Meiboom and D. Gill, 1958, *Rev. Sci. Instrum.*, 29, 688–691.
[91] D. Suwelack, W. P. Rothwell, and J. S. Waugh, 1980, *J. Chem. Phys.*, 73, 2559–2569.
[92] W. P. Rothwell and J. S. Waugh, 1981, *J. Chem. Phys.*, 75, 2721–2732.
[93] H. Saitô, S. Tuzi, S. Yamaguchi, M. Tanio and A. Naito, 2000, *Biochim. Biophys. Acta,* 1460, 39–48.
[94] H. Saitô, S. Tuzi, M. Tanio and A. Naito, 2002, *Annu. Rep. NMR Spectrosc.*, 47, 39–108.
[95] S. Tuzi, A. Naito, and H. Saitô, 1994, *Biochemistry*, 33, 15046–15052.
[96] S. Tuzi, S. Yamaguchi, A. Naito, R. Needleman, J. K. Lanyi, and H. Saitô, 1996, *Biochemistry*, 35, 7520–7527.
[97] J. R. Lyerla and C. S. Yannoni, 1983, *IBM J. Res. Develop.* 27, 302.
[98] V. J. McBrierty, D. C. Douglass, and D. R. Falcone, 1972, *J. Chem. Soc. Farady Trans.* II, 68, 1051–1059.
[99] S. Tuzi, A. Naito, and H. Saitô, 1996, *Eur. J. Biochem.*, 239, 294–301.
[100] T. Yamanobe, M. Tsukahara, T. Komoto, J. Watanabe, I. Ando, I. Uematsu, K. Deguchi, T. Fujito and M. Imanari, 1988, *Macromolecules*, 21, 48–50.
[101] S. Tuzi, S. Yamaguchi, M. Tanio, H. Konishi, S. Inoue, A. Naito, R. Needleman, J. K. Lanyi, and H. Saitô, 1999, *Biophys. J.*, 76, 1523–1531.
[102] H. Saitô, J. Mikami, S.Yamaguchi, M. Tanio, A. Kira, T. Arakawa, K. Yamamoto, and S. Tuzi, 2004, *Magn. Reson. Chem.*, 42, 218–230.
[103] Y. Kawase, M. Tanio, A. Kira, S. Yamaguchi, S. Tuzi, A. Naito, M. Kataoka, J. K. Lanyi, R. Needleman, and H. Saitô, 2000, *Biochemistry*, 39, 14472–14480.
[104] H. Saitô, S. Yamaguchi, K. Ogawa, S. Tuzi, M. Márquez, C. Sanz, and E. Padrós, 2004, *Biophys. J*, 86, 1673–1681.

[105] S. Kimura, A. Naito, S. Tuzi and H. Saitô, 2001, *Biopolymers*, 58, 78–88.
[106] S. Yamaguchi, S. Tuzi,, M. Tanio, A. Naito, J. K. Lanyi, R. Needleman, and H. Saitô, 2000, *J. Biochem.(Tokyo)*, 127, 861–869.
[107] T. Arakawa, K.Shimono, S. Yamaguchi, S. Tuzi, Y. Sudo, N. Kamo, and H. Saitô, 2003, *FEBS Lett.*, 536, 237–240.
[108] S. Yamaguchi, K. Shimono, Y. Sudo, S. Tuzi, A. Naito, N. Kamo and H. Saitô, 2004, *Biophys. J.*, 86, 3131–3140.
[109] S. Yamaguchi, S. Tuzi, J. U. Bowie and H. Saitô, 2004, *Biochim. Biophys. Acta*, 1698, 97–105.

Chapter 4
MULTINUCLEAR APPROACHES

4.1. ^{31}P NMR

^{31}P nucleus with spin number of 1/2 and relatively higher sensitivity is a suitable probe to reveal conformation and dynamics of biopolymer systems containing phosphorus, including nucleic acids such as DNA, RNA, or phospholipids in biomembranes. Conventional CP technique for detection of NMR signals from rare spins such as ^{13}C or ^{15}N nuclei can be also utilized for ^{31}P nuclei in spite of 100% natural abundance, because line broadenings due to homonuclear dipolar interactions from neighboring nuclei are usually not necessary to be taken into account owing to a limited number of ^{31}P nuclei present. In particular, analysis of static ^{31}P NMR spectra under proton decoupling has proved to be very useful for analyzing local conformation of phosphodiester linkages for mechanically oriented nucleic acids and dynamic feature of lipid bilayers as viewed from choline head groups of phospholipids. Static ^{31}P NMR spectra of phospholipids in liquid crystalline phase are also available from conventional high-resolution solution NMR spectrometer because of motionally averaged CSA.

4.1.1. Nucleic Acids

RNAs are found only in two related conformations A and A′, which both belong to the A-family double-helical structures. In contrast, DNAs can adopt several other conformations depending on environmental condition such as counterion and relative humidity (RH).[1,2] The B-DNA form is obtained when DNA is fully hydrated (over RH 92%) as it is *in vivo* and takes a double helix with antiparallel strands and with Watson–Crick base pairs oriented roughly at right angles to the helix axis, while A-DNA with base pair tilt by 20° relative to the helix axis is obtained under dehydrated condition (RH 70–80%). Besides these right-handed double helices, a left-handed variety (Z-DNA) has been discovered.

^{31}P NMR is a very convenient tool to characterize these polymorphic structures of DNA as viewed from their local conformation and dynamics of phosphodiester backbone, as demonstrated.[3–5] As an illustrative example, Figure 4.1 shows the observed and simulated ^{31}P NMR spectra for oriented salmon sperm NaDNA fibers taking A form at RH 79% at goniometer angles between fiber direction and the applied magnetic field.[6] The A form of DNA fibers exhibits a typical spectral pattern for the axially oriented molecules with a single conformation: the parallel spectrum is a singlet, the 45° spectrum is a triplet, and the perpendicular spectrum is a bimodal pattern. In case of oriented materials along one axis such as DNA fibers, it is convenient to define the molecular axis system (*a, b, c*) and to use Euler angles α, β, and γ (Figure 4.2, left) instead of the direction cosines (Eq. (2.3)) for the expression of a relative orientation of the axis systems. The behavior of the chemical shielding tensor is conveniently expressed by Wigner rotation matrices $D^{(2)}_{ab}(\Omega)$. When rotational transformation from the laboratory frame (magnetic field) to the principal axis system are carried out, namely

$$\text{principal axis system} \xrightarrow[\Omega(\alpha,\beta,\gamma)]{} \text{molecular axis system}$$

$$\xrightarrow[\Omega(\psi,\theta,0)]{} \text{fiber axis system} \xrightarrow[\Omega(0,\phi,0)]{} \text{laboratory frame}$$

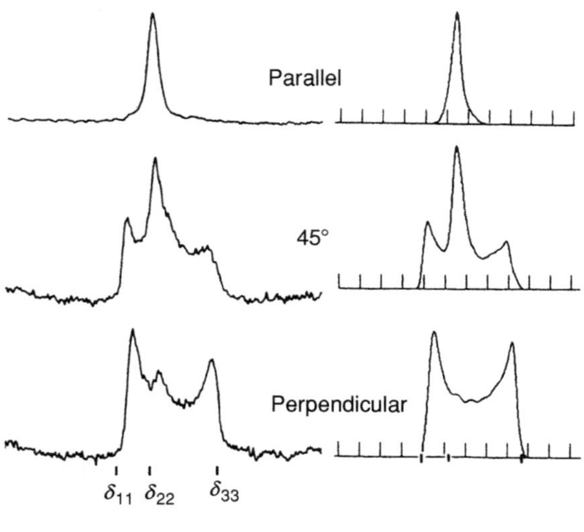

Figure 4.1. Observed and simulated ^{31}P NMR spectra of oriented fibers of the A-form DNA at goniometer angles between the fiber direction and the magnetic field as indicated (from Shindo *et al.*[6]).

then the observed chemical shift, $\sigma^{obsd}(\phi,\alpha)$, is given by[5,7]

$$\rho_{20}^{lab} = \sqrt{3/2}(\sigma^{obsd} - (1/3)\mathrm{Tr}\sigma_{ii})$$
$$\rho_{20}^{lab} = \Sigma_{m,n,p} D_{p0}^{(2)}(0,\phi,0) D_{mp}^{(2)}(\varphi,\theta,0) D_{nm}^{(2)}(\alpha,\beta,\gamma)\rho_{2n} \quad (4.1)$$

where $D_{mn}^{(2)}(\alpha,\beta,\gamma)$ is the Wigner rotation matrix of the second rank, ρ_{2n} is the irreducible tensor of chemical shielding anisotropy, and the notations of the Euler angles correspond to those in Figure 4.2. The principal axis system is defined by β and γ with respect to the molecular axis system (a, b, c). The fiber axis is taken to be identical with the goniometer axis system (Figure 4.2, right). In case of the phosphodiesters of DNA, the principal axis system of the chemical shielding tensor of ^{31}P nuclei is reasonably related to the atomic coordinates of the phosphodiester,[8,9] so that the x-axis is taken as the axis normal to the O–P–O plane, the y-axis as the bisector of the vectors of two P–O bonds, and the z-axis as normal to the above two axes. By use of the observed principal values,[3,5] $\delta_{11} = 83$ ppm, $\delta_{22} = 22$ ppm, and $\delta_{33} = -110$ ppm relative to trimethyl phosphate in aqueous solution as an external reference, the phosphodiester orientation represented by Euler angles β and γ was chosen so as to meet the chemical shifts of the peaks in the spectra from the fibers oriented parallel, at 45°, and perpendicular to the magnetic field. Note that these δ values are converted from the original σ values by definition of Eq. (2.5). The line broadening in the simulated spectra was made by considering the imperfect alignment of the molecules with a standard deviation $\langle\theta\rangle = 12$ and a 4 ppm width from other contributions. The observed β and γ angles thus obtained are 70° and 52°, respectively, and close to those (77° and 79°) from X-ray fiber diffraction.[10]

The B-form DNA is always observed at high RHs (89–98%). However, it is very difficult to predict the static geometry of the B form directly from the ^{31}P NMR spectra, because the observed spectra were strongly perturbed by the

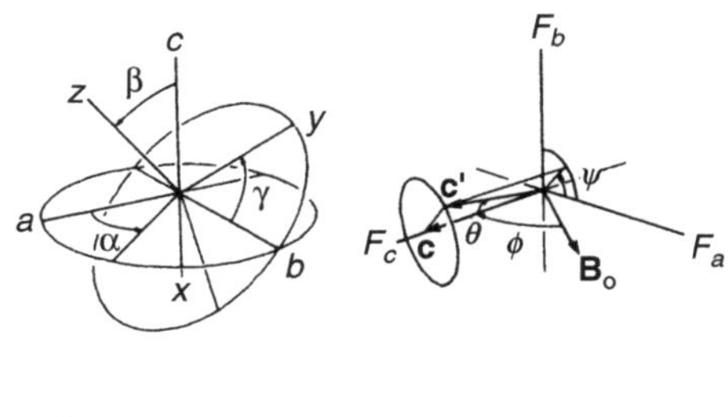

Figure 4.2. Left: Coordinate system *(a,b,c)* represents the molecular frame and axis *c* coincides with the helical axis of DNA as a reference frame. Right: Misalignment of a DNA molecule from the fiber axis F_c; vector **c** lies along fiber axis and **c'** represents the vector of the helical axis of misaligned DNA which makes an angle θ with respect to the fiber axis (from Shindo et al.[6]).

molecular motions of DNA.[6] The observed parallel spectra under high humidity conditions were found to be usually broad compared with those of the A-form DNA fiber, in spite of the fact that any motion must result in the motional narrowing of the spectral lines.[4–6] Out of several factors causing the line broadening, the contributions from the multiplicity in the backbone conformation and the imperfect alignment of the molecules within the fiber sample were found to be predominant.[3] ^{31}P NMR spectra observed for NaDNA fibers at seven different goniometer angles at RH 98% (the oriented B-form DNA) were simulated by taking into account the rotational diffusion about the helical axis as shown in Figure 4.3.[11] Rotational diffusion about the helical axis was found to perturb the spectral line shapes most strongly, and its constants were 1.5×10^4 and 5.0×10^4/s for DNA fibers at 92% and 98% RHs, respectively.[11] Two sets of simulated spectra, corresponding to the restricted rotation about the helical axis (A) and the rotational diffusion model (B) show an essentially good agreement both in the line shape and in the peak positions with those of the observed spectra. The broad ^{31}P NMR lines of the B form of natural DNA suggested that the phosphodiesters have a considerable dispersion in their orientation relative to the helical axis. If such a variation of the backbone conformation was induced by difference in base sequence, synthetic polynucleotide with a known sequence would emphasize a unique structure

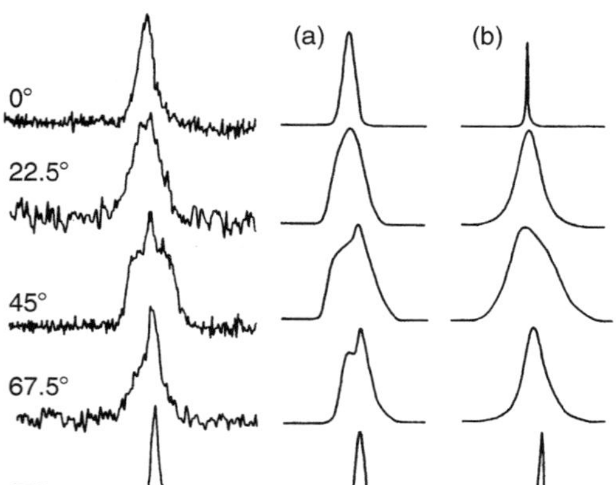

Figure 4.3. Dipolar decoupled ^{31}P NMR spectra (80.7 MHz) of NaDNA fibers at 98% RH at various goniometer angles as indicated. The observed spectra were compared with those calculated from two models, restricted rotation (a) and rotational diffusion (b). Parameters used for model A are 220° and 90° for rotational amplitude $\Delta\eta$ and its standard deviation $\langle\Delta\eta\rangle$ from $\Delta\eta$, respectively. Diffusion constant used for model B is $N = 1000$ (i.e., $D = 5.1 \times 10^4$/s) (from Fujiwara and Shindo[11]).

reflected in the spectral pattern. In fact, Shindo and Zimmerman[12] showed that poly (dAdT) fibers at 98% RH exhibit two well-defined peaks for the parallel spectrum depending upon two sequences, ApT and TpA, instead of the single line in the native DNA. Therefore, it was suggested that irregularity of the backbone conformation is ubiquitous for natural DNA.[4]

DiVerdi and Opella[13] showed that the ^{31}P chemical shielding tensor of DNA in solid fd bacteriophage is indistinguishable from that of single-stranded or double-stranded DNA in the absence of proteins; therefore the ^{31}P chemical shift does not show evidence of structural changes in DNA upon incorporation into the virus. The broad line width of fd in solution is due to static chemical shift anisotropy that is not motionally averaged, as illustrated in Figures 4.4 (a) and (b).[13a] These results indicate that DNA packaged inside fd is immobilized by the coat proteins. The oriented fd spectrum in Figure 4.4 (c) is identical with that for unoriented fd in

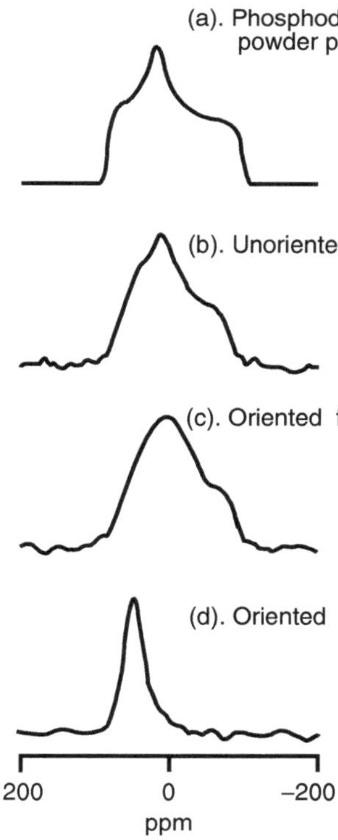

Figure 4.4. ^{31}P NMR spectra of filamentous virus fd and Pf1 at 60.9 MHz: (a) calculated powder pattern based on the spectrum of unoriented frozen fd ($\delta_{11} = 81$, $\delta_{22} = 15$, and $\delta_{33} = -100$ ppm); (b) unoriented fd spectrum; (c) magnetic field oriented fd spectrum; (d) magnetic field oriented Pf1 spectrum (from Cross *et al.*[13a]).

Figure 4.4 (b). This is in contrast to the ^{15}N chemical shift comparison of ^{15}N-labeled fd to be described in Section 9.6.2 (Figure 9.13), where the oriented fd sample gives a spectrum characteristic of aligned N–H bonds in the protein. This means that the fd virus is oriented by the magnetic field, but the phosphodiester backbone of DNA is not. The phosphates of Pf1 DNA are therefore oriented so that the chemical shift tensor axis of δ_{11} is approximately parallel to the filament axis and the magnetic field (Figure 4.4 (d)). As to an RNA virus, it was demonstrated that the ^{31}P CSA powder pattern of a stationary, unoriented solution of RNA in tobacco mosaic virus (TMV) shows the RNA to be immobilized by the coat protein–RNA interactions, since the principal values ($\delta_{11} = 83$, $\delta_{22} = 25$, $\delta_{33} = -108$ ppm relative to external 85% H_3PO_4) are essentially the same as those of a static phosphodiester group.[14] There are three peaks in the isotropic ^{31}P NMR spectrum obtained with MAS, indicating three distinct phosphate environments. There are also three peaks in the ^{31}P NMR spectrum from a magnetically oriented TMV solution as in filamentous bacteriophage, indicating three distinct phosphate orientations.

4.1.2. Phospholipids in Biomembranes[15–18]

Figure 4.5 illustrates the manner of schematic orientation for phosphodiester group in the head groups of phospholipids.[15] ^{31}P chemical shift of such phospholipids in biomembranes depends on the orientation and dynamics of the group with respect to the applied magnetic field of the spectrometer. Rigid phosphodiester moiety of a membrane lipid, if any, yields the ^{31}P powder pattern with the three principal components, δ_{11}, δ_{22}, and δ_{33}, which spans some 190 ppm as shown in Figure 4.6(a). Such a phosphodiester moiety in biomembrane, however, is not always rigid at ambient temperature but undergoes several types of motions as illustrated in Figures 4.6(b) and (c). Suppose, for simplicity, that this motion is about the 1 axis, thus averaging δ_{22} and δ_{33}:

$$\begin{aligned}\delta_{\perp} &= \int_0^{2\pi} (\delta_{22}\cos^2\theta + \delta_{33}\sin^2\theta)\mathrm{d}\theta \Big/ \int_0^{2\pi} \mathrm{d}\theta \\ &= (\delta_{22} + \delta_{33})/2,\end{aligned} \tag{4.2}$$

where θ is the angle between the applied magnetic field and the 2 axis. Thus the chemical shift will have the same value for the field anywhere in the 2–3 plane but a different value when the field is perpendicular to the plane; the

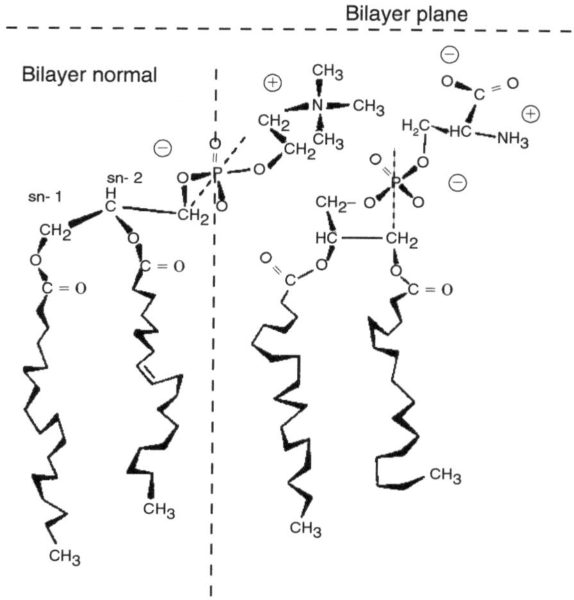

Figure 4.5. Representation of some typical lipids found in biological membranes: phosphatidylcholine (left) and phosphatidylserine (right) (from Smith and Ekiel[15]).

chemical shift tensor is axially symmetric. The chemical shift expected when the field is parallel to the unique axis 1 is labeled $\sigma_{\parallel}$, whereas that expected for the field perpendicular to this axis is $\sigma_{\perp}$. The resultant powder pattern is demonstrated in Figure 4.6(b).

The situation most relevant to biological membranes is that rapid motion of limited amplitude of the 1 axis of the phosphodiester moiety occurs, while retaining the rapid motion about the 1 axis (Figure 4.6(c)). The 1 axis now moves in a cone, and there is partial averaging of the former $\delta_{\parallel}$ and $\delta_{\perp}$ to yield new, smaller effective values, $\delta'_{\parallel}$ and $\delta'_{\perp}$. The effective tensor still has axial symmetry, but the reduced CSA, $\Delta\delta = \delta'_{\parallel} - \delta'_{\perp}$, can be easily determined by the edges of the experimental spectrum, and is a measure of orientation and average fluctuation of the phosphate segment. Since the molecular fixed chemical shielding tensor is not axially symmetric[8,19], the molecular interpretation of $\Delta\delta$ is more complicated, and requires two order parameters[20] instead of one for deuterium NMR described below. Even without a detailed molecular interpretation, $\Delta\sigma$ may be used as a convenient measure for the comparison of head group motion in different lipid bilayers.

An interesting property of ^{31}P NMR is also its sensitivity to lipid polymorphism: if the geometry of the lipid phase changes from lamellar

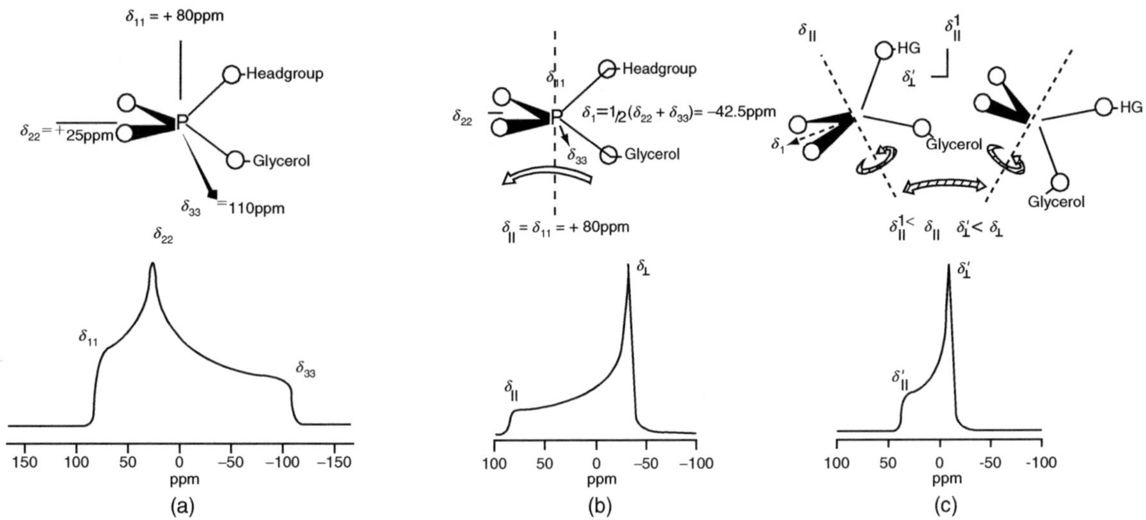

Figure 4.6. Various possible motional states of the phosphodiester moiety of a membrane lipid and the expected ^{31}P NMR spectra: (a) static phosphodiester; (b) ordered phosphodiester, rapid axial rotation; (c) disordered phosphodiester, rapid axial rotation. In the case of motional averaging of the chemical shift tensor to axial symmetry, $\delta_{\parallel}$ refers to the chemical shift for the external magnetic field parallel to the unique axis, and $\delta_{\perp}$ to that for the field in the equatorial plane (from Smith and Ekiel[15]).

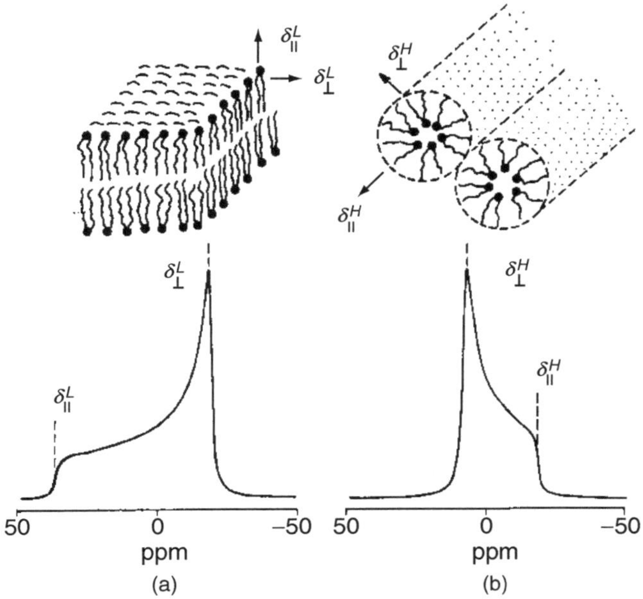

Figure 4.7. Representation of the (a) bilayer and (b) hexagonal (H_{11}) phases formed by membrane lipids and their expected ^{31}P NMR spectra (from Smith and Ekiel[15]).

(Figure 4.7 (a)) to hexagonal (Figure 4.7 (b)), ^{31}P chemical shift anisotropy $\Delta\delta$ changes its sign and is reduced by exactly a factor of two as shown below. The cylinders in a hexagonal phase have a very small radius, and therefore lateral diffusion about the cylinder axis can cause further averaging of tensor components. The unique axis of the system now becomes the axis of the cylinder, and thus we label its chemical shift component $\delta_{\parallel}^{H}$. However, noticing that along this axis the field would be roughly normal to the fatty acyl chains, we would expect the value of this chemical shift $\delta_{\parallel}^{H}$ to be similar to $\delta_{\perp}^{L}$. On the other hand, $\delta_{\perp}^{H}$ will be an average of $\delta_{\perp}^{L}$ and $\delta_{\parallel}^{L}$, owing to rapid motion around the cylinder axis. Hence

$$\delta_{\parallel}^{H} = \delta_{\perp}^{L}, \tag{4.3}$$

$$\delta_{\perp}^{H} = (\delta_{\parallel}^{L} + \delta_{\perp}^{L})/2, \tag{4.4}$$

and

$$\Delta\delta^{H} = -\Delta\delta^{L}/2. \tag{4.5}$$

The net result of this is that the ^{31}P powder pattern for the hexagonal phase has a $\Delta\sigma$ that is roughly half that of a corresponding lamellar phase, and a shape that has the buildup of intensity owing to the axial component on the other side of the chemical shift zero.

For relatively small, spherical vesicles ($R < 10{,}000$ Å), two diffusion processes contribute to the averaging of the spectra: rapid Brownian tumbling of the entire vesicle (characterized by the diffusion coefficient D_t) and lateral diffusion of lipids around the vesicle (with diffusion coefficient D_{diff}). The correlation time τ_c is given by the equation

$$1/\tau_c = (6/R^2)(D_t + D_{\text{diff}}), \tag{4.6}$$

where R is the radius of a vesicle, $D_t = kT/8\pi R\eta$, and η is the viscosity.[21] Using such a motional model and Freed's theory for the motional dependence of line shapes,[22] Burnell *et al.*[23] were able to simulate the experimental spectra of dioleoylphosphatidylcholine vesicles for different temperatures and viscosities of the medium (Figure 4.8). As seen from Eq. (4.6), the effectiveness of the averaging process will increase with decreasing size of the vesicle.

4.2. ^{2}H NMR

As described in Section 2.2, ^{2}H NMR is a very useful means to determine the orientation angle θ of a given C–H vector with respect to the applied magnetic field by means of Eq. (2.30), if magnetically oriented crystalline sample is examined. If this C–^{2}H vector undergoes rotational motion about the rotational axis that makes the angle α and β with the applied magnetic field and the C–^{2}H bond, respectively, as illustrated in Figure 4.9, the quadrupole splitting as defined by Eq. (2.31) is rewritten by

$$\Delta\nu_Q = (3/8)(e^2qQ/h)\ (3\cos^2\alpha - 1)(3\cos^2\beta - 1), \tag{4.7}$$

because

$$3\cos^2\theta - 1 = 1/2(3\cos^2\alpha - 1)\ (3\cos^2\beta - 1). \tag{4.8}$$

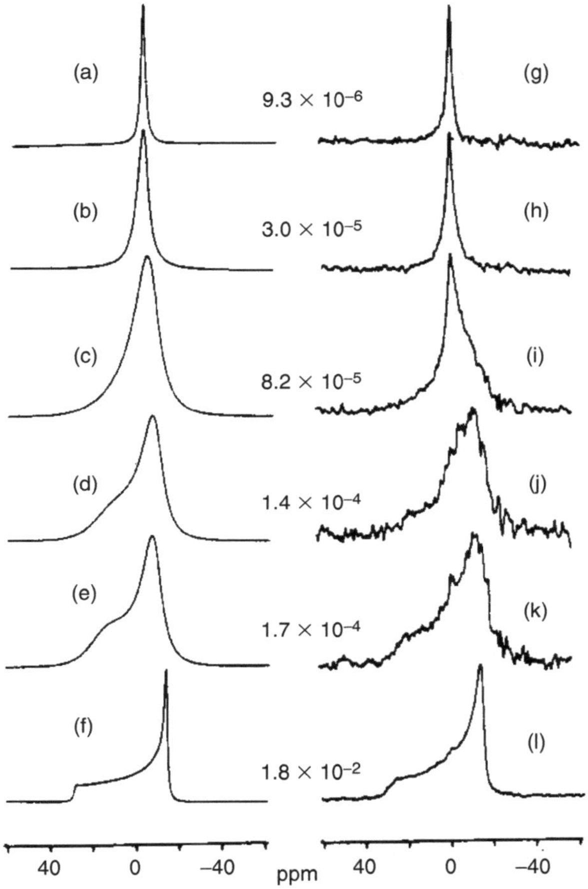

Figure 4.8. Simulated (a–f) and experimental (g–l) 80 MHz ^{31}P NMR spectra of dioleoylphosphatidylcholine vesicles ($R = 980$ Å) for different values of τ_c (in seconds) as indicated. The spectra were simulated using $\delta_{\parallel} - \delta_{\perp} = 3550$ Hz (43.8 ppm). The experimental spectra were measured under the following conditions: (g) 60°C; (h) 25°C; (i) 0°C; (j) −10°C; (k) −15°C; (l) 30°C, unsonicated liposomes (from Burnell *et al.*[23]).

The corresponding quadrupole splitting for unoriented powder sample is:

$$\Delta\nu_{Q,\text{powder}} = (3/8)(e^2qQ/h)\,(3\cos^2\beta - 1), \tag{4.9}$$

because the α angles are randomly distributed.

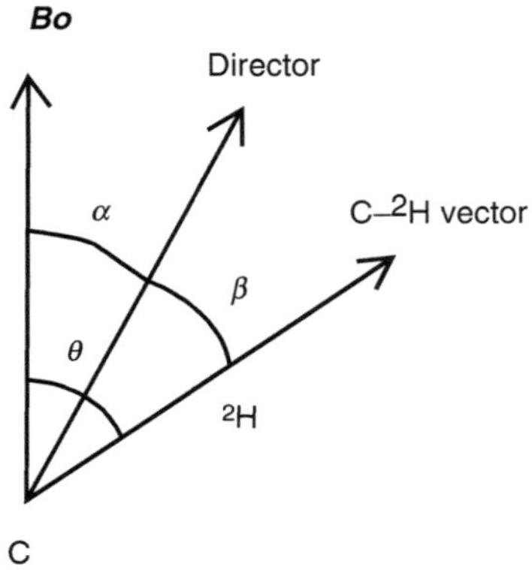

Figure 4.9. Rotation of C–^{2}H vector about the director axis and their relative orientation to the static magnetic field.

4.2.1. Magnetically Oriented System

Obviously, ^{2}H NMR analysis of ^{2}H-labeled preparation provides one unique data as to how C–^{2}H vector under consideration is oriented to the applied magnetic field based on Eq. (2.30), if single crystal or mechanically or magnetically oriented systems are employed, as will be described in Section 5.2. Undoubtedly, preparation of large single crystalline sample for current NMR study is not practical as far as proteins are concerned. Further, the latter approach using mechanically or magnetically oriented system for a ^{2}H NMR study has not always been fully explored, probably because its peak separation for individual residues based on chemical shift (equivalent to ^{1}H chemical shift) is not sufficient as compared with that of other nuclei such as ^{13}C or ^{15}N nuclei frequently utilized for this purpose.

Instead, microcrystals of a variety of paramagnetic heme proteins, suspended in ≈90% saturated $(NH_4)_2SO_4$, may be perfectly aligned by an intense static external magnetic field due to the larger anisotropy in the magnetic susceptibility of the protein caused by the paramagnetic center.[24,25] The magnetic ordering method permits the recording of ''single crystal'' NMR spectra from microcrystalline array of proteins that cannot be prepared in large enough quantities for NMR spectroscopy. For this purpose, they prepared ^{2}H-methylmethionine labeled myoglobin samples (at Met-55 and Met-13) by reaction with deuterated methyl iodide by treatment with mercaptoethanol. As compared with the ^{2}H NMR powder pattern arising from unoriented crystalline powder (Figure 4.10 (a)), a very narrow line spectrum having $\Delta\nu_Q = 53.6$ kHz is obtained when the microcrystals (Figure 4.10 (a)) are resuspended in ^{2}H-depleted 90% saturated ammonium sulfate (pH 6.3)

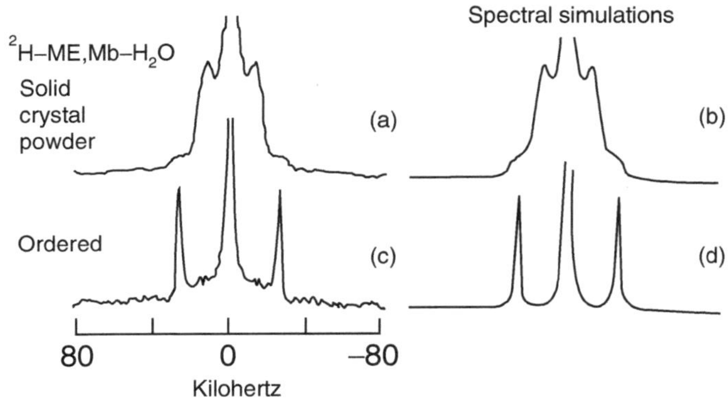

Figure 4.10. ^{2}H NMR spectra of ^{2}H-methionine labeled sperm whale myoglobin as a solid hydrated crystalline powder (a) and magnetically ordered sample (c), and their corresponding spectral simulations (b and d) (from Rothgeb and Oldfield[24]).

(Figure 4.10 (c)).[24] This spectrum has a $\Delta\nu_Q$ much greater than the $\approx$ 40 kHz expected for only methyl group rotation to be described below and does not have the appearance of a normal powder pattern. From the result in Figure 4.10 (c), it follows that β angle for at least one methionine S_δ–C_ε vector is $17.5 \pm 2°$, as shown in Figure 4.10 (d). The results of sample freezing and additional pH dependence experiments, in which crystal orientation is changed by means of a spin-state transition indicates that a second methionine resonance lies under the intense central HOD resonance. Thus, β for this group is $54.7 \pm 2°$. Further, ^{2}H NMR spectra of [*meso*-α,β,γ,δ-$^2H_4$], [methyl-1,3-$^2H_6$], and [methylene-6,7-β-$^2H_4$]heme-labeled aquoferrimyoglobin microcrystals from *Physeter catodon* in the magnetically oriented system permit partial determination of the static organization of the heme, and the results obtained are in good agreement with those obtained using X-ray crystallography.[25]

4.2.2. Order Parameters for Liquid Crystalline Phase

Liquid crystalline molecules consisting of biomembrane or lipid bilayer are characterized by their tendency to align the constituent rod-like molecules parallel to the long axes. The rotation of liquid crystalline molecules about the unique axis, defined as the director axis, occurs with frequencies in the range of 10^7–10^{10} Hz, whereas those perpendicular to the long axes are restricted. This means that the angle β in Eqs. (4.7–4.9) is not always static, but this angle is now defined by the instantaneous angle between the C–^{2}H bond and the direction of the bilayer normal, and $<\ldots>$ denotes the time average. Therefore, ^{2}H quadrupole splitting is determined by the time-averaged value as defined by an order parameter characterized by[26,27]

$$\Delta\nu_Q = (3/4)\,(e^2qQ/h)\,S_{\mathrm{CD}}\,(3\cos^2\alpha - 1) \tag{4.10}$$

$$\text{and } S_{\mathrm{CD}} = (1/2)\left\langle 3\cos^2\beta - 1\right\rangle. \tag{4.11}$$

The manner of orientation of the director axis with respect to the applied magnetic field is readily determined, if ^{2}H-labeled single crystal, mechanically, or magnetically oriented samples are available. Figure 4.11 illustrates ^{2}H NMR spectra of an illustrative example from oriented bilayer of soap system in which total of approximately 20 mg of material, 1,1-dideuteriooctanol (30 wt.%), sodium octanoate (34 wt.%), and water (36 wt.%) were sandwiched between a stack of 20 glass plates of $20 \times 6.6 \times 0.15$ mm.[26] Note that the angle α in Eq. (4.10) corresponds to the angle θ in the angular dependence of the observed separation of the peaks in Figure 4.11.

The order parameter S_{CD} is also available from unoriented, multilamellar dispersion of lipid bilayer as demonstrated in Eq. (4.12),

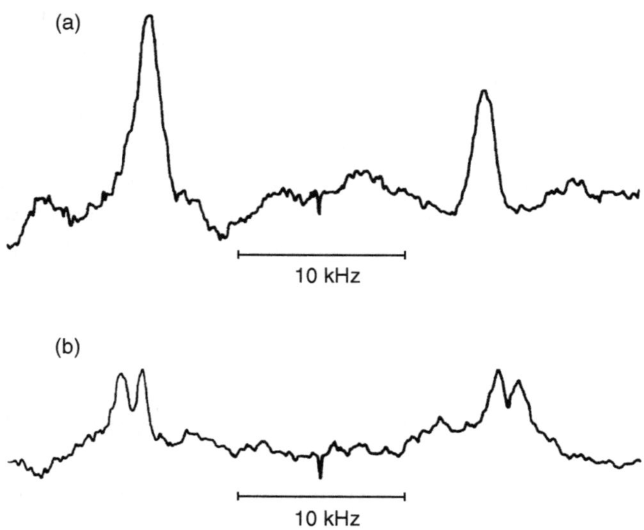

Figure 4.11. ^{2}H NMR spectra of oriented bilayers of 1,1-dideuteriooctanol, sodium octanonate, and water in stacked in glasses (from Seelig and Niederberger[26]).

$$\Delta\nu_{Q,\text{powder}} = (3/4)\,(e^2qQ/h)\,S_{\text{CD}}, \tag{4.12}$$

although prior knowledge about the relative orientation of the director vector to the bilayer normal is unavailable from the powder pattern spectra alone. In fact, it is known that any C–^{2}H fragment from fatty acyl chains in a biological membrane or lipid bilayer is known to undergo instantaneous, anisotropic fluctuation motion with respect to the director, parallel to the bilayer normal. Accordingly, the average orientation of the C–^{2}H bond is essentially perpendicular to the bilayer normal; hence it is reasonable to assume that S_{CD} is negative. The segment direction is given by the normal to the plane spanned by the two C–^{2}H bonds of a methylene unit. The segmental order parameter S_{mol} is defined by

$$S_{\text{mol}} = (1/2)\langle 3\cos^2\varsigma - 1\rangle, \tag{4.13}$$

where ς is the angle between the segment direction and the bilayer normal. Therefore, S_{mol} can be derived from S_{CD} by means of the following relation:

$$S_{\text{mol}} = S_{\text{CD}}[(3\cos^2 90^\circ - 1)/2]^{-1} = -2\,S_{\text{CD}}. \tag{4.14}$$

In the case of the terminal methyl group, the following transformation leads to

$$S_{\text{mol}} = S_{\text{CD}}[(3\cos^2 109.5^\circ - 1)/2]^{-1} \approx -3S_{\text{CD}}. \tag{4.15}$$

In fact, Figure 4.12 (a) and (b) demonstrate the dependence of the degree of the order parameters of fatty acyl chains in deuterated dipalmitoyllecithin at nine different carbon atoms at 57°C and 65°C, respectively, together with the data from ESR spin-label experiments.[28] In a similar manner, Figure 4.12 (c) illustrates the similar plots of the segmental order parameters vs. position of deuteration for specifically deuterated stearic acid probes intercalated in egg lecithin lamellar dispersion (●) at 30°C, and (■) at 55°C, together with the data from ESR.[29] With exception of C-5 in dipalmitoylphosphatidylcholine (DPPC), the order parameters from ESR are appreciably lower than those obtained by ^{2}H NMR spectra. In both cases, the deuterium order parameter is approximately constant for positions 2–10 ($S_{\text{mol}} \approx 0.45$ at 41°), and drops rapidly with further distance from the carboxyl groups. A rise in temperature decreases the order parameter. Further, the constancy of the ^{2}H order parameters for the first 10 carbon atoms may be due to the higher probability of kinks (g^+tg^-), leaving the hydrocarbon chains essentially parallel to each other.

The effective length of the hydrocarbon chain in the bilayer $\langle L\rangle$ is given by

$$\begin{aligned}\langle L\rangle &= \Sigma_i\langle l\rangle_i = \Sigma_i 1.25\langle\cos\varsigma\rangle_i\\ &= 1.25\,[n - 0.5\,\Sigma_i(1 - (S_{\text{mol}})_i)/1.125],\end{aligned} \tag{4.16}$$

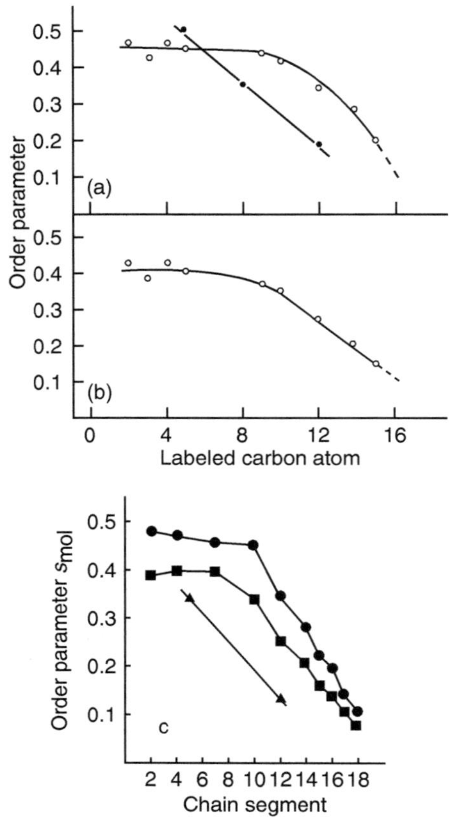

Figure 4.12. Comparison of the segmental order parameters obtained by ^{2}H NMR spectra of specifically deuterated lipids and electron spin resonance of the corresponding lipids labeled with the same position with 2-*spiro*-4,4-dimethyloxazolidinyl-3-oxyl moiety. (a) DPPC with the fatty acyl chains labeled with deuterium (o) or nitroxide spin labels (●) at 41°C,[28] (b) at 50°C, (from Seelig and Seelig [28]). (c) egg phoshatidyl choline containing fatty acids labeled with deuterium (●, 30°C; ■, 55°C) or nitroxide spin labels (▲, 30°C) (from Stocton *et al.*[29]).

where $\langle l \rangle_i$ is the average length of the ith segment and n is the number of segments.[28,29] The segmental length in the extended chain is 1.25 Å. This finding indicates that DPPC bilayer at liquid crystalline phase estimated on the basis of the ^{2}H NMR at 50°C is approximately 11.2 Å shorter than the same bilayer with completely extended chains.[28]

In a similar manner, the average chain length of stearic acid intercalated in egg lecithin bilayer is shorter by 6.15 Å at 30°C and 6.75 Å at 55°C, respectively, as compared with the length of the extended chain.[29]

These results are fairly in good agreement with those obtained by X-ray diffraction.

4.2.3. Side-Chain Dynamics of Amino Acid Residues

^{2}H NMR spectra as expressed by Eq. (4.10) could be substantially modified in the presence of motions depending upon their type, when several types of motions are present about the axis of motional averaging. Here, β is the angle between the axis of motional averaging and the principal axis of the electric field gradient tensor and α is the angle between the axis of motional averaging and B_0. It is expected that a variety of side-chain motions, including twofold jump, continuous diffusion, threefold rotation and/or flip for phenyl and methyl groups in ^{2}H-labeled fibrous, globular, or membrane proteins, if any, would substantially modify ^{2}H NMR spectral pattern due to the resulting modified electric field tensor and asymmetric parameter. This means that a closer examination of ^{2}H NMR spectral pattern provides one a clue as to side-chain dynamics of amino acid residues.

Figure 4.13(c) and (d) illustrate how ^{2}H NMR spectra of [$^2H_5$]phenylalanine are modified by fast (>>10^5 Hz) continuous diffusion or two-fold jump of phenyl ring about the C_γ–C_ζ axis, respectively, as compared with rigid lattice (static) powder spectrum, consistent with the simulated ^{2}H NMR spectrum with e^2qQ/h of 181 kHz and η of 0.05 (Figure 4.13(a) and (b))[30,31]. A quadrupole splitting of $\Delta\nu \approx 17$ kHz is predicted from Eq. (4.9) (Figure 4.13(c)), provided that phenyl ring would undergo rapid rotational diffusion about C_γ–C_ζ axis in which case the $C_{\delta 1,\delta 2,\varepsilon 1,\varepsilon 2}$–H vectors would be at a bond angle 60° to the axis of motional averaging. By contrast, a twofold jump for phenylalanine motion, whereby the aromatic ring executes the twofold reorientational flips about C_γ–C_ζ, predicts a very different ^{2}H NMR spectrum as shown in Figure 4.13(d). Assuming that motion is fast compared to the breadth of the rigid quadrupole interaction, a motionally averaged tensor with $\eta_{\text{eff}} = 0.66$ was calculated based on dependence on the effective asymmetry parameter as a function of the bond angle by the model of Soda and Chiba[32] for the reorientation about the twofold axis. The resulting spectra now contain a new sharp narrow feature, corresponding to the separation between the singularities in the powder pattern, having $\Delta\nu_{Q1} \approx 30.3$ kHz (Figure 4.13(d)). The splittings of the singularity, step, and edge as a function of breadth (ν_Q) and asymmetry parameter (η or η_{eff}) are as follows:

$$\Delta\nu_{Q1} = 1/2\ \nu_Q (1 - \eta)$$

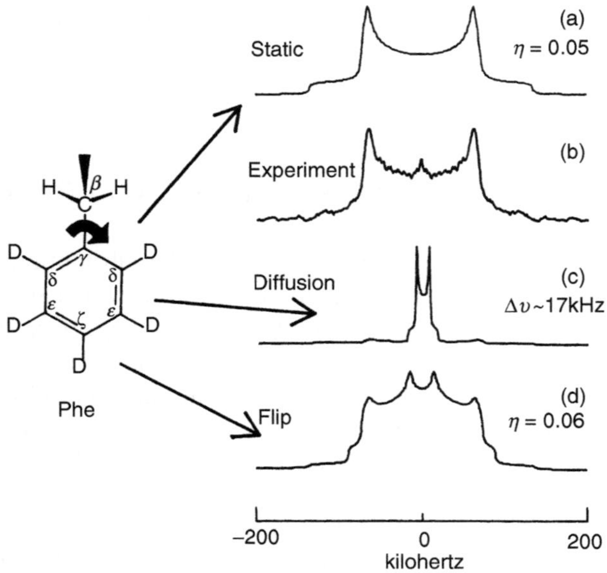

Figure 4.13. Simulated and experimental ^2H NMR spectra of [$^2H_5$]phenylalanine. (a) Simulated ^2H NMR spectrum of [$^2H_5$]phenylalanine as a rigid polycrystalline solid with the quadrupole coupling constant of 180 kHz and $\eta = 0.025$. (b) Experimental spectrum of [$^2H_5$]phenylalanine at 25°C. (c) Simulated line shape in which the phenyl ring is assumed to undergo rapid rotational diffusion about the C_γ–C_ζ axis. (d) Simulated line shape in which the phenyl ring is assumed to undergo rapid twofold flip motions about the C_γ–C_ζ axis (from Oldfield *et al.* [31]).

$$\Delta\nu_{Q2} = 1/2\ \nu_Q\,(1 + \eta) \tag{4.17}$$

$$\Delta\nu_{Q3} = \nu_Q$$

Similar results for the calculated spectral pattern in the presence of such motions was also presented by Gall *et al.*[33] and Rice *et al.*[34] based on the method of the coherent averaging of chemical shift tensor.[35,36] In the latter,[36] it was demonstrated that in chemically exchanging system the quadrupole echo pulse sequence introduces spectral distortion due to power roll-off due to the finite radio frequency pulse widths.[37] As a result, fast-limit spectra are not observed until the rate exceeds 10^7 Hz because of the echo distortion, as illustrated for crystalline [tyrosine-3,5-$^2H_5$]-labeled [Leu^5]enkephalin[34] in Figure 4.14, instead of the breadth of the quadrupole interaction (10^5 Hz) mentioned above. Flipping rate of the phenyl ring of tyrosine as fast as 10^4–

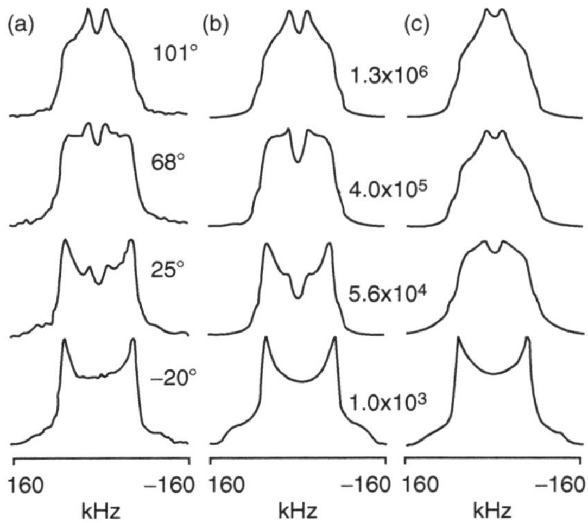

Figure 4.14. (a) ^{2}H quadrupole echo spectra of polycrystalline [tyrosine-3,5-$^{2}H_2$][Leu^5]-Leu-enkephalin as a function of temperature. (b) Simulations of the spectra using a twofold jump model including corrections for power roll-off and for intensity distortions produced by the quadrupole echo. The jump rate at each temperature is indicated in the Figure. (c) Line shapes at the same jump rate but without the echo distortion corrections (from Rice *et al.*[34]).

10^5 Hz is thus determined by the line-shape analysis based on the treatment of ^{2}H NMR spectra for such twofold jump was performed at temperature between −20°C and 101°C.

In a similar manner, it is straightforward to distinguish the manner of motion for aliphatic side chains for a variety of proteins, among possible several types of motions characterized by their ^{2}H NMR powder pattern, as demonstrated for simulated spectra of [5-$^{2}H_3$]leucine methyl groups (Figure 4.15), as a reference to spectral interpretation for more complicated ^{2}H NMR spectra of [$^{2}H_{10}$]Leu-labeled collagen.[38,39] ^{2}H NMR spectrum of methyl groups from polycrystalline [$^{2}H_{10}$]leucine appears with $\Delta\nu_Q \approx 40$ kHz even at −45°C as a doublet (Figure 4.15(c)) owing to the presence of rapid C_3 rotation, instead of the static spectra shown in Figure 4.15(d).[39] In addition, leucine can have its side chain in the two stable configuration, and jump between them would cause spectral change as illustrated in Figure 4.15(b). It is also plausible for there to be two separate rotational motions in long aliphatic side chains and the combined effect is shown in Figure 4.15(a).

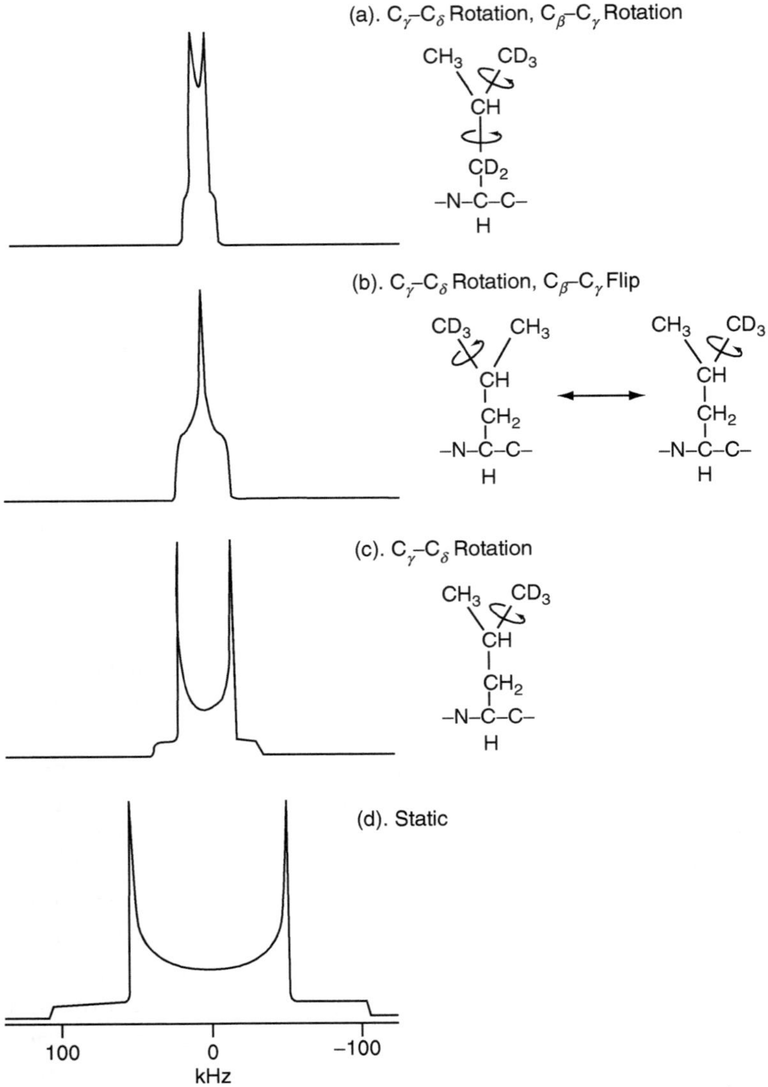

Figure 4.15. Calculated ^2H powder pattern of [5-$^2H_3$]leucine in the presence of various types of motions (from Opella[39]).

4.2.4. Determination of Oriented Spectrum from Powder Sample (DePaking)

^2H powder pattern NMR spectrum of perdeuterated lipid bilayers, for instance, from DPPC in excess water at 45°C exhibits a complicated spectral pattern arising from several superimposed Pake-doublets from individual ^2H-labeled positions of fatty acyl chains, as illustrated in Figure 4.16(a).[40] It is very difficult to identify and assign specific peak in this spectrum, and powder pattern simulations with so many parameters such as splitting, intensity, and linewidth for each peak are rather tricky.[41] Because of the local axial symmetry of the lamellar liquid crystalline phase, it is possible to extract the complete distribution of quadrupolar splittings directly from the powder pattern spectrum.[42] The oriented or "de-Paked" spectrum of Figure 4.16(b) was derived from the powder spectrum in Figure 4.16(a). Measurement of the amplitude and frequencies of the many components of the "de-Paked" spectrum is easier and more accurate than for the original powder spectrum, since most of the overlap between components has been removed.

In practice, Bloom *et al.*[42] denote the oriented spectrum for $\theta = 0$ by the normalized line shape function $F_0(\omega)$ and its powder spectrum by $G(\omega)$, where ω is expressed relative to ω_0 as a multiple of an appropriate coupling constant so the entire spectrum falls well within the range $-1 < \omega < +1$. We adopt the convention that $F_0(\omega) = 0$ for negative values of ω, so that the powder spectrum is related to $F_0(\omega)$ by

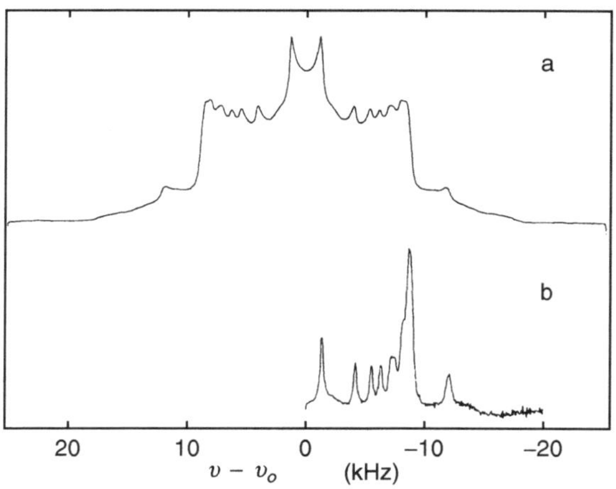

Figure 4.16. (a) ^2H NMR spectrum of perdeuterated DPPC/water (50:50 wt%) at 45°C. (b) The "de-Paked" spectrum obtained from (a) showing explicitly the distribution of quadrupolar splittings (from Davis[40]).

$$G(\omega) = \int_0^1 [3x(x+2\omega)]^{-1/2} F_0(x) I(x,\omega) \mathrm{d}x, \tag{4.18}$$

where $I(x,\omega) = 1$ for $-x/2 < w \leq x$ and $I(x,\omega) = 0$ outside these limits. For symmetrical powder spectra such as quadrupolar interaction, the experimental powder spectrum is identified with $[(G(\omega) + G(-\omega))/2]$ as a consequence of restriction of $F_0(\omega)$ to positive values of ω. From numerical calculation, it is possible to extract the oriented spectrum from the powder spectrum which is in close agreement with the original $F_0(\omega)$.

4.3. ^{17}O NMR

The oxygen atom is one of the important atoms constituting hydrogen-bonding structure in polypeptides and proteins. The significance of solid state ^{17}O NMR study for polypeptides in the solid state has been increasingly recognized. Nevertheless, the number of studies in this field is very small. This comes from very weak sensitivity for solid state ^{17}O NMR measurement that has two reasons. One is that the ^{17}O nucleus possesses a very low natural abundance of only 0.037%. Another is that the ^{17}O is a quadrupolar nucleus because the nuclear spin quantum number (I) is $-5/2$, and thus ^{17}O signals are broadened by nuclear quadrupolar effects in solid. On the other hand, solution ^{17}O NMR spectroscopy has been successfully employed to elucidate a number of structural problems in organic chemistry,[43–47] because ^{17}O signal becomes very sharp due to the disappearance of quadrupolar interaction by isotropic fast reorientation in solution. When the oxygen atom in the carbonyl group of various kinds of compounds is directly associated with formation of a hydrogen bond, the carbonyl ^{17}O NMR signal often moves to large upfield (low-frequency) shifts.[48,49] From these experimental results, solution ^{17}O NMR has been established as a means for investigating hydrogen-bonded structural characterization. From such situations, it can be expected that solid state ^{17}O NMR provides deep insight into understanding hydrogen-bonding structure in solid polypeptides.[50–56]

4.3.1. Measurement of Solid State ^{17}O NMR

Static ^{17}O CP-NMR experiments operating at 67.8 MHz (500 MHz for ^{1}H) at room temperature are described. The ^{17}O labeled polypeptide samples are contained in a cylindrical rotor made of silicon nitride. In CP method,

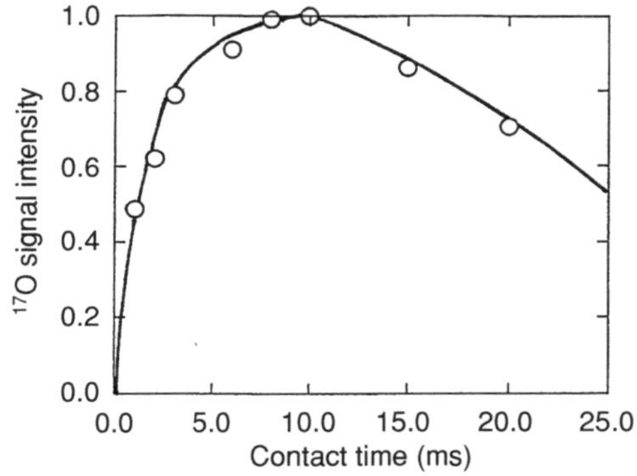

Figure 4.17. A plot of the ^{17}O peak intensities of 6% ^{17}O labeled Gly*-Gly against CP contact time.[50]

$Mg(^{17}OH)_2$ was used for ^1H–^{17}O CP matching ($\gamma_H B_{1H} = 3\gamma_O B_{1O}$, where γ_X and B_{1X} are the gyromagneto ratio and rf field for *X* nucleus, respectively.[57] The ^1H $\pi/2$ pulse length is 5 μs and the ^{17}O $\pi/2$ pulse length is 5 μs for solid sample (which corresponds to 15 μs for liquid sample such as water). The ^1H decoupling field strength is 50 kHz and repetition time is 5 s.

To determine the optimum contact time for ^{17}O nucleus in polypeptides, static ^{17}O CP-NMR of 6% ^{17}O labeled Gly*-Gly was recorded by changing the contact time as shown in Figure 4.17. The carbonyl oxygen in the amide group does not link with hydrogen atom, but the amide proton in hydrogen-bond is close to the oxygen. This amide proton induces the CP for the ^{17}O nucleus. The ^{17}O signal intensities are followed by a function of the CP contact time t by Eq. (3.12).[57,58] The optimum contact time is obtained as 9 ms, and CP time (T_{OH}) and $T_{1\rho}^{H}$ are obtained as 2.5 ms and 30 ms, respectively. The obtained T_{OH} for the peptide is much longer compared with an inorganic solid such as $AlO^{17}OH$ whose T_{OH} is 0.018 ms.[57] The ^{17}O chemical shifts are calibrated through external liquid water ($\delta = 0$ ppm).

4.3.2. Static ^{17}O NMR Spectra

As mentioned above, the ^{17}O possesses a quadrupolar moment. Thus, static ^{17}O spectrum contains information on quadrupolar interactions, and so the central transition signal (−1/2, 1/2) becomes broad by the second-order perturbation. A static ^{17}O NMR signal contains eight kinds of NMR parameters such as the nuclear quadrupolar constant (e^2qQ/h), which comes from the electrostatic interaction between the nuclear quadrupolar electric moment (eQ) and the electric field gradient tensor ($e\boldsymbol{q}$), the asymmetry parameter (η), and the principal values of chemical shift tensor (δ_{11}, δ_{22}, δ_{33}) (where $\delta_{11} > \delta_{22} > \delta_{33}$). The orientation of the principal values of chemical shift tensor in the quadrupolar frame of reference is defined by the three Euler angles (α, β, γ) as shown in Figure 2.3(a). The spectral pattern depends on these parameters. Thus, these eight parameters can be determined by superimposing the theoretical line shape[59–62] onto the observed spectrum.

In the static spectral pattern, the resonance line appears at a frequency of ω by the sum of the quadrupolar interaction term (ω_Q) and the chemical shift term ($\omega_{\rm cs}$) as

$$\omega = \omega_Q + \omega_{\rm cs} \tag{4.19}$$

The quadrupolar interaction term (ω_Q) is expressed by

$$\omega_Q = (\omega_q^2/6\omega_0)\,(I(I+1) - 3/4)\,[A\,(\phi)\cos^4\theta + B\,(\phi)\cos^2\theta + C(\phi)] \tag{4.20}$$

in which

$$\omega_q = 3\,e^2qQ/2I(2I-1)\,h, \tag{4.21}$$

$$A(\phi) = -27/8 - (9/4)\eta\cos 2\phi - (3/8)\eta^2\,\cos^2 2\phi, \tag{4.22}$$

$$B(\phi) = 30/8 - \eta^2/2 + 2\eta\cos 2\phi - (3/4)\eta^2\,\cos^2 2\phi, \tag{4.23}$$

and

$$C(\phi) = -3/8 + \eta^2/2 + (1/4)\eta\,\cos 2\phi - (3/8)\eta^2\cos^2 2\phi. \tag{4.24}$$

The chemical shift term ($\omega_{\rm cs}$) is expressed as

$$\begin{aligned}\omega_{\rm cs} = \delta_0 + (\delta'/2)\,\{&(3\cos^2\theta - 1)[(3\cos^2\beta - 1)/2 - (\eta'/2)\sin^2\beta\cos 2\gamma] \\ &+ \sin 2\theta\cos(\phi+\alpha)[-(3/2)\sin 2\beta - \eta'\sin\beta\cos\beta\cos 2\gamma] \\ &+ \sin 2\theta\sin(\phi+\alpha)(\eta'\sin\beta\sin 2\gamma) \\ &+ (1/2)\sin^2\theta\cos 2(\phi+\alpha)[3\sin^2\beta - \eta'\cos 2\gamma\,(1+\cos^2\beta)] \\ &+ (1/2)\sin^2\theta\sin 2(\phi+\alpha)(2\eta'\cos\beta\sin 2\gamma)]\}\end{aligned} \tag{4.25}$$

in which the angles θ and ϕ are the spherical angles describing the orientation of the Z-axis of the principal axis system.

$$\delta_0 = (\delta_{11} + \delta_{22} + \delta_{33})/3 \tag{4.26}$$

$$\delta' = \delta_{33} - \delta_0$$

and

$$\eta' = (\delta_{22} - \delta_{11})/(\delta_{33} - \delta_0)$$

The static spectral pattern can be calculated by integration over possible orientation.

As mentioned above, it can be expected that static ^{17}O CP-NMR spectral pattern is greatly influenced by the NMR frequency. In Figure 4.18, ^{17}O CP NMR spectral patterns of Gly*-Gly (6% ^{17}O labeled) observed at 36.6 MHz (270 MHz for ^{1}H), 54.2 MHz (400 MHz for ^{1}H), and 67.8 MHz (500 MHz for ^{1}H) are shown. It is seen that the ^{17}O spectrum at 36.6 MHz consists of two splitting signal groups. Their separation is decreased with increased NMR frequency, and the two split signals coalesce at 67.8 MHz. This means that such a splitting at the central transition (see Figure 5.14 also) comes from the second-order quadrupolar interaction (for example, see Figure 2.13 for $I = 3/2$ nuclei) and if one wants to obtain an undistorted ^{17}O signal, ultra-high-frequency NMR spectrometer is needed (see Figure 5.14). By computer simulation using some different frequency NMR spectrometers as mentioned above, we can obtain eight kinds of NMR parameters with high reliability.

4.3.3. MAS Spectral Pattern

In rare spins such as ^{13}C, ^{15}N, etc, high-resolution solid state NMR spectra are usually recorded by the standard technique of CP between rare spin and abundant spin (usually ^{1}H), combined with MAS under dipolar decoupling. If the spinning rate ν_r is less than the width of CSA ($\Delta\delta = |\delta_{11} - \delta_{33}|$), then the resulting spinning sidebands (rotational echo) flank the center band. For example, in ^{13}C spectra, the presence of such splitting sidebands is prominent in carbonyl carbons, whose CSA values are as large as 10 kHz. The ^{17}O CP-MAS NMR spectra of poly(L-alanine) [$(Ala)_n$] with the α-helix form (Figure 4.19) at 36.6 MHz and 67.8 MHz are shown (Figure 4.20) to clarify the effect of MAS on the spectral pattern. It is apparent that the spectral range for the MAS spectrum becomes effectively much narrower depending upon ν_Q^2/ν_L in Eq. (2.38) (see Figure 2.14) than that for the CP spectra (Figure 4.20). This means that extremely fast MAS rate is needed on ^{17}O nucleus with high-speed MAS and then the interval between the main peak and sidebands splitting becomes large, and the sideband intensity is reduced. Therefore, high-frequency NMR measurement is needed for a ^{17}O MAS

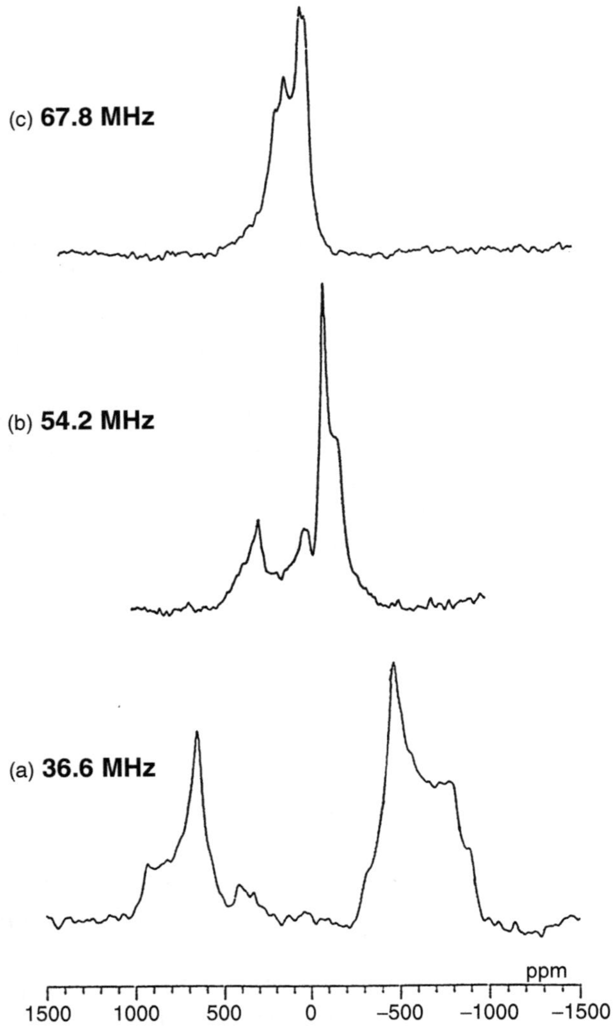

Figure 4.18. Static ^{17}O CP spectra of Gly*-Gly at 36.6 (a), 54.2 (b) and 67.8 (c) MHz.[50]

experiment simultaneously with high-speed MAS. In Figure 4.21, the ^{17}O MAS NMR spectra of $(Ala)_n$[degree of polymerization (DP) = ca. 100] with an α-helix form (a) and $(Ala)_n$ (DP = 5) with a β-sheet form (b) shown at 108.6 MHz and at 25 kHz MAS rate, where the upper is theoretically simulated spectrum taking into account 25 kHz MAS rate and the bottom is the observed spectrum. In the spectra, the central band signal was

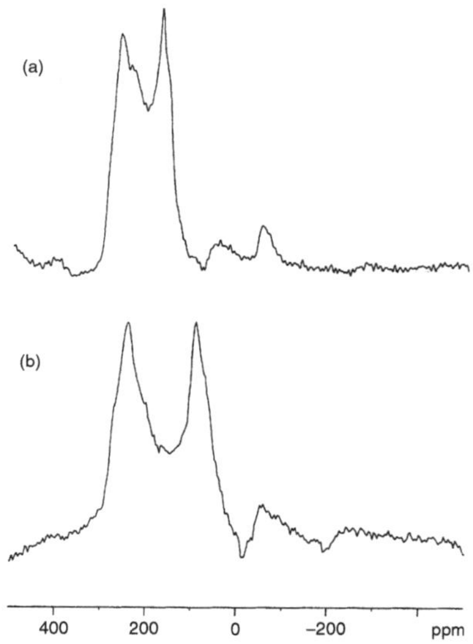

Figure 4.19. ^{17}O CP-MAS NMR spectra of $(Ala)_n$ (with DP = ca. 100) at 67.8 MHz (a) and 54.2 MHz (b).[50]

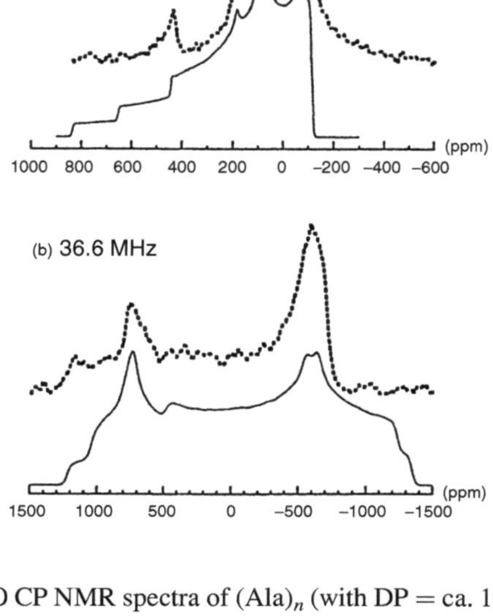

Figure 4.20. Static ^{17}O CP NMR spectra of $(Ala)_n$ (with DP = ca. 100) at 67.8 MHz (a) and 36.6 MHz (b).[52]

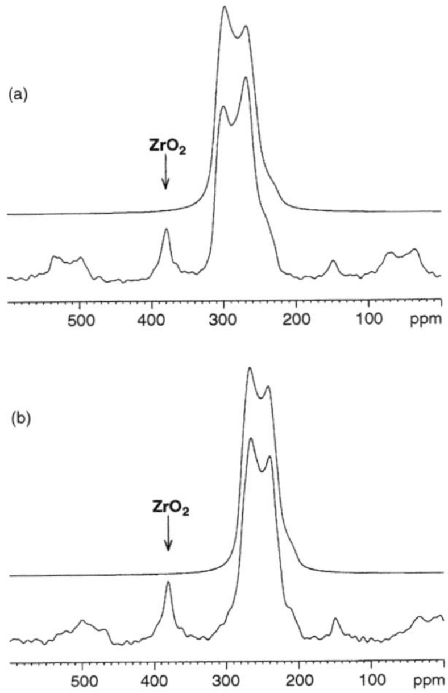

Figure 4.21. ^{17}O CP MAS NMR spectra of $(Ala)_n$ (with DP = ca. 100) with an α-helix form (a) and of $(Ala)_n$ (with DP = ca. 5) with a β-sheet form (b) at 108.6 MHz and at MAS rate of 25 kHz. The upper traces are theoretically simulated spectra and the bottom traces are the observed ones.[52]

completely separated from the sideband signals. It is seen from these spectra that the ^{17}O signal for the α-helix form appears at lower field that that for the β-sheet form.

References

[1] W. Saenger, 1983, *Principles of Nucleic Acid Structure*, Springer-Verlag, New York.
[2] S. B. Zimmerman, 1982, *Annu. Rev. Biochem.*, 51, 395–427.
[3] H. Shindo, J. B. Wooten, B. H. Pheiffer, and S. B. Zimmerman, 1980, *Biochemistry*, 19, 518–526.
[4] H. Shindo, 1984, in *Phosphorus-31 NMR, Principles and Applications*, D. G. Gorenstein, Ed., Academic Press, New York, pp. 401–422.
[5] B. T. Nall, W. P. Rothwell, and J. S. Waugh, 1981, *Biochemistry*, 20, 1881–1887.
[6] H. Shindo, T. Fujiwara, H. Akutsu, U. Matsumoto, and Y. Kyogoku, 1985, *Biochemistry*, 24, 887–895.

[7] M. Mehring, 1983, in *Principles of High Resolution NMR in Solids*, Second Revised Edition, Springer-Verlag, Berlin, Heidelberg, New York.
[8] S. J. Kohler and M. P. Klein, 1976, *Biochemistry*, 15, 967–973.
[9] J. Herzfeld, R. G. Griffin, and R. A. Harberkorn, 1978, *Biochemistry*, 17, 2711–2718.
[10] S. Arnott, R. Chandrasekaran, D. L. Birdsall, A. G. W. Leslie, and R. L. Ratliff, 1980, *Nature*, 283, 743–745.
[11] T. Fujiwara and H. Shindo, 1985, *Biochemistry*, 24, 896–902.
[12] H. Shindo and S. B. Zimmerman, 1980, *Nature*, 283, 690–691.
[13] J. A. DiVerdi and S. J. Opella, 1981, *Biochemistry*, 20, 280–284.
[13a] T. A. Cross, P. Tsang, and S. J. Opella, 1983, *Biochemistry*, 22, 721–726.
[14] T. A. Cross, S. J. Opella, G. Stubbs, and D. L. D. Caspar, 1983, *J. Mol. Biol.*, 170, 1037–1043.
[15] I. C. P. Smith and I. H. Ekiel, 1984, in *Phosphorus-31 NMR, Principles and Applications*, D. G. Gorenstein, Ed., Academic Press, New York, pp. 447–475.
[16] R. G. Griffin, 1981, *Methods Enzymol.*, 72 (part D), pp. 108–174.
[17] J. Seelig, 1978, *Biochim. Biophys. Acta*, 515, 105–140.
[18] P. R. Cullis and B. de Kruijff, 1979, *Biochim. Biophys. Acta*, 559, 399–420.
[19] R. G. Griffin, 1976, *J. Am. Chem. Soc.*, 98, 851–853.
[20] W. Niederberger and J. Seelig, 1976, *J. Am. Chem. Soc.*, 98, 3704–3706.
[21] M. Bloom, E. E. Burnell, M. I. Valic, and G. Weeks, 1975, *Chem. Phys. Lett.*, 14, 107–112.
[22] G. H. Freed, G. V. Bruno, and C. F. Polnaszek, 1971, *J. Phys. Chem.*, 73, 3385–3399.
[23] E. E. Burnell, P. R. Cullis, and B. de Kruijff, 1980, *Biochim. Biophys. Acta*, 603, 63–69.
[24] T. M. Rothgeb and E. Oldfield, 1981, *J. Biol. Chem.*, 256, 1432–1446.
[25] R. W. K. Lee and E. Oldfield, 1982, *J. Biol. Chem.*, 257, 5023–5028.
[26] J. Seelig and W. Niederberger, 1974, *J. Am. Chem. Soc.*, 96, 2069–2072.
[27] J. Seelig, 1977, *Q. Rev. Biophys.*, 10, 353–418.
[28] A. Seelig and J. Seelig, 1974, *Biochemistry,* 13, 4839–4845.
[29] G. W. Stockton, C. F. Polnaszek, A. P. Tulloch, F. Hasan, and I. C. P. Smith, 1976, *Biochemistry*, 15, 954–966.
[30] R. A. Kinsey, A. Kintanar, and E. Oldfield, 1981, *J. Biol. Chem.*, 256, 9028–9036.
[31] E. Oldfield, R. A. Kinsey, and A. Kintanar, 1982, *Methods Enzymol.*, 88, 310–325.
[32] G. Soda and T. Chiba, 1969, *J. Chem. Phys.*, 50, 439–455.
[33] C. M. Gall, T. A. Cross, and S. J. Opella, 1982, *Proc. Natl. Acad. Sci. USA*, 79, 101–105.
[34] D. M. Rice, R. J. Witterbort, R. G. Griffin, E. Meirovitch, E. R. Stimson, Y. C. Meinwald, J. H. Freed, and H. A. Scheraga, 1981, *J. Am. Chem. Soc.*, 103, 7707–7710.
[35] R. G. Barnes and J. W. Bloom, 1972, *J. Chem. Phys.*, 57, 3082–3086.
[36] M. Mehring, R. G. Griffin, and J. S. Waugh, 1971, *J. Chem. Phys.*, 55, 746–755.
[37] H. W. Spiess and H. Sillescu, 1981, *J. Magn. Reson.*, 42, 381–387.
[38] L. S. Batchelder, C. E. Sullivan, L. W. Jelinski, and D. A. Torchia, 1982, *Proc. Natl.Acad. Sci. USA*, 79, 386–389.
[39] S. J. Opella, 1986, *Methods Enzymol.*, 131, 327–361.
[40] J. H. Davis, 1983, *Biochim. Biophys. Acta*, 737, 117–171.
[41] J. H. Davis, 1979, *Biophys. J.*, 27, 339–358.

[42] M. Bloom, J. H. Davis, and A. L. Mackay, 1981, *Chem. Phys. Lett.*, 80, 198–202.
[43] D. W. Boykin, Ed., 1991, ^{17}O *NMR Spectroscopy in Organic Chemistry*, CRC Press, Boca Raton, FL, pp. 1–325.
[44] L. Baumstark and D. W. Boykin, 1991, ^{17}O NMR Spectroscopy: Applications to Structural Problems in Organic Chemistry, *in Advances in Oxygenated Processes,* Vol. III, A. L. Baumstark Ed., JAI Press, Greenwich, CT, p. 141.
[45] W. G. Klemperer, 1983, in *The Multinuclear Approach to NMR Spectroscopy*, J. B. Lambert and F. G. Riddell, Eds., D. Reidel, Dordrecht, Holland, pp. 245–260.
[46] J. P. Kintzinger, 1983, in *Newly Accessible Nuclei*, Vol. 2, P. Laszlo, Ed., Academic Press, New York, p.79.
[47] S. Kuroki, K. Yamauchi, I. Ando, A. Shoji, and T. Ozaki, 2001, *Curr. Org. Chem.*, 5, 1001–1016.
[48] A. L. Baumstark, M. Dotrong, R.R. Stark and D.W. Boykin, 1988 *Tetrahedron Lett*, 29, 2143–2146.
[49] D. W. Boykin and A. Kumar, 1992, *J. Heterocycl. Chem.*, 29, 1–4.
[50] S. Kuroki, A. Takahashi, I. Ando, A. Shoji, and T. Ozaki, 1994, *J. Mol. Struct.*, 323, 197–208.
[51] A. Takahashi, S. Kuroki, I. Ando, T. Ozaki, and A. Shoji, 1998, *J. Mol. Struct.*, 442, 195–199.
[52] S. Kuroki, K. Yamauchi, H. Kurosu, S. Ando, I. Ando, A. Shoji, and T. Ozaki, 1999, *ACS Symp. Ser.*, 732, 126–137.
[53] K. Yamauchi, S. Kuroki, I. Ando, A. Shoji, and T. Ozaki, 1999, *Chem. Phys. Lett.*, 302, 331–336.
[54] S. Kuroki, I. Ando, A. Shoji, and T. Ozaki, 1992, *J. Chem. Soc., Chem. Commun.*, 433–434.
[55] K. Yamauchi, S. Kuroki, and I. Ando, 2000, *Chem. Phys. Lett.*, 324, 435–439.
[56] S. Kuroki, S. Ando, and I. Ando, 1995, *Chem. Phys.*, 195, 107–116.
[57] T. H. Walter, G. L. Turner, and E. Oldfield, 1988, *J. Magn. Reson.*, 76, 106–120.
[58] S. R. Hartmann and E. L. Hahn, 1962, *Phys. Rev.*, 128, 2042–2053.
[59] R. Goc and D. Fiat, 1987, *Phys. Status Solidi.*, B140, 243–250.
[60] P. J. Chu and B. C. Gerstein, 1989, *J. Chem. Phys.*, 91, 2081–2101.
[61] J. T. Cheng, J. C. Edwards, and P. D. Ellis, 1990, *J. Phys. Chem.*, 94, 553–561.
[62] W. P. Power, R. E. Wasylishen, S. Mooibroek, B. A. Pettit, and W. Danchura, 1990, *J. Phys. Chem.*, 94, 591–598.

Chapter 5

EXPERIMENTAL STRATEGIES

5.1. Isotope Enrichment (Labeling)

As far as NMR signals from their major components are concerned, isotope enrichment is not always essential for NMR observation of fibrous proteins or polysaccharides which give rise to rather simple NMR spectra because of limited number of repeating residues. By contrast, it is difficult to distinguish individual NMR signals of respective residues either in globular or membrane proteins, in which their peaks are crowded because of increased numbers of amino acid residues as compared with fibrous proteins and their relative proportions are low. Accordingly, an appropriate isotope enrichment (labeling) by the most common ^{13}C or ^{15}N nuclei through biosynthesis by cell cultures or chemical synthesis is essential in order to enhance both their *peak intensities* and *selectivity* of their labeled signals from background signals arising from residues of natural abundance. Strictly speaking, this process should not be referred to as labeling in the case of radioisotope but to isotope-replacement or enrichment, because the extent of which is desirable as high as possible up to 100% for a sample to be measured by solid state NMR. The term ''labeling'' is also conventionally used for many instances, however.

5.1.1. Uniform Enrichment

Extensive isotope enrichment based on uniform isotope substitution (labeling) of proteins, using [$^{13}C_6$]glucose, [$^{13}C_2$]acetate, and/or $^{15}NH_4Cl$ as the only sources for ^{13}C or ^{15}N nuclei, seems to be naturally the most efficient and comprehensive means to prepare samples useful for solid state NMR as well as solution NMR,[1,2] as far as resulting signals could be well resolved and straightforwardly assigned to sites of individual residues. In fact, it turned out that uniform ^{15}N enrichment by $^{15}NH_4Cl$, which constitutes isolated ^{15}N spin network in a protein, has proved to be an excellent means to reveal 3D structures and dynamics of uniformly ^{15}N-labeled fibrous or membrane proteins in mechanically oriented lipid bilayer to the applied magnetic field, based on the data of ^{15}N chemical shifts/^{15}N–H dipolar interactions, and ^{15}N chemical shift tensor, respectively.[3–6] On the contrary, spectral analysis of uniformly ^{13}C-labeled proteins seems to be not always straightforward means because of inherent problems arising from high dense ^{13}C spin network. Pauli *et al.*[7] prepared uniformly ^{13}C-, ^{15}N-labeled microcrystalline sample of 62-residue, α-spectrin SH_3 domain by overexpression from *Escherichia coli*, using a minimal medium based on M9 salts, supplemented with 2 g [u-^{13}C]glucose and 1.0 g $^{15}NH_4Cl$ per liter of medium. They showed the well-resolved ^{13}C and ^{15}N peaks, in which the individual backbone and side-chain ^{13}C NMR signals were readily assigned by 2D MAS ^{15}N–^{13}C and ^{13}C–^{13}C correlation spectroscopy at 17.6 T, although their NMR signals from the first six N-terminal and the last C-terminal residues are absent.[7] For lyophilized preparations as discussed in Section 3.2, it should be anticipated that additional line broadening arises from chemical shift dispersion in the presence of conformationally heterogeneous contribution. For this reason, crystalline samples, in favorable case, containing hydrated water molecules are highly desirable for solid state NMR.

5.1.2. Selective Enrichment by [1-^{13}C]Glucose or [2-^{13}C]Glycerol

Uniformly ^{13}C-labeled proteins, however, may pose several difficulties in recording solution and solid state NMR. First, the indirect one-bond spin couplings as well as dipolar ^{13}C–^{13}C interactions causes severe line broadenings for fully ^{13}C-labeled proteins in the solid state NMR. The former type of line broadening cannot be removed even by high-speed MAS (>25 kHz), in contrast to the case of the latter type. In addition, the weak ^{13}C–^{13}C dipolar couplings arising from the remote ^{13}C–^{13}C distance are obscured by the strong dipolar couplings due to spin diffusion within the dense ^{13}C spin network and by the dipolar truncation mechanism.[8] Even in solution NMR, analysis of side-chain dynamics by ^{13}C spin relaxation is complicated by contributions from ^{13}C–^{13}C scalar and dipolar couplings. To overcome the latter complication in solution NMR, the ^{13}C label using [1-^{13}C] glucose or [2-^{13}C]glycerol as the sole carbon sources has been introduced to isolate groups in an alternating ^{12}C–^{13}C–^{12}C-labeling pattern.[9] Taking this labeling strategy, Hong and Jakes[10] and Hong[11] prepared selectively and extensively ^{13}C-labeled proteins suitable for solid state NMR. They showed that for amino acids synthesized in the linear part of the biosynthetic pathways, [1-^{13}C] glucose preferentially labels the ends of the side chains, while [2-^{13}C]glycerol labels the C_α of 10 amino acid residues, including Gly, Cys, Ser, Ala, Leu, Val, His, Phe, Tyr, and Trp, which are suitable for measurements of the backbone torsion angles simultaneously, as well as C_β of Val residue, as in Figure 5.1. Typically, they used a modified M9 medium containing, per liter, 1 g $^{15}NH_4Cl$, 4 g [2-^{13}C]glycerol, 3 g KH_2PO_4, 6 g Na_2HPO_4, 1 mM $MgSO_4$, the unlabeled amino acids Asp, Asn, Arg, Gln,

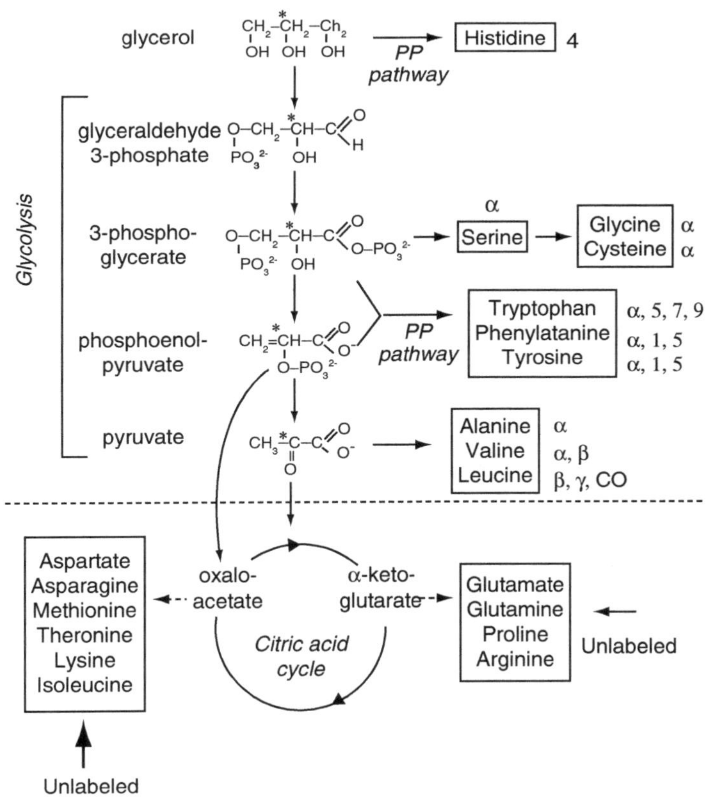

Figure 5.1. Amino acid biosynthetic pathways in bacteria, utilizing glycerol as the main carbon source. Selectively ^{13}C-labeled hydrophobic amino acid residues are largely produced from the glycolytic and pentose phosphate (PP) pathways. The growth media is supplemented with unlabeled amino acid products of the citric acid cycle (from Hong and Jakes[10]).

Glu, Ile, Lys, Met, Pro, and Thr at 150 μg/mL each, and 100 μg/mL ampicillin. They showed that the maximum labeling level is up to 50%. This approach yielded high sensitivity and spectral resolution, as demonstrated for 25-kDa colicin Ia channel domain in the absence of lipids. They expressed the protein with an amino-terminal His_6 tag from pKSJ120-containing *E. coli* BL21(DE3). The resulting spectra turned out to be remarkably simple for a 25-kDa protein. They showed two pairs of coupled ^{13}C spins, Val C_α/C_β and Leu C_β/C_γ, which are very convenient for the signal assignment. This approach was successfully used for the above-mentioned SH3 domain to determine the secondary structure, based on 286 inter-residue ^{13}C–^{13}C and six ^{15}N–^{15}N restraints determined by proton-driven spin diffusion (PDSD) spectra.[12]

5.1.3. Spectral Broadening Caused by Shortened Spin–Spin Relaxation Times of Dense ^{13}C Spin Network

As discussed already in Section 3.1, surface residues involved in loops of membrane proteins such as fully hydrated bacteriorhodopsin (bR) are very flexible at ambient temperature, even if they are embedded in naturally occurring 2D crystals (purple membrane; PM), in contrast to the case of crystalline globular proteins. Undoubtedly, it should be taken into account that this kind of flexibility is essential to exhibit its specific biological function. Nevertheless, it should be cautioned that spectral resolution of ^{13}C-labeled proteins might be desperately deteriorated, if a ^{13}C–^{13}C sequence is present in the dense ^{13}C spin networks, and this portion undergoes fluctuation motions with frequency from 10^8 to 10^2 Hz, as manifested from the observed ^{13}C CP-MAS NMR spectra of [1,2,3-$^{13}C_3$]Ala-labeled bR from PM,[13] as demonstrated in Figure 5.2. Surprisingly, the spectral resolution of the CP-MAS NMR spectra turned out to be hopelessly poor for densely ^{13}C-labeled bR, although the well-resolved signals are partially visible only from the corresponding DD-MAS NMR spectra of the carbonyl region. This is mainly caused by their shortened spin–spin relaxation times due to the presence of increased number of relaxation pathways modulated by their fluctuation motions through a number of ^{13}C–^{13}C homonuclear dipolar and scalar spin interactions, etc. In contrast, this sort of broadening does not occur for [3-^{13}C]- or [1-^{13}C]Ala-labeled bR, constituting an isolated spin system in the absence of the direct ^{13}C–^{13}C sequence, as illustrated in Figure 2.7.

Even in the absence of such problem arising from the dense spin network, it should be anticipated that ^{13}C NMR signal from an expected site is not always

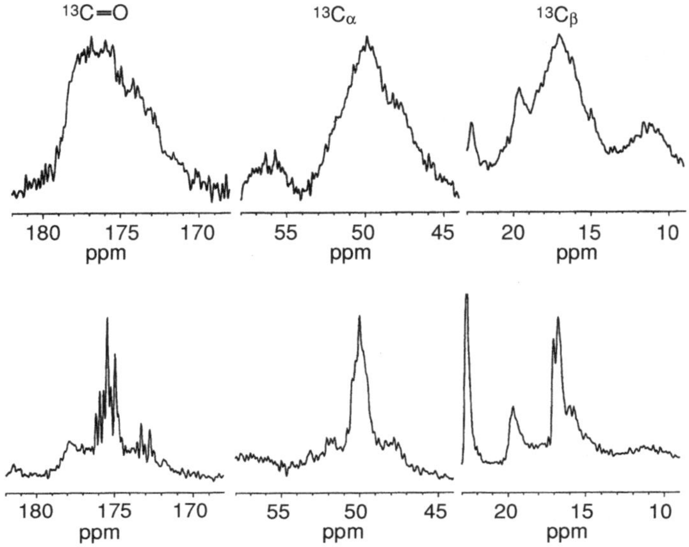

Figure 5.2. ^{13}C CP-MAS NMR (upper traces) and DD-MAS NMR spectra (lower traces) of ^{13}C=O, C_α and C_β carbons from [1,2,3-$^{13}C_3$]Ala-labeled bacteriorhodopsin.[13]

fully visible, if motional frequency of internal fluctuation, if any, is interfered with either frequency of proton decoupling or MAS as encountered for a variety of membrane proteins at ambient temperature.[14–17] This is possible for a variety of fully hydrated membrane proteins labeled by [3-^{13}C]Ala, [1-^{13}C]Gly, Ala, or Val, etc., as discussed already in Section 3.2. This kind of information, however, is very useful to locate residues in which frequencies of fluctuation motion with timescale of millisecond or microsecond of biological significance interfer with frequencies of either proton decoupling or MAS, once ^{13}C NMR peaks of interest are site-directly assigned to certain amino acid residues located at such particular site in advance.

5.1.4. Site-Directed (or amino acid specific) ^{13}C Enrichment

To avoid this complication, it is advised to try an alternative site-directed (or amino acid specific) enrichment in which one of the unlabeled amino acid residues is replaced by a selectively ^{13}C-labeled amino acid such as [3-^{13}C]Ala, [1-^{13}C]Ala, Val, etc. in the growth TS medium[18] for bR from

Halobacterium salinurum or M9 medium for a variety of membrane proteins expressed from *E. coli* (Table 5.1).[19] It is also possible to combine two different kinds of amino acid residues such as [3-^{13}C]Ala and [1-^{13}C]Val simultaneously because of no overlap of peaks and no direct ^{13}C–^{13}C connectivity. Obviously, this approach is the simplest means to incorporate one or two kinds of amino acid residue(s) as an isolated spin system to proteins, which is essential for observation of sharp NMR signals for membrane proteins undergoing fluctuation motions with a variety of frequencies depending upon their respective environment in 2D crystal, as demonstrated in Figure 2.7. In principle, all the ^{13}C signals thus obtained could be readily ascribed to amino acid residues incorporated, unless otherwise metabolic conversion from incorporated residues to the others. Metabolic conversion, however, from Asp to Glu or Trp residues should be also taken into account when one aims to prepare [4-^{13}C]Asp-labeled bR, in which this label is useful for probes for pH titration or salt bridge-like bonds.[20,20a,b] No such conversion, however, has been observed when amino acids such as hydrophobic Ala and Val are incorporated into bR. Nevertheless, background ^{13}C signals from unlabeled amino acid residues cannot be completely ignored as demonstrated for [1-^{13}C]Gly-labeled bR.

5.1.5. Isotope Enrichment in Primary Cell Cultures

In addition to the above-mentioned biosynthetic approaches utilizing bacterial cells, several attempts have been also made to utilize primary cultured cells as a means for biosynthetic incorporation of isotope labels. To prepare ^{13}C- or ^{2}H-labeled collagen,[21] calvaria parietal bones from 25 dozen 17-day-old chick embryos are preincubated at 38–39 °C in a humidified 5% CO_2/air atmosphere for two 2-h periods in order to deplete the free amino acid pool.[21] Both incubations use 250 mL of Eagle's minimal essential medium, deficient in the amino acid of interest supplemented, per 100 mL, with the following: 10 mg of glycine, 10 mg of L-proline, 5–20 mg of the ^{13}C- or ^{2}H-labeled amino acid of interest, L-ascrobic acid (sodium salt), 10–50 μCi of the ^{14}C- or ^{3}H-labeled amino acid, and 5000 units of penicillin/streptomycin. After the two 2-h preincubation periods, the calvaria are incubated for two 48-h periods in 250 mL of the above media, to which 5 mg of β-aminopropionitrile fumarate are added per 100 mL. (β-Aminopropionitrile fumarate inhibits collagen cross-linking.) Levels of incorporated labeled compounds were 14–85%, depending on the type of incorporated amino acids.

Table 5.1. TS and M9 media for amino acid specific enrichment.

TS medium (pH 6.6, 2-L scale)					
NaCl	500 g				
$MgSO_4 \cdot 7H_2O$	40 g				
KCl	2 g				
KNO_3	200 mg				
KH_2PO_4	300 mg				
K_2HPO_4	300 mg				
Sodium citrate $2H_2O$	1 g				
L-Ala	430 mg	L-Arg	800 mg	L-Asp	450 mg
L-Cys	100 mg	L-Glu	2600 mg	Gly	120 mg
L-His	300 mg	L-Ile	440 mg	L-Leu	1600 mg
L-Lys	1700 mg	L-Met	370 mg	L-Phe	260 mg
L-Pro	100 mg	L-Ser	610 mg	L-Thr	500 mg
L-Tyr	400 mg	L-Trp	50 mg	L-Val	1000 mg
Glycerol	2 g = 0.1% (g/mL)				
Metal salt solution	2 mL				
[Metal salt solution consisting of: 100 mL (0.1 N HCl)]					
$CaCl_2 \cdot 2H_2O$	700 mg				
$CuSO_4 \cdot 5H_2O$	5 mg				
$FeCl_2 \cdot 4H_2O$	230 mg				
$ZnSO_4 \cdot 7H_2O$	44 mg				
$MnSO_4 \cdot H_2O$	30 mg				
M9 medium (pH 7.4, l-L scale)[19]					
$Na_2HPO_4 \cdot 12H_2O$	15.1 g				
KH_2PO_4	3 g				
NaCl	0.5g				
NH_4Cl	1.0 g				
Nonlabeled amino acids (except for Trp) each 100 mg					
Trace metal solution 1.0 mL, consisting of:					
[Concentrated trace metal solution (100-mL scale)]					
4 mM $ZnSO_4 \cdot 7H_2O$	115 mg				
1 mM $MnSO_4 \cdot 4\text{–}5H_2O$	23.2 mg				
4.7 mM H_3BO_3	29.1 mg				
0.7 mM $CuSO_4 \cdot 5H_2O$	17.5 mg				
2.5 mM $CaCl_2 \cdot 2H_2O$	36.8 mg				
1.8 mM $FeCl_2 \cdot 4H_2O$	35.8 mg				
Adding the followings, after autoclaving:					
1 M $MgSO_4$	3 mL				
1 M $CaCl_2$	150 μL				
20% glucose	10 mL (2g/L)				
Trp (100 mg/10 mL)	10 mL (100 mg/mL)				
^{13}C-labeled amino acid (100 mg/10 mL)	10 mL (100 mg/L)				
Antibiotics (the amount varies depending upon protein to be overexpressed and bacterium)					

Perry *et al.*[22] used primary cultures of neonatal rat smooth muscle cells (NRSMC) which are isolated from the aortae of newborn Sprague–Dawley rats and grown in a mixture which includes fetal bovine serum and other standard growth medium components, including Dulbecco's modified Eagle's medium (DMEM) and nonessential amino acids. After about 1 week, the cells reach confluency and initiate abundant elastin synthesis. The primary source of glycine is contained in the DMEM. Therefore, DMEM is purchased "without glycine" and the isotopically enriched glycine ([1-^{13}C]Gly, 99%; [2-^{13}C]Gly, 99%; or [^{15}N]Gly, 98%) is supplemented into the medium at the appropriate concentration (30 mg Gly/L media) via a prepared nonessential amino acid mixture. The label is introduced immediately after seeding of the cells to ensure maximal incorporation of the label. After 6 to 7 weeks of growth, the mixture of cells, with elastin, growth medium, and other components of the matrix is harvested. Insoluble elastin is purified using the cyanogens bromide method.

Asakura and coworkers prepared ^{13}C-labeled *Bombyx mori* silk fibroin through the use of an artificial diet supplemented with [1-^{13}C]Gly or Ala.[23–25] Isotopes were incorporated into cocoons of *B. mori* by feeding 20 mg of [1-^{13}C]labeled amino acid along with 27 g of an artificial diet (silk Mate 2M, Nippon Nosan Kogyo Co., Tokyo) to 5-, 6- or 7-day-old silkworm larvae in the fifth instar stage. *In vitro* production of [^{15}N]Gly- and [^{15}N]Ala-enriched *B. mori* silk fibroin was performed with posterior silk glands isolated from 5-day-old silkworm larvae in the fifth instar stage. For preparation of ^{13}C, ^{15}N-labeled spider silk is to be used for rotational echo double resonance (REDOR) experiments, *Nephila clavipes* spiders (central Florida) were fed with DMEM which had been fortified with 1.3% (w/v) ^{15}N Ser, 0.56% (w/v) [1-^{13}C]Leu, and 0.56% [2-^{13}C]Leu.[26] Spiders were hand fed before, during, and after silking and once per day on nonsilking days. Silk was collected three times per week at an extraction rate of 2 cm/s for 45 min per session, with constant monitoring to ensure inclusion of only dragline fibers. A total of 57 mg of dragline silk was harvested from these 18 spiders. Kümmerlen *et al.*[27] prepared ^{13}C-labeled fibroin from spiders from *Nephila madagascariensis* kept at a low diet (of Tenebrio mealworms) supplemented with [1-^{13}C] Ala or Gly in aqueous solution. The isotope enrichment of the amino acids exceeded 98%, and in the resulting silk 60% enrichment was found.

Isotopically labeled peptides are usually synthesized by a solid-phase method using an automated peptide synthesizer.[27a] Fmoc-amino acids for this purpose are available from commercial sources. Isotopically labeled Fmoc-amino acids, however, are synthesized from 9-fluomethyl succinimidyl carbonate (Fmoc–Osu) following the method of Paquet.[27b] The crude peptides are purified using HPLC with a reversed phase column.

5.2. Assignment of Peaks

Site-specific assignment of well-resolved ^{13}C or ^{15}N NMR peaks, if any, to respective amino acid residues is prerequisite to reveal local conformation and dynamics of biopolymers under consideration. Before proceeding to the *site-specific assignment of peaks*, it is worthwhile to attempt *regio-specific assignment of peaks* for membrane proteins, if any, to distinguish signals of immobilized portions of the transmembrane helices from those of rather flexible N- or C-terminus and loops by comparing signals recorded by CP-MAS experiment with those recorded by DD-MAS experiment, respectively, as illustrated already in Figure 2.7 for bR (see also Figure 2.8 for amino acid sequence by taking into account of the secondary structure). Such assignment of peaks is very important step to be able to draw any reliable conclusion from observed NMR data, even though it is too laborious and time-consuming. In this section, we briefly outline the following three types of approaches to this end: (1) *regio-specific assignment of peaks*; (2) *site-specific assignment of peaks based on site-directed mutagenesis and proteolysis*, and (3) *sequential assignment*. The second and third experiments are applied to samples of site-directed (or amino acid specific) and uniform enrichment, respectively.

5.2.1. Regio-Specific Assignment of Peaks for Membrane Proteins

As an illustrative example, we have demonstrated how ^{13}C CP-MAS and DD-MAS NMR signals of fully hydrated [3-^{13}C]Ala-labeled bR are well-resolved (up to 12 peaks) at ambient temperature and can be site-specifically assigned to 29 individual Ala residues, as shown in Figure 2.7. Naturally, the three intense signals marked by gray in the ^{13}C DD-MAS spectrum are suppressed in the corresponding CP-MAS NMR spectrum, although the spectral feature of the rest is unchanged.[28] These partly resolved peaks at 17.20 ppm (Ala 235 and 103), 16.88 ppm (Ala 240–246 and 84), and 15.91 ppm (Ala 228 and 233) are from the Ala residues in the C-terminal α-helix neighboring the Pro residue (Ala 235), the C-terminal tail taking the random coil (Ala 240, 244–246), and the C-terminal α-helix (Ala 228 and 233) protruding from the transmembrane surface, respectively. This assignment is consistent with the data of the conformation-dependent ^{13}C chemical shifts available from the data of model polypeptides.[29] Obviously, these selectively suppressed peaks in the CP-MAS NMR spectrum arose from time averaging of dipolar interactions due to acquisition of very rapid or almost isotropic motions of the order of 10^{-8} s at the residues of the C-terminal end. Therefore, the comparison of the DD-MAS and CP-MAS NMR spectra provides the most convenient means to distinguish the freely fluctuating C- or N-terminal tail from the transmembrane α-helices and the interhelical loops.[28] The residues 245–248 and 231–244 are also readily confirmed by examination of the absence of Ala C_β peaks from [3-^{13}C]Ala–bR due to successive cleavage by carboxypeptidase A and papain, respectively.[30]

In contrast, the ^{13}C spin–lattice relaxation times of the C_α and C=O signals from the transmembrane α-helices and loops are substantially longer (10–20 s) than the repetition time of 4–8 s, conventionally used for recording DD-MAS NMR spectra.[13] Accordingly, recording the ^{13}C DD-MAS NMR spectra on the [1-^{13}C] or [2-^{13}C]amino acid-labeled bR turned out to be suitable for locating signals from the C- or N-terminal residues undergoing fluctuation motions, whereas the CP-MAS NMR approach is suitable for detecting the ^{13}C NMR signals of the transmembrane α-helices and interhelical loops. Consequently, quite different spectral features can be obtained from the ^{13}C DD-MAS (a) and CP-MAS (b) NMR spectra of [1-^{13}C]Ala–bR, as illustrated in Figure 5.3. There appears, however, no ^{13}C NMR signal from the loop region, because the two signals located at the loop region are unequivocally ascribed to the Ala 235 next to the Pro in the C-terminal α-helix and the random coil peak at the C-terminal end. This situation arose from the fact that the ^{13}C signal from the loop region is completely suppressed from the spectral region owing to failure of the attempted peak narrowing by MAS caused by interference of an incoherent motional frequency of the order of 10^4 Hz with the coherent frequency of MAS,[31] although this timescale is not sufficient to suppress the ^{13}C NMR signals of [3-^{13}C]Ala–bR (Figure 2.7), as confirmed by counting the number of individual carbon numbers in the deconvoluted spectra.[32] In the latter, peaks could be suppressed by motions with frequency of 10^5 Hz interfered with the proton decoupling frequency as encountered for [3-^{13}C]Ala–bR, bacterioopsin (bO),[33] or blue membranes.[28] This situation, however, is worsened when the ^{13}C NMR spectra of [2-^{13}C]Ala-labeled bR are recorded at ambient temperature: many ^{13}C NMR signals from the transmembrane α-helices are suppressed by interference of the motional frequency with magic angle frequency (10^4 Hz), in addition to the completely suppressed peaks from the loop region.[13] On the contrary, ^{13}C NMR signals from Val residues in the loops are almost fully visible from [1-^{13}C]Val-labeled bR, as described below.

^{13}C NMR signals of the ground state and M-intermediate of [4-^{13}C]Asp-labeled bR[20,20a,b] yielded changes in protonated/deprotonated aspartic acids from the shift positions.

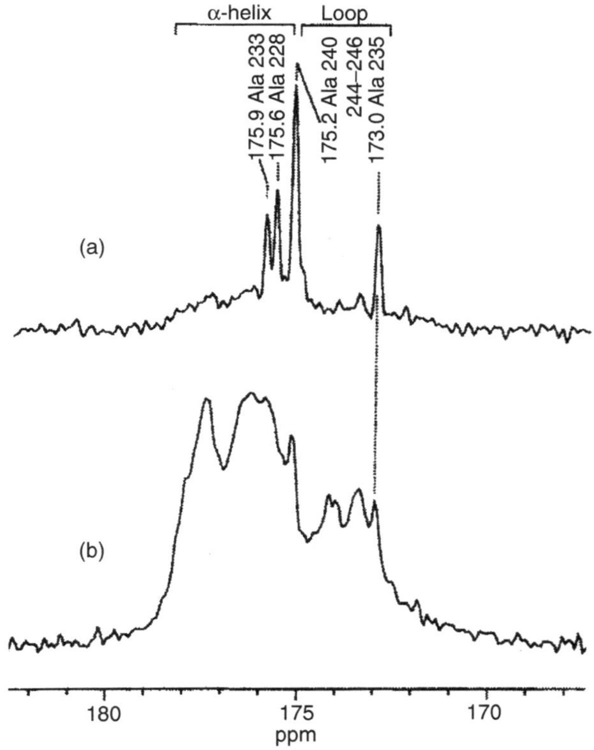

Figure 5.3. ^{13}C DD-MAS (upper trace) and CP-MAS (lower trace) of [1-^{13}C]Ala-labeled bacteriorhodopsin.[13]

5.2.2. Site-Specific Assignment of Peaks Based on Site-Directed Mutagenesis

The first step for this procedure is to locate the reduced peak in ^{13}C NMR spectra of the site-directed mutant, which lacks the amino acid residue of interest as compared with those of wild type. For instance, the Ala C_β NMR signals of Ala 196 and 126 of [3-^{13}C]Ala-labeled bR are straightforwardly assigned to the peaks whose intensities are significantly reduced in the site-directed mutants A196G and A126G, respectively, as shown in Figure 5.4,[28] since no additional spectral change was noted by introduction of this sort of site-directed mutagenesis. In many instances, however, a more complicated spectral pattern can arise owing to induced local conformational changes caused by introduced site-directed mutagenesis, as in A103C and A39V.[31]

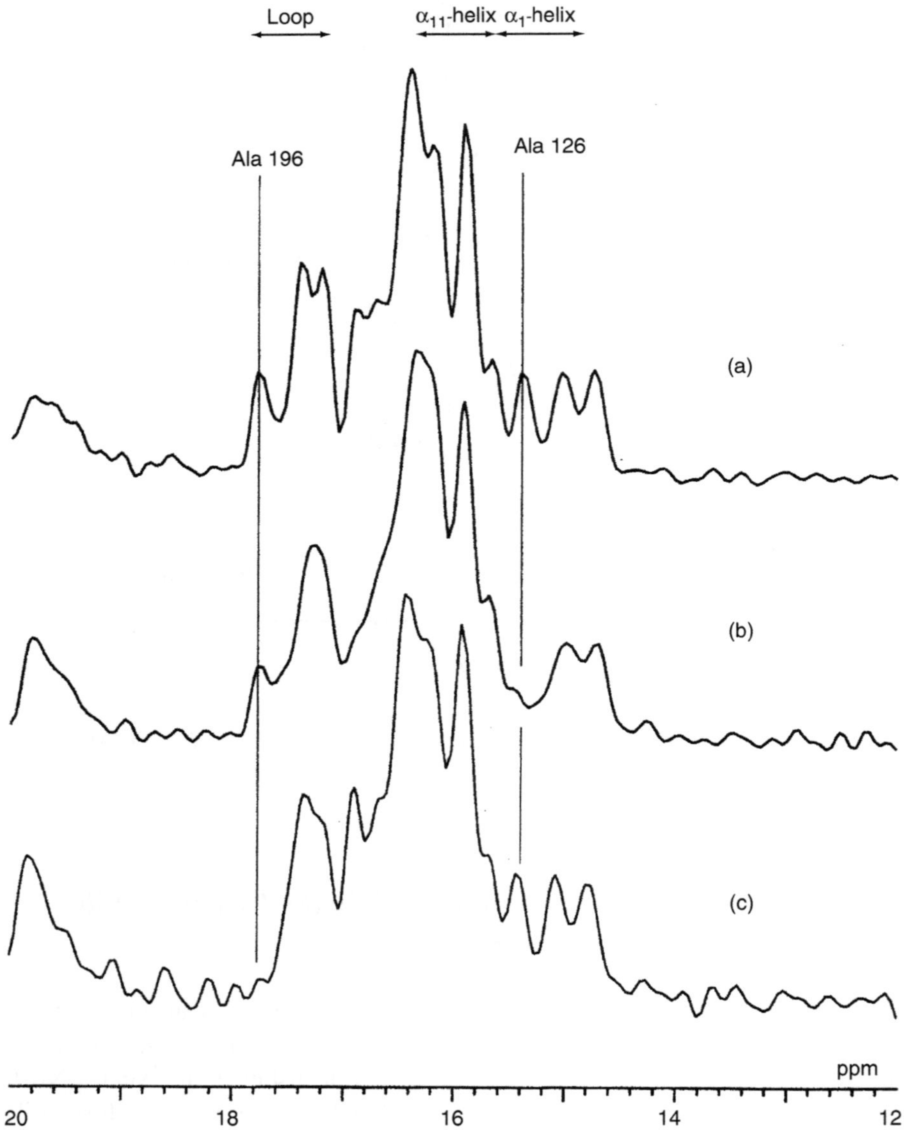

Figure 5.4. Comparison of the ^{13}C CP-MAS NMR spectra of [3-^{13}C]Ala-labeled bacteriorhodopsin (a) and A126G (b) and A196G (c) mutants.[28]

Even in such situations, this approach is still effective, as far as such accompanied conformational changes remain local and their sites can be identified. This type of peak assignment is made easier if resulting superimposed signals from other residues can be preferentially removed. If such perturbed signals are due to a residue located near a site at surface to which Mn^{2+} ion can bind,[34] the resulting accelerated transverse relaxation times due to the presence of the Mn^{2+} ions result in substantial line broadening, leading to suppressed peaks under the condition of high-resolution NMR.[35] The upper bound of the interatomic distances between the ^{13}C nuclei in bR and the Mn^{2+} bound to the hydrophilic surface to cause suppressed peaks by the presence of Mn^{2+} was estimated as 8.7 Å.[34] This results in peak broadening to 100 Hz, according to the Solomon–Bloembergen equation.[34–36] As demonstrated in Figure 5.5(a), the peak from the difference spectrum between wild type and A215G at 16.20 ppm is unequivocally ascribed to the single peak C_β signal of Ala 215, free from superimposed signals from other

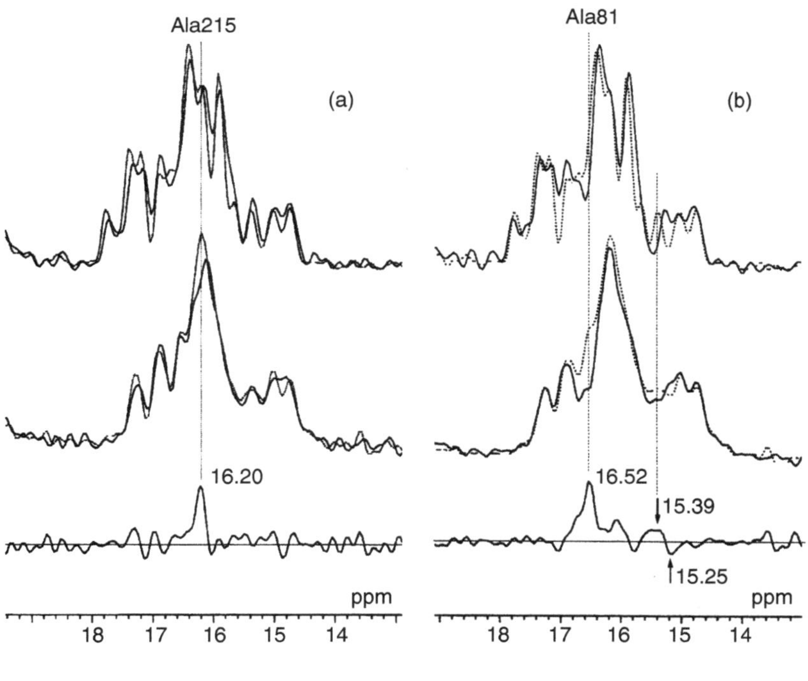

Figure 5.5. ^{13}C CP-MAS NMR spectra of [3-^{13}C]Ala-labeled A215G (a) and A81G (b) in the absence (top solid traces) and presence (middle solid traces) of Mn^{2+} ion. The corresponding spectra of [3-^{13}C]Ala-labeled wild-type bacteriorhodopsin are superimposed on the top and middle traces as dotted traces. Difference spectra between the wild type and mutants of bR in the presence of Mn^{2+} ion are shown in the bottom traces.[34]

Ala residues located at the cytoplasmic surface in the presence of the Mn^{2+} ion. In a similar manner, the signal at 16.52 ppm is assigned to the C_β signal of Ala 81, although the accompanied dispersion signals arise from local conformational changes in the transmembrane α-helix, induced by replacement of Ala 81 (Figure 5.5(b)). The assigned [3-^{13}C]Ala-labeled peaks of bR thus obtained are summarized in Table 5.2. Here, distinction is made for α_I and α_{II}-helices as viewed from the peak position of Ala C_β signals, as will be discussed in the Section 6.4.

In a similar manner, the ^{13}C NMR signals of Val residues from [1-^{13}C]Val-labeled bR, located in the loop regions such as Val 34 (A–B loop), 69 (B–C loop), 101 (C–D loop), 130 (D–E loop), and 199 (F–G loop), can be assigned[37] with reference to the data of assigned peaks using enzymatic cleavage and corresponding site-directed mutants (Figure 5.6). Therefore, they can be utilized as very convenient diagnostic probes to examine a plausible change in the conformation and dynamics of the cytoplasmic and extracellular loops of bR as a function of pH, ionic strength, temperature, etc. The assigned peaks for [1-^{13}C]Val-labeled bR, so far obtained by comparison of ^{13}C chemical shifts of a variety of mutants,[37–40] V29A, V34A, V49A, V101A,V130A, V151A, V167L, V179M, V180A, V187L, V199A V213A, V217A with those of wild type, are summarized in Table 5.2.

Table 5.2. Assigned ^{13}C chemical shifts for [3-^{13}C]Ala and [1-^{13}C]Val residues in [3-^{13}C]Ala-, [1-^{13}C]Val-labeled bacteriorhodopsin.[37]

		Chemical shifts	Ala residues	Val residues
[3-^{13}C]Ala	Loop	17.78	196	
	Loop	17.36	160	
	α_{II}-Helix	17.27	184	
	Loop, α_{II}-helix	17.19	103, 235	
	Random coil, α_{II}-helix	16.88	240, 244–246, 84	
	α_{II}-Helix	16.38	39,168	
	α_{II}-Helix	16.52	81	
	α_{II}-Helix	16.20	215	
	α_{II}-Helix	16.14	53	
	α_{II}-Helix	15.92	51	
	α_{II}-Helix	15.91–15.67	228,233	
	αI-Helix	15.02	126	
[1-^{13}C]Val	Loop, α-helix	171.07		101, 199
	Loop, α-helix	171.99		49, 130
	Loop	172.84		34,69
	α-Helix	173.97		151,167,180
	α-Helix	174.60		136,179,187
	α-Helix	174.99		217
	α-Helix	177.04		29,213

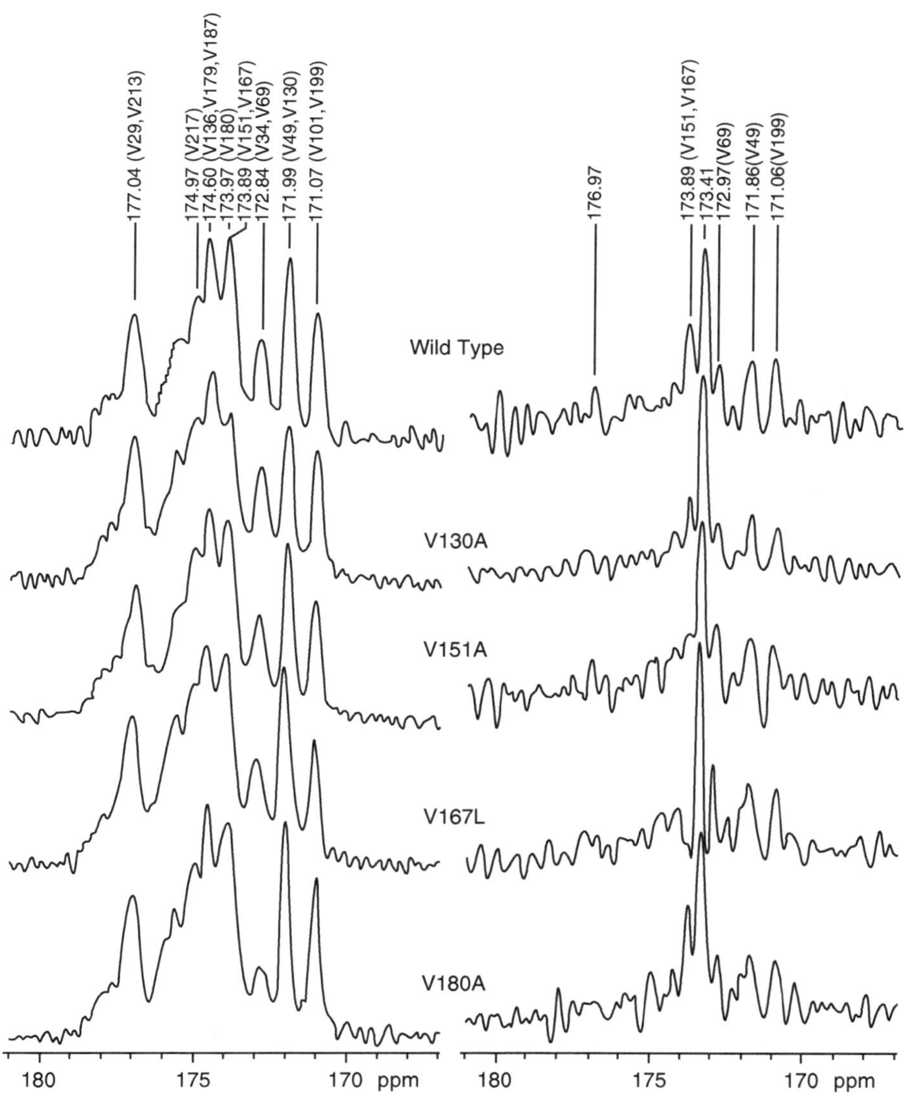

Figure 5.6. ^{13}C CP-MAS (left) and DD-MAS (right) spectra of [1-^{13}C]Val-labeled wild type, V130A, V151A, V167L, and V180A mutants[37].

Interestingly, ^{13}C NMR signals of bR from surface area (within 8.7 Å from the surface) turned out to be partially or almost completely suppressed for [1-^{13}C]Gly-, Ala-, Leu-, Phe-, and Trp-labeled preparations, owing to the presence of low-frequency conformational fluctuation present in these residues which interfers with frequency of MAS. In contrast, such conformational space allowed for fluctuation motion may be limited to a very narrow area for Val or Ile residues with bulky side-chains at C_α, together with limited χ_1 rotation around the C_α–C_β bond as shown by C_α–$C_\beta H(X)(Y)$ where X and Y are substituents on C_β. Accordingly, the above-mentioned low-frequency, residue-specific backbone dynamics may be present for Gly, Ala, Leu, Phe, and Trp residues, because backbone dynamics in these systems could be coupled with a possible rotational motion of the χ_1 angle around the C_α–C_β bond, as represented schematically by the C_α–$C_\beta H_2$-Z, where Z is H, isopropyl, phenyl, or indole group. For this reason, Val and Ile residues are the most appropriate probe to examine conformation and dynamics of amino acid residues of bR located at surface area. In a similar manner, carbonyl ^{13}C signals for Trp-, Ile-, and Pro-labeled bR were site-specifically assigned by repeating this procedure to compare the ^{13}C chemical shifts of wild type with those of corresponding site-directed mutants,[37] although several such carbonyl ^{13}C NMR signals from surface areas could also be suppressed for Gly, Ala, Leu, Phe, and Trp residues, as mentioned earlier. Nevertheless, it is true that carbonyl ^{13}C NMR signals of Val and Ile residues are exceptionally fully visible and most suitable probes to reveal conformation and dynamics of bR. In addition, the assigned peaks for [1-^{13}C]Trp-, Ile-, and Pro-labeled bR based on their respective mutants were summarized in Table 5.3.

It is cautioned, however, that this approach cannot be efficiently utilized for mutants in which any single mutation at a certain key position causes global conformational change as encountered for the ^{13}C NMR spectra of [3-^{13}C]Ala-labeled D85N mutant of bR.[41] Caution is also prerequisite for

Table 5.3. Assigned carbonyl ^{13}C chemical shifts of bacteriorhodopsin for Trp, Ile, and Pro residues (ppm from TMS) based on site-directed mutants.[37]

	Residue number	Chemical shift	Local conformation
Trp	182	175.8	α-Helix
	189	174.9	α-Helix
Ile	4	173.5	Random coil
	122	175.4	α-Helix
Pro	50	174.3	α-Helix
	91	176.0	α-Helix
	186	177.7	α-Helix

several ^{13}C-labeled preparations in which certain signals from surface areas are partially or completely suppressed when fluctuation frequency, if any, is interfered with frequency of MAS in the order of 10^4 Hz, as mentioned earlier.

5.2.3. Sequential Assignment

Undoubtedly, sequential assignment of resonance peaks is essential when uniformly ^{13}C or ^{15}N-labeled proteins are recorded by solid state NMR. In solution ^{1}H NMR, such sequential assignment based on *J*-coupling is a routine procedure to reveal 3D structure based on solution NMR as studied by multidimensional experiments such as correlated spectroscopy (COSY) and total correlated spectroscopy (TOCSY). For ^{13}C NMR in solution, ^{13}C–^{13}C through-bond correlation is available from incredible natural abundance double quantum transfer experiment (INADEQUATE) experiment.[42] To avoid significant loss of signal intensity to sidebands, the MAS frequency for solid state NMR must be at least comparable to or preferably higher than the anisotropy of chemical shift tensors. Further, the MAS frequency must be chosen carefully to avoid accidental matching of rotational-resonance conditions. For this reason, it is preferable to use a MAS frequency that is larger than the total width of the spectrometer. This is the reason why fast MAS frequency larger than 15 kHz is required for the purpose of sequential assignment of signals in the solid state.

5.2.3.1. Through-space connectivities

PDSD experiment has been a standard means to obtain homonuclear COSY by means of *through space* dipolar interactions with the nearest neighbors, for either pairs of natural abundant nuclei or uniformly labeled preparations.[43] Nevertheless, there exist severe line broadenings due to the presence of highly dense ^{13}C spin network for uniformly ^{13}C-labeled preparations, as discussed in Section 5.1. A resolution enhancement by homonuclear ^{13}C–^{13}C decoupling was proposed[44] in order to reduce line width up to 100 Hz. This is because the one-bond ^1J ($^{13}C_\alpha$–^{13}CO) couplings have values of ≈50 Hz and all the other aliphatic ^{13}C–^{13}C couplings are typically in the range of 30–40 Hz, which amount to the line width of 120 Hz. The proposed decoupling scheme, shown in Figure 5.7, allows one to remove from a specified spectral region all the multiplet splittings

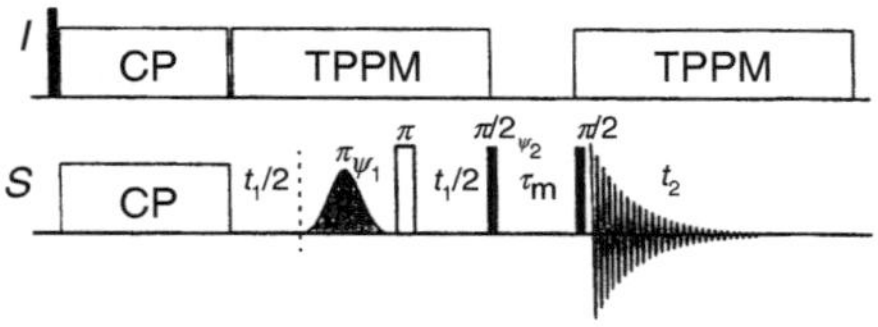

Figure 5.7. Pulse sequence to correlate the ^{13}C chemical shifts by PDSD with homonuclear decoupling in ω_1. All pulses except the shaped ones are high power (from Straus *et al.*[44]).

due to nuclei with resonance frequencies outside the specified spectral range. A pair of π pulses is placed in the middle of the evolution period t_1: a suitably shaped band-selective π pulse applied to the selected spectral region is followed by a nonselective π pulse. The chemical shifts of the out-of-band spins are refocused by the nonselective pulse, whereas the spins with resonance frequencies in the selected band experience no net rotation due to the effective 2π rotation. The experiment, combined with a standard spin-diffusion mixing period, τ_m, produces a 2D spectrum in which the J splittings are removed exclusively from the ω_1 domain while they are still present in ω_2. For this purpose, selective DANTE pulses were utilized in the center of the indirect evolution period.[45] Further, resolution enhancement is achieved[46] by using transition-selective excitation and spin-state selective polarization transfer using zero-quantum (ZQ) solid state NMR mixing sequences. It was shown that PDSD experiments with long mixing time (4 s) are helpful[47] for confirming the assignment of the protein backbone ^{15}N resonances and as an aid in the amide proton assignment. In the latter, phase-modulated Lee–Goldburg (PMLG) irradiation was used during proton evolution to suppress strong ^{1}H homonuclear dipolar interactions.[48]

A homonuclear through-space correlation experiment[49] was developed by employing dipolar recoupling enhancement through amplitude modulation (DREAM)[50,51] scheme for adiabatic dipolar transfer, which yields a complete assignment of the C_α and aliphatic side-chain ^{13}C resonances for uniformly ^{13}C, ^{15}N-labeled cyclic decapeptide antamide, *cyclo*-(Val1-Pro2-Pro3-Ala4-Phe5-Phe6-Pro7-Pro8-Phe9-Phe10), as illustrated in Figure 5.8(a).

5.2.3.2. Through-bond connectivities

In general, it appears that through-bond connectivities of ^{13}C chemical shifts are more reliable than through-space connectivities, in order to establish an

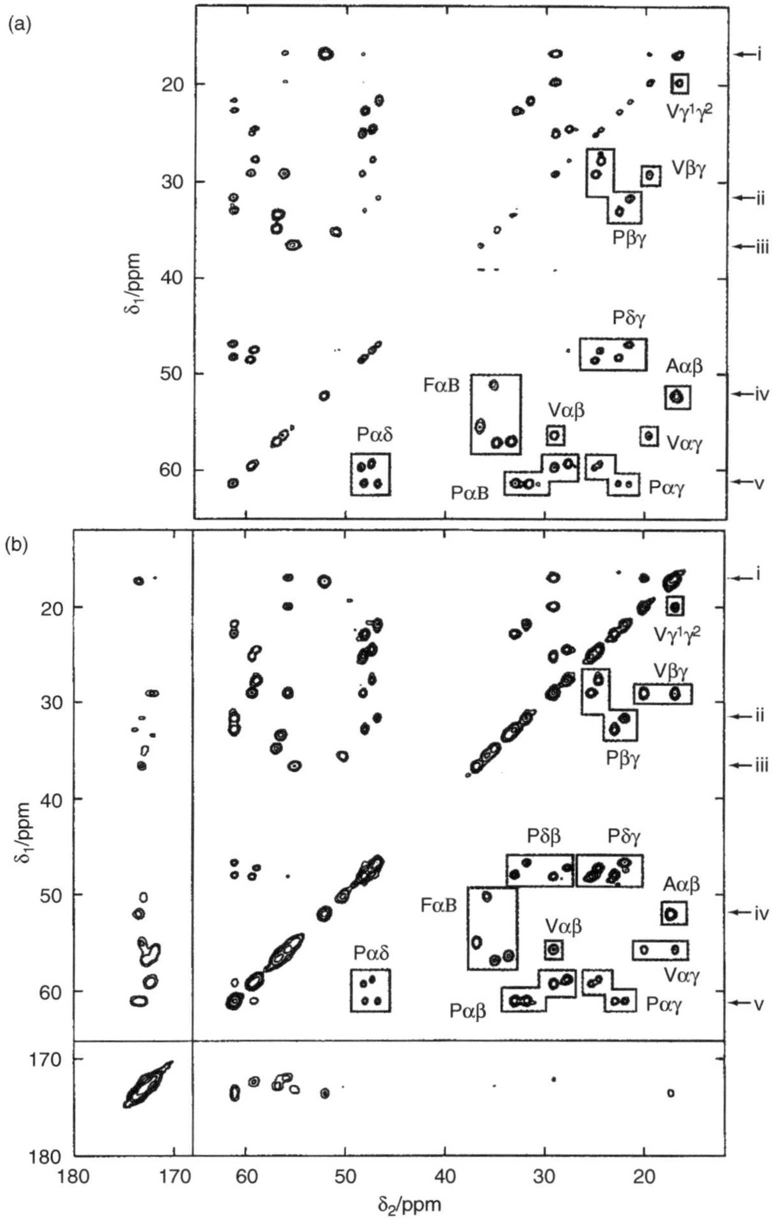

Figure 5.8. ^{13}C homonuclear shift-correlation spectra of cyclic decapeptide [U-^{13}C,^{15}N]antamide at a MAS frequency of $\omega_r/2\pi = 30\,\text{kHz}$. (a) Spectrum obtained with the DREAM sequence for dipolar transfer with a mixing time of 7 ms. The total acquisition time was 3.5 h. Only the region containing C_α and aliphatic side-chain resonances is shown. (b) Spectrum obtained with the $P91_{12}^{1}$ TOBSY for transfer via the *J*-couplings. The total acquisition time was 7 h (from Detken et al.[49]).

unambiguous assignment of peaks. The same information as DREAM can be obtained from total through-bond correlation spectroscopy (TOBSY) experiment[53–55] to be described below. A 2D DREAM correlation experiment of a uniformly labeled amino acid maps out the complete connectivity. Further, a dipolar INADEQUATE with a large double quantum spectral width was proposed to identify ^{13}C connectivity pattern for spectral assignment for a uniformly ^{13}C-labeled amino acid and an extensively ^{13}C-labeled protein.[52] The achieved transfer efficiency of up to 76% was obtained at a MAS frequency of 26.67 kHz[55] based on the initial transverse carbon polarization created by cross polarization via adiabatic passage through Hartmann–Hahn condition (APHH-CP),[56] as illustrated for 2D ^{13}C–^{13}C correlation spectrum of cyclic decapeptide [U-^{13}C, ^{15}N]antamide (Figure 5.8b). Scalar-coupling-driven, uniform-sign cross-peak double-quantum-filtered correlation spectroscopy (UC2QF COS) was introduced[57] as an effective through-bond correlation spectroscopy for disordered solids.

The INADEQUATE experiment is a well-known liquid state technique to establish direct scalar connectivity in the ^{13}C skeleton.[42] As in solution NMR, directly bonded ^{13}C resonances are identified by the fact that a common double quantum frequency is generated (Figure 5.9).[58–60] Homonuclear carbon–carbon through-bond correlations are available from the ordinary (Figure 5.9(a)) and refocused INADEQUATE (Figure 5.9(b)) experiments, yielding anti-phase and in-phase line shapes, respectively, on solid samples where the line widths greatly exceed the value of the scalar coupling. The refocused INADEQUATE experiment leads to a sufficiently

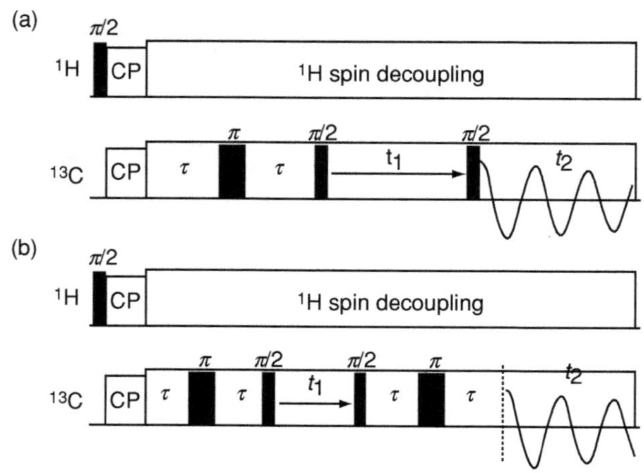

Figure 5.9. Pulse sequence for solid-state INADEQUATE (a) and refocused INADEQUATE (b) (from Lesage *et al.*[59]).

efficient excitation of double quantum coherence, because experiments yielding anti-phase line shapes have previously been considered as impractical in ordinary organic solids because of signal cancellation due to line width. This technique was applied to assign ^{13}C NMR signal of 11% ^{13}C-labeled cellulose. Quite interestingly, the observed ^{13}C NMR spectrum clearly demonstrates the presence of two different types of glucopyranose residues in the unit cells, although 1D spectrum is too broad to obtain any conclusion. Consistent with this finding, two sets of the ^{13}C–^{13}C through-bond correlations are also available[61] for individual glucose residues from highly crystalline cellulose I_α from natural abundant *Cladophora* and I_β from tunicate: both cellulose I_α and I_β contain two magnetically nonequivalent glucose residues in the unit cells, as will be discussed in more detail in Section 10.1.3.

5.2.3.3. ^{15}N–^{13}C and ^{15}N–^{13}C–^{13}C correlation

For the assignment of fully ^{13}C–^{15}N-labeled proteins in the solid state, a variety of the combination of intra-residue and inter-residue ^{13}C–^{15}N heteronuclear correlation experiment with ^{13}C–^{13}C spin diffusion have been proposed, as illustrated in Figure 5.10, together with the abbreviated

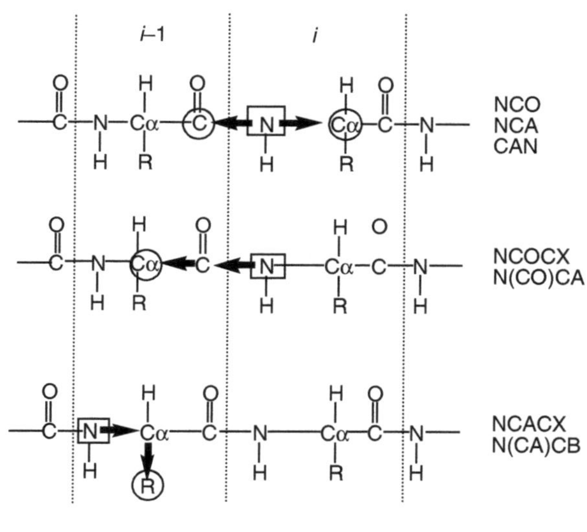

Figure 5.10. Schematic representation of several types of ^{15}N–^{13}C connectivities.

names.[49,62–67] The intra-residue (NCA or CAN) and inter-residue (NCO) two-bond correlations are achieved by direct cross polarization[49,62] with or without double APHH-CP[56] and band-selective spectrally induced filtering in combination with cross polarization (SPECIFIC CP) transfer,[64,67,68] respectively. Unlike the conventional CP experiment,[69,70] SPECIFIC CP is characterized by low- to medium-size rf field that facilitates polarization transfer whenever the effective field of rf irradiation (ω_1, ω_2) and chemical shift offsets (Ω_1, Ω_2) of heteronuclear spin pair fulfill the following (ZQ, $n = +1, +2$) recoupling condition:

$$\sqrt{\omega_1^2 + \Omega_1^2} - \sqrt{\omega_2^2 + \Omega_2^2} = n\omega_R \tag{5.1}$$

Using a slow rf amplitude modulation for one of the two rf fields (e.g., $\omega_1(t)$), the recoupling condition of Eq. (5.1) can be fulfilled for a defined range of chemical shift offsets. Further, theoretical considerations on how to select the applied rf fields ω_1 and ω_2 or the resonance offsets Ω_1 and Ω_2 in Eq. (5.1) and how to maximize the transfer efficiency for a given MAS frequency ω_R can be found.[68] This concept was applied to obtain NCA and NCO spectra of the precipitated α-spectrin SH3 domain (7.2 kDa, 62 amino acids) from a $(NH_3)_2SO_4$-rich solution, containing exclusively N_i-$C\alpha_i$ and N_i-CO_{i-1} cross-peaks (Figure 5.11), respectively, which provide a high-resolution finger print of the protein.[68] In Figure 5.11 top, the NCO cross-peaks yield sequential information, as indicated with the residue contributing to the nitrogen chemical shift first, followed by the residue that contributes to the carbonyl chemical shifts.

The inter- and intra-residue connectivities by ^{13}N–^{13}C–^{13}C correlation spectroscopy such as NCOCX (or N(CO)CA) and NCACX (or N(CA)CB) sequences (Figure 5.10) have been developed by extending the above-mentioned NCA and NCO sequences with ^{13}C–^{13}C through-space polarization transfer[49] using the DREAM and the rotational-resonance tickling (R^2T),[71–73] through-bond transfer[49] by the TOBSY, ^{13}C–^{13}C mixing unit,[37] with a double quantum mixing unit to achieve carbon–carbon transfer,[67] REDOR,[63] or RFDR[63,66,74] mixing, or by soft-triple resonance.[65] As an illustrative example of NCOCA spectrum based on APPH-CP and R^2T sequence,[49] a series of 2D slices at constant ^{15}N frequency from a 3D NCOCA spectrum are shown in Figure 5.12. For each amide nitrogen N_k, the following peaks are expected: $C_{k\alpha}$ and C_{k-1}=O diagonal peaks (direct correlations), and C_{k-1}=O–$C_{k-1,\alpha}$ and $C_{k\alpha}$–C_k=O cross-peaks (relayed N–C correlations). Indeed, each of the displaced slices contains just these four well-separated peaks due to the good resolution along the ^{15}N dimension. Peaks at δ_3 frequencies are found again in the corresponding slices of the next or previous residues, respectively. Thus,

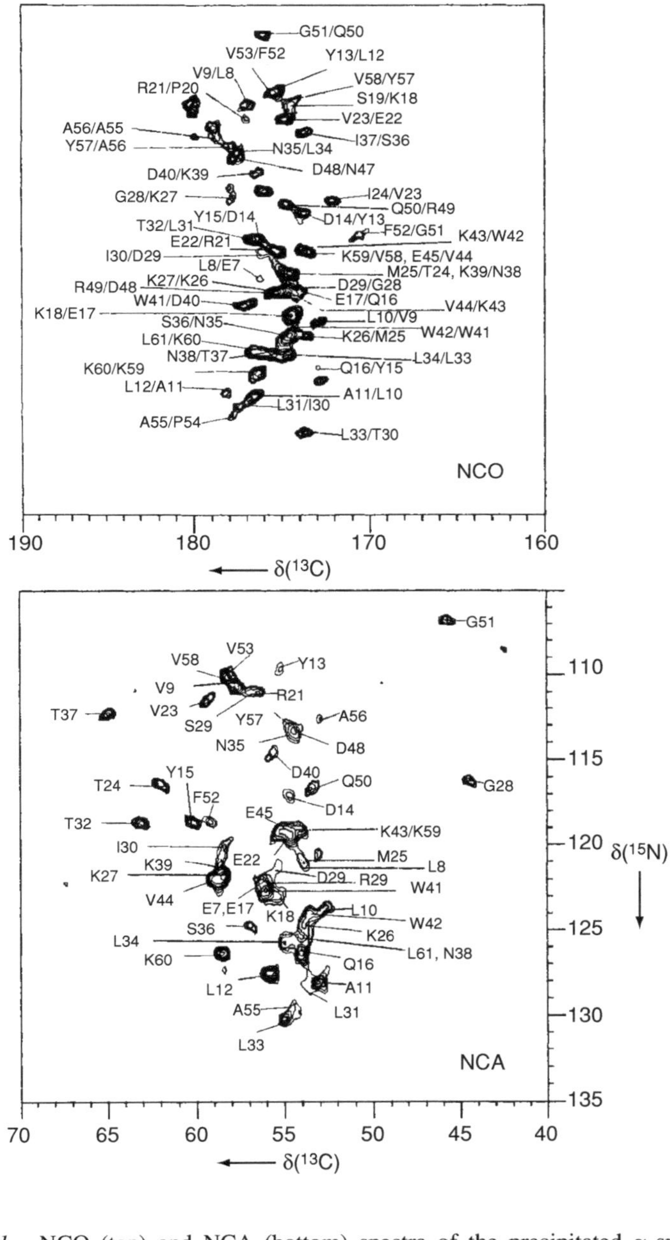

Figure 5.11. NCO (top) and NCA (bottom) spectra of the precipitated α-spectrin SH3 sample recorded with the band-selective SPECIFIC CP. The spinning frequency was 12 kHz. Assigned signals are indicated with the respective residue type and number. In the spectrum at the left, the NCO cross-peaks yielding sequential information are indicated with the residue contributing to the nitrogen chemical shift first, followed by the residue that contributes to the carbonyl chemical shift (from Pauli *et al.*[68]).

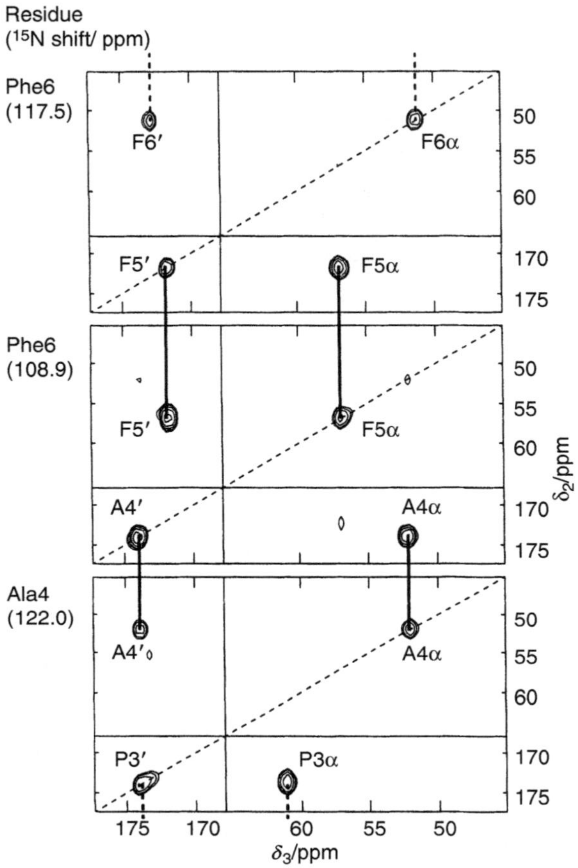

Figure 5.12. Slices of a 3D ^{15}N–^{13}C correlation spectrum of antamide for constant ^{15}N chemical shift. After 3D Fourier transformation, slices were extracted at the ^{15}N chemical shifts of Ala4, Phe5, and Phe6, respectively. Each slice contains diagonal peaks of the carbons directly attached to the corresponding ^{15}N nucleus, and cross-peaks from these carbons to the next carbons along the backbone (from Detken *et al.*[49]).

by connecting the slices as indicated by the bold vertical lines, an independent sequence-specific assignment can be completed.

The assignment of the peaks by this way is nearly complete for the SH3 domain residues 7–61, while the signals of the N- and C-terminal residues 1–6 and 62, respectively, outside the domain boundaries are not detected in the MAS spectra. One reason for this may be the flexibility of the N-terminus, which could be on a timescale that interferes with the proton decoupling frequency. Another explanation for the missing signals relates to the occurrence of heterogeneous broadening, which may result from a multitude of conformers that are ''frozen out'' upon precipitation during the sample preparation, which leads to increased chemical shift dispersion. To make situation worse, it should be also anticipated that spectral resolution of ^{13}C NMR signals in the dense ^{13}C spin network, undergoing fluctuation motions from 10^8 to 10^2 Hz, might desperately deteriorate owing to shortened spin–spin relaxation times due to the presence of increased number of relaxation pathways, as illustrated for [1,2,3-$^{13}C_3$]Ala-labeled bR (Figure 5.2). In fact, backbone and side-chain correlations involving N- and C-terminal and loop segments outside the domain other than the transmembrane α-helices are significantly attenuated in the ^{13}C NMR signals of [U-^{13}C, ^{15}N]-labeled membrane–protein complex, LH2 light-harvesting complex from purple photosynthetic bacteria, in detergent at −50 °C.[65,75] In addition, ^{13}C NMR signals of several residues in the transmembrane α-helices are also missing.

Therefore, it should be taken into account that sequential assignment of signals is feasible only for rigid, uniformly ^{13}C, ^{15}N-labeled proteins. Otherwise, considerable loss of ^{13}C NMR signals should be anticipated, especially when ^{13}C NMR signals from membrane proteins at ambient temperature were attempted.

5.3. Ultra-High Field and Ultra-High Speed MAS NMR

5.3.1. Ultra-High Field NMR

It has been recognized that a high magnetic field has an advantage for resolution and sensitivity enhancement. Although dipolar interaction and spin–spin interactions do not increase as the magnetic field is increased, chemical shift values are proportional to the magnetic field in the frequency unit. Naturally, this causes resolution enhancement as the magnetic field is increased. However, when the line width dominates in the inhomogeneous distribution of conformations such as amorphous samples, resolution enhancement is not efficient as that expected in the high field. It is also possible to increase sensitivity as the magnetic field is increased. One of the factors of the sensitivity enhancement is the Boltzmann distribution that brings about a large magnetization for the samples in the high magnetic field. Actually, *S/N* ratio can be expressed as

$$S/N = N_s V_s \frac{(h\nu_0)^2}{(4kT)^{3/2}} \frac{\nu_0}{B_0} \frac{1}{\Delta\nu} \sqrt{\frac{\mu_0 \pi}{V_c}}, \tag{5.2}$$

for a typical electric circuit,[76] where, N_s is the number of spins, V_s is the sample volume, $\Delta\nu$ is the intrinsic line width, and V_c is the coil volume. One can easily realize that an *S/N* ratio is proportional to the square of ν_0. Although size of coil is dramatically decreased for a circuit designed in the high field and hence sample volume is decreased, the *S/N* ratio is inversely proportional to the root of coil volume. Thus, one can still expect to enhance the sensitivity with a great amount by using ultra-high field magnet. In addition, recent developments of a microprobe show that a great sensitivity enhancement is achieved because of the significant increase of rf field at the sample.[77] In the solid state NMR, a higher field NMR is demanding because material application with quadrupolar nuclei and complicated biologically active molecules such as membrane proteins are important targets to solve the structures to understand the properties of the molecules. High-field NMR spectroscopy may have advantages to observe high-resolution resonance lines for half-integer transitions because a second-order perturbation term is greatly attenuated.[78] In case of complicated biological molecules, high magnetic field also improves the resolution and sensitivity, which allow to determine the high-resolution structures of molecules. In the ultra-high field NMR, it is possible to observe high-resolution ^{1}H NMR signals by using ultra-fast MAS NMR alone without using any multiple-pulse sequence.[79]

5.3.1.1. Reduction of the second-order quadrupole interaction

The second-order quadrupolar effect to the central transition $\nu(1/2 \leftrightarrow -1/2)$ of half-integer spins is proportional to the inverse of Larmor frequency as described in Eq. (2.39). This situation could be visualized in Figure 2.4.[80]

Therefore, enhancement in both resolution and sensitivity is expected with a higher magnetic field. This is the major importance in terms of applicability of solid state NMR to crystalline and amorphous inorganic materials containing quadrupolar nuclei such as ^{11}B, ^{17}O, ^{23}Na, ^{25}Mg, ^{27}Al, ^{69,71}Ga, and ^{87}Pb, especially for compounds with severe second-order quadrupolar shift causing the line broadening.

Various types of 1D and 2D experiments have been developed to overcome the second-order quadrupolar broadening by double rotation

(DOR)[81] or spreading spectra in 2D with DAS,[82] MQ MAS,[83] and satellite transitions and magic-angle spinning (STMAS)[84] experiments. The 2D approach relies on the same principle of averaging out the second-order quadrupolar effect by correlating different transitions of quadrupolar spins. The development of these methods has considerably extended the applicability of solid state NMR to study the structures and properties of organic and inorganic materials in crystalline, amorphous, or glassy states.

5.3.1.2. Resolution enhancements in quadrupolar nuclei

Solid state NMR spectra of quadrupolar nuclei are usually broadened by the second-order quadrupole effects. Without any correction, the magnetic field homogeneity and stability add about 5 ppm extra line broadening. This broadening becomes small compared to the broadening from large quadrupole couplings in a higher magnetic field (see Eq. (2.38)). The ultra-high field can reduce the second-order quadrupolar effect to improve spectral resolution. The solid-state ^{17}O MAS NMR spectra for crystalline η-P_2O_5 taken at different magnetic field strengths, 9.6, 14.1, and 19.6 T, are shown in Figure 5.13.[80] The reduction in the breadth of the central transition with increasing magnetic field strength is clearly seen in Figure 5.13 and indicative of the inverse dependence of the Larmor frequency over the second-order quadrupolar coupling.

It was demonstrated that the possibility of using ultra-high magnetic field, 25 T resistive and 40 T hybrid magnets to reduce the second-order quadrupolar line broadening such that high-resolution spectra can be obtained using simple excitation schemes for keeping the quantitative interpretation of obtained spectra.[78] Aluminoborate $9Al_2O_3 + 2B_2O_3(A_9B_2)$ samples were used for the field-dependent demonstration. The compound has four different sites: one AlO_4, two AlO_3, and double intensity AlO_6 sites. Figure 5.14 displays the spectra of A_9B_2 obtained at 14, 19.6, 25, and 40 T. Until 14 T, severe spectral overlap makes the fitting of all the four ^{27}Al sites difficult. Consequently, chemical shift and quadrupolar coupling parameters could not be reliably measured from just 1D spectra. At 19.6 T, a field close to the maximum magnetic field available with superconducting magnet, the AlO_6 line begins to separate from the others. At 25 T, the AlO_4 no longer overlaps with the two AlO_5 sites that are still partially overlapped. At 40 T, the second-order quadrupolar shift and broadening are nearly wiped out, and most of the residual line width of about 5 ppm comes from the time fluctuation of the principal field. All four sites are well-resolved and positioned close to their

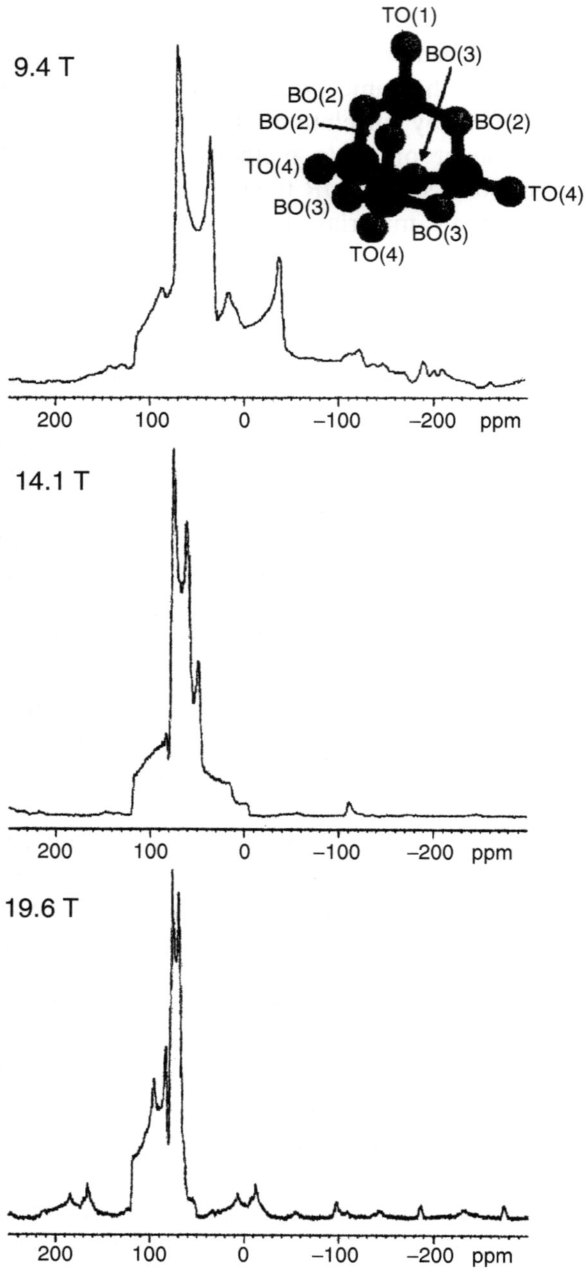

Figure 5.13. Solid-state ^{17}O MAS NMR spectra for crystalline η-P_2O_5 as a function of magnetic field strength. The inset is the adamantoid structure of η-P_2O_5, with TO and BO showing the terminal and bridging oxygenes, respectively (from Zhang *et al.*[80]).

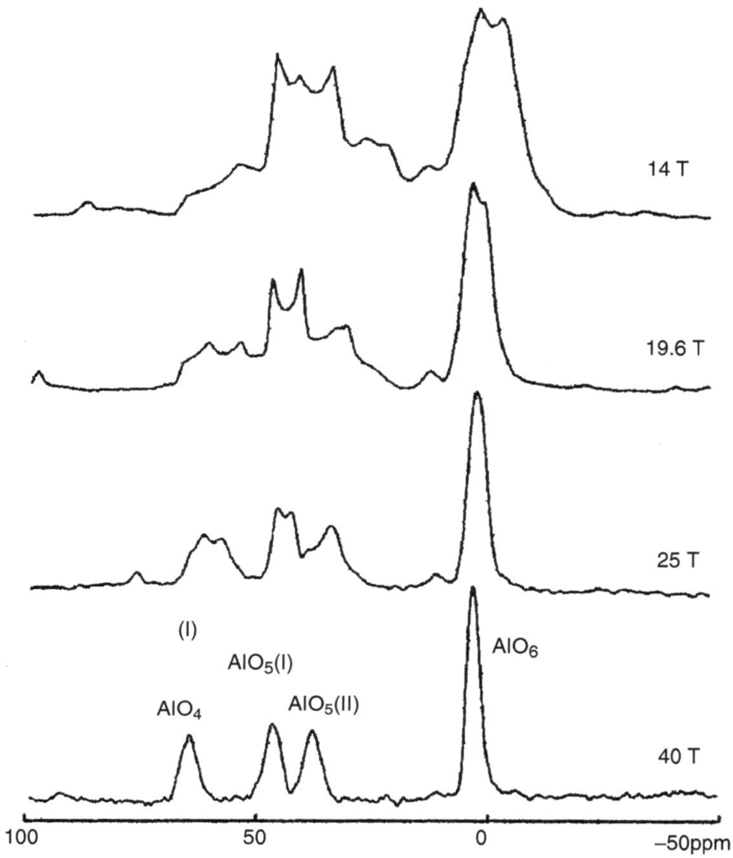

Figure 5.14. ^{27}Al MAS spectra of A_9B_2 compound as a function of magnetic field strength from 14 to 40 T (from Gan *et al.*[78]).

individual chemical shifts with near Gaussian line shapes and total intensities representing directly their abundance in the sample (1:1:2). At this ultra-high field, the ^{27}Al central transition spectrum becomes nearly free of the second-order quadrupolar effect and appear in a way similar to that of a spectrum of $I = 1/2$ spins. With reduced line width and gain in spectral sensitivity by high field, high-resolution spectra can be acquired quickly and reliably, and most important, the quantitative nature of NMR spectra is retained in comparison with that obtained with the 2D methods. The only drawback of the spectral simplification and higher resolution is the loss of information on the quadrupolar coupling similar to switching from a static spectrum to a MAS spectrum in the case of $I = 1/2$ spin.

5.3.2. Ultra-High Speed MAS Under High Field

In a higher field, chemical shift anisotropy in the frequency unit will be proportionally increased and this causes many sideband lines. To suppress the sideband lines, one can increase the spinning frequency of MAS. In 1999, 2.5-mm rotors, specified up to 33–35 kHz, became commercially available. Speed of up to 50 kHz with rotors of about 2-mm diameter was demonstrated in 1999 in an academic laboratory.[85] Several artifacts accompany the high-speed MAS experiments. The sample is subjected to a stress gradient, ranging from 0 to about 10^6 g (gravity) at the periphery. This may, for example, broaden lines of quadrupole nuclei due to piezoelectric effects. Another potential complication may be the sample heating. Friction with the surrounding atmosphere becomes quite noticeable at high speeds. As the NMR community paid attention to that,[86] the surface temperature heats up to as a quadratic function of the speed. The isotropic shift of Pb in lead nitrate is a convenient thermometer for solid state NMR.[87]

5.3.2.1. Rotor design

The experimental history of high-resolution NMR in solids has been a quest for higher spinning speed. Principally two different rotor designs have been used. First experiments used a conical rotor design, exploiting Bernoulli forces to hold the rotor in the coil.[88] Better stability and higher speeds have been achieved by a cylindrical design of the rotor[89] and modified by using double gas bearing system, pioneered by the group of Lippmaa.[90] The design principles are relatively well mastered by now and several implementations have claimed a rotor surface speed exceeding the speed of sound in air. However, practical applications are limited to subsonic speeds by the dependable strength of rotor material. The presence of the magnetic field limits the choice of materials to nonconductive ceramics or polymer compounds.

5.3.2.2. Homonuclear decoupling under fast MAS

The most intriguing high-speed application is the possibility of suppressing homonuclear dipolar interactions. Homonuclear dipolar line broadening in a rigid system of hydrogen atoms may exceed 50 kHz. Under MAS, the residual line width is only a weak function of spinning speed, decreasing linearly with first power of the speed. Multiple-pulse sequences have been developed to refocus dipolar dephasing periodically. This refocusing has been demonstrated to work also in combination with a slow sample rotation.[91,92] For rigid systems the multiple-pulse approach can offer resolution of a few tenths of ppm. In the case of plastic crystals and the presence of fast molecular motion, the multipulse approach becomes principally less effective. It may also be technically difficult to retain stability over a sufficiently long measurement period. Figure 5.15 presents ^{1}H MAS only spectra of camphor, where *J*-coupling becomes visible at 50 kHz. Moving to the higher fields, fast rotation starts to provide informative spectra also for rigid systems such as alanine crystals, as shown in Figure 5.16.[79] The relative resolution grows proportionally with the field strength, since residual line broadening retains its absolute value which depends mostly on the rotation speed. Both advantages combined bring a virtually quadratic

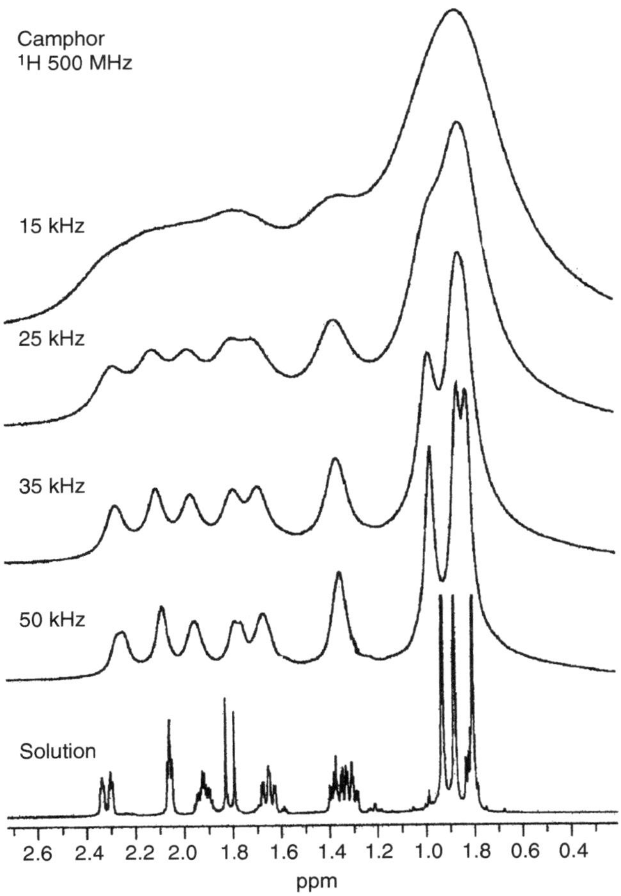

Figure 5.15. ^{1}H NMR spectra of camphor as a function of MAS frequency from 15 to 50 kHz. Bottom show a solution state ^{1}H NMR spectrum.[79]

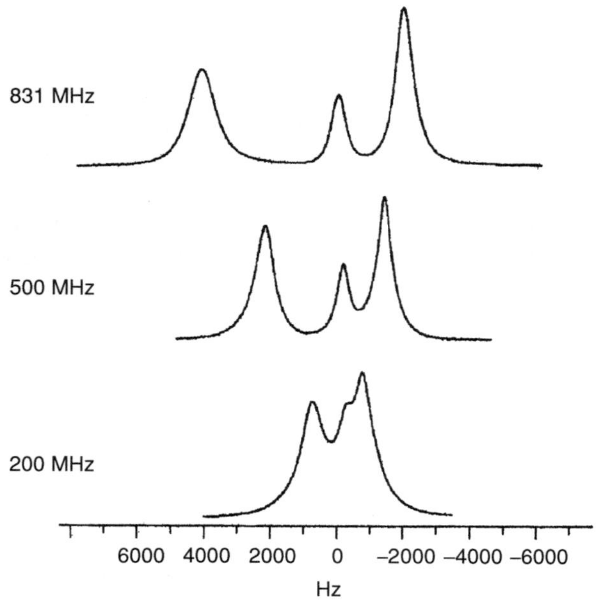

Figure 5.16. ^{1}H NMR spectra of alanine at 200, 500, and 831 MHz. MAS frequency is 45 kHz.[79]

improvement in resolution. However, a subtle deviation from this trend can already be observed in higher field spectra (831 MHz). Although the resolution is better, the absolute line width is clearly larger than 500 MHz. Here a natural, susceptibility-induced broadening starts to compete with the residual dipolar line width and no further substantial resolution enhancement is possible. The case of high speed as a method of improving resolution may actually turns out to be the only alternative for measurements at high fields, because several experimental parameters start to complicate the application of the multiple-pulse sequences. Firstly, it is technically more demanding to generate the required rf field strength. Secondly, the multiple-pulse sequences inherently lose efficiency with expansion of the spectral width. This is particularly true for ^{19}F spectroscopy. A qualitative improvement can be observed when both high field and rotational speed are combined. For several experimental situations, resolution enhancement under MAS may not only be a matter of convenience, but also imperative in order to free the pulse space for specific tasks. High rotation speeds also allow exchange of averaging order in decoupling experiments. A very low amplitude proton rf field is then needed for well-resolved carbon spectra.

5.3.2.3. Heteronuclear decoupling under fast MAS

Achieving high spectral resolution is an important prerequisite for the application of solid state NMR to biological molecules. High spectral resolution allows to resolve a large number of resonance and leads to higher sensitivity. Among other things, heteronuclear spin decoupling is one of the important factors, which determine the resolution of a spectrum. Continuous-wave rf irradiation leads only in a zeroth-order approximation to a full decoupling of heteronuclear spin system in solid under MAS. In a higher-order approximation, a cross-term between the dipolar coupling tensor and the chemical-shielding tensor is reintroduced, providing a scaled coupling term between the heteronuclear spins. In strongly coupled spin systems, this second-order recoupling term is partially averaged out by the proton spin-diffusion process, which leads to exchange-type narrowing of the lines by proton spin flips. This process can be described by a spin-diffusion type super operator, allowing the efficient simulation of strong coupled spin systems under heteronuclear spin decoupling. Low-power continuous-wave (CW) decoupling at fast MAS frequency offers an alternative to high-power irradiation by reversing the order of the averaging process. At fast MAS frequencies, low-power CW decoupling leads to significant narrower lines than high-power CW decoupling while at the same time reducing the power dissipated in the sample by several order of magnitude. The best decoupling is achieved by multiple-pulse sequences at high rf fields and under fast MAS. Two such sequences, two-pulse phase-modulated decoupling (TPPM)[93] and X-inverse-X (XiX) decoupling,[94] are proposed to be efficient decoupling pulse sequences.

The best decoupling results are achieved by high-power decoupling using multiple-pulse sequence. Both TPPM and XiX decoupling show obvious improvements in line widths and line heights compared to CW decoupling. The peak height for XiX decoupling is increased by an additional 29% over TPPM, while the line width is only slightly reduced, namely from 33 (TPPM) to 31 Hz (XiX).[95] The reason for this apparent contradiction is that the broad ''foot'' at the base of the TPPM-decoupling spectrum is reduced under XiX decoupling. Such an increase in peak height by 29% is equivalent to a decrease of the measurement time by 40% to obtain the same *S/N* ratio.

5.3.2.4. Cross polarization efficiency under fast MAS

Heteronuclear CP, combined with MAS, is a useful technique in solid state NMR to improve the sensitivity of nuclei with low gyromagnetic ratio (e.g., ^{13}C, ^{15}N). Polarization from abundant nuclei with a large gyromagnetic ratio can be transferred to insensitive species by the Hartmann–Hahn CP (HHCP) sequence, which consists of rf fields simultaneously applied to both nuclear species. For an efficient transfer, the matching condition $\omega_{1S} = \omega_{1I}$ has to be fulfilled where $\omega_{1I} = \gamma_I B_{1I}$ and $\omega_{1S} = \gamma_S B_{1S}$ are the rf field strengths applied to the sensitive or insensitive nuclei, respectively.[70,95] The insensitivity to a mismatch of the Hartmann–Hahn matching conditions is an important quality measured for a CP scheme.

For static sample, the CP matching profile consists of a broad unstructured peak centered at $\omega_{1S} - \omega_{1I} = 0$ with a width comparable to the homonucleus *I*-spin line width.[96] Sample spinning does not significantly influence the polarization transfer as long as the MAS frequency is slow compared to the homogeneous line width of the abundant *I* spins.

The efficiency of very fast CP-MAS was first described by Stejskal *et al.*[97] The matching profile, centered at $\omega_{1S}-\omega_{1I}$ is split into narrow sidebands separated by the rotation frequency ω_r, leading to the new matching conditions $\omega_{1S} = \omega_{1I} + f\omega_r$, where $f = 0, \pm 1, \pm 2, \ldots$.[98] The sidebands in the matching profile are resolved whenever the MAS frequency exceeds the width of the abundant *I*-spins resonance line and heteronuclear dipolar coupling frequency. To obtain better CP efficiency with high-speed MAS, a pulse sequence is desired which improves the efficiency of the polarization transfer on the central and, in addition, broader the matching condition to render the transfer less sensitive to deviation from the exact Hartmann–Hahn condition.

Barbara and Williams[99] and Wu and Zilm[100] have presented a pulse sequence (SPICP) with rotor-synchronized 180° phase shift on both channels during the mixing time. Peersen *et al.*[101,102] have proposed a variable-amplitude CP scheme (VACP or RAMP-CP) where the rf field of the *I*-spins (or *S*-spins) is stepped (slowly compared to the MAS frequency) through a number of amplitude values in the vicinity of the original Hartmann–Hahn condition, VACP leads to a sequential transfer for the different spin packets in the sample, each of which is expected to experience efficient center or sideband cross polarization for at least one of the amplitude values. Recently it is demonstrated that rotor synchronized amplitude-modulated CP (AMCP) spin-lock sequence can provide efficient polarization transfer at the Hartmann–Hahn condition and under suitable conditions, can also lead to a broadening of the Hartmann–Hahn condition.[103,104]

References

[1] L.-Y. Lian and D. A. Middleton, 2001, *Prog. Nucl. Magn. Reson. Spectrosc.*, 39, 171–190.

[2] N. K. Goto and L. E. Kay, 2000, *Curr. Opin. Struct. Biol.*, 10, 585–592.

[3] G. C. Leo, L. A. Colnago, K. G. Valentine, and S. J. Opella, 1987, *Biochemistry*, 26, 854–862.
[4] F. M. Marassi, A. Ramamoorthy, and S. J. Opella, 1997, *Proc. Natl. Acad. Sci. USA*, 94, 8551–8556.
[5] F. M. Marassi and S. J. Opella, 2000, *J. Magn. Reson.*, 144, 150–155.
[6] J. Wang, J. Denny, C. Tian, S. Kim, Y. Mo, F. Kovacs, Z. Song, K. Nishimura, Z. Gan, R. Fu, J. R. Quine, and T. A. Cross, 2000, *J. Magn. Reson.*, 144, 162–167.
[7] J. Pauli, M. Baldus, B. van Rossum, H. de Groot, and H. Oschkinat, 2001, *Chembiochem*, 2, 272–281.
[8] S. Kiihne, M. A. Mehta, J. A. Stringer, D. M. Gregory, J. C. Shiels, and G. P. Drobny, 1998, *J. Phys. Chem. A*, 102, 2274–2282.
[9] D. M. LeMaster and D. M. Kushlan, 1996, *J. Am. Chem. Soc.*, 118, 9255–9264.
[10] M. Hong and K. Jakes, 1999, *J. Biomol. NMR*, 14, 71–74.
[11] M. Hong, 1999, *J. Magn. Reson.*, 139, 389–401.
[12] F. Castellani, B. van Rossum, A. Diehl, M. Schubert, K. Rehbein, and H. Oschkinat, 2002, *Nature*, 429, 98–102.
[13] S. Yamaguchi, S. Tuzi, K. Yonebayashi, A. Naito, R. Needleman, J. K. Lanyi, and H. Saitô, 2001, *J. Biochem. (Tokyo)*, 129, 373–382.
[14] H. Saitô, K. Yamamoto, S. Tuzi, and S. Yamaguchi, 2003, *Biochim. Biophys. Acta*, 1616, 127–136.
[15] T. Arakawa, K. Shimono, S. Yamaguchi, S. Tuzi, Y. Sudo, N. Kamo, and H. Saitô, 2003, *FEBS Lett.*, 536, 237–240.
[16] S. Yamaguchi, K. Shimono, Y. Sudo, S. Tuzi, A. Naito, N. Kamo, and H. Saitô, 2004, *Biophys. J.*, 86, 3131–3140.
[17] S. Yamaguchi, S. Tuzi, J. U. Bowie, and H. Saitô, 2004, *Biochim. Biophys. Acta*, 1698, 97–105.
[18] H. Onishi, E. McCance, and N. E. Gibbons, 1965, *Can. J. Microbiol.*, 11, 365–373.
[19] H. Kandori, K. Shimono, Y. Sudo, M. Iwamoto, Y. Shichida, and N. Kamo, 2001, *Biochemistry*, 40, 9238–9246.
[20] M. Engelhard, B. Hess, D. Emeis, G. Metz, W. Kreutz, and F. Siebert, 1989, *Biochemistry*, 28, 3967–3975.
[20a] G. Metz, F. Siebert, and M. Engelhard, 1992, *Biochemistry*, 31, 455–462.
[20b] G. Metz, F. Siebert, and M. Engelhard, 1992, *FEBS Lett.*, 303, 237–241.
[21] D. A. Torchia, 1982, *Methods Enzymol.*, 82, 174–186.
[22] A. Perry, M. P. Stypa, J. A. Foster, and K. K. Kumashiro, 2002, *J. Am. Chem. Soc.*, 124, 6832–6833.
[23] M. Demura, M. Minami, T. Asakura, and T. A. Cross, 1998, *J. Am. Chem. Soc.*, 120, 1300–1308.
[24] T. Asakura, R. Sakaguchi, M. Demura, T. Manabe, A. Uyama, K. Ogawa, and M. Osanai, 1993, *Biotechnol. Bioeng.*, 41, 245–252.
[25] T. Asakura, H. Yamada, M. Demura, and M. Osanai, 1990, *Insect Biochem.*, 20, 261–266.
[26] C. A. Michal and L. W. Jelinski, 1998, *J. Biomol. NMR*, 12, 231–241.
[27] J. Kümmerlen, J. D. van Beek, F. Vollrath, and B. H. Meier, 1996, *Macromolecules*, 29, 2920–2928.
[27a] L. E. Cammish and S. A. Kates, 2000, in *F-moc Solid Phase Peptide Synthesizer*, Chapter 13, W. C. Chan and P. D. White, Ed., Oxford University Press, Oxford.
[27b] A. Paquet, 1982, *Can. J. Chem.*, 60, 976.

[28] S. Tuzi, S. Yamaguchi, M. Tanio, H. Konishi, S. Inoue, A. Naito, R. Needleman, J. K. Lanyi, and H. Saitô, 1999, *Biophys. J.*, 76, 1523–1531.
[29] H. Saitô, 1986, *Magn. Reson. Chem.*, 24, 835–852.
[30] S. Tuzi, A. Naito, and H. Saitô, 1994, *Biochemistry*, 33, 15046–15052.
[31] W. T. Rothwell and J. S. Waugh, 1981, *J. Chem. Phys.*, 75, 2721–2732.
[32] S. Tuzi, S. Yamaguchi, A. Naito, R. Needleman, J. K. Lanyi, and H. Saitô, 1996, *Biochemistry*, 35, 7520–7527.
[33] S. Yamaguchi, S. Tuzi, M. Tanio, A. Naito, J. K. Lanyi, R. Needleman, and H. Saitô, 2000, *J. Biochem. (Tokyo)*, 127, 861–869.
[34] S. Tuzi, J. Hasegawa, R. Kawaminami, A. Naito, and H. Saitô, 2001, *Biophys. J.*, 81, 425–434.
[35] I. Solomon, 1955, *Phys. Rev.*, 99, 559–565.
[36] N. Bloembergen, 1957, *J. Chem. Phys.*, 27, 572–573.
[37] H. Saitô, J. Mikami, S. Yamaguchi, M. Tanio, A. Kira, T. Arakawa, K. Yamamoto, and S. Tuzi, 2004, *Magn. Reson. Chem.*, 42, 218–230.
[38] H. Saitô, S. Tuzi, M. Tanio, and A. Naito, 2002, *Annu. Rep. NMR Spectrosc.*, 47, 39–108.
[39] M. Tanio, S. Inoue, K. Yokota, T. Seki, S. Tuzi, R. Needleman, J. K. Lanyi, A. Naito, and H. Saitô, 1999, *Biophys. J.*, 77, 431–442.
[40] H. Saitô, R. Kawaminami, M. Tanio, T. Arakawa, S. Yamaguchi, and S. Tuzi, 2002, *Spectroscopy*, 16, 107–120.
[41] Y. Kawase, M. Tanio, A. Kira, S. Yamaguchi, S. Tuzi, A. Naito, M. Kataoka, J. K. Lanyi, R. Needleman, and H. Saitô, 2000, *Biochemistry*, 39, 14472–14480.
[42] A. Bax, R. Freeman, and T. A. Frenkiel, 1981, *J. Am. Chem. Soc.*, 103, 2102–2104.
[43] N. M. Szeverenyi, M. J. Sullivan, and G. E. Maciel, 1982, *J. Magn. Reson.*, 47, 462–475.
[44] S. K. Straus, T. Bremi, and R. R. Ernst, 1996, *Chem. Phys. Lett.*, 262, 709–715.
[45] T. I. Igumenova and A. E. McDermott, 2003, *J. Magn. Reson.*, 164, 270–285.
[46] L. Duma, S. Hediger, B. Brutscher, A. Böckmann, and L. Emsley, 2003, *J. Am. Chem. Soc.*, 125, 11816–11817.
[47] B.-J. van Rossum, F. Casellani, J. Pauli, K. Rehbein, J. Hollander, H. J. M. de Groot, and H. Oschkinat, 2003, *J. Biomol. NMR*, 25, 217–223.
[48] E. Vinogradov, P. K. Madhu, and S. Vega, 1999, *Chem. Phys. Lett.*, 314, 443–450.
[49] A. Detken, E. H. Hardy, M. Ernst, M. Kainosho, T. Kawakami, S. Aimoto, and B. H. Meier, 2001, *J. Biomol. NMR*, 20, 203–221.
[50] R. Verel, M. Baldus, M. Ernst, and B. H. Meier, 1998, *Chem. Phys. Lett.*, 287, 421–428.
[51] R. Verel, M. Ernst, and B. H. Meier, 2001, *J. Magn. Reson.*, 150, 81–99.
[52] M. Hong, 1999, *J. Magn. Reson.*, 136, 86–91.
[53] M. Baldus and B. H. Meier, 1996, *J. Magn. Reson.*, A, 121, 65–69.
[54] E. H. Hardy, R. Verel, and B. H. Meier, 2001, *J. Magn. Reson.*, 148, 459–464.
[55] E. H. Hardy, A. Detken, and B. H. Meier, 2003, *J. Magn. Reson.*, 165, 208–218.
[56] S. Hediger, B. H. Meier, and R. R. Ernst, 1995, *Chem. Phys. Lett.*, 240, 449–456.
[57] R. A. Olsen, J. Struppe, D. W. Elliott, R. J. Thomas, and L. J. Mueller, 2003, *J. Am. Chem. Soc.*, 125, 11784–11785.
[58] A. Lesage, C. Auger, S. Caldarelli, and L. Emsley, 1997, *J. Am. Chem. Soc.*, 119. 7867–7868.
[59] A. Lesage, M. Bardet, and L. Emsley, 1999, *J. Am. Chem. Soc.*, 121, 10987–10993.

[60] M. Bardet, L. Emsley, and M. Vincendon, 1997, *Solid State Nucl. Magn. Reson.*, 8, 25–32.

[61] H. Kono, T. Erata, and M. Takai, 2003, *Macromolecules*, 36, 5131–5138.

[62] D. Sakellariou, S. P. Brown, A. Lesage, S. Hediger, M. Bardet, C. A. Meriles, A. Pines, and L. Emsley, 2003, *J. Am. Chem. Soc.*, 125, 4376–4380.

[63] S. K. Straus, T. Bremi, and R. R. Ernst, 1998, *J. Biomol. NMR*, 12, 39–50.

[64] M. Hong, 1999, *J. Biomol. NMR*, 15, 1–14.

[65] T. A. Egorova-Zachernyuk, J. Hollander, N. Fraser, P. Gast, A. J. Hoff, R. Cogdell, H. J. M. de Groot, and M. Baldus, 2001, *J. Biomol. NMR*, 19, 243–253.

[66] N. S. Astrof and R. G. Griffin, 2002, *J. Magn. Reson.*, 158, 157–163.

[67] T. Igumenova, A. J. Wand, and A. E. McDermott, 2004, *J. Am. Chem. Soc.*, 126, 5323–5331.

[68] J. Pauli, M. Baldus, B. van Rossum, H. de Groot, and H. Oschkinat, 2001, *Chembiochem*, 2, 272–281.

[69] M. Baldus, A. T. Petkova, J. Herzefeld, and R. G. Griffin, 1998, *Mol. Phys.*, 95, 1197–1207.

[70] S. R. Hartmann and E. L. Hahn, 1962, *Phys. Rev.*, 128, 2042–2053.

[71] A. Pines, M. G. Gibby, and J. S. Waugh, 1973, *J. Chem. Phys.*, 59, 569–590.

[72] K. Takegoshi, K. Nomura, and T. Terao, 1995, *Chem. Phys. Lett.*, 232, 424–428.

[73] K. Takegoshi, K. Nomura, and T. Terao, 1997, *J. Magn. Reson.*, 127, 206–216.

[74] P. R. Costa, B. Q. Sun, and R. G. Griffin, 1997, *J. Am. Chem. Soc.*, 119, 10821–10830.

[75] A. J. VanGammeren, F. B. Fulsberger, J. G. Hollander, and H. J. M. de Groot, 2005, *J. Biomol. NMR*, 31, 279–293.

[76] U. Haeberlen, 1991, Solid State NMR in High and Very High Magnetic Fields, in *NMR Basic Principles and Progress*, vol. 25, P. Doehl, E. Fluch, H. Gunther, R. Kosfeld, and J. Seelig, Eds., Springer-Verlag, Berlin, p. 143.

[77] K. Yamauchi, J. W. G. Janssen, and A. P. M. Kentgens, 2004, *J. Magn. Reson.*, 167, 87–96.

[78] Z. Gan, P. Gor'kov, T. A. Cross, A. Samoson, and D. Massiot, 2002, *J. Am. Chem. Soc.*, 124, 5634–5635.

[79] A. Samoson, 2002, *Encl. Nucl. Magn. Reson.*, 9, 59–64.

[80] H.-J. Behrens and B. Schnablel, 1982, *Physica*, 114B, 185-190

[81] A. Samoson, E. Lippmaa, and A. Pines, 1988, *Mol. Phys.*, 65, 1013–1018.

[82] K. Mueller, B. Sun, G. C. Chingas, J. Zwanziger, T. Terao, and A. Pines, 1990, *J. Magn. Reson.*, 86, 470–487.

[83] L. Frydman and J. S. Harwood, 1995, *J. Am. Chem. Soc.*, 117, 5367–5368.

[84] Z. Gan, 2000, *J. Am. Chem. Soc.*, 122, 3242–3243.

[85] A. Samoson, T. Tuherm, and J. Past, 2001, *J. Magn. Reson.*, 149, 264–267.

[86] T. Mildner, M. Ernst, and D. Freude, 1995, *Solid State Nucl. Magn. Reson.*, 5, 269–271.

[87] A. Bielecki and D. P. Burum, 1995, *J. Magn. Reson.*, A116, 215–220.

[88] E. R. Andrew, A. Bradbury, and D. G. Eades, 1958, *Nature*, 182, 1659.

[89] I. J. Lowe, 1959, *Phys. Rev. Lett.*, 2, 285–287.

[90] E. Lippmaa, M. Alla, A. Salumac, and T. Tuherm, 1981, U.S. Patent 42, 543,773.

[91] E. Lippmaa, A. Samoson, and M. Magi, 1986, *J. Am. Chem. Soc.*, 108, 1730–1735.

[92] C. E. Bronnimann, B. I. Hawkins, M. Zhang, and G. E. Maciel, 1988, *Anal. Chem.*, 60, 1743–1750.

[93] A. E. Bennett, C. M. Rienstra, M. Auger, K. V. Lakshmi, and R. G. Griffin, 1995, *J. Chem. Phys.*, 103, 6951–6958.
[94] A. Detken, E. H. Hardy, M. Ernst, and B. H. Meier, 2002, *Chem. Phys. Lett.*, 356, 298–304.
[95] M. Ernst, 2003, *J. Magn. Reson.*, 162, 1–34.
[96] M. H. Levitt, D. Suter, and R. R. Ernst, 1986, *J. Chem. Phys.*, 84, 4243–4255.
[97] E. O. Stejskal, J. Schaefer, and J. S. Waugh, 1977, *J. Magn. Reson.*, 28, 105.
[98] B. H. Meier, 1992, *Chem. Phys. Lett.*, 188, 201–207.
[99] T. M. Barbara and E. H. Williams, 1992, *J. Magn. Reson.*, 99, 439–442.
[100] X. Wu and K. W. Zilm, 1993, *J. Magn. Reson.*, A104, 154–165.
[101] O. B. Peersen, X. Wu, I. Kustanovich, and S. O. Smith, 1993, *J. Magn. Reson.*, A104, 334–339.
[102] O. B. Peersen, X. Wu, and S. O. Smith, 1994, *J. Magn. Reson.*, A106, 127–131.
[103] S. Hediger, B. H. Meier, and R. R. Ernst, 1993, *Chem. Phys. Lett.*, 213, 627–635.
[104] S. Hediger, B. H. Meier, and R. R. Ernst, 1995, *J. Chem. Phys.*, 102, 4000–4011.

Chapter 6

NMR CONSTRAINTS FOR DETERMINATION OF SECONDARY STRUCTURE

6.1. Orientational Constraint

Secondary structures of peptides and proteins can be revealed by a variety of structural constraints available from solid state NMR such as (1) orientational constraints, (2) distances, (3) torsion angles, and (4) conformation-dependent displacements of peaks. Magnetically or mechanically oriented samples with respect to the applied magnetic field are required for measurements of orientational angles which characterize the orientation of anisotropic interactions with respect to the applied magnetic field, although NMR measurements of unoriented samples using MAS are sufficient to examine the other constraints such as distances, torsion angles, and conformation-dependent chemical shifts. 3D structure can thus be determined by obtaining the numerous structural constraints mentioned above.

6.1.1. Single Crystalline System

Orientational constraints are derived from observations of a wide range of orientation-dependent anisotropic nuclear interactions such as dipolar interaction, CSA, and quadrupolar interaction in an oriented system such as single crystal sample. The following several types of characteristic angles for a peptide plane (see Figure 6.1) can be determined by solid state NMR as orientational constraints, depending upon the type of interactions.[1,2] The dipolar interaction between two spins, for instance, of amide nitrogen or an α carbon in a peptide linkage to nearby proton can be treated as an isolated spin pair. In an oriented system, the frequency splitting of the doublet ($\Delta\nu_D$) has the angular dependence described by

$$\Delta\nu_D = \nu_{\parallel}(3cos^2\theta - 1), \tag{6.1}$$

where θ is the angle between the direction of the applied magnetic field and the bond vector connecting the two nuclei, $2\nu_{\parallel}$ is the maximal frequency splitting for the dipolar coupling, which is observed when the internuclear vector is parallel to the field direction. Accordingly, θ^{H-N} and $\theta^{C'-N}$ in Figure 6.1 are readily determined by analysis of the N–H and C′–N dipole–dipole interactions, and $\nu_{\parallel}$ can be readily calculated from the relevant gyromagnetic ratios, γ_i, and the distance between the two nuclei, r, using the following equation:

$$\nu_{\parallel} = h\gamma_A\gamma_B(1/r^3) \tag{6.2}$$

In practice, it is necessary to use C–H or N–H bond lengths based on NMR measurements for evaluation of $\nu_{\parallel}$ values, while C–N and C–C bond lengths determined by diffraction method can be used.

For spin-1/2 nuclei such as ^{13}C and ^{15}N, the observed, static ^{13}C chemical shifts exhibit angular dependence due to the presence of CSA as shown in Eq. (2.6). δ_{ii} are the principal values of the CSA tensor which have been determined for several model peptides for ^{13}C[3–7] and ^{15}N[7–10] sites. In oriented samples, the chemical shift interaction gives a single resonance line for each site with a characteristic frequency for each orientation.

The ^{14}N quadrupole interaction for the amide site is useful for determining the two angles α and β, which describe the orientation of the principal axes of the EFG tensor at the nitrogen atom in the peptide plane with respect to the magnetic field. The resulting observed ^{14}N quadrupole splitting between the two frequencies (Eq. (2.29)) gives the angular information described in the following equation:

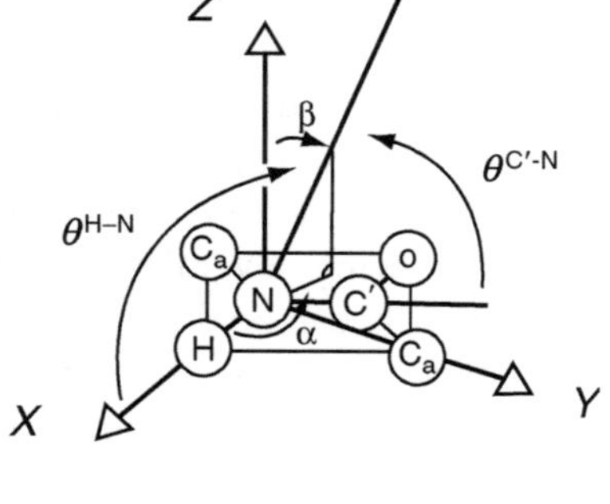

Figure 6.1. A peptide plane with the molecular axis system, X, Y, Z, defined. The angles α, β, θ^{H-N}, and $\theta^{C'-N}$ are shown for an arbitrary direction of the magnetic field vector, B_0 (from Opella and Stewart[1]).

$$\Delta\nu_Q = (3/4)(e^2qQ/h)[(3\cos^2\beta - 1) + \eta\sin^2\beta\cos 2\alpha] \tag{6.3}$$

where e^2qQ/h is the ^{14}N quadrupole coupling constant and η is the asymmetry parameter. ^2H quadrupole interactions are also useful for determining the peptide backbone structure in the oriented systems. For a deuterium bonded to carbon, however, the EFG tensor is axially symmetric with the largest principal component along with the C–^{2}H bond,

$$\Delta\nu_Q^{C-D} = (3/4)(e^2qQ/h)(3\cos^2\beta - 1), \tag{6.4}$$

where e^2qQ/h is the ^2H quadrupole coupling constant, which is approximately 180 kHz, and β is the angle between the C_α–H bond and the magnetic field. This angular dependence is the same as found for the ^1H–^{13}C dipolar interactions.

In principle, 3D structures of any proteins can be determined by a series of sufficient numbers of such orientational constraints available from analysis of the anisotropic nuclear interactions, when this approach is sequentially applied to respective peptide units.[1,2] In fact, this approach has been successfully applied to reveal secondary structure of small peptides. Nevertheless, this approach is not always practical for larger proteins, because requirement of larger single crystalline samples for NMR studies is more stringent than that for X-ray diffraction. In the latter, crystals with size of microns can be used together with very strong X-ray source available from synchrotron radiation (SR).

6.1.2. Mechanically Oriented System

Instead of the single crystalline system, revealing the secondary structure of ^{13}C- or ^{15}N-labeled peptides and proteins is also feasible for magnetically oriented proteins such as filamentous bacteriophage particles of fd coat protein oriented spontaneously with the direction of the strong applied magnetic field[11] (see Section 9.6.3) or incorporated preparations into highly anisotropic environment of mechanically or magnetically oriented lipid bilayers.[12,13] In the latter, lipids can be macroscopically oriented by attaching the lipid–water dispersions on flat glass plates because the lipid orientation is achieved by mechanical shearing forces. Alternatively, lipid molecules themselves can be oriented spontaneously to the magnetic field because of their intrinsic diamagnetic anisotropy. The mechanically oriented system incorporated into anisotropic environment of lipid bilayers is schematically drawn in Figure 6.2(a). To prepare the mechanically oriented

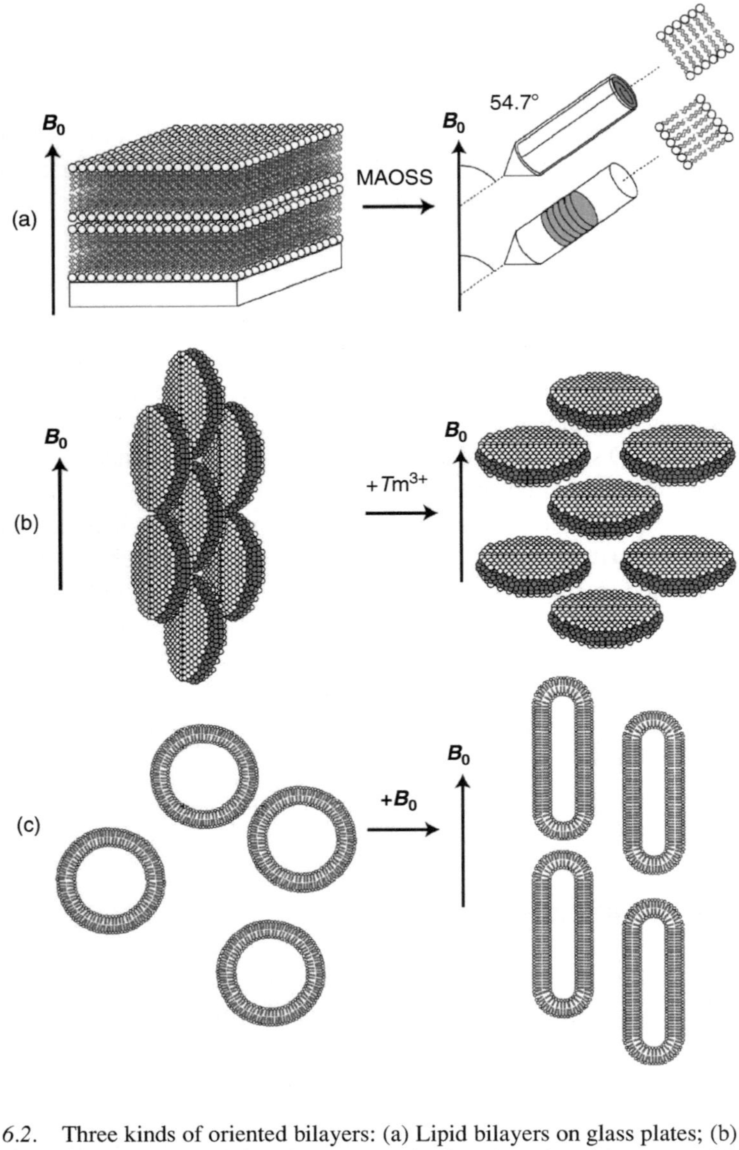

Figure 6.2. Three kinds of oriented bilayers: (a) Lipid bilayers on glass plates; (b) bicelles; (c) (left) large unilamellar vesicle, (right) magnetically oriented vesicle system (MOVS).

samples, methanol solution of ^{15}N-labeled gramicidin A (gA), consisting of 15 alternating L and D amino acids with both end groups blocked, and DMPC (in a 1:8 molar ratio) was dried[14] on glass microscope coverslips having dimensions of 5.8 × 12.0 mm:

HOC – Val[1]–Gly – Ala – D – Leu – Ala – D – Val – Val – D–
Val – Trp – D – Leu[10]–Trp – D – Leu – Trp – D – Leu–
Trp[15]–$NHCH_2CH_2OH$

Fourteen microliters of methanol solution containing 3.8 mg of gA and DMPC was spread on each coverslip; 28 of the coverslips were then stacked in a 12-mm-long section of square tubing with a 6-mm inner dimension, and enough deionized water was added to achieve 70% hydration. The ends of the tube were sealed with glass plates and a quick-setting epoxy. Orientation-dependent displacement of ^{31}P chemical shifts can be visualized by changing the angle θ between the bilayer normal and the applied magnetic field (Figure 6.3). The uniformity of the sample orientation is described by the mosaic spread (θ), a measure of the variation in the orientation of the bilayer normal with respect to the mean orientation of the bilayer normal, as measured by ^{31}P NMR spectra of hydrated DMPC bilayers containing gA. The asymmetry in the width at half-height for the 0° orientation spectrum can be used to estimate a maximum value for θ of $\pm 3°$. Alternatively, the vesicle suspension of fd coat protein was distributed over the surface of 40 glass cover slides with dimensions $11 \times 20 \times 0.07$ nm.[15] After allowing the bulk water to evaporate, the slides were stacked and placed in a sealed chamber together with a saturated ammonium phosphate solution, which

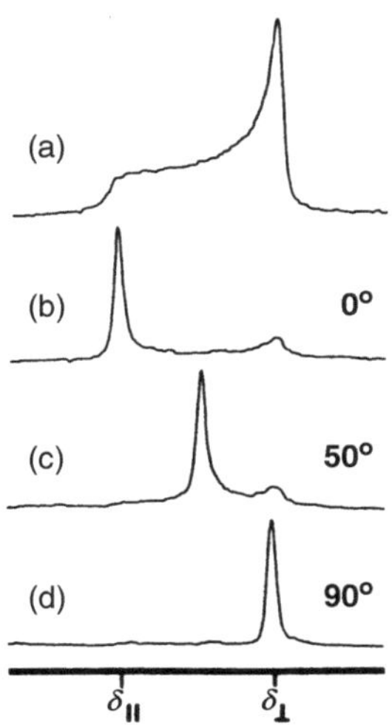

Figure 6.3. ^{31}P NMR spectra of hydrated DMPC bilayers containing gramicidin. (a) Unoriented, (b–d) spectra of samples oriented as indicated by the approximate angle of the bilayer normal with respect to the magnetic field (from Nicholson *et al.*[14]).

provided a 93% relative humidity atmosphere. Instead of the glass cover slides, much thinner polymer films such as a hydrophobic polymer, made of Halar (a copolymer of ethylene and chlorotrifluoroethylene) and a hydrophilic polycarbonate film were utilized.[16] Halar and polycarbonate are 25 and 2-μm thick compared with 150 μm for glass cover slides, and 60 layers of polymer could be easily stacked in the NMR tube.

6.1.2.1. *Gramicidin A channel*

High-resolution dimeric structure of the cation channel conformation of gramicidin A was elucidated solely on the basis of their amino acid sequence and 144 orientational constraints,[17–19] based on ^{15}N chemical shifts, ^{15}N–^{1}H and ^{15}N–^{13}C dipolar interactions, and ^{2}H quadrupolar interactions from uniformly aligned, isotopically labeled samples in DMPC bilayers. In calculating an initial molecular structure of the polypeptide backbone, they assumed that the peptide planes have an ω torsion angles of 180°. Consequently, the orientation of the planes is described by two vectors, the N–H and N–C_1 bonds. This can be achieved by the orientational constraints of the respective dipolar interactions as defined in Eq. (6.1). Once the orientation of a pair of adjacent planes is determined with respect to the magnetic field, one can take advantage of the tetrahedral geometry about the shared C_α carbon and generate the relative orientation of the planes. For the ^{15}N chemical shift, each structural constraint does not yield a unique structural solution, because observed chemical shift δ_{obs} does not uniquely define the orientation of the chemical shift tensor and the molecular frame (MF) with respect to the applied magnetic field which are expressed by the two angles α and β.[20] Hence, these NMR observables are structural constraints rather than unique determinants of structure. All of these conformations result in the same polypeptide fold and hydrogen-bonding pattern in gramicidin channel.[19] This represents an initial structure to which can be added the side chains largely defined by ^{2}H quadrupolar-derived constraints.

The folding motif for the two analytically derived structures are the same: they are both single-stranded helices with a right-handed helical sense and six to seven residues per turn. Both structures have the same hydrogen-bonding pattern. Because of the β-type torsion angles, the repeating unit in these helices is a dipeptide. One of these planes is essentially parallel to the channel axis and the other is tipped by about 20°. One structure has the amide protons tipped toward the channel lumen, and the other structure has the carbonyl oxygens tipped toward the channel lumen and available for solvating the cations in the channel. Only this latter conformation is consistent with the exclusive cationic functional role of gramicidin A. The set of 10

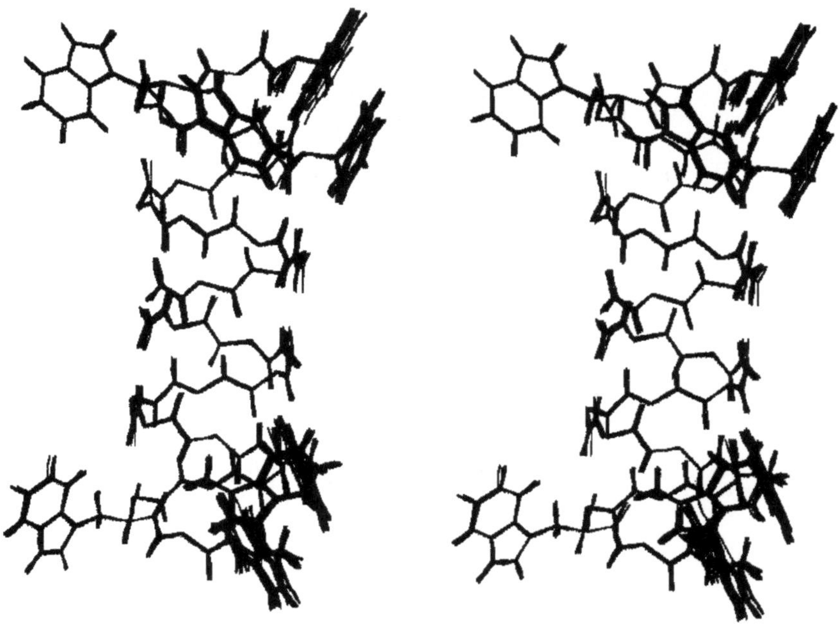

Figure 6.4. Stereo view of a set of 10 computationally refined structures of gramicidin A. The indole N–H groups are clustered at the bilayer surface, and the NH_2-terminus is buried at the bilayer center (from Ketchem *et al.*[17]).

computationally refined structures of gA as shown in Figure 6.4 illustrates a well-defined conformation resulting from constraints with very small error bars.

6.1.2.2. PISEMA and PISA wheel

The heteronuclear dipolar interaction is generally the dominant factor in determining line shapes, line widths, and relaxation parameters in NMR spectra of spin-1/2 nuclei in solids. This is serious when one extends the above-mentioned approach to larger molecular systems because of expected tremendous signal overlappings. To overcome this problem, a multidimensional solid state NMR called polarization inversion spin exchange at magic angle (PISEMA) experiment[21,21a,b] was developed (Figure 6.5, top) based on a flip-flop LG (Lee–Goldburg phase- and frequency-switched) pulse sequence,[22,23] which is used to spin-lock the *I* spins along the magic angle to suppress the homonuclear dipolar interaction. As a result, line widths in

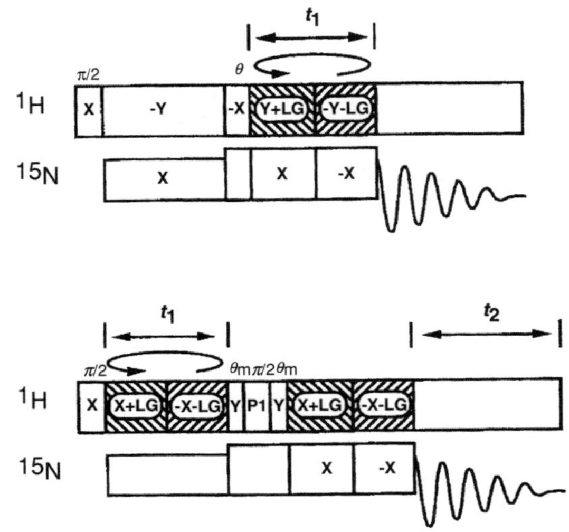

Figure 6.5. Pulse sequences of PIESMA (top) and HETCOR (bottom). X, -X, Y, -Y specify quadrature phases. +LG, −LG specify positive and negative frequency that fulfill the Lee–Goldburg condition. θ corresponds to 35.3° pulse and θ_m corresponds to a 54.7° pulse (from Wu *et al.*[21], Ramamoorthy *et al.* [21b]).

the dipolar dimension are reduced by more than an order of magnitude compared to a conventional separated-local-field (SLF) spectrum obtained without *I–I* homonuclear decoupling during the t_1 interval as well as the *I–S* heteronuclear decoupling by continuous irradiation of *S* spins. Here, LG would be at 180° in the opposite direction. By extending this approach to 3D solid state NMR correlation, the line-narrowing in the ^{1}H chemical shift and ^{1}H–^{15}N dipolar coupling dimensions was achieved.[24] A 2D ^{1}H chemical shift/^{15}N chemical shift heteronuclear correlation (HETCOR)[24a] spectrum (Figure 6.5, bottom) was obtained by fixing the spin exchange at the magic angle (SEMA) period, which is incremented as t_2 in the pulse sequence that generates a 3D correlation spectrum.[24,25] The 2D PISEMA spectrum of uniformly ^{15}N-labeled fd coat protein[15] as shown in Figure 6.6 has many resolved resonances, each of which is characterized by a ^{1}H–^{15}N dipolar coupling frequency and a ^{15}N chemical shift. The background sites that are mobile on the timescales of the ^{1}H–^{15}N dipolar and ^{15}N chemical shift interactions have resonances with zero dipolar and isotropic chemical shift frequencies. The line widths observed in 1D spectral slices taken from the 2D spectra are 1.2 ppm for the ^{1}H chemical shift, 300 Hz for the ^{1}H–^{15}N dipolar coupling, and 3 ppm for the ^{15}N chemical shift dimensions. This enables the measurement of the ^{1}H chemical shift, ^{1}H–^{15}N dipolar coupling, and ^{15}N chemical shift frequencies for all amide sites in the protein.

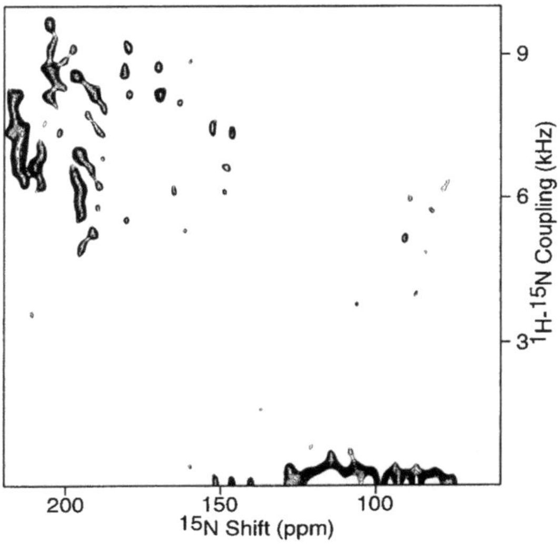

Figure 6.6. 2D $^{1}H–^{15}N$ dipolar coupling/^{15}N chemical shift PISEMA spectrum of an oriented sample of uniformly ^{15}N-labeled fd coat protein in phospholipid bilayers (from Marassi *et al.*[15]).

The characteristic "wheel-like" patterns, in 2D $^{1}H–^{15}N$ dipolar coupling/^{15}N chemical shift PISEMA spectra of uniformly ^{15}N-labeled samples in oriented bilayers, reflect helical wheel projections of residues in both transmembrane and in-plane helices and hence provide direct indices of secondary structure and topology of membrane proteins in phospholipids bilayers without resonance assignments.[26] No structural information, however, is available for the sites of the N- and C-terminal residues with isotropic resonances because of complete motional averaging of dipolar interactions. The PISEMA spectra calculated for the full range of possible orientations of an ideal 19-residue α-helix, with 3.6 residues per turn and identical backbone dihedral angles ($\phi = -65°$ and $\psi = -40°$) for all residues are shown in Figure 6.7.[26] When the helix axis is parallel to the bilayer normally all of the amide sites have an identical orientation relative to the direction of the applied magnetic field and therefore all of the resonances overlap with the same $^{1}H–^{15}N$ dipolar coupling and ^{15}N chemical shift frequencies. Tilting the helix away from the membrane normally breaks the symmetry, introducing variations in the orientations of the amide NH bond vectors relative to the field direction. This is manifest in the spectra as dispersions of both the $^{1}H–^{15}N$ dipolar coupling and the ^{15}N chemical shift frequencies, as shown in Figure 6.7((b)–(j)).

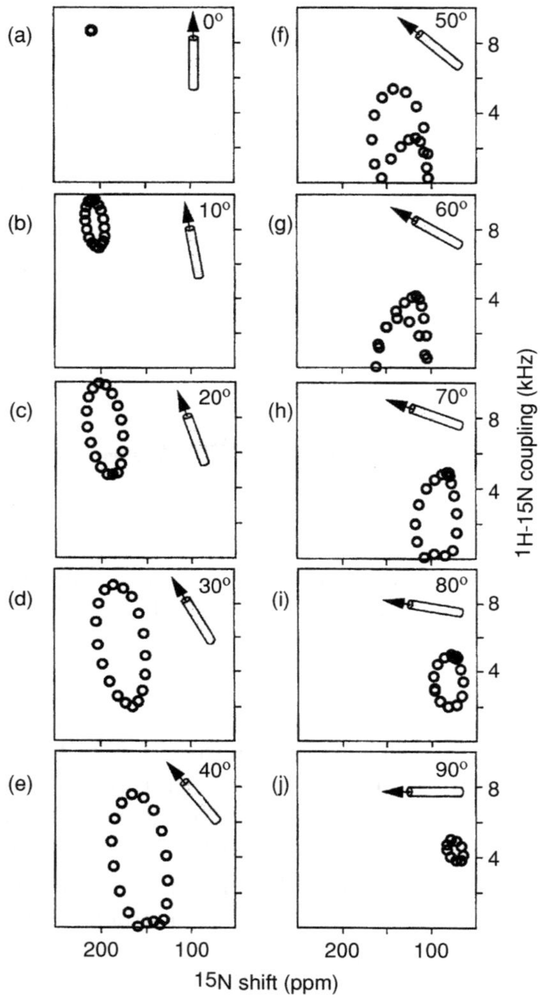

Figure 6.7. PISEMA spectra calculated for a 19-residue α-helix with 3.6 residue per turn and uniform dihedral angles ($\phi = -65°$ and $\psi = -40°$) at various tilt angles relative to the bilayer normal. (a) 0°; (b) 10°; (c) 20° (d) 30°; (e) 40°; (f) 50°; (g) 60°; (h) 70°; (i) 80°; and (j) 90°. The principal values and molecular orientation of the ^{15}N tensor ($\delta_{11} = 64$ ppm, $\delta_{22} = 77$ ppm, $\delta_{33} = 217$ ppm, δ_{33}, $\angle\delta_{33}$ NH $= 17°$ (from Marassi and Opella[26]).

By connecting the resonances within each dipolar transition of the dipolar dimension of ^{15}N PISEMA spectrum of multiple and single site labeled preparations of M2-TMP (transmembrane peptide of influenza A viral M2 protein) in hydrated lipid bilayers oriented with the bilayer normal parallel to the magnetic field direction, a mirror image pair of ‘‘PISA wheel’’ (polarity index slant angle) is immediately apparent as illustrated in Figure 6.8.[27] Calculating the ^{15}N anisotropic chemical shift of an ideal α-helical structure ($\phi = -65°$, $\psi = -40°$) and ^{15}N–^{1}H dipolar interactions for different helical tilts, τ, resulted in angle of 38° ± 3°, consistent with the tilt angle calculated from a set of assigned chemical shifts and from this set of assigned PISEMA results. However, the conclusion here is achieved without any use or need for resonance assignments. An algorithm for fitting protein structure to PISEMA spectra was described[28] and its application to helical proteins in oriented samples was demonstrated using both simulated and experimental results. Although the algorithm can be applied in an "assignment-free" manner to spectra of uniformly labeled proteins, the precision of the structure fitting was improved by the addition of assignment information, for example the identification of resonances by residue type from spectra of selectively labeled proteins. In addition, a simple, qualitative approach was proposed for determination of membrane protein secondary structure, including β-strands associated with membranes, and topology in lipid bilayers based on PISEMA and HETCOR spectra.[29] A "shotgun" NMR approach relies solely on the spectra from one uniformly and several selectively labeled ^{15}N samples, and on the fundamental symmetry properties of PISA wheels to enable the simultaneous sequential assignment of resonances and the measurement of the orientationally dependent

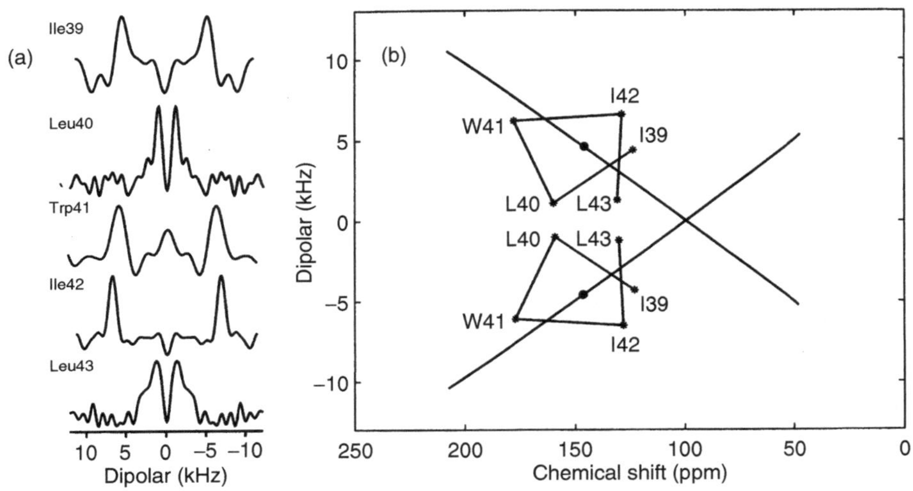

Figure 6.8. (a) Dipolar splittings observed from PISEMA spectra of multiple and single site labeled preparations of M2-TMP in hydrated lipid bilayers aligned with the bilayer normal to the parallel to the magnetic field direction. (b) Display of the dipolar splittings (*) at their observed chemical shift. The resonances are connected in helical wheel fashion (from Wang *et al.*[27]).

frequencies.[30] Shotgun NMR short-circuits the laborious and time-consuming process of obtaining complete sequential assignments prior to the calculation of a protein structure from the NMR data by taking advantage of the orientational information inherent to the spectra of aligned proteins. A total of five 2D $^1H/^{15}N$ PISEMA spectra, from one uniformly and four selectively ^{15}N-labeled samples, were sufficient to determine the structure of the membrane-bound form of the 50-residue major pVIII coat protein of fd filamentous bacteriophage.

6.1.2.3. Magic angle oriented sample spinning (MAOSS)

MAS NMR was applied to a uniformly aligned biomembrane samples by stacking glass plates with membrane sample and mounting onto a 7-mm MAS rotor, as illustrated in Figure 6.2 (a), right.[31] The spectral line width of multibilayer caused by orientational defects (''mosaic spread'') was easily removed by this MAOSS experiment to yield improved spectral resolution. For instance, the dramatic resolution improvement for protons, which is achieved in a lipid sample at only 220 Hz spinning speed in a 9.4 T field, is slightly better than any data published to date using ultra-high fields (up to 17.6 T) and high-speed spinning (14 kHz).

If the spinning rate ω_r is smaller than the anisotropic spin interactions such as CSA or quadrupole interaction, spinning sidebands will appear equally spaced by ω_r about the isotropic resonance. Therefore, orientational information of peptides or proteins in lipid bilayers or biomembranes is available from the analysis of the resulting ^2H, ^{13}C, or ^{15}N spinning sidebands of corresponding labeled samples. This approach was applied to study the orientation of the deuterated methyl group in [18-C^2H_3]-retinal in oriented bR in both the photocycle ground state (bR_{568}) and in the photo intermediate state M_{412} (see Figure 6.9).[32] To simulate ^2H MAOSS spectrum, it is necessary to rewrite the quadrupolar interaction in Eq. (2.23), using the Wigner rotation matrix after necessary reference frame rotations, from the principal axis system (PAS) to the MF, from their frame to the director frame (DF), the DF to the MAS frame (RF), and finally into the laboratory frame, together with the appropriate sets of the Euler angles Ω:

$$\mathrm{PAS} \xrightarrow[\Omega_{\mathrm{PM}}(\alpha_{\mathrm{PM}},\beta_{\mathrm{PM}},0)]{} \mathrm{MF} \xrightarrow[\Omega_{\mathrm{DR}}(\alpha_{\mathrm{MD}},\beta_{\mathrm{MD}},\gamma_{\mathrm{MD}})]{} \mathrm{DF} \xrightarrow[\Omega_{\mathrm{DR}}(0,0,0)]{} \mathrm{RF}$$

$$\xrightarrow[\Omega_{\mathrm{RL}}(\omega_r t,\theta,\psi)]{} \mathrm{LF}$$

The first rotation of the methyl group causes the unique z-axis of Z_P of the PAS of the EFG tensor to coincide with the C–C vector. The orientation of

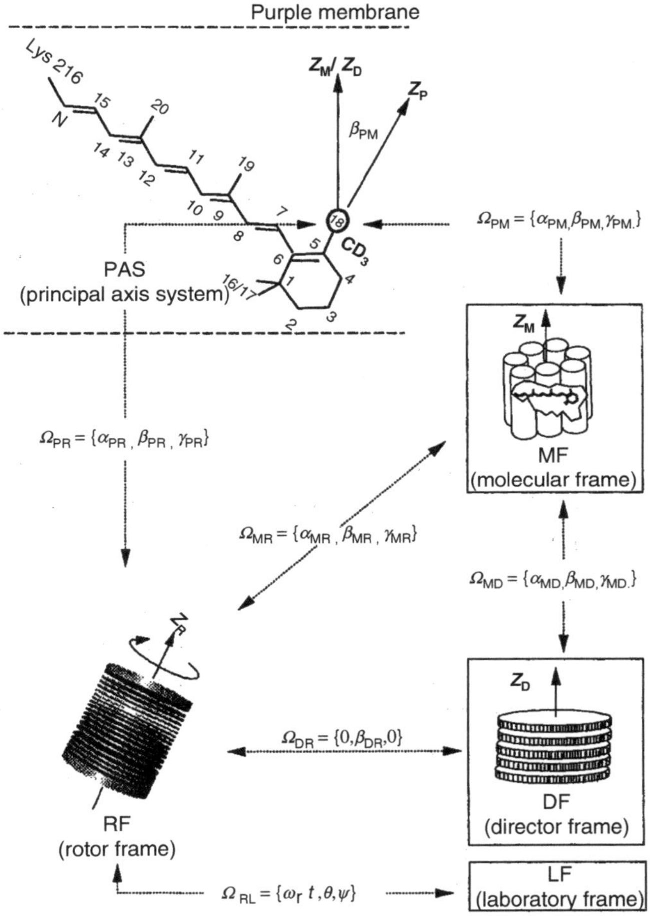

Figure 6.9. Method used to determine tilt angles for labeled groups by the MAOSS NMR method (from Glaubitz *et al.*[32]).

the axis Z_P with respect to the protein long axis Z_M is to be determined (β_{PM}). We define the z-axis Z_M of our MF as the average vector of all the seven protein helix long axes, which would be identical to the sample director Z_D in a perfectly aligned sample ($\beta_{MD} = 0°$). The DF is identical in our experimental setup with the RF, i.e., $Z_D \parallel Z_R(\beta_{DR} = 0°)$. The parallel to the sample director is given by $Z_D(\beta_{DR} = 0°)$. The RF Z_R axis is tilted with respect to the LF by the magic angle θ. Sample rotation makes a time-dependent set of Euler angles $\Omega_{RL}(\omega_r t, \theta, \psi)$ necessary (Figure 6.9).

The dependence of the ^{2}H MAS sideband pattern on the tilt angle β_{PM} in a well-oriented system ($\Delta\beta = 0°$) is shown in Figure 6.10(a) for relative

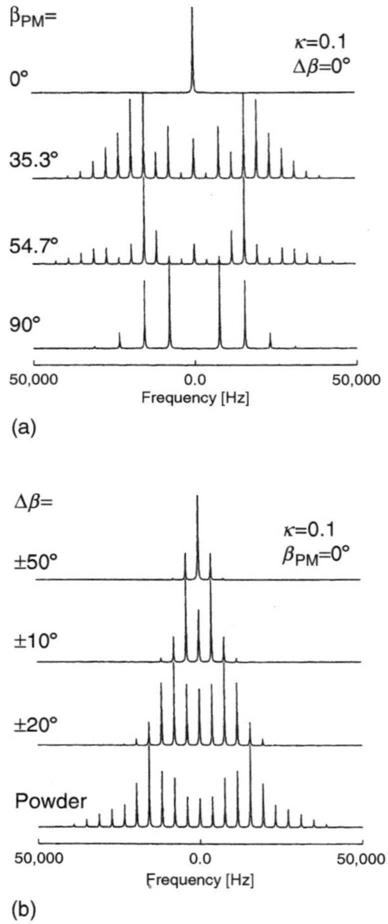

Figure 6.10. Simulations of ^{2}H MAS spectra for different tilt angles β_{PM} of the C–CD_3 vector with respect the MF in a perfectly aligned sample ($\Delta\beta = \pm 0°$) (a), and for the particular vector orientation of $\beta_{PM} = 0°$ with different degree of disorder $\Delta\beta$ (b). The relative spinning speed κ is the ratio of ω_r and quadrupole coupling (from Glaubitz *et al.*[32]).

spinning speed $\kappa = \omega_r/\chi_Q$. As discussed before, $\beta_{PM} = 0°$, only the isotropic central resonance remains, while for $\beta_{PM} = 90°$ all odd-numbered sidebands vanish. C-CD_3 orientations can now be determined by fitting these simulated sideband patterns to experimental MAOSS spectra. The effect of which different degrees of disorder ($\Delta\beta$) would have on a spectrum corresponding to a particular orientation is shown in Figure 6.10(b). In order to analyze the obtained function, a χ^2 merit function

$$\chi^2(\beta_{PM},\Delta\beta) = \Sigma_N\{[I_N^{ex} - I_N^{sim}(\beta_{PM},\Delta\beta)]/\Delta I_N^{ex}\}^2 \quad (6.5)$$

is minimized for the tilt angle β_{PM} and mosaic spread $\Delta\beta$ with I_N^{ex} and I_N^{sim} being experimental and simulated spectral intensities. A global minimum for bR_{568} was found at $\beta_{PM} = 36°$ and a mosaic spread $\Delta\beta$ between $\pm 0°$ and $\pm 8°$.[32] In contrast, the data for M_{412} photo intermediate state yields a tilt angle of $\beta_{PM} = 22°$ with $\Delta\beta$ between $\pm 10°$ and $\pm 18°$. Further, the tilt angle of bacteriophage coat protein M13 embedded in lipid bilayers was determined to be $20° \pm 10°$ around residues 29–31 based on sideband analysis by ^{13}C MAOSS NMR.[33]

6.1.3. Magnetically Oriented System

In general, lipid molecules themselves can be aligned spontaneously to the applied magnetic field because of their diamagnetic anisotropy. The interaction of the diamagnetic anisotropy is given by

$$F = -(1/2)B_0^2\{\chi_\perp + (\chi_\parallel - \chi_\perp)\cos^2\theta\} \quad (6.6)$$

where F is the interaction energy and $\chi_\perp$ and $\chi_\parallel$ are the perpendicular and parallel components of the magnetic susceptibility, respectively, and θ is the angle between $\chi_\parallel$ and the magnetic field B_0. Therefore, the orientation energy of the lipid bilayer with N lipids is expressed as

$$N\Delta F = -(1/2)N\Delta\chi B_0^2 \quad (6.7)$$

where $\Delta F = F(\theta = 0°) - F(\theta = 90°)$ and $\Delta\chi = \chi_\parallel - \chi_\perp$. If N is a small number, the orientation energy is not sufficient to align the macrodomain of the lipid bilayers, because the thermal energy, kT, tends to disturb the alignment. In the case of phospholipids bilayer, however, a macrodomain containing 10^6 lipid molecules can have sufficient energy for spontaneous alignment in the magnetic field, as demonstrated in the following three types of magnetically aligned systems: bilayered discoidal mixed micelles (bicelles)[34–36] or elongated lipid bilayers (magnetically oriented vesicle system; MOVS).[37]

6.1.3.1. Bicelles

Bicelles are lipid-detergent aggregates of DMPC with certain detergents[38,39] either short-chain phosphatidylcholine, dihexanoylphosphatidylcholine

(DHPC), or a zwitter ion bile salt derivative, CHAPSO. The function of the short-chain molecules is to coat the edges of the bilayered sections to protect the longer phospholipids chains from exposure of water. Bicelles possess great potential as model membrane for structural and dynamic studies of membrane proteins. Bicellar size varies as a function of the molar ratio [DMPC]/[DHPC] (with diameter $\approx$10–100 nm). Magnetically oriented phase is readily formed by the negative magnetic anisotropy of the lipid acyl chain when 2 < [DMPC]/[DHPC] < 6 (Figure 6.2(b)).[38,40] The phospholipids bicelles may be used as membrane mimetics for structural studies in both the anisotropic and the isotropic phases. In fact, a possibility of this system to mimic bilayer environment was explored[36] for studies of membrane proteins: resonances from reconstituted proteins by solution NMR were moderately broad but much less so than those from protein reconstituted into multibilayers oriented by mechanical method. The bicelles are shown to be oriented in a magnetic field by the fact that the acyl chain tends to orient perpendicular to the magnetic field if a large number of lipid molecules are ordered in the liquid crystalline phase possessing a sufficient degree of magnetic anisotropy to align the lipid bilayers along the magnetic field direction, as described above. This orientation ($\beta_{nl} = 90°$) is defined by an order parameter $S_{zz} = \langle 1/2(3\cos^2\beta_{nl} - 1)\rangle = -1/2$. For bilayer constituents undergoing rapid axially symmetric orientation with respect to the normal, the only disadvantage is that the anisotropic shifts and splittings will be scaled by a factor of 1/2. The addition of moderate amounts of paramagnetic ions, however, has the effect of flipping phospholipids bicelles from $S_{zz} \approx -1/2$ to $S_{zz} \approx 1$, as manifested from the increased ^{2}H quadrupolar splittings for ^{2}H NMR spectra of DMPC-d_{54} in the presence of Eu^{3+}, Er^{3+}, Tm^{3+}, and Yb^{3+} ions (Figure 6.2(b), right).[41,42] These oriented ^{2}H NMR spectra are naturally consistent with those from oriented spectra calculated as ''dePaked'' spectra from powder patterns, as shown in Figure 4.16. For the lanthanides, the origin of the paramagnetic shift in high-resolution NMR spectra is predominantly dipolar, and for an effectively uniaxially symmetrical ligand field, the relation between magnetic susceptibility anisotropy and direction of shift is given (in ppm) by Kurland and McGarvey[43] as

$$\Delta\nu/\nu_0 = -\Delta B/B = (\Delta\chi/3r^3)(3\cos^2\beta - 1) \tag{6.8}$$

where β is the angle between the principal axis of χ and the vector **r** between the ion and the nuclear spin, and it is readily seen that a nuclear spin located on or near the symmetry axis of the complex will experience deshielding, i.e., a downfield shift, when $\Delta\chi > 0$. In fact, suitable lanthanide ions are those with positive anisotropy of their magnetic susceptibility, namely Eu^{3+}, Er^{3+}, Tm^{3+}, and Yb^{3+}, as illustrated in Figure 6.11.

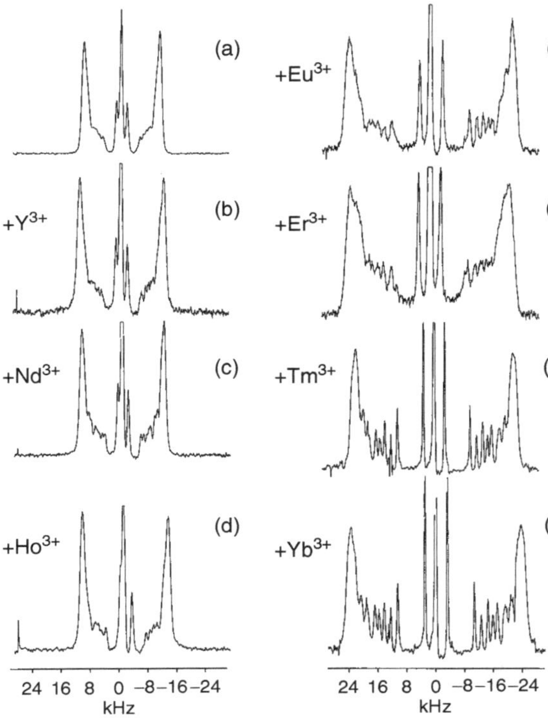

Figure 6.11. ^{2}H NMR spectra recorded at 37 °C of DMPC-d_{54} in magnetically aligned DMPC/DHPC bicellar samples with 80% water (w/v) and [DMPC]/[DHPC] = 4.6. (a), undoped sample; (b)–(h), from samples containing the indicated lanthanide ion in a ratio of Ln^{3+}/DMPC = 0.10. Note that spectrum *(f)* was obtained from a sample to which a small amount (DMPC/DMPA = 15:1) of dimyristoyl phosphatidic acid was added (from Prosser *et al.*[42]).

In the absence of special paramagnetic agents, phospholipid bicelles orient with their normal perpendicular to the magnetic field. The observed quadrupole splitting for a methylene deuteriums in an alkyl chain may be expressed[42] as

$$\Delta = \Delta_p S_{l\acute{n}} S_{\acute{n}n} S_{nm} S_{mp} \tag{6.9}$$

where $\Delta_p = (3/2)(e^2qQ/h)$ and $S_{ij} = (1/2)\langle 3\cos^2\beta_{ij} - 1\rangle$ represents an orientational order parameter linking the coordinate axis systems i and j to one another. Thus we need (1) average orientation, $\beta_{l\acute{n}} = 90°$, of the bicelle normal with respect to the magnetic field in the laboratory frame, l, for negatively ordered bicelles; (2) the angle, $\beta_{\acute{n}n}$, between the average bilayer normal, $\acute{n}$, and the instantaneous, or local, bilayer normal, n; (3) the orientation β_{nm} of a molecular axis m with respect to n; and (4) the local angle, β_{mp}, between the molecular axis and the principal axis p of the quadrupolar tensor along the C–D bond. If we choose the molecular axis as the normal to the plane of the CD_2 group, $\beta_{mp} = 90°$. Using a (vibrationally averaged) quadrupole coupling constant $e^2qQ/h = 168$ kHz, we then find that the observed splitting (in kHz) $\Delta = 63S_{\acute{n}m}$, where $S_{\acute{n}m} = S_{\acute{n}n}S_{nm}$ defines an internal order parameter. The internal order parameter, $S_{\acute{n}m}$, is seen to be approximately 0.32 for all the following three bicelles, which is slightly less than the internal order parameter $S_{\acute{n}m} = 0.39$ observed for multilamellar hydrated phospholipids bilayers without surfactant.[44] They are neutral bicelles prepared from DMPC and DHPC, and acidic bicelles, containing DHPC and DMPC with 25% of the DMPC replaced with, respectively, DMPS (1,2-dimyristoyl-n-glycero–3-[phosphor-L-serine]) or DMPG (1,2-dimyristoyl-n-glycero–3-[phospho-rac-(1-glycerol)]). Consequently, if we assume that the extents of phospholipids motion in bicelles and multibilayers are the same, we find that the bicellar order, expressed by $S_{\acute{n}m}$, is reduced by 15–20% relative to that in pure DMPC bilayers. They ascribed this decrease to bicellar ''wobble,'' defined more precisely as ''diffusion-in-a-cone'' of the bicelle normal.[45] Using the cone model, for which $S_{\acute{n}m} = 1/2\cos\beta_0(1 + \cos\beta_0)$, the reduction in order corresponds to a cone half-angle $\beta_0 = 28 - 32°$ for the [DMPC]/[DHPC] = 3.5 bicelles with diameter ≈500 Å. The inclusion of 25% DMPS or DMPG into phosphatidylcholine bicelles resulted in a moderate increase in the ordering of myristoylated peptide, Myr-d_{27} –GNAAAAKKGSEQES (Cat 14), the N-terminal 14-residue peptide from the catalytic subunit of cAMP-dependent protein kinase A relative to the bicelle normal as viewed from its ^{2}H NMR spectra. This is presumably because of favorable electrostatic interactions between the phospholipids headgroups and the two lysines in positions 7 and 8. Successful preparation of acidic bicelles was achieved by careful adjustment of lipid composition, pH, and ionic strength.

6.1.3.2. Magnetically oriented vesicle system

Magnetic ordering of lipid bilayers has been reported for pure and mixed phosphatidylcholine bilayers,[46–50] including melittin–phospholipid systems.[37,51–53] A DMPC bilayer containing a moderately high concentration of melittin (DMPC:melittin = 10:1 molar ratio) is subject to lysis and fusion at temperatures lower and higher than gel to liquid crystalline phase transition temperature, T_c, respectively.[37] The magnetically aligned, elongated vesicles are formed at a temperature above T_c as viewed from ^{31}P NMR and microscopic observation (Figure 6.2(c), right). Therefore, it was suggested that the elongated bilayer vesicles in which most of the surface area of the bilayers is oriented parallel to the magnetic field are formed rather than discoidal bilayers. This is because a large magnetic anisotropy can be induced owing to the negative magnetic anisotropy of the acyl chain axes. As pointed out already, the mechanically oriented lipid bilayer system in which the acyl chain can be aligned perpendicular to the glass plate is an excellent system for NMR study since it allows any orientation of the glass plate with respect to the magnetic field. However, this system is not always suitable for fully hydrated samples (up to 50% w/w at most), and the NMR sensitivity is inevitably low owing to the low filling factor of the glass plates in the NMR coil. On the contrary, this MOVS is suitable for studies of fully (or excess) hydrated samples of biological significance.

Instead, orientation of peptides bound to the spontaneously, magnetically oriented lipid bilayers can be examined by looking at the CSA of the carbonyl carbon in the backbone in the peptide chain under the condition of fully hydrated state. In particular, when an α-helix rotates about the helical axis, it is possible to determine the orientation angle of the α-helical axis with respect to the bilayer normal.[54] The tilt angle of the α-helix with respect to the average helical axis can be determined by comparing the anisotropic patterns of the carbonyl carbons of consecutive amino acid residues which form the α-helix with the chemical shift values of the corresponding magnetically aligned state, in a similar manner to that of PISEMA or PISA wheel described earlier. This relies on the fact that the angle between the consecutive peptide planes is 100° in the case of an ideal α-helix. For this purpose, it is essential to consider the $^{13}C{=}O$ chemical shift tensor to evaluate the orientation of the averaged α-helical axis of peptides bound to the magnetically oriented lipid bilayers in the case where the C=O axis is parallel to the helical axis.[35] The α-helical axis is thus defined by the polar angle α and β with respect to the principal axes of the $^{13}C{=}O$ chemical shift tensor, as shown in the top of Figure 6.12. When the averaged α-helix is inclined by θ to the static magnetic field, the observed ^{13}C chemical shift values, δ_{obs}, for the rotating α-helix in the magnetically oriented state obtained from the static experiment can be given by

$$\begin{aligned}\delta_{obs} = \delta_{iso} + (1/4)(3\cos^2\theta - 1)[(3\cos^2\beta - 1)(\delta_{33} - \delta_{iso}) \\ + \sin^2\beta\cos 2\alpha(\delta_{11} - \delta_{22})]\end{aligned} \tag{6.10}$$

When $\theta = 0^\circ$, the α-helical axis is considered to be parallel to the static magnetic field. In that direction, δ_{obs} is denoted as $\delta_{\parallel}$ and is given by

$$\begin{aligned}\delta_{\parallel} = \delta_{iso} + (1/2)[(3\cos^2\beta - 1)(\delta_{33} - \delta_{iso}) \\ + \sin^2\beta\cos 2\alpha(\delta_{11} - \delta_{22})]\end{aligned} \tag{6.11}$$

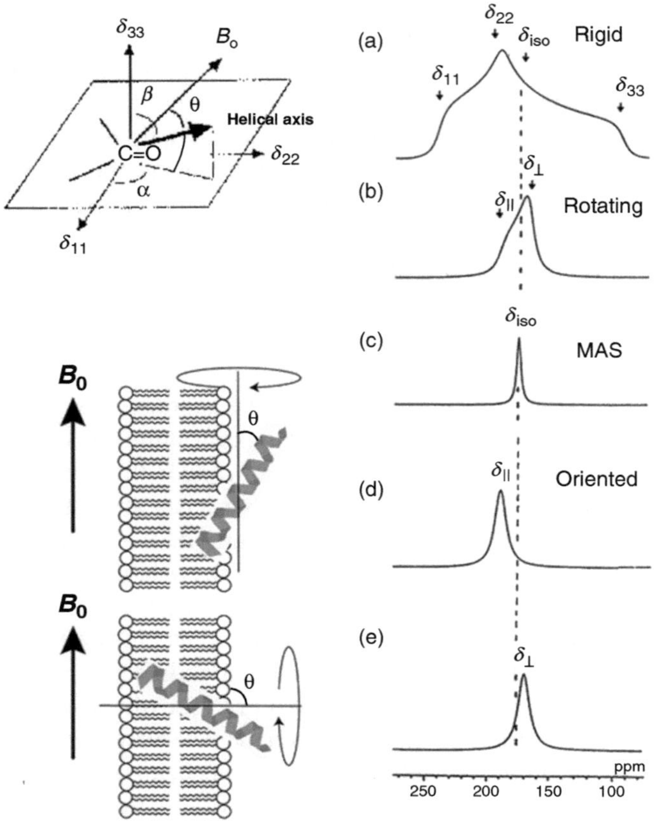

Figure 6.12. Distribution of the principal axes of the ^{13}C chemical shift tensor of the C=O group, helical axis, and static magnetic field, B_0, and ^{13}C NMR spectral patterns of the C=O carbons corresponding to the orientation of the α-helix with respect to the surface of the magnetically oriented lipid bilayers. Simulated spectra were calculated using $\delta_{11} = 241$, $\delta_{22} = 189$, and $\delta_{33} = 96$ ppm for the rigid case (a), rotation about the helical axis (slow MAS) (b), fast MAS (c), magnetic orientation parallel to the magnetic field (d), and the direction perpendicular to the magnetic field (e) (from Naito *et al.*[37]).

When $\theta = 90^\circ$, the α-helix is perpendicular to the static magnetic field, and then δ_{obs} corresponds to $\delta_{\perp}$ and is expressed as

$$\delta_{\perp} = \delta_{iso} - (1/4)[(3\cos^2\beta - 1)(\delta_{33} - \delta_{iso}) + \sin^2\beta\cos 2\alpha(\delta_{11} - \delta_{22})] \qquad (6.12)$$

Using Eqs. (6.11) and (6.12), the following equation is obtained:

$$\delta_{obs} = \delta_{iso} + (1/3)(3\cos^2\theta - 1)(\delta_{\parallel} - \delta_{\perp}) \qquad (6.13)$$

This relation makes it possible to predict that the α-helix is parallel to the magnetic field when δ_{obs} is displaced low field until $\delta_{\parallel}$, and the α-helix is perpendicular to the magnetic field when δ_{obs} is displaced upfield until $\delta_{\perp}$, as shown in Figure 6.12(d) and (e), respectively, for the case where $\delta_{\parallel}$ appears at lower field than $\delta_{\perp}$. The orientation of the α-helical axis with respect to the lipid bilayer surface, θ, can be determined by using Eq. (6.13) after the δ_{iso}, δ_{obs}, and $\delta_{\parallel} - \delta_{\perp}$ values are obtained from the fast MAS, static, and slow MAS experiment, respectively.

^{13}C NMR spectra for a variety of [1-^{13}C]amino-acid labeled melittin in DMPC bilayers were recorded under the slow MAS condition with spinning frequency of 100 ± 10 Hz to evaluate their respective $\delta_{\parallel} - \delta_{\perp}$ values,[37] as illustrated in Figure 6.13. Note that the anisotropy $|\delta_{\parallel} - \delta_{\perp}|$ of [1-^{13}C]Gly^3-melittin is much larger than that of [1-^{13}C]Gly^{12}-melittin in the slow MAS experiment. In the axially symmetrical powder pattern for Gly^3, $\delta_{\perp}$ appeared at lower field than $\delta_{\parallel}$. In contrast, $\delta_{\perp}$ in the axially symmetrical pattern for the [1-^{13}C]Ile^{20}-melittin appeared at higher field than $\delta_{\parallel}$ and the anisotropy was again much larger than that of [1-^{13}C]Gly^{12}-melittin. Anisotropies $|\delta_{\parallel} - \delta_{\perp}|$ for [1-^{13}C]Val^5- and Leu^{16}-melittin were also observed to be very small. The observed isotropic chemical shifts for these labeled portions are involved in the α-helical structure with reference to those of the conformation-dependent ^{13}C chemical shifts to be described in Section 6.4. This means that melittin undergoes rotation or reorientation of the whole α-helical rod about the average helical axis rather than the helical axis. Further, the angle for θ in the α-helical region around Gly^3 and Val^5 turns out to be nearly 90°, indicating that the α-helical rod of the N-terminal region is inserting into the lipid bilayers parallel to the bilayer normal. Similarly, the α-helical rod of the C-terminal region is inserted into the bilayers, making an angle of ~90° with the bilayer plane, which is again parallel to the bilayer normal.

In a similar manner, opioid peptide dynorphin A (1–17) is shown to strongly interact with DMPC bilayers to cause fusion and lysis across the phase transition temperature between the gel and liquid crystalline state ($T_c = 24$ °C) and results in subsequent magnetic ordering at a temperature above T_c.[54] Interestingly, the elongated vesicles are formed in dynorphin A (1–17)–DMPC systems with the long axis parallel to the magnetic field, when the bilayers are placed in the high magnetic field. Because the proportion of the powder pattern in ^{31}P NMR spectra is very low, the long axis is much longer than the short axis in the elongated vesicles. On the other hand, the shorter dynorphin A (1–13)–DMPC system did not exhibit magnetic ordering. Therefore, highly oriented DMPC bilayers containing dynorphin A (1–17) can be used to study the conformation and dynamics of membrane proteins and peptides, as a simple spontaneously aligned system to

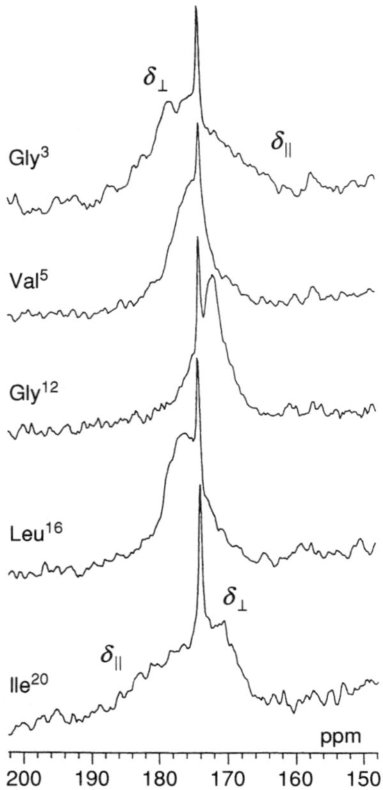

Figure 6.13. ^{13}C NMR spectra of carbonyl carbons for a variety of ^{13}C-labeled melittin bound to DMPC bilayers in the slow MAS condition. The sharp signals at 174 ppm indicate the C=O groups of DMPC (from Naito *et al.*[37]).

the applied magnetic field, instead of a mechanically oriented system using glass plates and bicelles. In fact, [1-^{13}C]Ala^{14}-labeled transmembrane peptide A (6–34) of bR in DMPC bilayer was spontaneously aligned to the applied magnetic field in the presence of dynorphin A (1–17) (dynorphin: DMPC = 1:10).[55] This magnetically aligned system turned out to be persistent even at 0 °C as viewed from ^{31}P NMR spectra of the lipid bilayers, after this peptide was incorporated into this system [A(6–34):dynorphin:DMPC = 4:10:100]. It was found from the ^{13}C NMR spectra of [1-^{13}C]Ala^{14}-labeled transmembrane peptide A (6–34) that the helical axis of A (6–34) is oriented parallel to the bilayer normal irrespective of the presence or absence of the reorientation motion about the helical axis at a temperature above the lowered gel to liquid crystalline phase transition.

6.1.3.3. Magnetically aligned proteins

There are several proteins that are spontaneously aligned to the applied magnetic field. Paramagnetic heme proteins suspended in buffer are perfectly aligned to the magnetic field due to the larger anisotropy in the magnetic susceptibility of the protein by the paramagnetic center, as already described in Section 4.2.1. In addition, nearly all of the α-helix in the coat protein of filamentous bacteriophages is extended approximately parallel to the filamentous axis. The resulting larger net diamagnetic anisotropy in the viral solution tends to align the particle to the applied magnetic field.

As a result, solid state NMR methods yield high-resolution spectra of concentrated solutions of bacteriophage particles. This subject will be discussed later in Section 9.6.3.

6.2. Interatomic Distance

6.2.1. Dipolar Interaction and its Recoupling Under MAS

When samples are rotating about an axis inclined to θ_m from the static magnetic field in the CP MAS or DD-MAS experiments as shown in Figure 2.3(b), θ is time-dependent and the $3\cos^2\theta - 1$ term of the dipolar interaction (see Eq. (2.1)) can be expressed as a function of time as follows:

$$\begin{aligned} 3\cos^2\theta(t) - 1 = {} & 1/2(3\cos^2\theta_m - 1)(3\cos^2\beta - 1) \\ & + \frac{3}{2}\sin^2\theta_m \sin^2\beta \cos(\alpha + \omega_r t) \\ & + \frac{3}{2}\sin^2\theta_m \sin^2\beta \cos 2(\alpha + \omega_r t), \end{aligned} \tag{6.14}$$

where α is the azimuthal angle and β is the polar angle defined by the internuclear vector with respect to the rotor axis. ω_r is the angular velocity of the rotor. When θ_m is the magic angle, the first term in Eq. (6.14) vanishes and is given by

$$3\cos^2\theta(t) - 1 = \sqrt{2}\sin^2\beta\cos\omega_r t + \sin^2\beta\cos^2(\alpha + \omega_r t). \tag{6.15}$$

Two frequencies due to the *I–S* dipolar interaction are expressed in angular velocity units as follows:

$$\omega_D(\alpha,\beta,t) = \pm\frac{D}{2}[\sin^2\beta\cos^2(\alpha + \omega_r t) - \sqrt{2}\sin^2\beta\cos(\alpha + \omega_r t)]. \tag{6.16}$$

Therefore, the dipolar interaction under the MAS condition is a function of time. This term can be null after taking an average over the rotor period as follows:

$$\begin{aligned}\overline{\omega_{\mathrm{D}}(\alpha,\beta,t)} &= \frac{1}{T_{\mathrm{r}}}\int_0^{T_{\mathrm{r}}} \omega_{\mathrm{D}}(t)\mathrm{d}t \\ &= 0,\end{aligned} \tag{6.17}$$

where T_r is the rotor period. This fact indicates that the dipolar interaction cannot affect the line shape of the center peak except for the intensities of the sidebands and difficult to obtain under the MAS conditions.

Accurate interatomic distances can be evaluated from the dipolar interactions which are normally sacrificed under the condition of high-power decoupling and MAS techniques.[56,57] A considerable improvement has been established to recouple the dipolar interaction by either introducing rf pulses synchronized with the MAS rotor period[58] or adjusting the rotor frequency as the difference of the chemical shift values of two isotopically labeled homonuclei.[59] Rotational echo double resonance (REDOR) [56,58] was explored to recouple the relatively weak heteronuclear dipolar interactions under the MAS condition by applying a π pulse synchronously with the rotor period. Consequently, the transverse magnetization cannot be refocused completely at the end of the rotor cycle, leading to a reduction of the echo amplitude. The extent of the reduction of the echo amplitude as a function of the number of rotor periods depends on the strength of the heteronuclear dipolar interaction. This method is extensively used to determine the relatively remote interatomic distance of 2–8 Å. When a number of isolated spin pairs are involved in a REDOR dephasing, the REDOR transformation could be useful in determining the interatomic distances because it yields single peaks in the frequency domain for each heteronuclear coupling strength.[60,61] This approach is, however, not applicable for the case where the observed nuclei in REDOR are coupled with multiple nuclei. The rotational resonance (RR) phenomenon[59,62] is a recoupling of the homonuclear dipolar interaction under the MAS condition. When the rotor frequency is adjusted to multiple of the difference frequency of the chemical shift values of two different resonance lines, line broadening and acceleration of the exchange rate of the longitudinal magnetization are observed. These effects depend strongly on the magnitude of the homonuclear dipolar interaction.

REDOR and RR methods have been most extensively explored, although several approaches to determine the interatomic distances in solid molecules have been proposed: transferred echo double resonance (TEDOR)[63] is a similar method used to determine the heteronuclear dipolar interactions by observing the buildup of echo amplitude. The magnetization in this method is transferred from one nucleus to the other through the heteronuclear dipolar interaction. It is, therefore, useful to eliminate naturally abundant background signals. Dipolar recovery at the magic angle (DRAMA)[64] is used to recouple the homonuclear dipolar interaction which is normally averaged out by MAS by applying $90°_x$ and $90°_{-x}$ pulses synchronously with the rotor period and hence interatomic distances between the two homonuclei can be determined. Since DRAMA strongly depends on the offset of the carrier frequency, Sun *et al.*[65] developed melding of spin-locking and DRAMA (MELODRAMA) by combining DRAMA with a spin-lock technique. This technique reduced the offset effect. Simple excitation for the dephasing of rotational echo amplitude (SEDRA)[66] and RF driven dipolar recoupling (RFDR)[67] are techniques used to apply a π pulse synchronously with the rotor period. Dipolar decoupling with a windowless sequence (DRAWS) is a pulse sequence using phase-shifted, windowless irradiation applied synchronously with sample spinning.[68] Using this pulse sequence, accurate interatomic distances can be determined for the case where the coupled spins have large chemical shift anisotropies and large difference in isotropic chemical shifts as well as the case of same isotropic chemical shifts. These techniques are also applied to determine the homonuclear dipolar interaction under the MAS condition. Because these techniques are not sensitive to the MAS frequency and offset effects, they will be useful to determine the dipolar interaction using multidimensional NMR for multiple site labeled systems.[65] It has not been fully evaluated, however, how accurately interatomic distances can be determined by these methods. As an alternative approach to evaluate molecular structure based on interatomic distances, methods of determining torsion angles have been proposed using magnetization or coherence transfer through a particular bond[69–72] or spin diffusion between isotopically labeled nuclei,[73] as will be described in Section 6.3.3.

6.2.2. REDOR

6.2.2.1. Simple description of the REDOR experiment[56]

Transverse magnetization which processes about the static magnetic field due to the dipolar interaction under the MAS condition moves back to the same direction at every rotor period because the integration of ω_D over one rotor period is zero. Consequently, the rotational echo signals are refocused at every rotor period. When a π pulse is applied to the S nucleus, which is coupled with the I nucleus, in one rotor period, this pulse plays a role to invert the precession direction of the magnetization of the observed I nucleus. Consequently, the magnetization vector of the I nucleus cannot move back to the same direction after one rotor period. Therefore, the amplitude of the echo intensity decreases (Figure 6.14(a)). The extent of the reduction of the rotational echo amplitude yields the interatomic distances. To evaluate the REDOR echo amplitude theoretically, one has to consider the averaging precession frequency in the presence of a π pulse at the center of the rotor period over one rotor cycle as follows:

$$\begin{aligned}\overline{\omega_{\mathrm{D}}(\alpha,\beta,t)} &= \pm\frac{1}{T_{\mathrm{r}}}\left(\int_{0}^{T_{\mathrm{r}}/2}\omega_{\mathrm{D}}\mathrm{d}t - \int_{T_{\mathrm{r}}/2}^{T_{\mathrm{r}}}\omega_{\mathrm{D}}\mathrm{d}t\right)\\ &= \pm\frac{D}{\pi}\sqrt{2}\sin 2\beta\sin\alpha\end{aligned} \tag{6.18}$$

Therefore, the phase angle, $\Delta\Phi(\alpha,\beta)$, for the N_{c} rotor cycle can be given by

$$\Delta\Phi(\alpha,\beta) = \overline{\omega_{\mathrm{D}}(\alpha,\beta)}N_{\mathrm{c}}T_{\mathrm{r}}, \tag{6.19}$$

where T_{r} is the rotor period. Finally, the echo amplitude can be obtained by averaging over every orientations as follows:

$$S_{\mathrm{f}} = \frac{1}{2\pi}\int_{\alpha}\int_{\beta}\cos\left[\Delta\Phi(\alpha,\beta)\right]\mathrm{d}\alpha\sin\beta\,\mathrm{d}\beta. \tag{6.20}$$

Therefore, the normalized echo difference, $\Delta S/S_{\mathrm{o}}$, can be given by

$$\begin{aligned}\Delta S/S_{\mathrm{o}} &= (S_{\mathrm{o}} - S_{\mathrm{f}})/S_{\mathrm{o}}\\ &= 1 - S_{\mathrm{f}}\end{aligned} \tag{6.21}$$

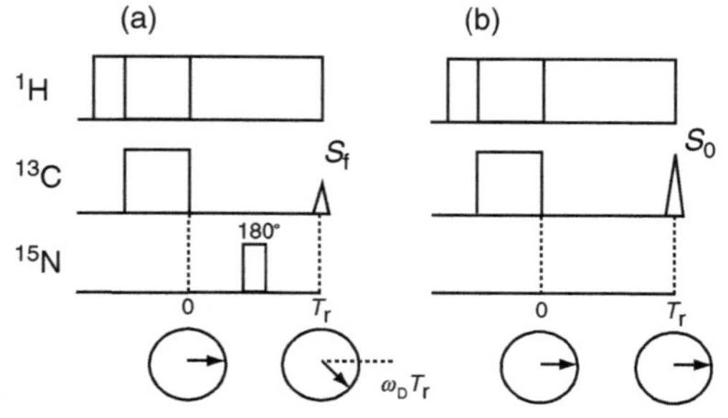

Figure 6.14. Pulse programs to observe rotational echo. (a) Transverse magnetizations do not refocus at the rotor period in the REDOR pulse sequence. (b) Transverse magnetizations refocus at the rotor period in the rotational echo pulse sequence.

Experimentally, REDOR and full echo spectra are acquired for a variety of N_cT_r values and the respective REDOR (S_f) and full echo (S_o) amplitudes are evaluated.

6.2.2.2. *Rotational echo amplitude calculated by the density operator approach*

The REDOR echo amplitude can be evaluated more rigorously by using density matrix operators and a pulse sequence for the REDOR experiment shown in Figure 6.15.[74] The time evolution of density operator, ρ_o, under the heteronuclear dipolar interaction during one rotor period can be considered by taking the pulse length into account. The average Hamiltonian in the rotating frame over one rotor period can be given by

$$\begin{aligned}
\overline{\mathcal{H}} &= \frac{1}{T_r}[\overline{\mathcal{H}_1(t)\tau} + \overline{\mathcal{H}_2(t)t_w} + \overline{\mathcal{H}_3(t)\tau}] \\
&= \frac{D}{4\pi}\{\sin^2\beta[\sin(2\alpha+\omega_r t_w) + \sin(2\alpha-\omega_r t_w) - 2\sin 2\alpha] \\
&\quad 2\sqrt{2}\sin 2\beta\left[\sin\left(\alpha+\frac{1}{2}\omega_r t_w\right) + \sin\left(\alpha-\frac{1}{2}\omega_r t_w\right) + 2\sin 2\alpha\right] \\
&\quad - \sin^2\beta[\sin(2\alpha+\omega_r t_w) + \sin(2\alpha-\omega_r t_w)]\frac{4\omega_r^2 t_w^2}{4\omega_r^2 t_w^2 - \pi^2} \\
&\quad + \sqrt{2}\sin 2\beta\left[\sin\left(\alpha+\frac{1}{2}\omega_r t_w\right) + \sin\left(\alpha-\frac{1}{2}\omega_r t_w\right)\right]\frac{2\omega_r^2 t_w^2}{\omega_r^2 t_w^2 - \pi^2}\Big\} I_z S_z \\
&\quad + \frac{D}{4\pi}\{\sin^2\beta[\cos(2\alpha+\omega_r t_w) + \cos(2\alpha-\omega_r t_w)]\frac{2\pi\omega_r t_w}{4\omega_r^2 t_w^2 - \pi^2} \\
&\quad - \sqrt{2}\sin 2\beta\left[\cos\left(\alpha+\frac{1}{2}\omega_r t_w\right) + \cos\left(\alpha-\frac{1}{2}\omega_r t_w\right)\right]\frac{2\pi\omega_r t_w}{\omega_r^2 t_w^2 - \pi^2}\Big\} I_z S_y \\
&= aI_z S_z + bI_z S_y,
\end{aligned} \tag{6.22}$$

where the same notations as in Eq. (6.14) is used.

$\overline{\mathcal{H}_1}(t)$, $\overline{\mathcal{H}_2}(t)$, and $\overline{\mathcal{H}_3}(t)$ are the average Hamiltonians corresponding to the period shown in Figure 6.15. Pulse length, t_w, is also considered in the calculations for the analysis of the REDOR results. The density operator, $\rho(T_r)$, at T_r after evolution under the average Hamiltonian can be calculated as

$$\rho(T_r) = \exp(-i\overline{\mathcal{H}}T_r)\rho_o\exp(i\overline{\mathcal{H}}T_r), \tag{6.23}$$

where ρ_o is considered to be I_y after the contact pulse. Then finally the transverse magnetization at T_r can be given by

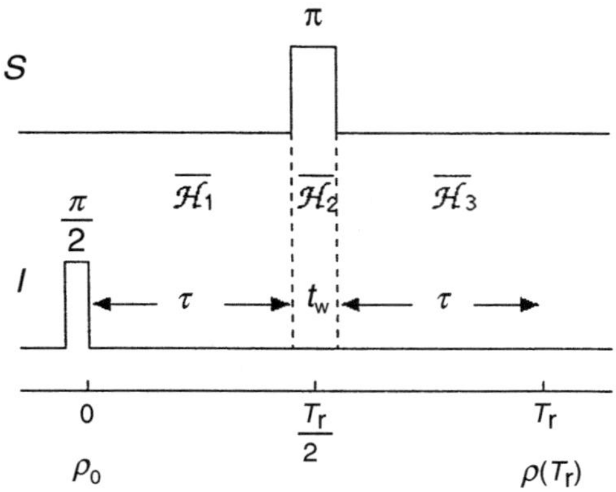

Figure 6.15. Pulse sequence and timing chart of REDOR experiment (from Naito *et al.*[74]).

$$\begin{aligned}\langle I_y(T_r)\rangle &= T_r\{\rho(T_r)I_y\}\\ &= \cos\left(\frac{1}{2}\sqrt{a^2+b^2}T_r\right).\end{aligned} \tag{6.24}$$

Echo amplitude in the powder sample can be calculated by averaging over every orientation as follows:

$$S_f = \frac{1}{2\pi}\int_\alpha\int_\beta \langle I_y(T_r)\rangle \sin\beta d\beta d\alpha \tag{6.25}$$

Therefore, the normalized echo difference, $\Delta S/S_o$, can be given by Eq. (6.21). When the length of t_w is zero, and Eq. (6.26) can be simplified as follows:

$$\langle I_y(T_r)\rangle = \cos\left(\frac{D}{\pi}\sqrt{2}\sin 2\beta \sin\alpha T_r\right). \tag{6.26}$$

In this case, Eq. (6.25) is equivalent to Eq. (6.21) in the case of $N_c = 1$.

6.2.2.3. *Echo amplitude in the three-spin system (S_I–I_I–S_2 system)*[75]

It is important to consider the case where the observed nucleus (I_1) is coupled with two other heteronuclei (S_1 and S_2). The Hamiltonian in the three-spin system can be given by

$$\mathcal{H}(t) = -\frac{\gamma_I \gamma_S h}{2\pi r_1^3}[3\cos^2\theta_1(t) - 1]I_{z_1}S_{z_1} - \frac{\gamma_I \gamma_S h}{2\pi r_2^3}[3\cos^2\theta_2(t) - 1]I_{z_1}S_{z_2}, \tag{6.27}$$

where r_1 and r_2 are the I_1–S_1 and the I_1–S_2 interatomic distances, respectively. $\theta_1(t)$ and $\theta_2(t)$ correspond to the angles between the magnetic field and the I_1–S_1 and the I_1–S_2 internuclear vectors, respectively. In the molecular coordinate system, the x-axis is along the I_1–S_1 internuclear vector, and the S_1–I_1–S_2 plane is laid in the x–y plane. The angle between I_1–S_1 and I_1–S_2 is denoted by ζ. The coordinate system is transformed from the molecular axis system to the MAS system by applying a rotation transformation matrix $R(\alpha,\beta,\gamma)$ with Euler angles α,β,γ followed by transforming from the MAS to the laboratory coordinate system by applying $R(\omega_r t,\theta_m,0)$. Finally, cos $\theta_1(t)$ and cos $\theta_2(t)$ are calculated as follows:

$$\cos\theta_1(t) = (\cos\gamma\cos\beta\cos\alpha - \sin\gamma\sin\alpha)\sin\theta_m\cos\omega t - (\sin\gamma\cos\beta\cos\alpha + \cos\gamma\sin\alpha)\sin\theta_m\sin\omega t + \sin\beta\cos\alpha\cos\theta_m$$

and

$$\begin{aligned}\cos\theta_2(t) = {} & [(\cos\gamma\cos\beta\cos\alpha - \sin\gamma\sin\alpha)\cos\zeta\sin\theta_m \\ & + (\cos\gamma\cos\beta\sin\alpha + \sin\gamma\cos\alpha)\sin\zeta\sin\theta_m]\cos\omega t \\ & - [(\sin\gamma\cos\beta\cos\alpha + \cos\gamma\sin\alpha)\cos\zeta\sin\theta_m \\ & + (\sin\gamma\cos\beta\sin\alpha - \cos\gamma\cos\alpha)\sin\zeta\sin\theta_m]\sin\omega t \\ & + \sin\beta\sin\alpha\cos\theta_m\sin\zeta + \sin\beta\sin\alpha\cos\theta_m\sin\zeta,\end{aligned} \tag{6.28}$$

where θ_m is the magic angle between the spinner axis and the static magnetic field, and ω_r is the angular velocity of the spinner rotating about the magic angle axis. The four resonance frequencies in the system are given by

$$\begin{aligned}\omega_{D_1} &= (D_1 - D_2)/2, \\ \omega_{D_2} &= (D_1 + D_2)/2, \\ \omega_{D_3} &= -(D_1 + D_2)/2, \\ \omega_{D_4} &= -(D_1 - D_2)/2.\end{aligned} \tag{6.29}$$

These dipolar transition frequencies are time-dependent and repeat the cycle in the spinning. In the REDOR pulse sequence, a π pulse is applied in the center of the rotor period. In this case, the averaged angular velocity over one rotor cycle for each resonance is given by

$$\overline{\omega_i(\alpha,\beta,\gamma,T_r)} = \frac{1}{T_r}\left(\int_0^{T_r/2}\omega_{Di}\mathrm{d}t\int_{T_r/2}^{T_r}\omega_{Di}\mathrm{d}t\right). \tag{6.30}$$

The phase accumulation after the N_c cycle is given by

$$\Delta\Phi_i(\alpha,\beta,\gamma,N_c,T_r) = \overline{\omega_i(\alpha,\beta,\gamma,T_r)}N_cT_r. \tag{6.31}$$

Finally, the REDOR echo amplitude after averaging over all Euler angles was calculated as

$$S_f = \frac{1}{8\pi^2}\sum_{i=1}^{4}\int_\alpha\int_\beta\int_\gamma[\cos\Delta\Phi_i(\alpha,\beta,\gamma,T_r)]\mathrm{d}\alpha\sin\beta\mathrm{d}\beta\mathrm{d}\gamma \tag{6.32}$$

The normalized echo difference, $\Delta S/S_o$, therefore strongly depends, not only on the dipolar couplings of I_1–S_1 and I_1–S_2 but also on the angle S_1–I_1–S_2.[75]

6.2.2.4. *Practical aspect of the REDOR experiment*

It is emphasized that accurate interatomic distances are prerequisite to achieve 3D structure of peptides, proteins, and macromolecules. A careful evaluation of the following several points is the most important step to obtain reliable interatomic distances by the REDOR experiment, although they have not always been seriously taken into account in the early papers. In practice, it is advisable to employ a standard sample such as [1-^{13}C,^{15}N]glycine,[74] whose C–N interatomic distance is determined to be 2.48 Å by a neutron diffraction study, to check that the instrumental conditions of a given spectrometer are correct before the experiment of a new sample.

As described in Section 6.2.2, the finite pulse length may affect the REDOR factor. In fact, this effect is experimentally observed and calculated using Eq. (6.23) as shown in Figure 6.16. The REDOR parameter, $\Delta S/S_o$, as measured for 20% [1-^{13}C,^{15}N]Gly, is plotted for the lengths of the ^{15}N π pulse of 13.0 μs for the experiment and 24.6 μs (chosen to satisfy 10% rotor cycle) as a function of N_cT_r with the calculated lines using the δ pulse length and finite lengths (13.0 and 24.6 μs) of the π pulse. It turns out, however, that the finite length of the ^{15}N π pulse does not significantly affect the REDOR effect provided that the pulse length is less than 10% of the rotor cycle at the rotor frequency of 4000 Hz. In a much faster spinning rate, rf irradiation occupies a significant fraction of the rotor period (10–60%). Thus the $\Delta S/S_o$ curve for REDOR deviate from that with ideal δ-function pulse.[76] Simultaneous frequency amplitude modulation (SFAM)[76a] uses rf fields

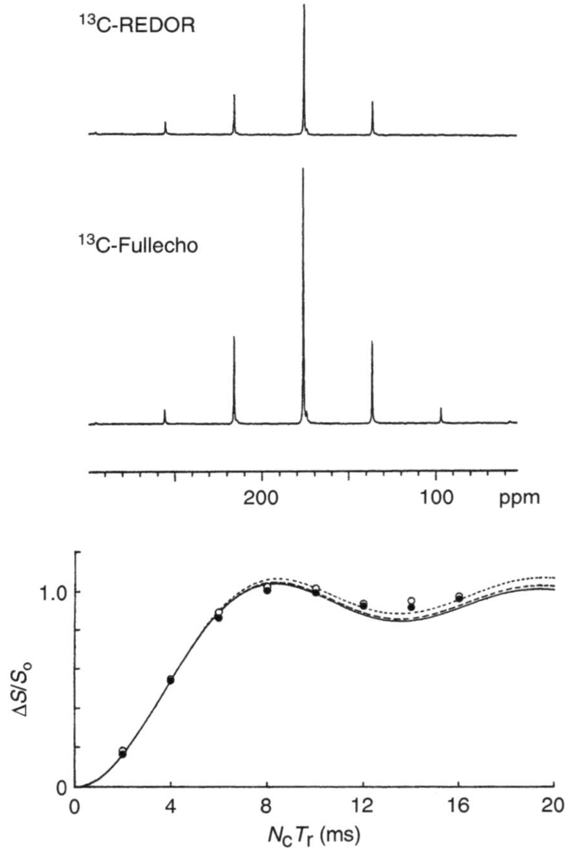

Figure 6.16. ^{13}C REDOR and full echo spectra of [1-^{13}C, ^{15}N]glycine as recorded by the rotor frequency of 4000 Hz and N_cT_r of 4 ms (top) and plots of the $\Delta S/S_o$ vs. N_cT_r (bottom). Solid and open circles denote the experimental points recorded using a ^{15}N π-pulse of 13.0 and 24.6 ms, respectively. Solid, broken, and dotted lines are calculated using the π pulses of δ, 13.0 and 24.6 ms, respectively, together with the C–N interatomic distance of 2.48 Å (from Naito *et al.*[74]).

whose amplitude and carrier frequency are sinusoidally modulated in order to overcome the shortcomings of REDOR arising from the finite pulse. Indeed, SFAM keeps more than 95% recoupling efficiency at the range of more than 85% rf effective field, whose range is twice that of REDOR.[76b]

For most spectrometers, it is very difficult to be free from *fluctuations of rf power* during the acquisition of REDOR experiments. It is, therefore, very important for the rf power to be stabilized after waiting a certain time period. If not, the π pulse cannot stay for a long time as the exact π pulse. Consequently, the REDOR factor is greatly decreased to yield relatively longer interatomic distances if the rf power changes. Compensation of instability of such rf power by the pulse sequence is, therefore, necessary to be free from long-term fluctuation of amplifiers. The *xy*-4 and *xy*-8 pulse sequences have been developed for this purpose and an *xy*-8 pulse is known to be the best sequence to compensate for the fluctuation of the rf power.[77]

Since the early stage of the REDOR experiment, contributions of natural abundance nuclei have been seriously considered as the major error source for the distance measurement.[78] The observed dipolar interaction modified by the presence of such neighboring naturally abundant nuclei was originally taken into account by simply calculating the $\Delta S/S_o$ value for isolated two pairs and adding proportionally to the natural abundant fraction.[78] Careful analysis of the three-spin system, however, indicates that this sort of simple addition of the fraction of the two-spin system may result in a serious overestimate of the natural abundance effect, to yield shorter distances.[74] The most accurate way to consider the natural abundance effect is, therefore, to treat the whole spin system as a three-spin system by taking into account the neighboring carbons in addition to the labeled pair. In practice, contributions from naturally abundant nuclei can be ignored[75] for ^{13}C REDOR but not for ^{15}N REDOR, because the proportion of naturally abundant ^{13}C nuclei is much higher than that of the ^{15}N nuclei.

^{13}C, ^{15}N-doubly labeled samples are usually used in the REDOR experiment to determine the interatomic distances between the labeled nuclei. More importantly, the dipolar interaction with the *labeled ^{15}N nuclei for the neighboring molecules* should be taken into account as an additional factor to contribute to the dipolar interaction of the observed pair under consideration. This contribution could be serious when the observed distance is quite remote, because there are many contributions from nearby nuclei. This effect can be completely removed by diluting the labeled sample with a sample of naturally abundant molecules. The sensitivity of the signals, however, has to be sacrificed if one wants to remove the effect completely as previously the sample is diluted to 1/49.[79] Instead, it is advised to evaluate the REDOR factors at the infinitely diluted condition by extrapolating the data by stepwise dilution of the sample (i.e., 60%, 30%, etc.) without losing sensitivity,[75] because a linear relationship between the REDOR factor and the dilution is ascertained by a theoretical consideration. Alternatively, the observed plots of $\Delta S/S_o$ values against the corresponding N_cT_r values for the sample without dilution can be fitted by a theoretical curve obtained from the dipolar interactions among three-spin systems, although the accuracy is not always improved to the level of the dilution experiment.

The transverse magnetization of the REDOR experiment decays as a function of the ^1H decoupling field.[80,81] Dipolar decoupling may be strongly interfered by molecular motion, if any, when the motional frequency is of the same order of magnitude as the decoupling field, and hence the transverse relaxation times are significantly shortened. In fact, it was found that the T_2 value of the carbonyl carbons in crystalline Leu-enkephalin are very short because of the presence of backbone motion.[82] This is a serious problem for the REDOR experiment, especially for the long distance pairs, because the *S/N* ratio is significantly deteriorated. In this case, it is worth considering measuring the ^{13}C REDOR signal under a strong decoupling field to elongate the transverse relaxation times. It is also useful to measure the distances at low temperature to be able to reduce the motional frequency. It is cautioned, however, that crystalline phase transition could be associated together with the freezing of the solvent molecule as encountered for a variety of enkephalin samples.[83]

In a commercial NMR spectrometer, the rotor is usually designed to allow the sample volume as large as possible in order to achieve better sensitivity. This arrangement, however, could cause B_1 inhomogeneity, which results in a broad distribution of the lengths of the 90° pulses. This problem is serious for the REDOR experiment in which a number of π pulses are applied. As a result, the pulse error can accumulate during the acquisition to give serious error. Particularly, the samples located at the top or bottom part of the sample rotor feel a quite weak rf field.[74,84] This causes a great reduction of the REDOR factor for sample which is filled all the way along in the sample rotor in a case of cylindrical rotor. This effect should be seriously taken into account prior to experiment with a commercial spectrometer. It is, therefore, strongly recommended to fill the sample only in the center part of the coil just like a multiple pulse experiment to be able to acquire accurate interatomic distances as much as possible by the REDOR method.

6.2.2.5. Distance measurements of multiple spin systems

It should be taken into account that a spin system for a doubly labeled sample should be generally treated as a multiple spin system (S–I_n system) as shown in Figure 6.17 in which S^* is the observed nuclei and the recoupling pulse is applied to the I^* nuclei. The ideal S–I system is available only when the doubly labeled preparation is infinitely diluted with unlabeled preparation (Figure 6.17(a)). In contrast, uniformly and singly labeled preparations can be treated as S_m–I_n and S–I_n systems, respectively (Figure 6.17(b) and (c)), although S^* in the latter is S nuclei of natural abundance. For this reason, it is advisable to utilize a variety of isotopically doubly labeled preparations to determine the 3D structure of the pentapeptide, Leu-enkephalin.[85] In addition, it is essential to make certain that all

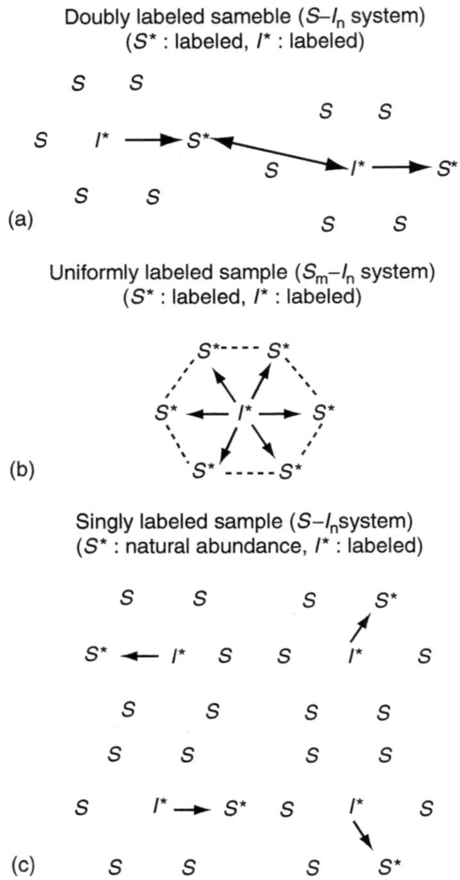

Figure 6.17. Schematic representation of multiple spin systems for (a) doubly labeled, (b) uniformly labeled, (c) singly labeled samples.

such preparations take the same polymorphic structure, because crystalline peptides under consideration may take different kinds of polymorphs depending on condition for crystallization, pH, temperature, etc.[74,83] The most convenient way to check this problem is to use the conformation-dependent ^{13}C chemical shifts as probes to certain crystalline forms, which are very sensitive to the presence of such polymorphic structures.[74,85,86]

In general, use of a uniformly labeled sample may be more attractive both for the sample preparation and also to avoid the problem of polymorphs.[87] This view turned to be not always true, however, because separation of a particular S–I interaction from S_m–I_n spin systems, consisting of many homo- and heteronuclear dipolar and scalar interactions, is not always as straightforward as anticipated. In practice, all of the following conditions should be taken into account to determine the accurate interatomic distances of a particular spin pair from the uniformly labeled sample. First, *J*-coupling among the observed *S* nuclei should be removed to reduce the phase twist due to the coherent evolution by *J*-coupling, when its magnitude is comparable to the dipolar coupling, as illustrated for uniformly [^{15}N, ^{13}C]-labeled ammonium L-glutamate monohydrate at $N_cT_r = 8$ ms (Figure 6.18).[88] This phase twist can be removed by using a shaped pulse as a *Z*-filter instead of a π pulse at the center position of the REDOR evolution period.[89] Second, the line widths of individual signals might be broadened due to strong homonuclear dipolar interaction with neighboring *S* nuclei, which should be eliminated by homonuclear decoupling pulses.[90] Thus, in the case of a uniformly labeled sample, combination of both homonuclear dipolar and scalar decoupling is required to simplify the S_m–I_n to S–I_n spin system. Third, relatively shortened transverse relaxation times (T_2) due to residual dipolar interactions reduce the transverse magnetization during the dipolar recoupling period in REDOR experiments. This makes it difficult to observe the REDOR effect for the nuclei with relatively long internuclear distances, because it is required to measure relatively long evolution times to observe slow dephasing of the REDOR signals. Fourth, REDOR factors against the

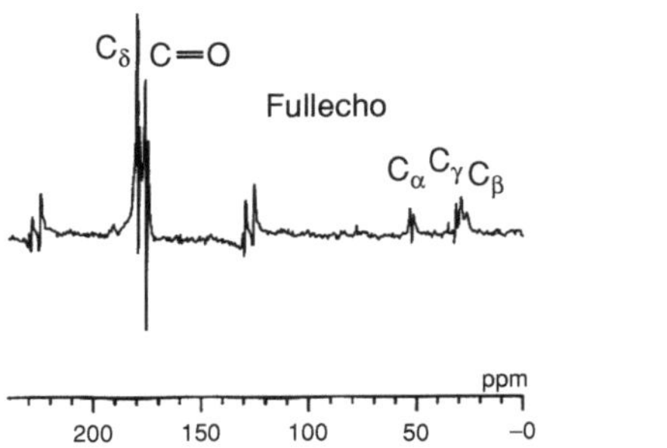

Figure 6.18. ^{13}C-REDOR and full echo spectra at $N_cT_r = 8$ ms for crystalline uniformly labeled ammonium [^{15}N]L-glutamate monohydrate (from Naito and Saito[88]).

dipolar evolution time should be analyzed using multiply coupled S–I_n spin systems rather than an isolated two-spin system[75,91,92] under the condition of homonuclear decoupling. Accordingly, the four kinds of obstacles mentioned above should be removed to determine a set of accurate interatomic distances simultaneously for a uniformly labeled sample.

Gullion and Pennington[93] proposed a θ-REDOR as an approach to determine the individual interatomic distances in a uniformly labeled sample by dividing S–I_n spin system into n number of isolated S–I two-spin systems. This makes the REDOR analysis simple and allows one to use the REDOR transformation to obtain the dipolar coupling frequencies of individual S–I spin couplings, although the method is not always easy to apply for measurement of a relatively long interatomic distance. Because ^{13}C–^{13}C isotropic J-couplings are typically 50–100 Hz for carbons separated by two bonds, ^{13}C echo trains for coupled carbons are attenuated and lose phase coherence after 15 ms. One approach to eliminate the effect of homonuclear ^{13}C isotropic J-coupling is to apply a tailored, frequency-selective REDOR (FSR) in which ^{13}C π pulse is applied to a single ^{13}C spin.[89,94] FSR experiment[94] can be applied to uniformly isotopically enriched samples, because the frequency of particular I–S spin pair is selected by applying simultaneous soft pulses. Individual internuclear distances are available, unless otherwise more than two observed nuclei are resonated accidentally at the same position. The selected spin then exhibits normal heteronuclear coupling to a distant ^{15}N, for example, while all other ^{15}N–^{13}C and all ^{13}C–^{13}C isotropic J-couplings involving the selected spin are suppressed. This approach has been used successfully to measure selected ^{15}N–^{13}C couplings within ^{13}C clusters.

A second approach is to combine ^{13}C π pulse (to refocus chemical shifts in Hahn echoes) with ^{13}C π/2 pulses (to refocus J-couplings in solid echoes). For an isolated ^{13}C–^{13}C pair, ^{13}C π pulses at T_r and $3T_r$, together with a π/2 pulse at $2T_r$, result in full refocusing at $4T_r$. For more than two coupled spins, however, the refocusing is incomplete. Nevertheless, the notion of combining Hahn echoes and solid echoes can be retained using just slightly more complicated pulse sequences. This sequence (called RDX for REDOR of $^{13}C_x$) has four π pulses and four π/2 pulses in $8T_r$. (The π pulse centered in the last two rotor cycles produces a Hahn echo that avoids beginning data acquisition coincident with a pulse; this pulse is not an integral part of the RDX sequence.) Under the RDX pulse sequence, the difference and full-echo ^{13}C{^{15}N} spectra of L-[$^{13}C_6$, α-^{15}N]histidine are all in phase. The labeled histidine was diluted 10-fold by natural abundance material and recrystallized in the free base form from water. Anti-phase components that evolve due to homonuclear scalar couplings and are not completely refocused are small for RDX, and with the sort of line widths typical in solid state NMR spectra, are not a problem. Multiple ^{15}N–^{13}C couplings are measured by RDX in the same 1D experiment. This approach therefore has a potential advantage over the use of frequency-selective π pulses that keep only regions of the spectrum in phase and require multiple experiments for multiple couplings, or with time-consuming 2D experiments that have no refocusing of isotropic J-couplings.

6.2.2.6. *Natural abundance ^{13}C REDOR experiment*[95]

A natural abundance REDOR experiment is performed because this method is free from both homonuclear dipolar and scalar interactions, as described earlier. This approach provides one the interatomic distances of number of isolated spin pairs simultaneously under high-resolution condition. Further, this approach is free from problem of the shortened T_2 values due to homonuclear ^{13}C spin interaction. Figure 6.19 shows the REDOR and full echo spectra of natural abundance ^{13}C nuclei of singly ^{15}N labeled crystalline ammonium [^{15}N] L-glutamate monohydrate (I) at $N_c T_r = 8$ ms. The assignment of peaks to the C_β, C_γ, C_α, C=O, and C_δ carbon nuclei, from the high- to the low-field, is also shown in Figure 6.19. The C=O (176.6 ppm) peak was distinguished from the C_δ (179.5 ppm) peak as a peak yielding stronger REDOR effect as shown in Figure 6.19. Similarly, the peak at 56.0 ppm was assigned to the C_α carbon nucleus, which gives the strongest REDOR effect among the three aliphatic ^{13}C peaks. In a similar manner, the peak at 30.5 ppm showed stronger REDOR effect compared with that of 35.2 ppm. Thus, the peaks at 35.2 and 30.5 ppm were assigned to the C_γ and C_β carbon nuclei, respectively.

The interatomic C–N distances between ^{15}N and ^{13}C=O, $^{13}C_\alpha$, $^{13}C_\beta$, $^{13}C_\gamma$, and $^{13}C_\delta$ carbon nuclei for I were determined with a precision of 0.15 Å, after the experimental conditions were carefully optimized. ^{13}C-REDOR factors for the three-spin system, $(\Delta S/S_o)_{CN1N2}$, and the sum of the two isolated two-spin system, $(\Delta S/S_o)^* = (\Delta S/S_o)_{CN1} + (\Delta S/S_o)_{CN2}$, were further evaluated by the REDOR measurements on isotopically diluted I in a controlled manner. Subsequently, the intra- and intermolecular C–N distances were separated by searching the minima in the contour map of the root mean square deviation (RMSD) between the theoretically and experimentally obtained $(\Delta S/S_o)^*$ values against the two interatomic distances, $r_{C\text{–}N1}$ and $r_{C\text{–}N2}$. When the intermolecular C–N distance ($r_{C\text{–}N1}$) of the particular carbon nucleus is substantially shorter than the intermolecular one ($r_{C\text{–}N2}$), the C–N distances between the molecule in question and the nearest neighboring molecules can also be obtained, although accuracy was lower. On the

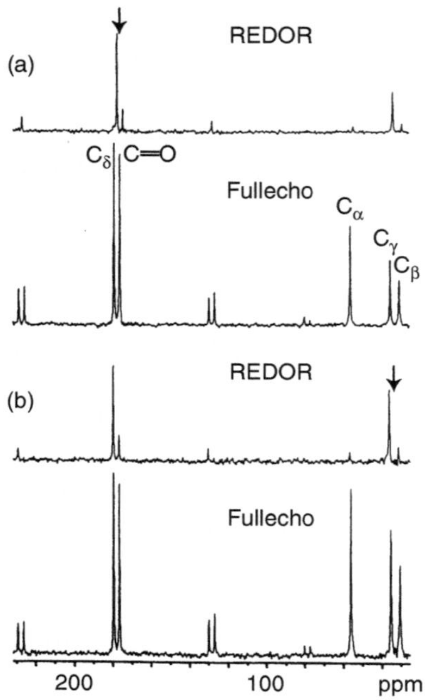

Figure 6.19. Naturally abundant ^{13}C-REDOR and full echo spectra at $N_c T_r = 8$ ms with two different carrier frequencies for crystalline ammonium [^{15}N]L-glutamate monohydrate. Blacked arrows indicate carrier frequency position at the center of: (a) C=O and C_δ resonances and (b) C_β and C_γ resonances (from Nishimura *et al.*[95]).

contrary, it was difficult to determine the interatomic distances in the same molecule when the intermolecular dipolar contribution is larger than the intramolecular one as in the case of C_δ carbon nucleus.

6.2.3. Rotational Resonance (RR)

6.2.3.1. Simple description of RR experiment[62]

In contrast to the REDOR experiment, homonuclear dipolar interactions can be recoupled in the RR experiment by adjusting the rotor frequency to be a multiple of the difference frequency of the isotropic chemical shift values of two chemically different homonuclear spins which is called rotational resonance conditions ($\Delta\omega_{iso} = n\omega_r$) (Figure 6.20). Under this condition, the energy level due to the sideband of one resonance becomes equal to that due to the center band of the other, and therefore mixing of two spin states occurs as indicated by the dotted arrow as shown in Figure 6.20. Consequently, the broadening of the line shape and the acceleration of the exchange rate of the longitudinal magnetization are observed. These rotational resonance phenomena are strongly dependent on the interatomic distance. One can, therefore, determine the interhomonuclear distance by the RR experiment. When the interatomic distance between the labeled nuclei is short, a characteristic line shape due to the mixing of the spin states can be observed. In contrast, when the interatomic distance is long, change of the signal intensity due to exchange of the longitudinal magnetizations can be observed.

The homonuclear dipolar interaction under the MAS condition has the Fourier components $\omega_{\mathrm{B}}^{(m)}$ at the frequency $m\omega_{\mathrm{r}}$ as follows:

$$\omega_{\mathrm{B}}^{(n)} = \sum_{n=-2}^{2} \omega_{\mathrm{B}}^{(m)} \exp(im\omega_{\mathrm{r}}t)\omega_{\mathrm{B}}(t) \tag{6.33}$$

The Fourier components in this equation can be further expanded by CSA as follows:

$$\omega_{\mathrm{B}}^{(n)} = \sum_{n=-2}^{2} \omega_{\mathrm{B}}^{(m)} a_{\Delta}^{(m-n)} \omega_{\mathrm{B}}^{(n)} \tag{6.34}$$

where $\omega_{\mathrm{B}}^{(n)}$ is the consequence of dipolar $\omega_{\mathrm{B}}^{(m)}$ and CSA $a_{\Delta}^{(n)}$. When the rotational resonance condition $\Delta\omega_{\mathrm{iso}} = n\omega_{\mathrm{r}}$ is fulfilled, $\omega_{\mathrm{B}}^{(n)}$ is the Fourier

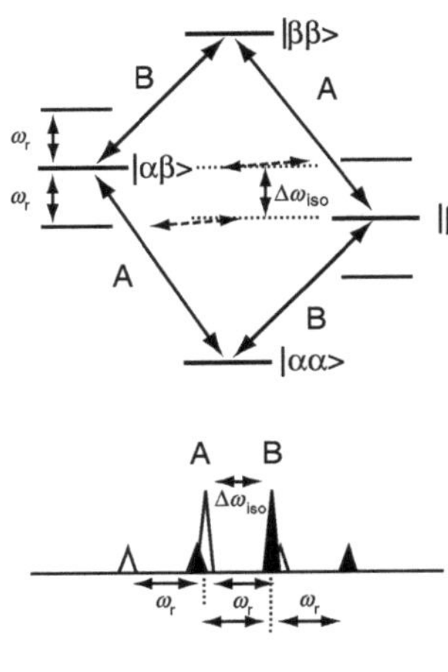

Figure 6.20. Energy level diagram in the homonuclear two-spin system under near rotational resonance condition.

component which is equal to the energy gap between two spin states and therefore the spin exchange or mixing of the wave functions becomes efficient. At the rotational resonance condition, the exchange rate R can be expressed as:

$$R^2 = r^2 - 4\left|\omega_{\mathrm{B}}^{(n)}\right|^2, \tag{6.35}$$

where $r = 1/T_2^{\mathrm{ZQ}}$ represents the zero quantum transverse magnetization rate and $\left|\omega_{\mathrm{B}}^{(n)}\right|$ implies the rotationally driven exchange rate. When $R^2 > 0$, the difference of the longitudinal two spins is given by

$$\langle I_z - S_z \rangle(t) = \mathrm{e}^{-rt/2}\left[\cosh(Rt/2) + \frac{r}{R}\sinh(Rt/2)\right], \tag{6.36}$$

when $R^2 < 0$,

$$\langle I_z - S_z \rangle(t) = \mathrm{e}^{-rt/2}\left[\cos(iRt/2) + \frac{r}{iR}\sin(iRt/2)\right]. \tag{6.37}$$

As is expected, T_2^{ZQ}, CSA and dipolar interaction are involved in the exchange rate of the longitudinal magnetization under the MAS condition. Therefore, the separation of the dipolar interaction is quite complicated as compared with the REDOR experiment.

6.2.3.2. *Practical aspect of RR experiment*

Experimentally, homonuclear dipolar interactions can be determined by measuring the extent of the exchange rate of the longitudinal magnetization as a function of mixing times τ_{m}.[84] A selective inversion pulse is used to invert one of the two resonances followed by the mixing time. One of the advantages of the RR experiment is that it can be performed by an ordinary double resonance spectrometer as long as the spinner speed can be controlled using a spinner frequency controller. It is advised to use $n = 1$ as the rotational resonance condition because the CSA is strongly affected when the n value is greater than 3. When a high field spectrometer is used, CSA increases proportionally to the frequency used. In that case, one has to include the CSA tensor for both nuclei in the analysis of the longitudinal magnetization. In the analysis of the RR experiment, one has to use the T_2^{ZQ} as discussed above. This value is difficult to determine experimentally. Practically, the T_2^{ZQ} value can be approximated from the single quantum relaxation time (T_2) of the two spins by the expression[96]

$$\frac{1}{T_2^{ZQ}} = \frac{1}{T_2^{I_1}} + \frac{1}{T_2^{I_2}} \tag{6.38}$$

$$T_2^{ZQ} = \frac{1}{\pi(\nu_{I_1} + \nu_{I_2})} \tag{6.39}$$

When the chemical shift difference is very small, it is difficult to perform the RR experiment because of the overlap of the resonance lines of the dipolar coupled nuclei, leading to difficulty in the analysis of the RR data. In this case, a rotational resonance experiment in the tilted rotating frame [dipolar-assisted rotational resonance (DARR)][97] can be used, because a much higher spinning speed can be adopted for the case of a smaller chemical shift difference in the system.

The natural abundance background signal can also affect the apparent amount of RR magnetization exchange. Consequently, the observed magnetization exchange rate then yields a smaller magnetization exchange rate than the observed one. Therefore, a rate smaller than the real rate is obtained, which results in an overestimate for the interatomic distance. Incomplete proton decoupling prevents magnetization exchange between the coupled spins because of the B_1 field inhomogeneity. It is therefore advised to irradiate a strong proton-decoupling field (>80 kHz) to a small size of sample in the rf coil to prevent B_1 inhomogeneity.

Rotational resonance experiments have been applied to determine internuclear distance in uniformly labeled molecules.[97,99] The rotational resonance conditions in uniformly ^{13}C-labeled threonine are sufficiently narrow to permit the measurement of five distances between the four carbon spins with an accuracy of better than 10% as shown in Figure 6.21. The polarization-exchange curves (Figure 6.21, bottom) are analyzed using a modified two-spin model consisting of the two active spins. Using this model, polarization-exchange curves were calculated in Liouville-space and one passive spin (spin 3). The passive spin is dipolar coupled to one of the active spins (spin 1). The modified model includes an additional offset in the final polarization, which comes from the coupling to the additional passive spins. The relative deviation of the distance extracted with two-spin model from the "true" distance is examined. As expected, accurate results for the internuclear distance r_{12} were obtained if the passive spin is far from rotational resonance, while the largest deviations occur when the passive spin is also close to a rotational resonance condition. The width of three unfavorable recoupling condition is of the size of the dipolar coupling to the passive spin (d_{13}) and becomes smaller for higher-order rotational resonance conditions. In addition, the J-couplings to the passive spins can contribute to the deviations in the fitted distances.

The analysis was carried out for a many-spin system as a pair of active spins and a set of passive spins. If the chemical shift differences between the

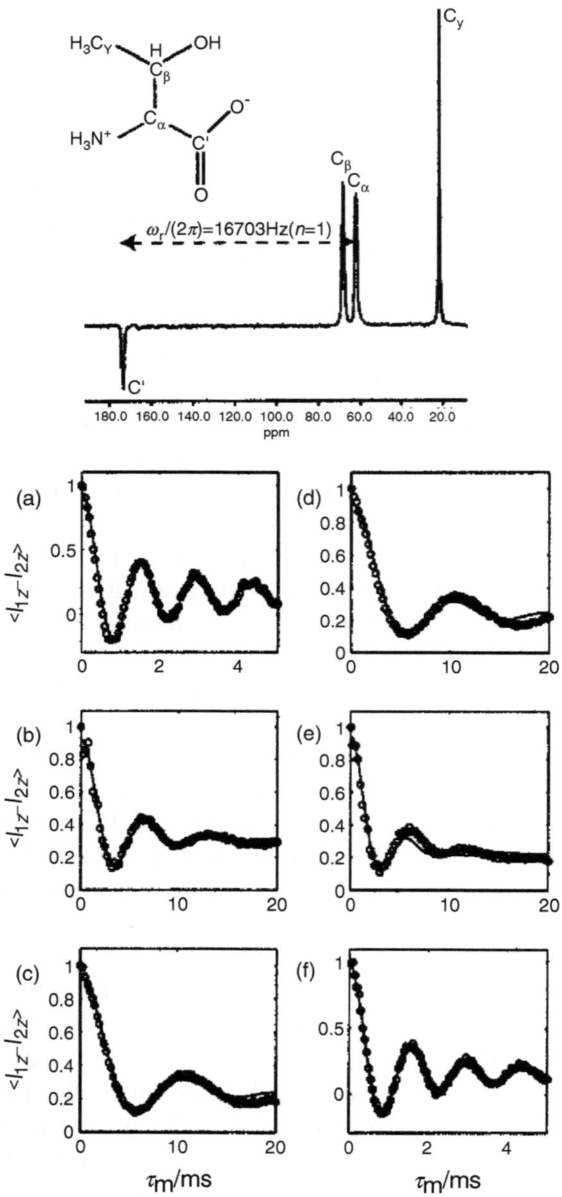

Figure 6.21. (Top) ^{13}C CP-MAS spectrum of L-threonin in the first time period (τ_m = 0 ms) in a rotational–resonance exchange curve between C′ and C_α with the C′ selectively inverted by a DANTE pulse train. (Bottom) Polarization-exchange curves for RR measurements in L-threonin as a function of the mixing time t_m. The measured data for the transfer from the C′ resonance to C_α (a), C_β (b), and C_γ (c), as well as between C_γ and C′(d), C_α (e), and C_β (f) (from Williamson *et al.*[98]).

active and the passive spins are not closer to a rotational resonance conditions, then the dipolar coupling constant between the active–passive spin pair, the presence of the passive spins only leads to a finite final polarization approached in the dynamics of the active spin pair and to a moderate systematic error in the internuclear distance.

The increased uncertainty in the distance, evaluated from the exchange curves in uniformly labeled samples using the modified two-spin model compared with that in selectively labeled samples, is often more than offset by the increase in the number of distances which can be measured from a single, uniformly labeled sample. The method described here can be looked as a homonuclear version of the frequency selective heteronuclear recoupling experiments in uniformly labeled methods (FSR).[94]

6.2.4. Structure Determination of Peptides and Proteins Based on Internuclear Distances

Internuclear distances were determined to elucidate the 3D structure of *N*-acetyl-Pro-Gly-Phe.[74] The carbonyl carbon of the $(i-1)$th residue and the amino nitrogen of the $(i+1)$th residue were labeled with ^{13}C and ^{15}N, respectively: [1-^{13}C]*N*-acetyl-Pro-[^{15}N]Gly-Phe (I), *N*-acetyl-[1-^{13}C]Pro-Gly-[^{15}N]Phe (II), and [1-^{13}C]*N*-acetyl-Pro-Gly-[^{15}N]Phe (III) were synthesized and the respective ^{13}C–^{15}N distances were determined as 3.24 $\pm$ 0.05, 3.43 $\pm$ 0.05, and 4.07 $\pm$ 0.05 Å, respectively, utilizing the REDOR factor obtained from the infinitely diluted state to prevent errors from the contributions of the neighboring labeled nuclei. No correction for the contributions of the naturally abundant nuclei turned out to be necessary. Surprisingly, these distances do not agree well with the values obtained by an X-ray diffraction study available when the initial part of this experiment was performed in spite of the expected accuracy, showing the maximum discrepancy between them to be 0.5 Å. This value seems to be unexpectedly larger than the experimental error estimated from the precision of the REDOR experiment ($\pm$0.05 Å). The reason why the discrepancy is unexpectedly larger was later explained by the fact that the crystal (orthorhombic) used for the REDOR experiments is different from that expected from literature data by X-ray diffraction study (monoclinic). To check the accuracy of the REDOR experiment, an X-ray diffraction study was performed on the same crystals used for the REDOR experiment. It was found that the distances from the newly examined crystalline polymorph (orthorhombic) agree well between REDOR and X-ray diffraction, within an accuracy of 0.05 Å.

Conformational maps based on the possible combinations for the torsion angles of Pro and Gly residues are calculated on the basis of the distance constraints thus obtained. Further, the difference of the chemical shifts between the C_β and C_γ carbons for the Pro residues $\Delta\beta_\gamma$ was used as a constraint to determine the ϕ value (−13°).[74] The ϕ angle of the Pro residue is in many instances restricted to −75°, which shows the minimum energy in the residue. Therefore, the torsion angles of the Pro residue were uniquely determined to be (−75°, −28°). Using these torsion angles, conformational maps for possible ϕ and ψ angles were again calculated. Finally two pairs of torsion angles were selected as (−112°, 48°) and (−112°, −48°). Minimization by a molecular mechanics calculation yielded the structure of the β turn I structure, as shown in Figure 6.22. It is found that the 3D structure of this peptide is well reproduced by a molecular dynamics simulation by taking into account all of the intermolecular interactions in the crystals, although a molecular dynamics calculation treating single molecular system does not always yield a correct picture as illustrated in Figure 6.22 (b), unless otherwise all the molecules in a unit cell are correctly taken into account.[74]

Elucidation of the 3D structure of an opioid peptide Leu-enkephalin crystal, Tyr-Gly-Gly-Phe-Leu grown from MeOH/H_2O mixed solvent, was performed by the REDOR method alone.[85] This seems to be an additional challenge for this technique to reveal the 3D structure of more complicated

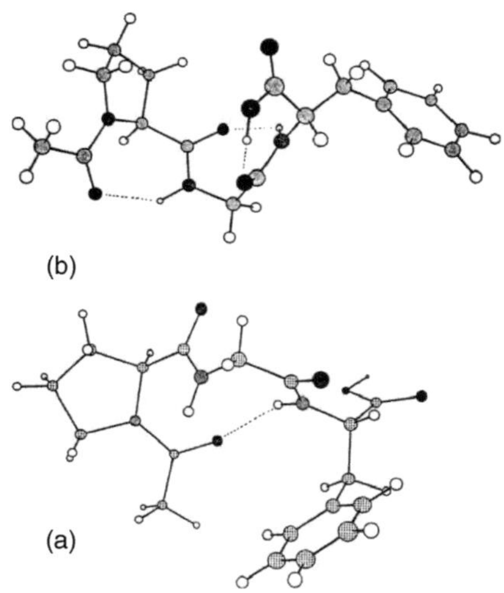

Figure 6.22. Optimized conformation of *N*-acetyl-Pro-Gly-Phe as obtained by the minimization of energy from the initial form as deduced from the REDOR experiment (a) and a "snapshot" of the conformation deduced by MD simulation *in vacuo* (at 100 K) (b) (from Naito *et al.*[74]).

systems. Six differently labeled Leu-enkephalin molecules were synthesized and the resulting interatomic distances were accurately determined. It turns out, however, that an extremely careful experiment is required to prevent unexpected crystalline conversion among polymorphs either by dehydration using a tightly sealed rotor or a phase transition occurring at ambient temperature. In other words, it is essential to guarantee the six differently labeled samples are all in the same crystalline polymorph by examination of the ^{13}C chemical shifts. Otherwise, meaningless data can be obtained without this precaution. When the distance data are converted to yield the necessary number of torsion angles, a unique combination of torsion angles in the corresponding conformational map is determined sequentially as shown in Figure 6.23.[85]

It is emphasized that accuracy within 0.1 Å is essential to arrive at unique set of torsion angles for each amino acid residue. It is also important to point out that the chemical shift data can be used for this purpose as additional constraints. A 3D structure was thus determined as shown in Figure 6.23. However, this structure is not the same as that previously determined by X-ray diffraction because one is dealing with a crystalline polymorph that is not fully explored. It is possible to figure out that the molecular packing in the crystals by looking at the intermolecular dipolar contribution. Since the intermolecular dipolar contributions in I and III are larger than that of II, a β-sheet is formed and one peptide interacts with two peptides as an anti-parallel β sheet as shown in Figure 6.23.

It should also be pointed out that, in order to isolate an S–I spin pair, it is essential to avoid any unnecessary errors from dipolar contributions of neighboring molecules from overall S–I_n spin system. This is generally encountered for a variety of peptides and the aforementioned dilution procedure turned out to be not required in the case of α-helical polypeptides as far as the measurement of the interatomic distance for $^{15}\mathrm{NH}(i)\cdots\mathrm{O}{=}^{13}\mathrm{C}(i+4)$ is concerned.[100] This is because the interhelical dipolar contribution is found to be negligible in view of the mutual packing of α -helical chains or random mutual orientation between isotopically labeled atoms. Further, it was found that in helical peptides, due to the manner of chain packing, this kind of measurement is feasible near the ambient temperature to yield 4.5 $\pm$ 0.1 Å, without any assumptions, for a variety of transmembrane α -helical peptides in lipid bilayers to mimic a biomembrane at a temperature below liquid crystalline–gel phase transition at which rotation of such peptides about the α-helical axis does not occur, in spite of the presence of rotational motions of lipid molecules, as viewed from the observation of three principal components of ^{13}C and ^{31}P chemical shift tensors.[101] As shown in this example, it is essential to ensure that molecular motion is reasonably ceased.

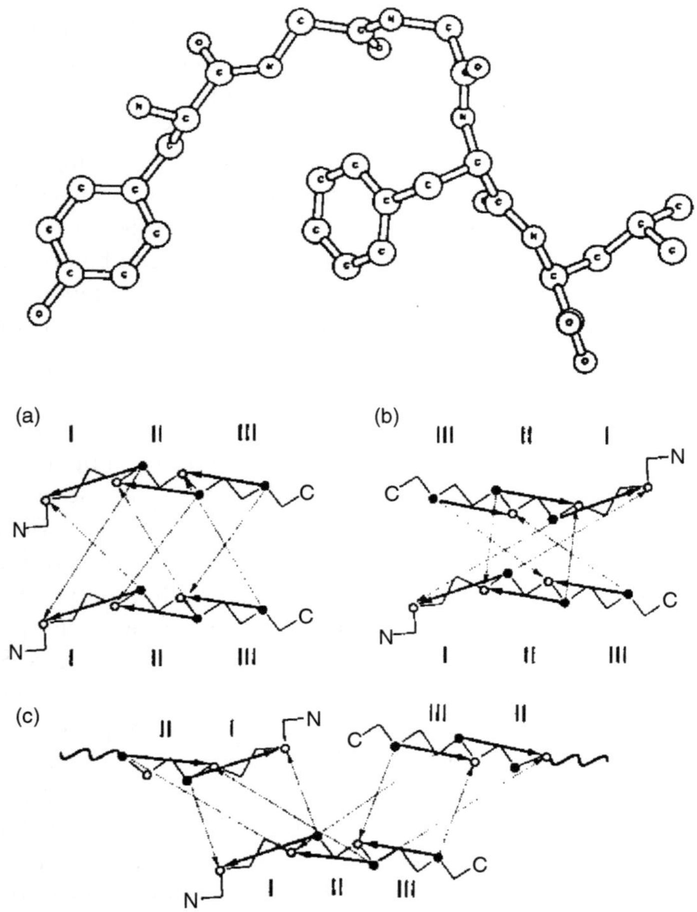

Figure 6.23. (Top) 3D structure of Leu-enkephaline dihydrate determined uniquely by the successive application of the conformation maps as well as additional constraints of the conformation-dependent ^{13}C chemical shifts. (Bottom) Three kinds of models for putative intermolecular packing as illustrated in the cartoons: (a) interaction through parallel β sheet form, (b) through an antiparallel β sheet, and (c) through interaction of three molecules. Solid and dotted arrows correspond with the dipolar interactions between intramolecular and intermolecular ^{13}C–^{15}N pairs, respectively (from Nishimura *et al.*[85]).

3D TEDOR NMR experiments have been employed for the simultaneous measurement of multiple carbon–nitrogen distances in uniformly ^{13}C, ^{15}N-labeled solids.[102] These approaches employ ^{13}C–^{15}N coherence transfer and ^{15}N and ^{13}C frequency labeling for site-specific resolution, and build on several 3D TEDOR techniques. The novel feature of the 3D TEDOR technique is specifically designed to circumvent the effects of homonuclear

^{13}C–^{13}C *J*-couplings on the measurement of weak ^{13}C–^{15}N dipolar couplings. The first experiment employs *z*-filter period to suppress the antiphase and MQ coherences and generates 2D spectra with purely absorptive peaks for all TEDOR mixing times. The second approach uses band-selective ^{13}C pulse to refocus *J*-couplings between ^{13}C spins within the selective pulse bandwidth and ^{13}C spins outside the bandwidth. The internuclear distances are extracted by using a simple analytical model, which accounts explicitly for multiple spin–spin couplings contributing to cross-peak buildup. The experiments are demonstrated in two U-^{13}C, ^{15}N-labeled peptides, *N*-acetyl-L-Val-L-Leu (N-ac-VL) and *N*-formyl-L-Met-L-Leu-L-Phe (N-f-MLF), where 20 and 26 ^{13}C–^{15}N distances up to ~5–6 Å were measured, respectively. Of the measured distances, 10 in N-ac-VL and 13 in N-f-MLF are greater than 3 Å and provide valuable structural constraints.

Selective recoupling techniques have also been applied to accurate measurements of multiple long-range ^{13}C–^{13}C and ^{13}C–^{15}N distances in the active site of light-adapted [U-^{13}C, ^{15}N] bR from PM at frozen state.[103]

6.3. Torsion Angles

Torsion angles (ϕ, ψ, and χ) are the important parameters to define the secondary structure of peptides and proteins for the sake of characterization of backbone and side-chain conformations (Figure 6.24). Of course, these torsion angles are indirectly determined by the above-mentioned distance constraints as revealed by REDOR or RR. In the former, however, it was demonstrated that accuracy in careful distance measurement with error less than $\pm$ 0.10 Å is essential to arrive at a unique set of the backbone torsion angles, by taking into account any conceivable source of errors.[85,88,104] It is not easy, however, to achieve such accuracy in distance measurement by

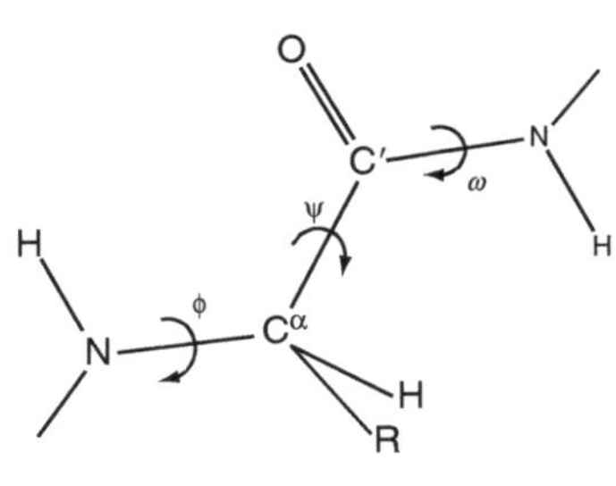

Figure 6.24. Definition of the torsion angles in peptides, polypeptides, and proteins.

RR, because it depends on parameters such as T_2^{ZQ} which is experimentally unavailable parameter. So far, several approaches have been explored to directly measure the torsion angles of interest.

6.3.1. Torsion Angles for H–C–C–H, ψ in Peptide and ψ and ψ' in Glycosidic Linkages

Double-quantum spectroscopy (DQSY), utilizing a sequence similar to INADEQUATE, was proposed for correlating the chemical shift tensors in pairs of bonded ^{13}C-labeled sites.[71] It provides relatively simple 2D spectral patterns as a result of eliminating the homonuclear dipolar coupling in one spectral dimension by suppression of the natural-abundance background signals, and by removal of the intense diagonal ridge that dominates related 2D exchange spectra. The experiment, which demonstrated on polyethylene isotopically labeled with dilute (≈4%) ^{13}C–^{13}C spin pairs, confirmed the all-*trans* structure in the crystalline region. This is a method to determine ψ angle in a doubly ^{13}C-labeled polypeptide. A method named relayed anisotropy correlation (RACO) was proposed for determining the mutual orientation of the two interaction tensors.[105] The 2D powder pattern was observed for [1,2-^{13}C] DL-alanine as a demonstration, to correlate the $^{13}C_1$ CSA and the $^{13}C_2$–^{1}H dipolar coupling via polarization transfer. The O–C_1–C_2–H dihedral angle as well as the orientation of the $^{13}C_1$ chemical shift tensor was determined from the spectrum.

The torsion angles of a doubly ^{13}C-labeled H–C–C–H moiety of ammonium hydrogen 2,2′-$^{13}C_2$-maleate and diammonium 2,2′-$^{13}C_2$-fumarate were determined based on the evolution of ^{13}C double-quantum coherence (2 QC) under the influence of magnetic field from the neighboring proton spins.[106] This method is called double-quantum heteronuclear local field (2Q-HLF) NMR and is based on the excitation of the double-quantum coherence by a sevenfold symmetric irradiation scheme called C7[107] between neighboring ^{13}C isotopic nuclear spin labels. By allowing this correlated quantum state to evolve in the presence of local fields from the bonded ^{1}H spins, it is possible to probe the correlations in these local fields and thereby the geometrical relationship of the bonded protons to the ^{13}C pair. Accordingly, the H–C–C–H torsion angles were measured with accuracies of $\approx \pm 20°$ in the neighborhood of the *cis*-conformation and $\approx \pm 10°$ in the neighborhood of the *trans*-conformation for AHM and DAF, respectively. Further, the H–C10–C11–H torsional angle for [10,11-$^{13}C_2$]*retinal*-labeled rhodopsin and polycrystalline all-*trans*-[10,11-$^{13}C_2$]retinal for calibration were determined.[108] The ^{13}C CP-MAS and 2Q-HLF spectra for the former are illustrated in Figure 6.25,

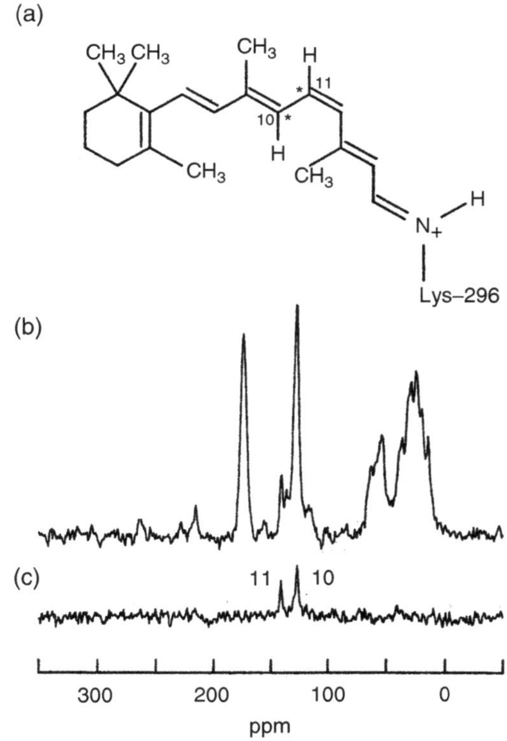

Figure 6.25. (a) Molecular structure of the retinylidene chromophore in rhodopsin. (b) ^{13}C CP-MAS NMR spectrum of membrane-bound bovine rhodopsin. (c) The signals from $^{13}C_2$-labeled retinal chromophore are selected by the double-quantum filtration, while the natural abundance background signals were suppressed (from Feng *et al.*[108]).

together with its molecular structure. The estimated HCCH torsion angle $|\phi|$ turns out to be 160° ± 10° for the former, significantly deviated from the planar 10–11-*s-trans* conformation, and 180° for the latter. In addition, H–C10–C11–H torsional angle of 10,11-$^{13}C_2$-rhodopsin was determined as 180° ± 25° for metarhodopsin-I for photointermediate after irradiation for about 20 h with focused light (250 W) from a slide projector at liquid nitrogen temperature.[109] Application of 2Q-HLF NMR was made to a sample of a $^{13}C_2$-labeled disaccharide, octa-*O*-acetyl-β,β'-thio-[1,1′-$^{13}C_2$]-trehalose,[110] in which the two ^{13}C spins are located on opposite sides of the glycosidic linkages. The evolution of the double-quantum coherences is found to be consistent with the solid-state conformation of the molecule, as determined by X-ray diffraction.

In the HCCH 2Q-HLF experiment, the spinning rate must be $\omega_r < \delta_d/3$, where δ_d is the multiple-pulse scaled ^1H–^{13}C dipolar coupling constant.[111] Therefore, these experiments are performed in the slow (3–4 kHz) spinning rate. Further, the time resolution with which data is gathered is limited by the length of the MREV pulse cycle, particularly if extension to uniformly labeled sample is desired.[111] The 3D NCCN 2Q-HLF approaches[111,112] as modified 2Q-HLF were proposed for direct means to estimate the torsion angle $\psi = (120^\circ - 180^\circ)$ of a ^{15}N–^{13}C–^{13}C–^{15}N molecular fragment in an isotopically labeled peptide. The principle of this method is similar to that of HCCH 2Q-HLF but this method requires a new experimental strategy, because the ^{15}N–^{13}C couplings are typically quite small and are readily averaged out by the MAS. In fact, $^{13}C_2$ 2QCs[111,112] created either by MELODRAMA[65] or C7[107] are allowed to evolve freely and strong rf pulses of flip angle π are applied at the ^{15}N Larmor frequency during evolution period. As in the REDOR technique, the π pulses inhibit the coherent averaging of the ^{15}N–^{13}C dipolar coupling by the magic angle rotation. This leads to dephasing of signal amplitudes of the $^{13}C_2$ 2QCs for [^{15}N, $^{13}C_2$-Gly]-[^{15}N-Gly]-Gly HCl under the influence of the heteronuclear local fields as a function of evolution interval, as shown in Figure 6.26.[112] The molecular torsional angle ψ is determined as $|\psi| = 162^\circ \pm 5^\circ$ [111,112] by comparing the experimental curves with numerical simulations, using the bond lengths, bond angles of the N–C–C–N moiety and N–C–C–N torsional angle taken from X-ray structure, and the decay time constant T_2^{2Q} (2.7 ms) of the $^{13}C_2$ double-quantum coherence from the calibration experiment.[112]

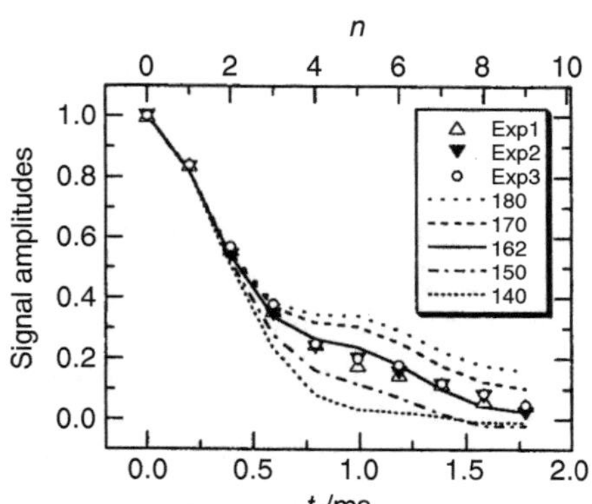

Figure 6.26. Signal amplitudes for [^{15}N, $^{13}C_2$-Gly]-[^{15}N-Gly]-Gly HCl as a function of evolution interval. Three different versions of ^{15}N π pulses were used: well-calibrated π pulses, of duration 8.0 μs (labeled Exp1); misset π pulses, of duration 7.3 μs (Exp2); composite π pulses with the central element having a duration 8.1 μs (Exp3) (from Feng *et al.*[112]).

In contrast to the situation found for β-sheet, the NCCN moiety in α-helical conformations is far from being planar, with the ψ values ranging between $-20°$ and $-70°$. Therefore, the $^{15}N^{13}C$–$^{13}C^{15}N$ experiment is outside its region of optimal sensitivity. To address this problem, a new experiment which correlates $^{1}H^{13}C_{\alpha}$ and $^{13}C'^{15}N$ dipolar interactions in the $^{1}H^{13}C_{\alpha}$–$^{13}C'^{15}N$ spin quartet was applied for *N*-formyl-[U-^{13}C,^{15}N]Met-Leu-Phe-OH (MLF).[113] An initial REDOR driven $^{13}C'$–^{15}N dipolar evolution period is followed by the C′ to C_{α} polarization transfer and by Lee–Goldburg cross polarization recoupling of the $^{13}C_{\alpha}{}^{1}H$ dipolar interaction. Simulation of resulting HCCN dephasing curves as a function of $\theta_{N-C'-C_{\alpha}-H}$ yields ψ angle by using the relation

$$\psi = \theta_{N-C'-C_{\alpha}-H} + 120°$$

The value of ψ extracted is close to that reported in diffraction studies for the methyl ester of MLF, *N*-formyl-[U-^{13}C,^{15}N]Met-Leu-Phe-OMe.

To simultaneously determine the ψ torsion angles in a uniformly ^{13}C, ^{15}N-labeled protein (α-spectrin SH3 domain), two different experiments were proposed by combination of the chemical shift correlation and NCCN dipolar correlation experiments.[114] The first NCCN experiment utilizes DQSY combined with the INADEQUATE type ^{13}C–^{13}C chemical shift correlation. The decay of the DQ coherence formed between $^{13}C'_i$ and $^{13}C_{\alpha i}$ spin pairs is determined by the "correlated" dipolar field due to $^{15}N_i$–$^{13}C_{\alpha i}$ and $^{13}C'_i-{}^{15}N_{i+1}$ dipolar couplings and is particularly sensitive to variations of the torsion angle in the regime $|\psi| > 140°$. This approach provides better signal-to-noise and is better compensated with respect to the homonuclear *J*-coupling effect. However, the ability of this experiment to constrain multiple ψ torsion angles is limited by the resolution of the $^{13}C_{\alpha}$–^{13}CO correlation spectrum. This problem is partially addressed in the second approach, which is an NCOCA NCCN experiment. In this case the resolution is enhanced by the superior spectral dispersion of the ^{15}N resonance present in the $^{15}N_{i+1}$–$^{13}C_{\alpha i}$ part of the NCOCA chemical shift correlation spectrum (see Section 5.2.3.3). For the case of the 62-residue α-spectrin SH3 domain, 13 ψ angle constraints were determined with the INADEQUATE NCCN experiment and 22 ψ constraints were measured in the NCOCA NCCN experiment.

It is possible to combine recoupling of chemical shift anisotropy (ROCSA)[115] to recouple ^{13}C CSA and $^{13}C_{\alpha}$–$^{1}H_{\alpha}$ dipolar dephasing by Lee–Golburg (LG) irradiation, to correlate the $^{13}C'$ CSA and the $^{13}C_{\alpha}$–$^{1}H_{\alpha}$ dipolar tensors.[116] The ψ angle in any single uniformly labeled residue can then be determined under fast MAS within ambiguities dictated by symmetry. $^{13}C' \rightarrow C_{\alpha}$ cross-peak intensities were measured in a series of 2D ^{13}C–^{13}C correlation spectra with variable ROCSA period and fixed LG period.

Intensities arising from the real and imaginary $^{13}C'$ transverse magnetization components after N ROCSA cycles are denoted ReN and ImN. Numerical simulation indicates that essential information about the ψ angle is contained in Re0 (or Rel), Im1, Re2, Im2, and Re3. Minimum χ^2 values for Ala 9 of the 17-mer peptide, MB(i+4)-EK, occur at $-25°$ and $-95°$, consistent with the expected $-40° \pm 15°$ for an α-helical conformation. Minimum χ^2 values for the labeled amyloid-forming peptide $A\beta_{11-25}$ labeled at Val 18, Phe 19, Phe 20, and Ala 21, occur at 150–160° and 80–90°, consistent with the expected $\psi = 140° \pm 20°$ for a β-stranded conformation.

6.3.2. ϕ Angles

By extending the HCCH 2Q-HLF approach, a method for determining the torsion angle ϕ in peptides[117] was proposed based on the measurement of the relative orientation of the N–H_N and C_α–H_α bonds, which is manifested in the rotational sideband spectrum of the sum and difference of the two corresponding dipolar couplings. The method exploits ^{15}N–^{13}C double-quantum and zero-quantum coherences, which evolve simultaneously under the N–H and C–H dipolar interactions, instead of the homonuclear ^{13}C–^{13}C double-quantum coherence. The magnitude of these dipolar couplings scaled by the proton homonuclear decoupling sequence are directly extracted from control experiments that correlate the dipolar interactions with the isotropic chemical shifts (DIPSHIFT, dipolar chemical shift). The peptide backbone torsion angle ϕ = C(O)–N–C_α–C(O) can be obtained from the relative orientation of the N–H_N and C_α–H_α dipolar tensors, whose unique axes are along the respective bonds. The corresponding angle H_N–N–C_α–H_α termed ϕ_H is directly related to ϕ according to $\phi_H = \phi - 60°$ for L-amino acid residue. Applied to ^{15}N-labeled *N*-acetyl-D,L-valine (NAV), best-fit simulation to the projected HMQ dipolar sideband spectrum of the C_α resonance yielded $\phi = -135°$, which agrees well with the X-ray crystal structure. Simulations indicate that the accuracy of the measured angle ϕ is within $\pm$ 10° when the N–H^N and C^α–H^α bonds are approximately antiparallel and $\pm$ 20° when they are roughly parallel. The effective doubling of the N–H dipolar coupling due to selective dipolar ''doubling'' by inserting rotor-synchronized 180° pulse results in greater intensities at the first- and second-order sidebands compared to the original experiment.[118] In such HNCH correlation experiments, the spectra for pairs of ϕ angles centered around $-120°$ and $+60°$ are identical. These degeneracies occur because the relative orientation of the unique axes of two dipolar tensors is characterized by only one parameter, the angle between the axes. The correlation involving the nonuniaxial ^{15}N chemical shift interaction is less prone to such degeneracies. Therefore, a correlation of ^{15}N CSA and $^{13}C_\alpha$–$^1H_\alpha$ coupling interaction yields the relative orientation of the two tensors and hence the H_N–N–C_α–H_α torsion angle with an angular resolution better than ±10°.[119] A correlation of ^{15}N chemical shift and $^{13}C_\alpha$–$^1H_\alpha$ dipolar tensor was utilized to reveal the peptide torsion angle ϕ.[120]

6.3.3. ϕ and ψ Angles

Two-dimensional NMR exchange spectroscopy with MAS permits the determination of the relative orientation of two isotopically labeled chemical groups with a molecule in an unoriented sample, thus placing strong constraints on the molecular conformation.[120] Structural information is contained in the amplitudes of cross-peaks in rotor-synchronized 2D MAS exchange spectra that connect spinning sideband lines of the two labeled sites. A new technique that enhances the sensitivity of 2D MAS exchange spectra to molecular structure, called orientationally weighted 2D MAS exchange spectroscopy, is introduced. The theory for calculating the amplitudes of spinning sideband cross-peaks in 2D MAS exchange spectra, in the limit of complete magnetization exchange between the labeled sites, is presented in detail. Experimental demonstrations of the utility of 2D MAS exchange spectroscopy in structural investigations of peptide and protein backbone are carried out on a model ^{13}C-labeled tripeptide, L-alanylglycylglycine. The dihedral angles ϕ and ψ that characterize the peptide backbone conformation at Gly–2 are obtained accurately from the orientationally weighted and unweighted 2D ^{13}C exchange spectra. Dual processing procedure of 2D MAS exchange spectra was also described by Tycko and Berger.[121]

Correlation of ^{13}C CSA tensors in double-quantum MAS experiments was proposed[122] in which the double quantum state is prepared using DRAWS[24] as DQDRAWS to determine the Ramachandran angles ϕ and ψ in a crystalline tripeptide, [1-^{13}C]Ala-[1-^{13}C]Gly-Gly and a 14 amino acid peptide Ac-LKKLLKL*L*KKLLKL, where * indicates a ^{13}C-label on the C_1 carbonyl. The experimental DQDRAWS projection data for crystalline AGG and a best-fit simulation ($\phi = -80°$, $\psi = 164°$), which is in very good agreement with the crystallographic torsion angles ($\phi = -83°$, $\psi = 169°$). In a similar manner, the experimental DQ data for the LK peptide are juxtaposed with a best-fit simulation ($\phi = -59°$, $\psi = -39°$) and several simulations that are based on common secondary structural motifs. The DQDRAWS spectrum shows a marked sensitivity to small variations in ψ.

A solid-state MAS NMR method was presented for measuring the NH_i-NH_{i+1} projection angle $\theta_{i,i+1}$ in peptides.[123] The experiment is applicable to uniformly ^{15}N-labeled peptides and is demonstrated on chemotactic tripeptide *N*-formyl-L-Met-L-Leu-L-Phe. The projection angle $\theta_{i,i+1}$ is directly related to the peptide backbone torsion angles ϕ_i and ψ_i. The method utilizes the T-MREV recoupling scheme to restore ^{15}N–^{1}H interactions, and proton-mediated spin diffusion to establish ^{15}N–^{15}N correlations. The T-MREV sequence is a MREV–8[124–126] derived, γ-encoded,[127] TC–5 type[128] multiple-pulse sequence that does not refocus the dipolar coupling after each rotor period and is shown to increase the dynamic range of the ^{15}N–^{1}H recoupling by γ-encoding, and permits an accurate determination of the recoupled NH dipolar interaction. Constraints of peptide backbone and side-chain conformation were demonstrated for 3D ^{1}H–^{15}N–^{13}C–^{1}H dipolar chemical shift, MAS NMR experiments.[129] In these experiments, polarization is transferred from ^{15}N[*i*] by ramped SPECIFIC cross polarization to the $^{13}C_\alpha[i]$, $^{13}C_\beta[i]$, and $^{13}C_\alpha[i\text{–}1]$ resonances and evolved coherently under the correlated ^{1}H–^{15}N and ^{1}H–^{13}C dipolar couplings. The resulting set of frequency-labeled ^{1}H^{15}N–^{13}C^{1}H dipolar spectra depend strongly upon the molecular torsion angles $\phi[i]$, $\chi[i]$, and $\psi[i-1]$. To interpret the data with high precision, they considered the effects of weakly coupled protons and differential relaxation of proton coherences via an average Liouvillian theory formalism for multispin clusters and employed average Hamiltonian theory to describe the transfer of ^{15}N polarization to three coupled ^{13}C spins ($^{13}C_\alpha[i]$, $C_\beta[i]$, and $^{13}C_\alpha[i-1]$). Degeneracies in the conformational solution space were minimized by combining data from multiple ^{1}H^{15}N–^{13}C^{1}H line shape and analogous data from other 3D ^{1}H–$^{13}C_\alpha$–$^{13}C_\beta$–^{1}H (χ_1), ^{15}N–$^{13}C_\alpha$–^{13}C′–^{15}N (ψ), and ^{1}H–^{15}N[*i*]–^{15}N[*i*+1]–^{1}H ($\phi+\psi$) experiments. For the uniformly ^{13}C,^{15}N-labeled solid tripeptide *N*-formyl-Met-Leu-Phe-OH, the combined data constraints a total of eight torsion angles (three ϕ, three χ_1, and two ψ).

Torsion angles of doubly ^{13}C-labeled polycrystalline sample of the tripeptide AlaGlyGly and a sextuply labeled lyophilized sample of the 17-residue peptide MB($i+4$)EK were determined by ^{13}C NMR.[130] The technique is applicable to peptides and proteins that are labeled with ^{13}C at two (or more) consecutive backbone carbonyl sites. Double-quantum (DQ) coherences are excited with a rf-driven recoupling sequence and evolve during a constant time t_1 period at the sum of the two anisotropic chemical shifts. The relative orientation of the two CSA tensors, which depends on ϕ and ψ backbone torsion angles, determines the t_1-dependence of spinning sideband intensities in the DQ-filtered ^{13}C MAS spectrum. Experiments and simulations show that both torsion angles can be extracted from a single data set. This technique, called DQCSA spectroscopy, may be especially useful when analyzing the backbone conformation of a polypeptide at a particular doubly labeled site in the presence of additional labeled carbons along the sequence.

6.4. Conformation-Dependent Chemical Shifts

6.4.1. Peptides and Proteins

Proteins and peptides are a major class of biopolymers or oligomers, which play an essential role for a wide variety of biological functions such as catalysis, transport, regulation, structural materials, etc. They consist of a repeating sequence of peptide units with 20 different kinds of substituents R at the C_α carbon (Figure 6.24). The secondary structure of these molecules is classified as right-handed α-helix (α_R-helix), left-handed α helix (α_L-helix), antiparallel and parallel β-sheets, 3_{10}-helix, etc. defined by a set of torsion angles (ϕ, ψ) in which the angles ϕ and ψ designate the extent of rotation about the N–C_α and C_α–C′ bonds, respectively. These secondary structures can be demonstrated in geometrically permitted areas for ϕ and ψ in the Ramachandran map.[131]

Chemical shift has long been treated as a ''stepchild'' in NMR structural studies of biological macromolecules.[132] Indeed, coupling constants as well as distance constraints available from NOE have been considered as the essential parameters for conformational characterization. This may be mainly because it has not been easy in solution NMR to recognize how ^{1}H, ^{13}C, or ^{15}N chemical shifts of polypeptides and proteins are specifically displaced depending upon their respective secondary structures, until solid state NMR data from a variety of polypeptides and fibrous proteins and recent burst of chemical-shift database from globular proteins in solution are available. In fact, for various linear model polypeptides frequently used such as polylysine, only an α-helical form is stable in aqueous solution, leaving the rest secondary structures, if any, under rapid conformational equilibrium including random coil, while β-sheet polypeptides are inevitably not soluble but present as precipitate. Conformation-dependent displacement of chemical shifts in solution, if any, tends to be masked by time-averaging process among various ordered and random coiled states.

6.4.1.1. *Helix-induced ^{13}C chemical shifts*

Initially, only a few examples about this subject were noticed as displaced peaks due to helix–coil transition of polypeptides in a review article of 1978 on ^{13}C NMR of peptides and proteins.[133] For instance, the ^{13}C chemical shifts of C_α, C_β, and C=O carbons of $(Lys)_n$ are appreciably displaced downfield by 1.7, 0.76, and 0.40 ppm, respectively, by going from random coil at neutral pH to α-helix at pH 10.2.[134] A possible contribution of the neutralization effect at the ε-amino acid group to the chemical shift of interest, however, should be anticipated for the observed conformation-dependent chemical shift recorded by the pH-induced helix–coil transition. To avoid this complication, it is more preferable to examine ^{13}C NMR spectral change of helix–coil transition of $(Lys)_n$ and $(Arg)_n$ induced by salts at neutral pH, which is caused by reduced electrostatic repulsion among the positively charged side chains due to partial screening of the positive charge in the presence of perchlorate or thiocyanate anion.[135] The resulting α-helix formed at neutral pH is free from contribution of such charge-induced perturbation to chemical shifts and suitable as a reference for the helical portion of protein: the downfield displacements of the C_α and C=O signals with respect to those of random coil form were observed in the α-helical state (2.8–3.1 and 2.0–2.4 ppm, respectively). In contrast, the upfield shift of the C_β signal (−0.6 ppm) is noted in the salt-induced conformational transition in contrast to the case of pH-induced transition. Indeed, the C_β ^{13}C NMR signals from several residues including Ala residue are significantly displaced upfield and can be used as convenient markers to the presence of the α-helical segments as found for histone H1 in aqueous solution and 2-chloroethanol solution.[136]

6.4.1.2. *Conformation-dependent ^{13}C chemical shifts from polypeptides in the solid state*

^{13}C chemical shifts of polypeptides, taking particular secondary structures such as α-helix, β-sheet, etc. in the solid state which are revealed by IR, X-ray diffraction, or other techniques can be conveniently used as reference data to determine local conformation at the site of respective amino acid residues of interest, because such ^{13}C chemical shifts in the solid are free from the time-averaging process due to conformational fluctuations as encountered in solution. More importantly, recording ^{13}C NMR spectra of polypeptides in the β-sheet form, which is sparingly soluble in aqueous solution is feasible only in the solid state. As an illustrative example, Figure 6.27 shows the ^{13}C CP-MAS NMR spectra of $(Ala)_n$ with various molecular weights defined by degree of polymerization (DP_n) or number-average degree of polymerization ($\overline{DP}_n$) being 16, 65, 700, and 2800 for PLA 1, 5, 50, and 200, respectively, and $Z\text{-}(Ala)_8\text{-}N(CH_2)_3CH_3$,[137] because their conformations vary with their chain lengths. The two kinds of conformations, α-helix or β-sheet form, are readily distinguished from the peak positions of the C_α, C_β, and carbonyl ^{13}C NMR signals. As summarized in Table 6.1, the ^{13}C chemical shifts of a variety of polypeptides are

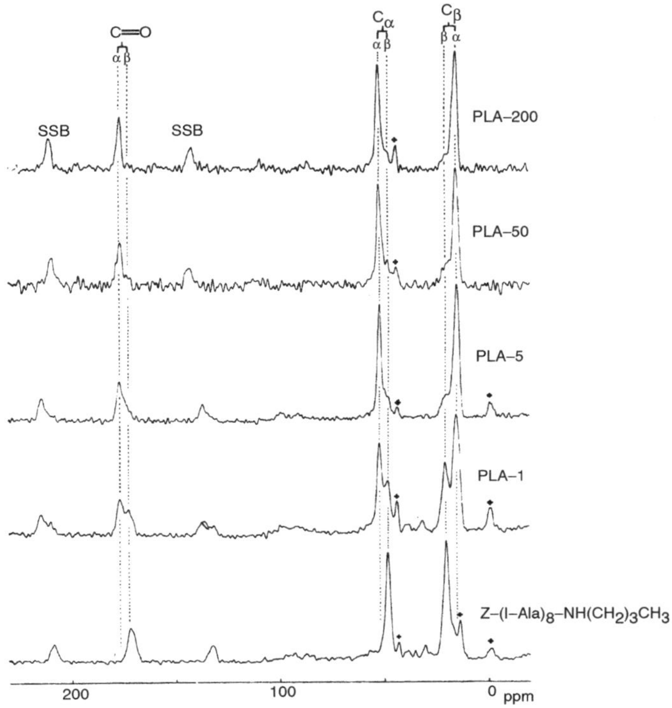

Figure 6.27. ^{13}C CP-MAS NMR spectra of $(Ala)_n$ with various degrees of polymerization. PLA-1, 16; PLA-5, 65; PLA-50, 700; PLA-200, 2800 (from Saitô *et al.*[137]).

appreciably displaced among the α-helix, β-sheet, and random coil forms.[138–143] The ^{13}C chemical shifts of random coil form in CF_3COOH or aqueous solution in the presence of a few drops of H_2SO_4 are resonated at the midpoint of the peak positions between the α-helix and β-sheet forms, reflecting the time-averaging among allowed conformations. The helix-induced shifts of $(Ala)_n$, $(Leu)_n$, $(Val)_n$, and $(Ile)_n$ with reference to those of random coil form are 0.5 to 2.8, −0.2 to −3.0, and 0.3 to 0.9 ppm for the C_α, C_β, and carbonyl carbons,[138,139] respectively, consistent with the data of the above-mentioned salt-induced α-helix for $(Lys)_n$ and $(Arg)_n$.[135]

It is noted, however, that the ^{13}C chemical shifts of the α-helix in some instances are very close to those of random coil as encountered for Ala and Leu residues (Table 6.1). This is also notable in the ^{13}C chemical shifts of membrane proteins as summarized in Table 5.1: the lowermost ^{13}C signal of Ala 84 (at 16.88 ppm) in [3-^{13}C]Ala-labeled bR located at the helix C is superimposed upon the peak position of the random coil. In a similar manner, the highermost [1-^{13}C]Val-labeled peak, Val 151, 167, and 180 at

Table 6.1. ^{13}C chemical shifts characteristic of the α-helix, β-sheet, and random coil forms (ppm from TMS).

Amino acid residues in reference polypeptides	C_α				C_β				C=O			
	α-Helix	β-Sheet	Random coil[a]	Δ[b]	α-Helix	β-Sheet	Random coil[a]	Δ	α-Helix	β-Sheet	Random coil[a]	Δ[b]
Ala	52.4	48.2	51.1	4.2	14.9	19.9	15.7	−5.0	176.4	171.8	176.1	4.6
	52.3	48.7		3.6	14.8	20.0		−5.2	176.2	171.6		4.6
	52.8	49.3		3.5	15.5	20.3		−4.8	176.8	172.2		4.6
Leu	55.7	50.5	55.2	5.2	39.5	43.3	39.7	−3.8	175.7	170.5	175.7	5.2
	55.8	51.2		4.6	43.7[c]	39.6		(4.1)	175.8	171.3		4.5
Val	65.5	58.4	61.2	7.1	28.7	32.4	31.7	−3.7	174.9	171.8	174.4	3.1
		58.2				32.4				171.5		
Ile	63.9	57.8	61.1	6.1	34.8	39.4	37.1	−4.6	174.9	172.7	175.8	2.2
		57.1				33.1				171.0		
Glu(OBzl)	56.4	51.2		5.2	25.6	29.0		−3.4	175.6	171.0		4.6
	56.8	51.1		5.7	25.9	29.7		−3.8	175.4	172.2		3.2
As(OPBzl)	53.4	49.2		4.2	33.8	38.1		−4.3	174.9	169.8		5.1
	53.6[d]				34.2[d]				174.9			
Lys[e]	57.4				29.9				176.5			
Lys (Z)	57.6	51.4		6.2	29.3	28.5		−0.8	175.7	170.4		5.3
Arg[e]	57.1				28.9				176.8			
Phe	61.3	53.2		8.1	35.0	39.3		−4.3	175.2	169.0		6.2
Met	57.2	52.2		5.0	30.2	34.8		−4.6	175.1	170.6		4.5
Gly		43.2								168.4		
		44.3								169.2		
									171.6[f]	168.5		3.1

[a]In CF_3COOH solution. A few drops of H_2SO_4 was added in the cases of $(Ile)_n$ and $(Leu)_n$.
[b]Difference in the ^{13}C chemical shifts of the α-helix form relative to those of the β-sheet form.
[c]This assignment should be reversed.
[d]Erroneously assigned from the left-handed α-helix.
[e]Data taken from the data of salt-induced α-helix in neutral aqueous solution.
[f]Average values from the data of polypeptides containing ^{13}C-labeled glycine residues.

173.97 ppm, are very close to the peak of random coil at 173.6 ppm. In such case, distinction of the α-helix form from the random coil form is readily feasible by examination of the peak-intensity recorded by the CP. It is expected that *^{13}C NMR signals from such α-helix form should be observed by CP-MAS NMR in view of its anisotropic environment in biomembranes, while ^{13}C signals from random coil portion are visible only by DD-MAS NMR* in view of its completely time-averaged dipolar contributions.

Instead, distinction of the peaks between the α-helix and β-sheet forms is more prominent in the solid state. The backbone C_α and C=O ^{13}C shifts of the α-helix (ordinary α-helix or α_I-helix) forms are significantly displaced downfield (by 3.5–8.0 ppm) with respect to those of the β-sheet forms, while the side-chain C_β peak of the α-helix form is displaced upfield (by 3.4–5.2 ppm) with respect to that of the β-sheet form. The differences of the ^{13}C chemical shifts for the C_α, C_β, and C=O carbons between the α-helix and β-sheet forms up to 8 ppm, Δ, do not vary strongly among a variety of amino acid residues (Table 6.1), although the absolute ^{13}C chemical shifts of the C_α and C_β carbons are strongly affected by the chemical structure of the individual amino acid residues. Further, it is demonstrated in Table 6.2 that the following additional five forms, left-handed α-helix (α_L-helix), α_{II}-helix, collagen-type triple helix, silk I, and random coil, can be readily distinguished with reference to the specific displacements of the ^{13}C chemical shifts of the Ala C_α, C_β, and C=O peaks.[141–144] The presence of the α_{II}-helix in $(Ala)_n$ was initially proposed by Krimm and Dwivedi[145] as the major form of the transmembrane α-helices present in bR as an alternative type of the ordinary α-helix, in which the amide plane might be tilted from the helical axis on the basis of anomalously high frequency absorption in the amide I band of IR spectra. The ^{13}C chemical shifts of the α_{II}-helix form, especially for Ala C_β signals, can be defined with reference to the ^{13}C chemical shifts of $(Ala)_n$ in hexafluoroisopropanol (HFIP) solution on the basis of IR.[146] It turned out, however, that the α_{II}-helix form should be recognized as dynamic picture, rather than the above-mentioned static picture, in membrane proteins as will be discussed in Section 6.4.1.3.

Table 6.2. Conformation-dependent ^{13}C chemical shifts of Ala residues (ppm from TMS).[144]

	α_I-Helix (α_R-helix)	α_{II}-Helix	α_L-Helix	β-Sheet	Collagen-like triple helix	Silk I	Random coil
C_α	52.4	53.2	49.1	48.2	48.3	50.5	50.1
C_β	14.9	15.8	14.9	19.9	17.6	16.6	16.9
C=O	176.4	178.4	172.9	171.8	173.1	177.1	175.2

The present database of the ^{13}C chemical shifts from a variety of polypeptides is consistent with that available from the calculated ^{13}C chemical shielding contour map, as manifested from those of the C_β carbon in *N*-acetyl-*N'*-methyl-L-alanine amide as a model for Ala residue for a variety of proteins and peptides (see Figure 3.1).[139] As to the ^{13}C chemical shifts for C_α and C=O carbons, however, hydrogen-bonding effect should be naturally taken into account to the ^{13}C chemical shift contour map. In any case, such relative displacements of the ^{13}C chemical shifts of polypeptides depending upon the secondary structures (Table 6.2) can be effectively used as a means to elucidate local conformations of the respective amino acid residues of any given proteins or peptides, since all the ^{13}C chemical shifts of amino acid residues adopting unfolded conformations in solution turn out to be independent of all neighboring residues except for the proline residue.[133] The ^{13}C chemical shifts of the Ala C_β taking a turned structure located at the loop region of bR occur at lower field than the peak of 16.88 ppm of the random coil.[143,144] Therefore, transferability of these parameters for the particular residues from the simple model system to more complicated proteins is excellent and can be applied to any types of proteins as reference, as far as the amino acid residues under consideration are virtually static (rigid) as in silk fibroin,[147,148] collagen,[149,150] and synthetic transmembrane peptides of bR in the solid.[151]

Further, Ando *et al.*[151a] demonstrated that the principal components of ^{13}C CSA tensor of polypeptides are also very sensitive to the secondary structure of amino acid residues in the solid state and can be utilized as a diagnostic tool for conformational characterization. This problem was later discussed by several workers.[151b–d] Of course, they are consistent with theoretical treatments as discussed already in Section 3.1.1.4.

6.4.1.3. *Conformation-dependent ^{13}C chemical shifts for membrane proteins*

It is noteworthy that secondary structure of *fully hydrated* membrane proteins or membrane-associated peptides under physiological condition are far from *static* in contrast to anticipation from the pictures available from crystalline studies at lower temperature. To support this view, the site-directed ^{13}C solid state NMR studies[143,144,152–154] revealed that membrane proteins in 2D crystal or embedded in lipid bilayers are very flexible at ambient temperature, undergoing various kinds of molecular motions with correlation times of the order of 10^{-2} to 10^{-8} s, depending upon the site under consideration. The assigned ^{13}C NMR peak positions of [3-^{13}C]Ala-, [1-^{13}C]Val-labeled bR as a typical membrane protein are summarized in

Table 5.2. It is noted that Val 199 and 49 ^{13}C NMR peaks of [1-^{13}C]Val-labeled bR resonated at the higher field positions than those of random coil form are ascribed to the loop region, in spite of their locations at the transmembrane α-helices,[155] because they are followed by a Pro residue and the carbonyl ^{13}C chemical shifts of amino acid residues in such circumstances are usually displaced upfield by 1.4–2.5 ppm.[156,157]

Further, it is notable that many of the ^{13}C NMR peaks from the transmembrane α-helices of [3-^{13}C]Ala-labeled proteins are resonated at the peak position of α_{II}-helix rather than α_{I}-helix form, as shown in Tables 6.1 and 6.2. The estimated proportion of the α_{II}-helix based on the carbonyl ^{13}C shift as referred to the peak positions of $(Ala)_n$ in HFIP solution, however, is much the less than that estimated from the Ala C_β signals described above.[158] This means that the ^{13}C chemical shifts of Ala C_β carbons for the α_{II} forms should not always be interpreted in terms of *static* picture as proposed on the basis of infrared spectral data[145] but dynamic picture as proposed.[158] In fact, ^{13}C NMR peaks of Ala C_β of the transmembrane peptides pointed toward lipid bilayers are much influenced by local fluctuation of the peptide unit in lipid bilayers but those of carbonyl groups are not, because they are not in direct contact with the lipids.[158] Therefore, it is more realistic to discuss the conformation and dynamics of more flexible membrane proteins or their fragments in lipid bilayers as viewed from their Ala C_β signals in terms of the *dynamics-dependent displacements of ^{13}C chemical shifts.*

The ^{13}C chemical shifts of Ala C_β taking a turned structure located at the loop region of bR occur at lower field than the peak of 16.88 ppm of the random coil.[143,144] Therefore, a view of the dynamic equilibrium should be implicitly taken into account for the interpretation of the ^{13}C chemical shifts of the loop and random coil forms. Indeed, the conformation of loop region of such membrane protein from fully hydrated species is obviously far from *static* as in a turn structure present in globular proteins, because ^{13}C NMR peaks of [1-^{13}C]Ala-labeled bR are suppressed by conformational fluctuation with 10^4 Hz[159] and the observed Ala C_β ^{13}C peak arises from time-averaged conformations. Surprisingly, the ^{13}C NMR signals of [3-^{13}C]Ala184 and Ala235 residues from the transmembrane α-helices are resonated at lower field than the peak position of random coil form (16.88 ppm), as illustrated in Table 5.2. As to the former, the torsion angles revealed by diffraction studies are deviated substantially from those of the normal α-helices depending on the kink angle at Pro186, if any, because Ala–184 is shown to be located at the hinge of the helix *F* caused by a lack of hydrogen bond between the carbonyl oxygen of Trp182 and the imido group of Pro186 (see Figure 2.8).[158] This is the reason why the ^{13}C NMR peak of Ala184 is resonated at unusually lower field region, in spite of the residue located at the transmembrane α-helices. This view is also justified because the ^{13}C signal of Ala C_β at Ala184 is displaced upfield toward the peak-positions of the normal α-helices, when this kinked structure is removed by mutation of Pro186→Ala.[158,160] The reason why the ^{13}C NMR signal of Ala235 of bR resonated at lower field is its location at the terminus of the C-terminal α-helix protruding from the membrane surface. Furthermore, the Ala84 ^{13}C signal is also accidentally superimposed at the peak-position of the random coil due to the presence of fluctuation motions of the transmembrane α-helix at the site close to the interfacial region.

6.4.1.4. *Database for the conformation-dependent chemical shifts from globular proteins*

Instead of utilizing ^{13}C chemical shifts from solid polypeptides, Spera and Bax[161] proposed an alternative approach based on database of ^{13}C chemical shifts from globular proteins: they proposed an empirical correlation of ^{13}C chemical shifts for 442 residues for which the ϕ and ψ are known with good precision by high-resolution X-ray work in which structures have been resolved at 1.0–2.2 Å. For this purpose, they excluded N- or C-terminals where no crystallographic coordinates were reported. Also excluded were regions with substantial disorder. The deviation from the random coil chemical shifts, often referred as secondary shift, of residue k is referred to as $\delta(\phi_k, \psi_k)$, where ϕ_k and ψ_k are the torsional angle of residue k. The average value $\Delta(\phi, \psi)$ of ^{13}C chemical shifts of the secondary structure as a function of the torsion angles were obtained by convolution of each of the experimental shifts, $\delta(\phi_k, \psi_k)$, with a Gaussian function,

$$\Delta(\phi,\psi) = \frac{\Sigma\delta(\phi_k,\psi_k)\exp[-((\phi-\phi_k)^2+(\psi-\psi_k)^2)/450]}{\Sigma\exp[-((\phi-\phi_k)^2+(\psi-\psi_k)^2)/450].} \tag{6.41}$$

Plots of these $\Delta(\phi, \psi)$ values for the C_α and C_β resonances are shown in Figure 6.28(a) and (b) , respectively. The $\Delta(\phi, \psi)$ surface is calculated from 442 $\delta(\phi_k, \psi_k)$ from the database used. Figure 6.28(c) shows the distribution of secondary shifts in α-helix and β-sheet. This is obviously consistent with the data from the polypeptides in the solid as summarized in Tables 6.1 and 6.2. For α-helix (119 residues) the average C_α secondary shift, $\delta_{C_\alpha} = 3.09 \pm 1.0$ ppm, and the average C_β secondary shift, $\delta_{C_\beta} = -0.38 \pm 0.85$ ppm. For β-sheet (126 residues), $\delta_{C_\alpha} = -1.48 \pm 1.23$ ppm and $\delta_{C_\beta} = 2.16 \pm 1.91$ ppm. It is notable that the chemical shift dispersion for these data collected from globular proteins in solution is surprisingly larger than that from the homopolypeptides in the solid as summarized in Table 6.1. Wishart *et al.*[162] found on the basis of chemical shift database for a variety of proteins that their ^{1}H, ^{13}C, and ^{15}N chemical shifts reveal strong correlations to protein

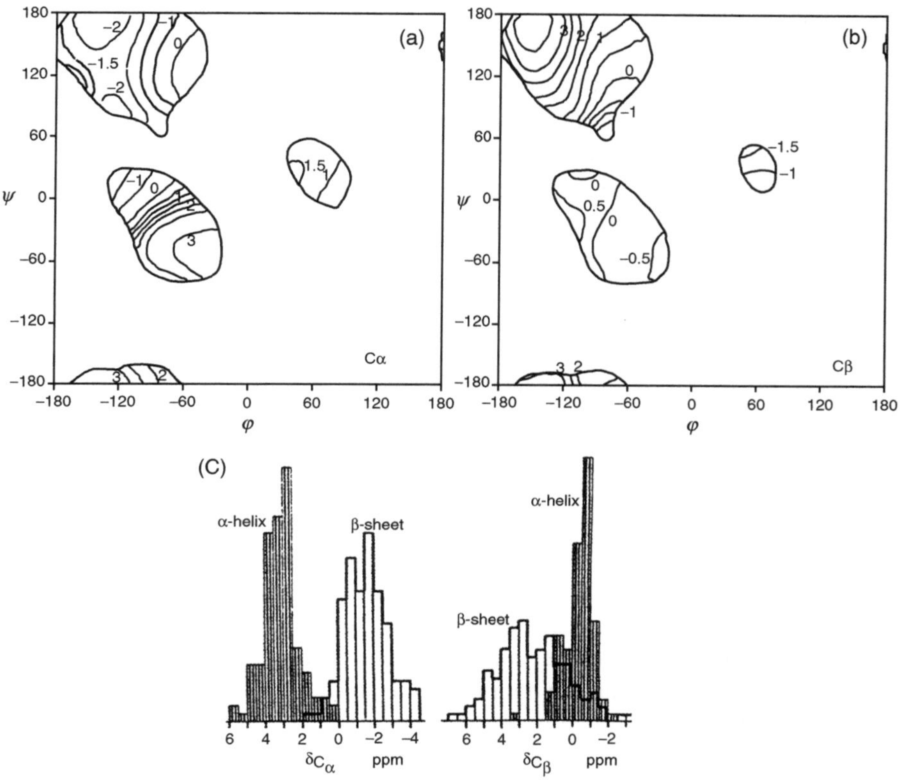

Figure 6.28. Contour plots of the average secondary shift $\Delta(\phi, \psi)$ of C_α (a) and C_β (b), resonances and histograms (c) of secondary shift distribution in α-helix and β-sheet (from Spera and Bax[161]).

secondary structure. In particular, all 20 naturally occurring amino acids experience a mean α-^{1}H upfield shift of 0.39 ppm (from the random coil value) when placed in a helical configuration. In a similar manner, the α-^{1}H chemical shifts is found to move downfield by an average of 0.37 ppm, when the residue is placed in a β-strand or extended configuration. As to the ^{13}C chemical shifts for C_α and C=O carbons, displacements of peaks for all 20 amino acid residues in the α-helix, β-sheet, and random coil environments are consistent with the data mentioned earlier. Currently, several additional sets of database are available from the data of globular proteins.[163–165]

The database from such solution NMR is, in many instances, utilized as a reference for the prediction of secondary structure of solid proteins and peptides recorded by the solid state NMR, although more preferable database suitable for this purpose are available, as demonstrated in

Tables 6.1 and 6.2. It seems to be worthwhile to comment that use of such database from solution NMR is not always sufficient as reference for interpretation of ^{13}C NMR spectra recorded by solid state NMR, because persistent errors remain due to difference of the chemical shift reference between the solid and solution NMR and also errors due to substantially higher dispersion of chemical shifts inevitable from solution NMR: First, such data are not always free from ambiguities in choosing appropriate chemical shift of the "random-coil" essential for the choice of the data collection for individual residues. Second, such chemical shifts are easily modified by averaging process of chemical shifts due to conformational fluctuation, if any, in certain residues. Third, chemical shift dispersion in the database from solution NMR might be substantially amplified when one include ^{13}C chemical shift data of certain residues of a variety of globular proteins in which local conformations are distorted to some extent, as manifested from "overlapping of peaks" between different conformations such as α-helix and β-sheet forms. Most seriously, it is well known that chemical shift references of a given molecule is not always observed at the same chemical shift position between the data of aqueous solution and solid, unless otherwise correction of bulk magnetic susceptibility is properly made for this system.[165a] Therefore, care should be taken to utilize database from solution NMR rather than the data from the solid state, because ambiguity in the chemical shift reference is present up to 1–2 ppm.

On the contrary, there is indeed no overlap (or crossing) in the ^{13}C chemical shift data available from solid state NMR between the α-helix and β-sheet and the peak-positions of the random coil form are exactly located at the border of these two types of conformers (Tables 6.1 and 6.2), although there still appear some anomalies in the data of bR (Table 5.2). Further, it is not necessary to be bothered by the problem of "random-coil" form, which is strongly dependent on the chemical shift reference. The conformation-dependent Ala ^{13}C chemical shift for β-sheet form in the solid are 48.2, and 171.8 ppm for C_α, and C=O carbons, respectively (Table 6.1). Nevertheless, these values from solution NMR are claimed to be 49.7 and 175.6 ppm![162] It is advised, therefore, to utilize the database from the solid state (Tables 6.1 and 6.2), if one aims to interpret the ^{13}C chemical shift data from solid state NMR.

6.4.1.5. ^{15}N chemical shifts

It is anticipated that ^{15}N chemical shifts are also sensitive to the secondary structure of proteins as demonstrated for ^{15}N-labeled homo- and copolypeptides in the solid state:[166–170] a small but significant downfield shift was observed for amide ^{15}N resonances in the β-sheets relative to those in the α-helices. Ashikawa *et al.*[170a] also showed that isotropic ^{15}N chemical shift or δ_{22} component are significantly displaced depending upon various conformations of poly(β-benzyl L-aspartate). A similar displacement of ^{15}N chemical shifts were also available from a variety of globular proteins in aqueous solution.[162] Shoji *et al.*[167] demonstrated that isotropic ^{15}N chemical shift depends not only on the conformation but also on the primary structure or probably on the higher order structure. Thus, the origin of ^{15}N chemical shifts is rather complex as compared with the above-mentioned ^{13}C chemical shifts. Therefore, it may be generally difficult to estimate the conformation of copolypeptides from the isotropic ^{15}N chemical shift values, although δ_{22} component of ^{15}N chemical shift tensor, which lies perpendicular to the H–N–C plane can be utilized as a good parameter to distinguish the α-helix from the β-sheet forms. In fact, ^{15}N chemical shifts are strongly influenced by side-chain bulkiness of residue $i-1$ affecting the backbone ^{15}N shift of the *i*th residue.[170] Further, it was also shown that ^{15}N chemical shifts are very sensitive to the manner of hydrogen bonding.[171–173] This means that there is a possibility that displacement of ^{15}N shift arising from hydrogen bond may surpass the effect due to conformational change. Therefore, it appears that isotropic ^{15}N chemical shifts alone are not always informative as a diagnostic tool for the conformational characterization, as compared with the above-mentioned ^{13}C chemical shifts.

6.4.2. Polysaccharides

In a similar manner, it is expected that such conformation-dependent displacement of ^{13}C chemical shifts could be readily visualized for a variety of polysaccharides, once one records their ^{13}C NMR spectra in aqueous solution. Such an attempt, however, is not always successful for a number of carbohydrates or polysaccharides in aqueous solution, because they are either insoluble in aqueous solution in the presence of hydrophobic intermolecular association or soluble as random coil as monomer without such molecular association. In gel state, however, the conformation-dependent displacements of ^{13}C chemical shifts were clearly visible: the appreciable downfield displacements of the C–1 and C–3 ^{13}C chemical shifts at the glycosidic linkages (2.8 and 3.2 ppm, respectively) are clearly seen in a gel-forming (1→3)-β-D-glucan, curdlan, as compared with those of aqueous solution of low molecular weight oligomers.[174]

The diversity of the secondary structure of polysaccharides arises generally from the variety of anomeric forms, α or β, from the variety of constituent residues such as glucose, mannose, etc. from the different possible glycosidic linkages such as 1→2, 1→3, 1→4, etc., and from sources of isolation or physical treatment. Naturally, such specific displacement of ^{13}C NMR peaks located at the carbons of the glycosidic linkages could be more clearly visualized, when one compares the ^{13}C NMR signals of various polymorphs of polysaccharides in the solid state.[175] Therefore, it is expected that secondary structures of various types of polysaccharides can be generally characterized by a set of the characteristic torsion angles (ϕ, ψ) at the glycosidic linkages. In fact, the C–1 and C–4 ^{13}C shifts of cycloamyloses are split into several peaks, reflecting differences in conformations in individual glucose residues.[176] In particular, the torsion angles of the symmetrical complexes of enclosing potassium acetate and sodium benzenesulfonate are $160° \pm 1.2°$ and $171.2° \pm 2°$ for ϕ and ψ, respectively, whereas those of the asymmetrical complexes enclosing H_2O, methanol, and 1-propanol are broadly distributed in the following two or three regions: $\phi = 160° \pm 2°$ and $169° \pm 7°$ and $\psi = -183° \pm 7°$, $-168° \pm 9°$, and $-150° \pm 5°$. The appearance of the two or three peaks is consistent with this distribution of the torsion angles. On the contrary, Ripmeester[177] showed that a better relationship is obtained by plotting the C–1 and C–4 signals with the ψ angles. In contrast, the C–1 and C–4 signals are well correlated with the ψ and ϕ angles, respectively. In addition, the C–6 signals were ascribed to either the g^-g^+ or g^+t forms as viewed from the C–6–O–6 orientation with respect to the C–4–C–5 and O–5–C–5 bonds.[178]

This kind of correlation for polysaccharides seems to be much obscured as compared with the case of polypeptides and proteins. This is because the variety of polymorphs in polysaccharides is limited and detailed molecular parameters available from fiber diffraction studies are not always sufficient. This may be the reason why it is difficult to obtain a general rule as to the conformation-dependent displacement of peak for polysaccharides as compared for a variety of model polypeptides, and globular and membrane proteins, as a convenient parameter to be able to transfer the data from model system to complex polysaccharide systems.

Nevertheless, it was demonstrated that three kinds of conformations in the (1→3)-β-D-glucan, triple-helix, single-helix, and single-chain, can be readily distinguished, on the basis of the ^{13}C chemical shifts of the C–1 and C–3 carbons, although additional peaks such as C–2 and C–4 located at the peak-positions of their neighbors are also displaced.[175] It is also noteworthy that the ^{13}C NMR line widths of the hydrated sample are substantially narrowed from the anhydrous preparation corresponding to the single-chain form. In general, broadened peaks in the solid state are ascribable to the superposition of slightly different conformations in view of the conformation-dependent displacement of peaks.

References

[1] S. J. Opella and P. L. Stewart, 1989, *Methods Enzymol.*, 176, 242–275.
[2] S. J. Opella and P. L. Stewart, 1987, *Q. Rev. Biophys.*, 19, 7–49.
[3] R. E. Stark, L. W. Jelinski, D. J. Ruben, D. A. Torchia, and R. G. Griffin, 1983, *J. Magn. Reson.*, 55, 266–273.
[4] T. G. Oas, J. Hartzell, T. J. McMahon, G. P. Drobny, and F. W. Dahlquist, 1987, *J. Am. Chem. Soc.*, 109, 5956–5962.
[5] C. J. Hartzell, M. Whitfield, T. G. Oas, and G. P. Drobny, 1987, *J. Am. Chem. Soc.*, 109, 5967–5969.
[6] Y. Wei, D.-K. Lee, and A. Ramamoorthy, 2001, *J. Am. Chem. Soc.*, 123, 6118–6126.
[7] N. Asakawa, S. Kuroki, H. Kurosu, I. Ando, A. Shoji, and T. Ozaki, 1992, *J. Am. Chem. Soc.*, 114, 3261–3265.
[8] G. S. Harbison, L. W. Jelinski, R. E. Stark, D. A. Torchia, J. Herzfeld, and R. G. Griffin, 1984, *J. Magn. Reson.*, 60, 79–82.
[9] T. G. Oas, C. J. Hartzell, F. W. Dahlquist, and G. P. Drobny, 1987, *J. Am. Chem. Soc.*, 109, 5962–5966.
[10] A. Shoji, S. Ando, S. Kuroki, I. Ando, and G. A. Webb, 1993, *Annu. Rep. NMR Spectrosc.*, 26, 55–98.
[11] A. C. Zeri, M. F. Mesleh, A. A. Nevzorov, and S. J. Opella, 2003, *Proc. Natl. Acad. Sci. USA*, 100, 6458–6463.
[12] T. A. Cross, 1994, *Annu. Rep. NMR Spectrosc.*, 29, 123–167.
[13] T. A. Cross, 1997, *Methods Enzymol.*, 289, 672–697.
[14] L. K. Nicholson, F. Moll, T. E. Mixon, P. V. LoGrasso, J. C. Lay, and T. A. Cross, 1987, *Biochemistry*, 26, 6621–6626.
[15] F. M. Marassi, A. Ramamoorthy, and S. J. Opella, 1997, *Proc. Natl. Acad. Sci. USA*, 94, 8551–8556.
[16] S. Augé, H. Mazarguil, M. Tropis, and A. Milon, 1997, *J. Magn. Reson.*, 124, 455–458.
[17] R. R. Ketchem, W. Fu, and T. A. Cross, 1993, *Science*, 261, 1457–1460.
[18] R. R. Ketchem, K.-C. Lee, S. Huo, and T. A. Cross, 1996, *J. Biomol. NMR*, 8, 1–14.
[19] R. R. Ketchem, B. Roux, and T. A. Cross, 1997, *Structure*, 5, 1655–1669.
[20] T. A. Cross, 1998, Studies in Physical and Theoretical Chemistry, in *Solid State NMR of Polymers*, vol. 84, I. Ando and T. Asakura, Eds., Elsevier, Amsterdam, pp. 218–235.
[21] C. H. Wu, A. Ramamoorthy, and S. J. Opella, 1994, *J. Magn. Reson., Ser. A*, 109, 270–272.
[21a] A. Ramamoorthy, Y. Wei, and D. K. Lee, 2004, *Annu. Rep. NMR Spectrosc.*, 52, 2–52.
[21b] A. Ramamoorthy, C.H. Wu, and S.J. Opella, 1999, *J. Magn. Reson.*, 140, 131–140.
[22] M. Lee and W. Goldburg, 1965, *Phys. Rev.*, 140, 1261–1271.

[23] A. Bielecki, A. C. Kolbert, and M. Levitt., 1989, *Chem. Phys. Lett.*, 155, 341–346.
[24] A. Ramamoorthy, C. H. Wu, and S. J. Opella, 1995, *J. Magn. Reson., Ser. B*, 107, 88–90.
[24a] P. Caravatti, L. Braunschweiler, and R. R. Ernst, 1983, *Chem. Phys. Lett.*, 100, 305–310.
[25] A. Ramamoorthy, C. H. Wu, and S. J. Opella, 1999, *J. Magn. Reson.*, 140, 131–140.
[26] F. M. Marassi and S. J. Opella, 2000, *J. Magn. Reson.*, 144, 150–155.
[27] J. Wang, J. Denny, C. Tian, S. Kim, Y. Mo, F. Kovacs, Z. Song, K. Nishimura, Z. Gan, R. Fu, J. R. Quine, and T. A. Cross, 2000, *J. Magn. Reson.*, 144, 162–167.
[28] A. A. Nevzorov and S. J. Opella, 2003, *J. Magn. Reson.*, 160, 33–39.
[29] F. M. Marassi, 2001, *Biophys. J.*, 80, 994–1003.
[30] F. M. Marassi and S. J. Opella, 2003, *Protein Sci.*, 12, 403–411.
[31] C. Glaubitz and A. Watts, 1998, *J. Magn. Reson.*, 130, 305–316.
[32] C. Glaubitz, I. J. Burnett, G. Gröbner, A. J. Mason, and A. Watts, 1999, *J. Am. Chem. Soc.*, 121, 5787–5794.
[33] C. Glaubitz, G. Gröbner, and A. Watts, 2000, *Biochim. Biophys. Acta*, 1463, 151–161.
[34] P. Ram and J. H. Prestegard, 1988, *Biochim. Biophys. Acta*, 940, 289–294.
[35] C. R. Sanders, B. J. Hare, K. P. Howard, and J. H. Prestegard, 1994, *Prog. Nucl. Magn. Reson.*, 26, 421–444.
[36] C. R. Sanders and G. C. Landis, 1995, *Biochemistry*, 34, 4030–4040.
[37] A. Naito, T. Nagao, K. Norisada, T. Mizuno, S. Tuzi, and H. Saitô, 2000, *Biophys. J.*, 78, 2405–2417.
[38] C. R. Sanders and J. P. Schwonek, 1992, *Biochemistry*, 31, 8898–8905.
[39] C. R. Sanders and J. H. Prestegard, 1990, *Biophys. J.*, 58, 447–460.
[40] R. R. Vold and R. S. Prosser, 1996, *J. Magn. Reson., Ser. B*, 113, 267–271.
[41] R. S. Prosser, S. A. Hunt, J. A. DiNatale, and R. R. Vold, 1996, *J. Am. Chem. Soc.*, 118, 269–270.
[42] R. S. Prosser, J. W. Hwang, and R. R. Vold, 1998, *Biophys. J.*, 74, 2405–2418.
[43] R. J. Kurland and B. R. McGarvey, 1970, *J. Magn. Reson.*, 2, 286–301.
[44] J. Struppe, E. A. Komives, S. S. Taylor, and R. R. Vold, 1998, *Biochemistry*, 37, 15523–15527.
[45] M. P. Warchol and W. E. Vaughan, 1978, *Adv. Mol. Relax. Interact. Processes*, 13, 317–330.
[46] F. Scholz, E. Boroske, and W. Helfrich, 1984, *Biophys. J.*, 45, 589–592.
[47] J. Seelig, F. Borle, and T. A. Cross, 1985, *Biochim. Biophys. Acta*, 814, 195–198.
[48] J. B. Speyer, P. K. Sripada, S. K. Das Gupta, G. G. Shipley, and R. G. Griffin, 1987, *Biophys. J.*, 51, 687–691.
[49] T. Brumm, C. Mops, C. Dolainsky, S. Bruckner, and T. M. Bayerl, 1992, *Biophys. J.*, 61, 1018–1024.
[50] X. Qiu, P. A. Mirau, and C. Pidgeon, 1993, *Biochim. Biophys. Acta*, 1147, 59–72.
[51] C. E. Dempsey and A. Watts, 1987, *Biochemistry*, 26, 5803–5811.
[52] C. E. Dempsey and B. Sternberg, 1991, *Biochim. Biophys. Acta*, 1061, 175–184.
[53] T. Pott and E. J. Dufourc, 1995, *Biophys. J.*, 68, 965–977.
[54] A. Naito, T. Nagao, M. Obata, Y. Shindo, M. Okamoto, S. Yokoyama, S. Tuzi, and H. Saitô, 2002, *Biochim. Biophys. Acta*, 1558, 34–44.
[55] S. Kimura, A. Naito, S. Tuzi, and H. Saitô, 2002, *Biopolymers*, 63, 122–131.
[56] T. Gullion and J. Schaefer, 1989, *Adv. Magn. Reson.*, 13, 57.

[57] E. Bennett, R. G. Griffin, and S. Vega, 1994, *NMR Basic Principles and Progress*, vol. 33, Springer-Verlag, New York, p. 1.
[58] T. Gullion and J. Schaefer, 1989, *J. Magn. Reson.*, 81, 196.
[59] D. P. Raleigh, M. H. Levitt, and R. G. Griffin, 1988, *Chem. Phys. Lett.*, 146, 71–76.
[60] K. T. Mueller, T. P. Jarvie, D. J. Aurentz, and B. W. Roberts, 1995, *Chem. Phys. Lett.*, 242, 535–542.
[61] T. P. Jarvie, G. T. Went, and K. T. Mueller, 1996, *J. Am. Chem. Soc.*, 118, 5330–5331.
[62] M. H. Levitt, D. P. Raleigh, F. Creuzet, and R. G. Griffin, 1990, *J. Chem. Phys.*, 92, 6347–6364.
[63] A. W. Hing, S. Vega, and J. Schaefer, 1992, *J. Magn. Reson.*, 96, 205.
[64] R. Tycko and G. Dabbagh, 1990, *Chem. Phys. Lett.*, 173, 461–465.
[65] B.-Q. Sun, P. R. Costa, D. Kocisko, P. T. Lansbury, Jr., and R. G. Griffin, 1995, *J. Chem. Phys.*, 102, 702–707.
[66] T. Gullion and S. Vega, 1992, *Chem. Phys. Lett.*, 194, 423–428.
[67] A. E. Bennett, R. G. Griffin, J. H. Ok, and S. Vega, 1992, *J. Chem. Phys.*, 96, 8624–8629.
[68] D. M. Gregory, D. J. Mitchell, J. A. Stringer, S. Kiihne, J. C. Shiels, J. Callahan, M. A. Mehta, and G. P. Drobny, 1995, *Chem. Phys. Lett.*, 246, 654–663.
[69] G. J. Boender, J. Raap, S. Prytulla, H. Oschkinat, and H. J. M. de Groot, 1995, *Chem. Phys. Lett.*, 237, 502–508.
[70] T. Fujiwara, K. Sugase, M. Kainosho, A. Ono, A. Ono, H. Akutsu, 1995, *J. Am. Chem. Soc.*, 117, 11351–11352.
[71] K. Schmidt-Rohr, *Macromolecules*, 1996, 29, 3975–3981.
[72] M. Baldus, R. J. Iuliucci, and B. M. Meier, 1997, *J. Am. Chem. Soc.*, 119, 1121–1124.
[73] D. P. Weliky and R. Tycko, 1996, *J. Am. Chem. Soc.*, 118, 8487–8488.
[74] A. Naito, K. Nishimura, S. Kimura, S. Tuzi, M. Aida, N. Yasuoka, and H. Saitô, 1996, *J. Phys. Chem.*, 100, 14995–15004.
[75] A. Naito, K. Nishimura, S. Tuzi, and H. Saitô, 1994, *Chem. Phys. Lett.*, 229, 506–511.
[76] C. P. Jaroniec, B. T. Tounge, C. M. Rienstra, J. Herzfeld, and R. G. Griffin, 2000, *J. Magn. Reson.*, 146, 132–139.
[76a] R. Fu, S. A. Smith, and G. Bodenhausen, 1997, *Chem. Phys. Lett.*, 272, 361–369.
[76b] K. Nishimura, R. Fu, and T. A. Cross, 2001, *J. Magn. Reson.*, 152, 227–233.
[77] T. Gullion and J. Schaefer, 1991, *J. Magn. Reson.*, 92, 439–442.
[78] Y. Pan, T. Gullion, and J. Schaefer, 1990, *J. Magn. Reson.*, 90, 330–340.
[79] J. R. Garbow and C. A. McWherter, 1993, *J. Am. Chem. Soc.*, 115, 238–244.
[80] D. Suwelack, W. P. Rothwell, and J. S. Waugh, 1980, *J. Chem. Phys.*, 73, 2559–2569.
[81] W. P. Rothwell and J. S. Waugh, 1981, *J. Chem. Phys.*, 74, 2721–2732.
[82] A. Naito, A. Fukutani, M. Uitdehaag, S. Tuzi, and H. Saitô, 1998, *J. Mol. Struct.*, 441, 231–241.
[83] M. Kamihira, A. Naito, K. Nishimura, S. Tuzi, and H. Saitô, 1998, *J. Phys. Chem. B*, 102, 2826–2834.
[84] O. B. Peersen, M. Groesbeek, S. Aimoto, and S. O. Smith, 1995, *J. Am. Chem. Soc.*, 117, 7228–7237.
[85] K. Nishimura, A. Naito, S. Tuzi, H. Saitô, C. Hashimoto, and M. Aida, 1998, *J. Phys. Chem. B*, 102, 7476–7483.

[86] H. Saitô, S. Tuzi, and A. Naito, 1998, *Annu. Rep. NMR Spectrosc.*, 36, 79–121.
[87] C. A. Michal and L. W. Jelinski, 1997, *J. Am. Chem. Soc.*, 119, 9059–9060.
[88] A. Naito and H. Saito, 2002, *Encl. Nucl. Magn. Reson.*, 9, 283–291.
[89] C. P. Jaroniec, B. A. Tounge, C. M. Rienstra, J. Herzfeld, and R. G. Griffin, 1999, *J. Am. Chem. Soc.*, 121, 10237–10238.
[90] J. Schaefer, 1999, *J. Magn. Reson.*, 137, 272–275.
[91] K. Nishimura, A. Naito, S. Tuzi, and H. Saitô, 1999, *J. Phys. Chem. B*, 103, 8398–8404.
[92] L. M. McDowell, C. A. Klug, D. D. Beusen, and J. Shaefer, 1996, *Biochemistry*, 35, 5395–5403.
[93] T. Gullion and C. H. Pennington, 1998, *Chem. Phys. Lett.*, 290, 88–93.
[94] C. P. Jaroniec, B. A. Tounge, J. Herzfeld, and R. G. Griffin, 2001, *J. Am. Chem. Soc.*, 123, 3507–3519.
[95] K. Nishimura, K. Ebisawa, E. Suzuki, H. Saitô, and A. Naito, 2001, *J. Mol. Struct.*, 560, 29–38.
[96] A. Kubo and C. A. McDowell, 1988, **J.** *Chem. Soc., Faraday Trans. 1*, 84, 3713–3730.
[97] K. Takegoshi, K. Nomura, and T. Terao, 1995, *Chem. Phys. Lett.*, 232, 424–428.
[98] P. T. F. Williamson, A. Verhoeven, M. Ernst, and B. H. Meier, 2003, *J. Am. Chem. Soc.*, 125, 2718–2722.
[99] A. Verhoeven, P. T. F. Williamson, H. Zimmermann, M. Ernst, and B. H. Meier, 2004, *J. Magn. Reson.*, 168, 314–326.
[100] S. Kimura, A. Naito, H. Saitô, K. Ogawa, and A. Shoji, 2001, *J. Mol. Struct.*, 562, 197–203.
[101] S. Kimura, A. Naito, S. Tuzi, and H. Saitô, 2002, *J. Mol. Struct.*, 602–603, 125–131.
[102] C. P. Jaroniec, C. Filip, and R. G. Griffin, 2002, *J. Am. Chem. Soc.*, 124, 10728–10742.
[103] C. P. Jaroniec, J. C. Lansing, B. A. Tounge, M. Belenky, and R. G. Griffin, 2001, *J. Am. Chem. Soc.*, 123, 12929–12930.
[104] A. Naito, S. Tuzi, and H. Saitô, 1998, *Solid State NMR of Polymers*, Chapter 2, I. Ando and T. Asakura, Eds., Elsevier, Amsterdam.
[105] Y. Ishii, T. Terao, and M. Kainosho, 1996, *Chem. Phys. Lett.*, 256, 133–140.
[106] X. Feng, Y. K. Lee, D. Sandström, M. Eden, H. Maisel, A. Sebald, and M. H. Levitt, 1996, *Chem. Phys. Lett.*, 257, 314–320.
[107] Y. K. Lee, N. D. Kurur, M. Helmle, O. G. Johannessen, N. C. Nielsen, and M. H. Levit, 1995, *Chem. Phys. Lett.*, 242, 304–309.
[108] X. Feng, P. J. E. Verdegem, Y. K. Lee, D. Sandström, M. Eden, P. Bovee-Geurts, W. J. de Grip, J. Lugtenburg, H. J. M. de Groot, and M. H. Levitt, 1997, *J. Am. Chem. Soc.*, 119, 6853–6857.
[109] X. Feng, P. J. E. Verdegem, M. Eden, D. Sandström, Y. K. Lee, P. H. M. Bovee-Geurts, W. J. de Grip, J. Lugtenburg, H. J. M. de Groot, and M. H. Levitt, 2000, *J. Biomol. NMR*, 16, 1–8.
[110] S. Ravindranathan, T. Karlsson, K. Lycknert, G. Widmalm, and M. H. Levitt, 2001, *J. Magn. Reson.*, 151, 136–141.
[111] P. R. Costa, J. D. Gross, M. Hong, and R. G. Griffin, 1997, *Chem. Phys. Lett.*, 280, 95–103.
[112] X. Feng, M. Edén, A. Brinkmann, H. Luthman, L. Eriksson, A. Gräslund, O. N. Antzutkin, and M. H. Levitt, 1997, *J. Am. Chem. Soc.*, 119, 12006–12007.
[113] V. Ladizhansky, M. Veshtort, and R. G. Griffin, 2002, *J. Magn. Reson.*, 154, 317–324.

[114] V. Ladizhansky, C. P. Jaroniec, A. Diehl, H. Oschkinat, and R. G. Griffin, 2003, *J. Am. Chem. Soc.*, 125, 6827–6833.

[115] J. C. Chan and R. Tycko, 2003, *J. Chem. Phys.*, 118, 8378–8389.

[116] J. C. Chan and R. Tycko, 2003, *J. Am. Chem. Soc.*, 125, 11828–11829.

[117] M. Hong, J. D. Gross, and R. G. Griffin, 1997, *J. Phys. Chem. B*, 101, 5869–5874.

[118] M. Hong, J. D. Gross, C. M. Rienstra, R. G. Griffin, K. K. Kumashiro, and K. Schmidt-Rohr, 1997, *J. Magn. Reson.*, 129, 85–92.

[119] R. Tycko, D. P. Weliky, and A. E. Berger, 1996, *J. Chem. Phys.*, 105, 7915–7930.

[120] M. Hong, J. D. Gross, W. Hu, and R. G. Griffin, 1998, *J. Magn. Reson.*, 135, 169–177.

[121] R. Tycko and A. Berger, 1999, *J. Magn. Reson.*, 141, 141–147.

[122] P. V. Bower, N. Oyler, M. A. Mehta, J. R. Long, P. S. Stayton, and G. P. Drobny, 1999, *J. Am. Chem. Soc.*, 121, 8373–8375.

[123] B. Reif, M. Hohwy, C. P. Jaroniec, C. M. Rienstra, and R. G. Griffin, 2000, *J. Magn. Reson.*, 145, 132–141.

[124] P. Mansfield, M. J. Orchard, D. C. Stalker, and K. H. B. Richards, 1973, *Phys. Rev.*, B7, 90–105.

[125] W. K. Rhim, D. D. Elleman, and R. W. Vaughan, 1973, *J. Chem. Phys.*, 59, 3740–3749.

[126] W. K. Rhim, D. D. Elleman, L. B. Schreiber, and R. W. Vaughan, 1974, *J. Chem. Phys.*, 60, 4595–4604.

[127] N. C. Nielsen, H. Bildsøe, H. J. Jacobsen, and M. H. Levitt, 1994, *J. Chem. Phys.*, 101, 1805–1812.

[128] J. D Gross, P. R. Costa, and R. G. Griffin, 1988, *J. Chem. Phys.*, 108, 7286–7293.

[129] C. M. Rienstra, M. Hohwy, L. J. Mueller, C. P. Jaroniec, B. Reif, and R. G. Griffin, 2002, *J. Am. Chem. Soc.*, 124, 11908–11922.

[130] F. J. Blanco and R. Tycko, 2001, *J. Magn. Reson.*, 149, 131–138.

[131] G. N. Ramachandran and V. Sasisekharan, 1968, *Adv. Protein Chem.*, 23, 283–337.

[132] L. Szilágyi, 1995, *Prog. Nucl. Magn. Reson. Spectrosc.*, 27, 325–443.

[133] W. Howarth and D. M. J. Liley, 1978, *Prog. Nucl. Magn. Reson. Spectrosc.*, 12, 1–78.

[134] H. Saitô and I. C. P. Smith, 1973, *Arch. Biochem. Biophys.*, 158, 154–163.

[135] H. Saitô, T. Ohki, M. Kodama, and C. Nagata, 1978, *Biopolymers*, 17, 2587–2599.

[136] H. Saitô, M. Kameyama, M. Kodama, and C. Nagata, 1982, *J. Biochem. (Tokyo)*, 233–241.

[137] H. Saitô, R. Tabeta, A. Shoji, T. Ozaki, and I. Ando, 1983, *Macromolecules*, 16, 1050–1057.

[138] T. Taki, S. Yamashita, M. Satoh, A. Shibata, T. Yamashita, R. Tabeta, and H. Saitô, 1981, *Chem. Lett.*, 1803–1806.

[139] I. Ando, H. Saitô, R, Tabeta, A. Shoji, and T. Ozaki, 1984, *Macromolecules*, 17, 457–461.

[140] H. Saitô, 1986, *Magn. Reson. Chem.*, 24, 835–852.

[141] H. Saitô and I. Ando, 1989, *Annu. Rep. NMR Spectrosc.*, 21, 209–290.

[142] H. Saitô, S. Tuzi, and I. Ando, 1998, *Annu. Rep. NMR Spectrosc.*, 36, 79–121.

[143] H. Saitô, S. Tuzi, M. Tanio, and A. Naito, 2002, *Annu. Rep. NMR Spectrosc.*, 47, 39–108.

[144] H. Saitô, S. Tuzi, S. Yamaguchi, M. Tanio, and A. Naito, 2000, *Biochim. Biophys. Acta*, 1460, 39–48.

[145] S. Krimm and A. M. Dwivedi, 1982, *Science*, 216, 407–408.
[146] J. R. Parrish, Jr. and E. R. Blout, 1972, *Biopolymers*, 11, 1001–1020.
[147] H. Saitô, M. Ishida, M. Yokoi, and T. Asakura, 1990, *Macromolecules*, 23, 83–88.
[148] M. Ishida, T. Asakura, M. Yokoi, and H. Saitô, 1990, *Macromolecules*, 23, 88–94.
[149] H. Saitô, R. Tabeta, A. Shoji, T. Ozaki, I. Ando, and T. Miyata, 1984, *Biopolymers*, 23, 2279–2297.
[150] H. Saitô and M. Yokoi, 1992, *J. Biochem. (Tokyo)*, 111, 376–382.
[151] S. Kimura, A. Naito, S. Tuzi, and H. Saitô, 2001, *Biopolymers*, 58, 78–88.
[151a] S. Ando, T. Yamanobe, I. Ando, A. Shoji, T. Ozaki, R. Tabeta, and H. Saitô, 1985, *J. Am. Chem. Soc.*, 107, 7648–7652.
[151b] Y. Wei, D.-K. Lee, and A. Ramamoorthy, 2001, *J. Am. Chem. Soc.*, 123, 6118–6162.
[151c] R. H. Havlin, D. D. Laws, H.-M. L. Bitter, L. K. Sanders, H. Sun, J. S. Grimley, D. E. Wemmer, A. Pines, and E. Oldfield, *J. Am. Chem. Soc.*, 123, 10362–10369.
[151d] M. Hong, 2000, *J. Am. Chem. Soc.*, 122, 3762–3770.
[152] T. Arakawa, K. Shimono, S. Yamaguchi, S. Tuzi, Y. Sudo, N. Kamo, and H. Saitô, 2003, *FEBS Lett.*, 536, 237–240.
[153] S. Yamaguchi, S. Tuzi, J. U. Bowie, and H. Saitô, 2004, *Biochim. Biophys. Acta*, 1698, 97–105.
[154] S. Yamaguchi, K. Shimono, Y. Sudo, S. Tuzi, A. Naito, N. Kamo, and H. Saitô, 2004, *Biophys. J.*, 86, 3131–3140.
[155] M. Tanio, S. Inoue, K. Yokota, T. Seki, S. Tuzi, R. Needleman, J. K. Lanyi, A. Naito, and H. Saitô, 1999, *Biophys. J.*, 77, 431–442.
[156] D. A. Torchia and J. R. Lyerla, 1974, *Biopolymers*, 13, 97–114.
[157] D. S. Wishart, C. G. Bigam, A. Holm, R. S. Hodges, and B. D. Sykes, 1995, *J. Biomol. NMR*, 5, 67–81.
[158] S. Tuzi, J. Hasegawa, R. Kawaminami, A. Naito, and H. Saitô, 2001, *Biophys. J.*, 81, 425–434.
[159] S. Yamaguchi, S. Tuzi, K. Yonebayashi, A. Naito, R. Needleman, J. K. Lanyi, and H. Saitô, 2001, *J. Biochem. (Tokyo)*, 129, 373–382.
[160] H. Saitô, S. Yamaguchi, H. Okuda, A. Shiraishi, and S. Tuzi, 2004, *Solid State Nucl. Magn. Reson.*, 25, 5–14.
[161] S. Spera and A. Bax, 1991, *J. Am. Chem. Soc.*, 113, 5490–5492.
[162] D. S. Wishart, B. D. Sykes, and F. M. Richards, 1991, *J. Mol. Biol.*, 222, 311–333.
[163] M. Iwadate, T. Asakura, and M. P. Williamson, 1999, *J. Biomol. NMR*, 13, 199–211.
[164] X.-P. Xu and D. A. Case, 2001, *J. Biomol. NMR*, 21, 321–333.
[165] D. S. Wishart, M. S. Watson, R. F. Boyko, and Brian D. Sykes, 1997, *J. Biomol. NMR*, 10, 329–336.
[165a] J. A. Pople, W. G. Schneider, and H. J. Berstein, 1959, *High-resolution Nuclear Magnetic Resonance*, McGraw-Hill, New York.
[166] H. G. Förster, D. Müller, and H. R. Kricheldorf, 1983, *Int. J. Biol. Macromol.*, 5, 101–105.
[167] A. Shoji, T. Ozaki, T. Fujito, K. Deguchi, and I. Ando, 1987, *Macromolecules*, 20, 2441–2445.

[168] A. Shoji, T. Ozaki, T. Fujito, K. Deguchi, S. Ando, and I. Ando, 1989, *Macromolecules*, 22, 2860–2863.

[169] A. Shoji, T. Ozaki, T. Fujito, K. Deguchi, S. Ando, and I. Ando, 1990, *J. Am. Chem. Soc.*, 112, 4693–4697.

[170] A. Shoji, S. Ando, S. Kuroki, I. Ando, and G. A. Webb, 1993, *Annu. Rep. NMR Spectrosc.*, 26, 55–98.

[170a] M. Ashikawa, A. Shoji, T. Ozaki, and I. Ando, 1999, *Macromolecules*, 32, 2288–2292.

[171] H. Saitô, K. Nukada, H. Kato, T. Yonezawa, and K. Fukui, 1965, *Tetrahedron Lett.*, 111–117.

[172] H. Saitô and K. Nukada, 1971, *J. Am. Chem. Soc.*, 93, 1072–1076,

[173] H. Saitô,Y. Tanaka, and K. Nukada, 1971, *J. Am. Chem. Soc.*, 93, 1077–1081.

[174] H. Saitô, T. Ohki, and T. Sasaki, 1977, *Biochemistry*, 16, 908–914.

[175] H. Saitô, M. Yokoi, and Y. Yoshioka, 1989, *Macromolecules*, 22, 3892–3898.

[176] H. Saitô, G. Izumi, T. Mamizuka, S. Suzuki, and R. Tabeta, 1982, *J. Chem. Soc., Chem. Commun.*, 1386–1388.

[177] J. A. Ripmeester, 1986, *J. Inclusion Phenom.*, 4, 129.

[178] R. P. Veregin, C. A. Fyfe, R. H. Marchessault, and M. G. Taylor, 1987, *Carbohydr. Res.*, 160, 41.

Chapter 7

DYNAMICS

It has been well recognized that flexibility is also essential for biopolymers such as proteins, peptides, nucleic acids, biomembranes, etc. in relation to their respective biological functions such as enzyme catalysis, recognition, signal transduction, transport, ligand binding, barrier of cells as well as environment for membrane proteins, etc. Nevertheless, the intrinsic beauty and remarkable detail of the drawings of protein structures so far revealed by X-ray crystallography would tempt one to convince that each protein atom is fixed in place owing to proposed rigid pictures. It is also true that such static view of protein structures is being replaced by a dynamic picture in which atoms are recognized to be in a state of constant motion at ordinary temperature for their respective biological activities. This is because the functions of molecules such as enzymes occur in real time, which may vary from 10^{-13} to 10^{-2} s depending upon types of motions, through fluctuations in structure.[1] Highly refined diffraction studies showed that small amplitude of motions can be revealed through temperature factors[2] as well as molecular dynamic (MD) simulation,[3] although the timescale of the latter approach is limited to up to a few hundred picoseconds (ps). Instead, such wide variety of motions can be very conveniently examined in the solid by means of a variety of solid state NMR techniques, together with their characteristic timescales most sensitively detected, as summarized in Figure 7.1 based on the arguments discussed so far.

In aqueous solution, the relaxation parameters available from NMR such as T_1, T_2, and NOE data are very sensitive to fast picosecond motions. By contrast, it has been recently pointed out that the presence of slow motions of backbone and side chains with timescale of millisecond and microsecond is more biologically relevant for a variety of globular proteins in relation to their specific activity including transient formation of ligand-binding competent states and transitions coupled to enzyme catalysis mentioned above than the fast motions detectable by relaxation parameters.[4–6] For this purpose, excess transverse relaxation rate (R_{ex}) caused by the exchange of

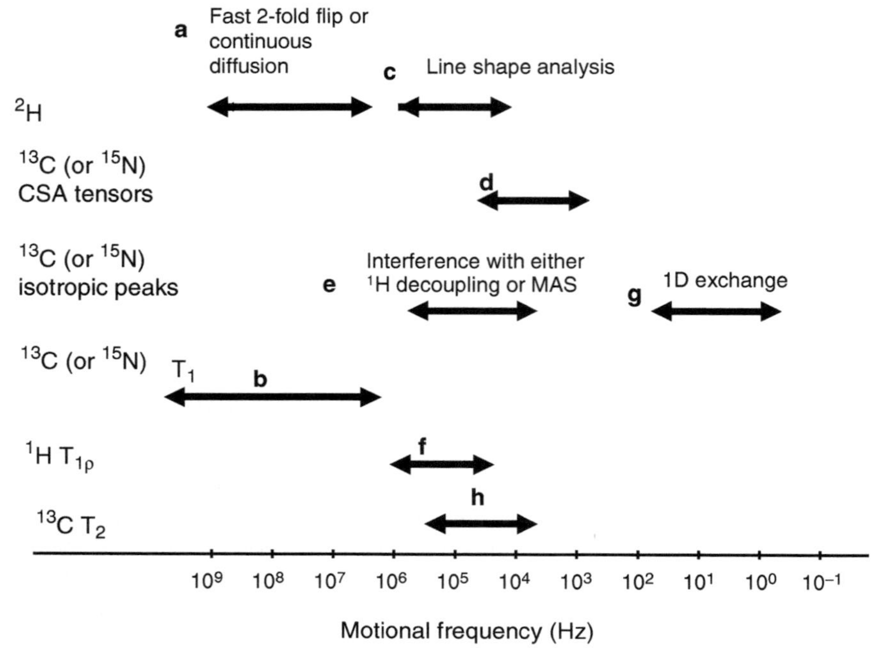

Figure 7.1. Timescale of motions detected by solid state NMR.

nuclei between different conformations/states with different chemical shifts can be nicely measured by applying a train of 180° rf pulses separated by a delays of length τ_{cp}, as in the CPMG experiment. This is because detection of chemical exchange process through the relaxation time is very sensitive to the presence of such slow motions.

Such slow motions are also very important for a variety of membrane proteins, as manifested from bR, a typical membrane protein, which is active as the simplest known light energy converting system, and undergoes global conformational changes during its photocycle: bR, K, L, M, N, and O are subsequently formed before recovery of the initial state, with lifetime of μs to ms, except for the very rapid conversion from bR to K intermediate with lifetime of ps.[7–9] Proton uptake from the cytoplasmic side occurs after N formation, while proton release to the extracellular space is an event at the time of M formation. Consistent with this view, it is also interesting to note that the activity of bR from PM is well related to the onset of low-frequency, large-amplitude anharmonic motions possible only under hydrated condition, as demonstrated by examination of hydrogen mean square displacement ($\langle u^2 \rangle$) determined by inelastic neutron scattering in dry PM (o) and in hydrated PM (■), as a function of temperature (Figure 7.2).[10] There is a

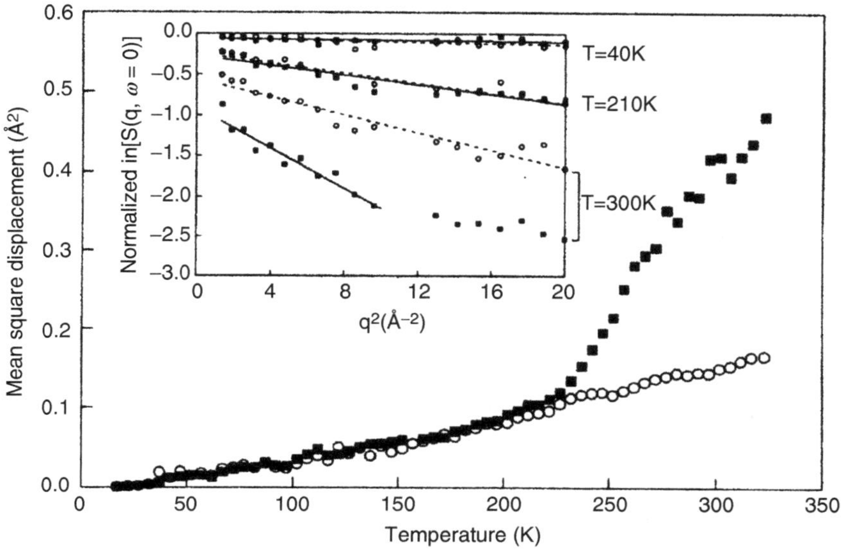

Figure 7.2. Hydrogen mean square displacements ($\langle u^2 \rangle$) in dry PM (o) and in hydrated (■) states as a function of temperature. Water contents were 0.02 and 0.55 g of ^2H_2O per g of bR (from Ferrand *et al.*[10]).

distinct difference between the wet and dry samples, beyond a dynamical transition at about 220–240 K, and the onset of the anharmonic, large-amplitude motions is possible in the latter. It appears that the "softness" of the membrane modulates the function of bR by allowing or not allowing the large-amplitude motions in the proteins.[10] Therefore, caution should be exercised to the extent of hydrated state of samples, which strongly influences the resulting protein dynamics when these preparations are examined by solid state NMR spectroscopy.

Several kinds of motions with different motional frequencies present in the solid, lipid bilayer, or fully hydrated membrane proteins can be readily distinguished by means of a variety of solid state NMR techniques as summarized in Figure 7.1: (1) fast motions with motional frequency > 10^6 Hz, (2) intermediate or slow motions with frequencies between 10^6 and 10^3 Hz, and (3) very slow motions with frequency < 10^3 Hz. The NMR techniques for this purpose are classified as: (1) ^{2}H NMR powder pattern with fast twofold flip–flop motions or continuous diffusion **a**, or ^{2}H, ^{13}C, or ^{15}N spin–lattice relaxation time T_1 in the laboratory frame **b**; (2) ^{2}H NMR line shape analysis of powder pattern **c**; ^{13}C or ^{15}N chemical anisotropy powder pattern **d**; motionally broadened or suppressed, isotropic ^{13}C NMR peaks due to interference of fluctuation frequency with proton decoupling or magic angle spinning frequencies **e**; ^{13}C or ^{1}H spin–lattice relaxation times in the rotating frame, **g**; or spin–spin relaxation times under proton decoupling, **h**; and (3) slow motion detected by **f**; ^{13}C NMR line shape analysis of isotropic peaks or 1D exchange.

7.1. Fast Motions with Motional Frequency>10^6 Hz

The presence of either fast continuous diffusion or twofold flip–flop motions of phenyl rings about the C_γ–C_ζ axis can be readily identified when ^2H NMR powder pattern spectra of ^2H-labeled Phe or Tyr residues are examined for a variety of proteins with reference to Figure 4.13(c) and (d). For instance, dynamic aspects of aromatic side chains such as Trp, Phe, or Tyr were examined by this approach. It turned out that ^2H NMR spectra of nearly all (9±2) of the 11 Tyr and (13±2) of the 13 Phe in bR gave rise to the spectral pattern characteristic of the static side chain at lower temperatures (−25 and −90°C), rapid twofold jump motions about the C_γ–C_ζ axis at ambient temperatures (37°C and 54°C) and isotropic signal at 95°C (Figure 7.3 for [$^2H_2$]Tyr-labeled bR).[12] Upon heating the sample at temperatures between ≈ 86°C and 92°C, the protein apparently unfolds, and a narrow isotropic peak is obtained. This is the consequence of the backbone motions arising from unfolding as well as rotation about the C_α–C_β axis. In contrast, there is no evidence for fast motion about C_α–C_β axis as manifested from ^2H NMR spectra of [γ-$^2H_6$]Val-labeled bR at any temperature.[11] In the case of Val-labeled bR, motion about C_β–C_γ is fast (>10^6 Hz) at all temperatures studied (down to 120 K) with the quadrupole splittings as large as 33–39 kHz with reference to Figure 4.16.

As illustrated in Figure 7.3, the ^2H NMR powder pattern of ^2H-labeled bR is changed to the isotropic peak, arising from the completely time-averaged quadrupole interaction for unfolded protein at such higher temperatures, resonated at the central peak at elevated temperature (95°C). The presence of such an isotropic ^2H NMR signal in native bR at ambient temperature, if any, could be utilized as a very convenient means to locate segments undergoing isotropic conformational fluctuation. Oldfield and coworkers recorded the intense isotropic signals at the central peak positions at ambient temperature besides the widely spread ordinary powder pattern spectra for a variety of ^2H NMR spectra of [α-$^2H_2$]Gly-, [γ-$^2H_6$]Val-, [δ-$^2H_3$]Leu-, [β-$^2H_2$]Ser-, [β,γ-$^2H_4$]Thr-, [α,β,γ,δ,ε-$^2H_9$]Lys-, [δ,δ,ε,ε,ζ-$^2H_5$]Phe-, [ε-$^2H_2$]Tyr-labeled bR, although

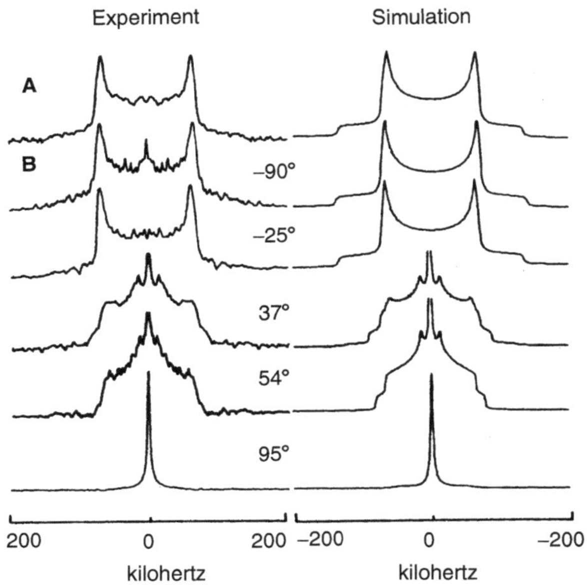

Figure 7.3. ^{2}H NMR spectra of [$^{2}H_2$]tyrosine (A) and [$^{2}H_2$]Tyr-bR from purple membrane at various temperatures (B) (from Kinsey *et al.*[12]).

no isotropic signal was visible from [$\delta_1,\varepsilon_3,\zeta_2,\zeta_3,\eta_2$-$^{2}H_2$]Trp-bR.[13] Consequently, they proposed a rather flexible picture of bR at surface residues: all amino acids outside the surface (surface residues) of bR are highly mobile on the timescale of ^{2}H NMR experiment, while all residues inside the surface are essentially crystalline. Their interpretation of the ''isotropic'' signal is not always convincing in view of the secondary structure as revealed by cryo-electron microscope or X-ray diffraction, however,[14–17] because the claimed isotropic averaging for residues located at the loop region is physically impossible that will lead to complete removal of the quadrupole interactions in view of their short ''lengths'' as the loops. This view is consistent with ^{13}C CP-MAS NMR studies on [3-^{13}C]Ala- or [1-^{13}C]Val-labeled bR, exhibiting characteristic peaks ascribable to loop regions, instead of random coil form,[18,19] with reference to the data of the conformation-dependent displacement of peaks.[20,21] Therefore, such isotropic motions as viewed from ^{2}H NMR may be possible for some Gly, Ser, Ala, and Thr residues at the terminal ends of the N- or C-terminus which are anchored at the membrane surface. In fact, Ala or Gly at this region take random coil form as proved by their respective conformation-dependent displacements of ^{13}C NMR peaks[22,23] and ^{13}C T_1's, which are significantly shortened as compared with those of the transmembrane α-helices.[24] Therefore, recording ^{13}C or ^{15}N NMR signals from such regions should be performed by DD-MAS instead of CP-MAS method, because those signals are usually selectively suppressed by the latter.

Sarkar *et al.*[25,26] measured ^{13}C T_1 and NOE at 15.09 and 62.98 MHz to reveal dynamics of [1-^{13}C]Gly- and [2-^{13}C]Gly-labeled collagen in hard and soft tissue. They assumed that molecular motion is primarily a consequence of reorientation about the long axis of the molecule in view of the highly organized form in the direction parallel to the molecular axis.[26] The root mean square fluctuation in azimuthal angle, γ_{rms}, is shown as[25]

$$(\gamma_{rms})^2 = \langle \gamma^2 \rangle = \int p(\gamma)(\gamma - \gamma_0)^2 d\gamma, \tag{7.1}$$

where γ_0 is the mean value of γ. This motion changes the orientation of the C_α–H vector relative to the applied magnetic field for [2-^{13}C]Gly-labeled collagen and modulates the strong ^{13}C–^{1}H dipolar coupling producing spin–lattice relaxation. The orientation averaged spin–lattice relaxation times $\langle 1/T_1 \rangle$ for the two C–H vectors of Gly making β_1 and β_2 angles with respect to the helical axis are

$$\langle 1/T_1 \rangle / (C_{\gamma_{rms}}{}^2) = (1/10)\omega_D^2 \times [g(\tau,\omega_I - \omega_S) + 3g(\tau,\omega_I) + 6g(\tau,\omega_I + \omega_S)], \tag{7.2}$$

where $C = 3(\sin^2\beta_1 + \sin^2\beta_2) = 5.28$ based on the atomic coordinate of the collagen model peptide (Pro-Pro-Gly)$_{10}$,[27] and $g(\tau,\omega) = \tau/(1 + \omega^2\tau^2)$. Therefore, plots of $(C\gamma_{rms}^2)\langle 1/T_1 \rangle^{-1}$ vs τ (Figure 7.4(a)) are the familiar curves obtained for isotropic motion in solution, together with the orientation averaged $\langle$NOE$\rangle$ in Figure 7.4(b). First, they obtained τ from the measured $\langle$NOE$\rangle$ together with Figure 7.4(b) and then γ_{rms} from the measured $\langle 1/T_1 \rangle$ and the definition of C. They showed that, at 22°C, τ is in the 1–5 ns range for all samples and γ_{rms} is 10°, 9°, and 5.5° for the non-cross-linked, cross-linked, and mineralized samples, respectively. At −35°C, γ_{rms} is less than 3° for all samples.

7.2. Intermediate or Slow Motions with Frequencies Between 10^6 and 10^3 Hz

As demonstrated in Figure 4.15, line shape analysis provides rate constant for twofold flip motion of the Tyr side chain in crystalline enkephalin about

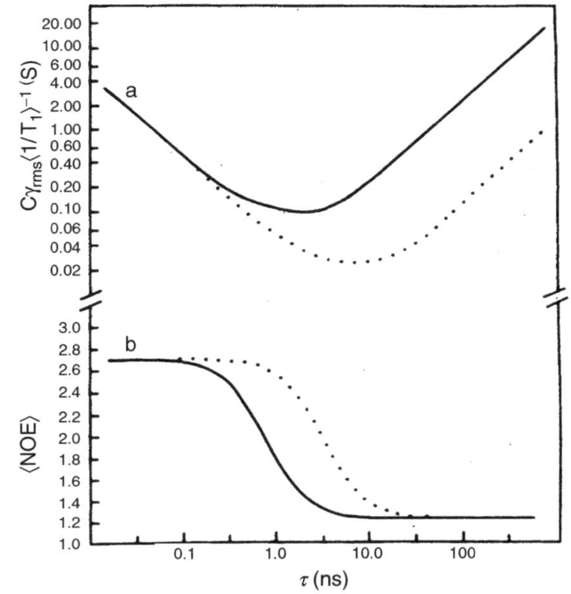

Figure 7.4. Averaged NMR parameters (a) $C\gamma^2_{rms}\langle 1/T_1\rangle^{-1}$ and (b) ⟨NOE⟩ at two fields of 62.98 (—) and 15.09 MHz (....) assuming small amplitude motions in the collagen azimuthal orientation (from Sarkar et al.[26]).

the C_β–C_γ axis for ^{2}H NMR spectra of [tyrosine-3,5-$^2H_2$] Leu^5-enkephalin as approximately 5 × 10^4 Hz at room temperature and that it increases to about 10^6 Hz at 101°C.[28] This frequency as viewed from [$^2H_5$]Phe^4-labeled Leu^5-enkephalin varies with crystalline structure as manifested from the data of 5.0 × 10^3, 3.0 × 10^4, 2.4 × 10^6 Hz depending upon crystals from H_2O, methanol/H_2O, and *N,N*-dimethylformamide/H_2O, respectively.[29]

The chemical shift tensors for the ^{13}C carbonyl[29–31] and ^{15}N amide nitrogen[31–35] atoms in peptide units have been determined for several model peptides. It was shown from a single crystalline study on [1-^{13}C]Gly-[^{15}N]Gly hydrochloride monohydrate that δ_{11} component of nearly axially symmetric ^{15}N chemical shift tensor is tilted 21° away from the N–H bond, although the orientation of δ_{33} and δ_{22} could not be determined.[32] Naito *et al.* determined the principal components and their directions of ^{15}N chemical shift tensors by recording proton-decoupled and ^{15}N–^{1}H dipolar coupled ^{15}N CP-static powder NMR spectra for a variety of peptides.[35] They showed the magnitude of δ_{22} (perpendicular to the peptide plane) and δ_{33} (parallel to the C–N bond) as illustrated in Figure 7.5. Reversal of the magnitudes of δ_{22} and δ_{33}, however, was noted

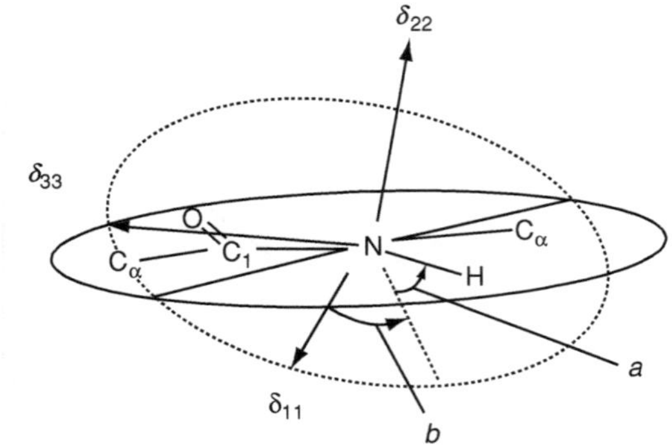

Figure 7.5. Directions of the principal components of the ^{15}N chemical shift anisotropy tensor. The values *a* and *b* are the polar angles of the N–H vector in the principal axis frame.[35]

for Gly-Pro-D-Leu-[^{15}N]Gly.[35] It is expected that backbones of small peptides are usually considered to be static compared with side chains in the solid state. In contrast, Naito *et al.*[34] showed that backbone dynamics of small peptides containing Gly-Gly residues can be examined by analyzing the ^{15}N NMR line shapes powder pattern and the ^{13}C T_2 under the proton decoupling. It was found that by lowering temperature from 40°C to −120° C, the δ_{11} and δ_{22} values of Gly[^{15}N]Gly were shifted to low and high field by 2.5 ppm, respectively, while the δ_{33} value was unchanged. The observed displacement of the principal values of the ^{15}N chemical shift tensor was interpreted by taking librational motions with small amplitudes into account. This librational motion was interpreted in terms of a two-site jump model about the δ_{33} axis that is very close to the $C_\alpha C_\alpha$, direction of the peptides with a jump angle of 17° with jump frequency higher than 1 kHz, as manifested from the simulated spectra demonstrated in Figure 7.6. The correlation times of the librational motions of the peptide plane was estimated as 2.3×10^{-4} s at ambient temperature by analyzing ^{13}C T_2 values (Eq. (3.9)) in the presence of an ^{1}H decoupling field of 50 kHz.[36] Here, it is noticed that the chemical shift anisotropies for ^{15}N amide and ^{13}C carbonyl nuclei are in the order of 10^3 and 10^4 Hz, respectively, in view of the breadth of respective tensors. This means that protein dynamics reflected in changes in CSAs is sensitive to the timescales of 10^{-3} to 10^{-4} s, respectively.

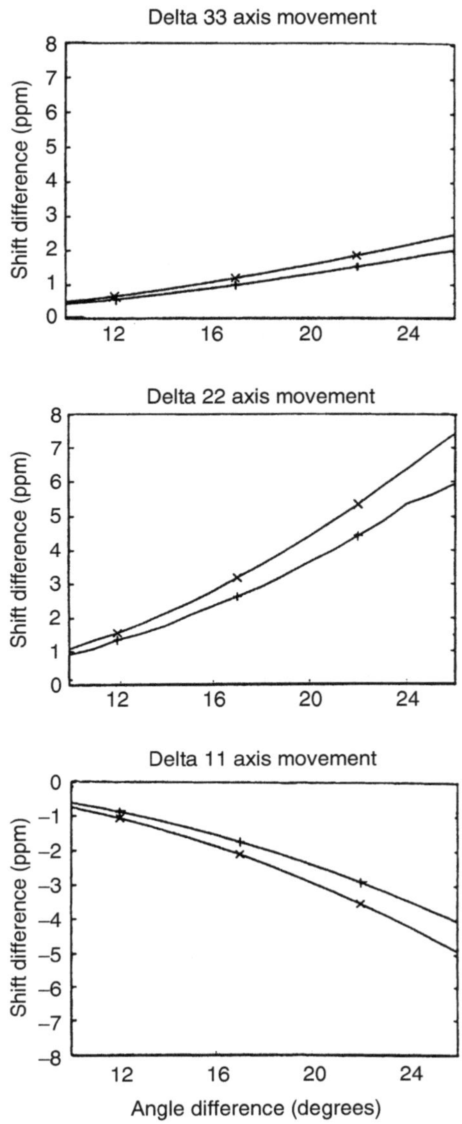

Figure 7.6. Plots of the shift difference in the principal values of ^{15}N chemical shift tensor of Gly[^{15}N]Gly against jump angles between states A and B for the jump frequencies of 1 kHz (x) and 1 MHz (+). The larger values in the vertical axis indicate the higher field shift compared with the rigid molecules.[34]

The averaged CSA tensor $\langle\sigma_H\rangle$ for [1-^{13}C]Gly-labeled collagen was calculated by[25]

$$\begin{aligned}\langle\sigma_H\rangle &= \int p(\gamma)\sigma_H \mathrm{d}\gamma \\ &= \sigma_H + (\Sigma\partial^2\sigma_H/\partial\gamma^2)\langle\gamma\rangle^2/2 + \cdots .\end{aligned} \tag{7.3}$$

If $\gamma_{\mathrm{rms}} \ll 0.7$ radian, it is not necessary to specify $p(\gamma)$ explicitly since the NMR line shape depends only upon γ_{rms}. In the case of a two-site model in which azimuthal orientations are equally populated, $\gamma_{\mathrm{rms}} = (\gamma_1 - \gamma_2)/2$. As illustrated in Figure 7.7, comparison of the ^{13}C NMR spectra of [1-^{13}C]Gly-labeled collagen from different tissues gave rise to the breadths of the ^{13}C CSA Δ and resulting γ_{rms} values. It should be noted that γ_{rms} values obtained from the line shape analysis of [1-^{13}C]Gly-collagen are three to four times larger than the values obtained by the relaxation measurements mentioned already. This is because slow motions of collagen fibrils having large amplitudes with

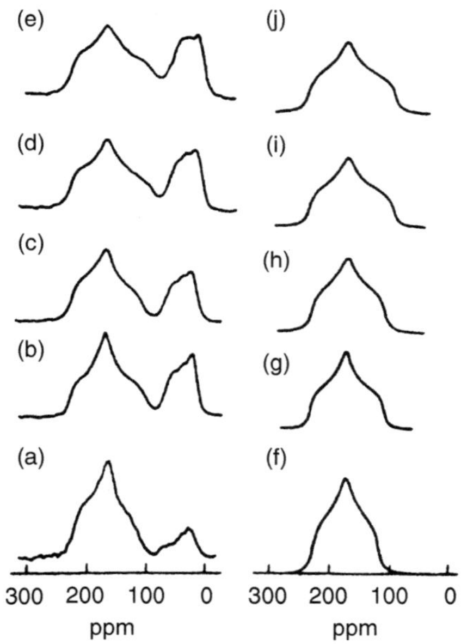

Figure 7.7. Comparison of the experimental spectra (a–e) from [1-^{13}C]Gly-labeled collagen with the line shapes calculated for the glycine carbonyl carbon (f–j). (a) Reconstituted collagen fibrils (22°C); (b) rat tail tendon collagen (22°C); (c) demineralized rat calvaria (22°C); (d) mineralized rat calvaria (22°C); (e) mineralized rat calvaria (−35°C); (f) $\Delta = 108$ ppm, $\gamma_{\mathrm{rms}} = 41°$; (g) $\Delta = 124$ ppm, $\gamma_{\mathrm{rms}} = 31°$; (h) $\Delta = 120$ ppm, $\gamma_{\mathrm{rms}} = 33°$; (i) $\Delta = 140$ ppm, $\gamma_{\mathrm{rms}} = 14°$; (j) $\Delta = 145$ ppm, $\gamma_{\mathrm{rms}} = 0°$ (Δ = frequency breadth of powder pattern; γ_{rms} = range of motion through azimuthal angle (from Sarkar *et al.*[25]).

correlation times less than 10^{-3} s are sensitively detected by line shape analysis, instead of the small amplitude motions with nanosecond correlation times as found by the relaxation measurements mentioned above. It was shown that ^{15}N chemical shift anisotropy powder pattern of the amide group in the peptide bonds is substantially modified either by isotropic motion or by a variety of libration motions about the principal axis system.[37] The phenyl ring dynamics of [$^2H_5$] Phe^4-labeled Leu^5- and Met^5- enkephalin crystals were examined as frequencies of 10^3–10^6 Hz by ^{2}H NMR.[38]

Figure 7.8 illustrates ^{2}H NMR spectra of [$^2H_{10}$]Leu-labeled collagen exhibiting marked temperature dependence, which is attributed to motion about the Leu side-chain bonds.[39] Only the signals of the methyl deuterium nuclei of Leu are included in the spectral analysis because of the low signal-to-noise ratio of the methine and methylene deuterium signals. The calculated spectra in the right-hand side (h–n) were obtained by using a two-site hop model in which the C_γ–C_δ bond axes are assumed to ''hop'' between the two predominant conformations revealed by X-ray diffraction separated by 108°–112°: g^+t conformation with $\chi_1 = 300°$, $\chi_2 = 180°$, and tg^-

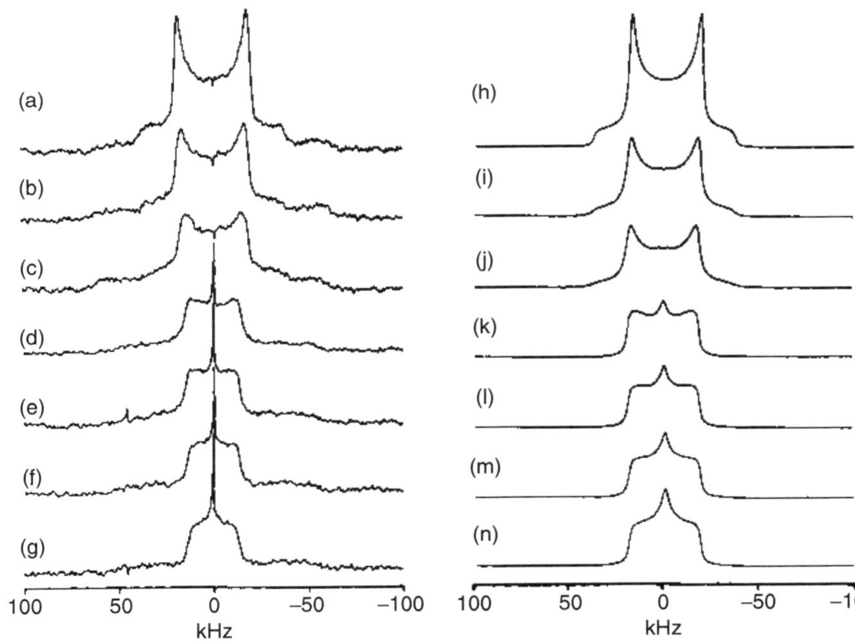

Figure 7.8. Experimental (a–g) and calculated (h–n) ^{2}H NMR spectra of [$^2H_{10}$]Leu-labeled collagen in equilibrium with 0.02 M Na_2HPO_4: (a), −85°C; (b), −43°C; (c), −18°C; (d), −6°C; (e), 1°C; (f), 15°C; (g), 30°C. In spectrum (h), $\kappa \leq 10^8$ rad/s; (i), $\kappa = 1.9 \times 10^4$ rad/s; (j), $\kappa = 3.1 \times 10^4$ *rad/s*, (k), $\kappa = 3.7 \times 10^5$ *rad/s*; (l), $\kappa = 6.3 \times 10^5$ *rad/s*; (m), $\kappa = 8.7 \times 10^5$ *rad/s*, (n), $\kappa = 1.2 \times 10^6$ rad/s (from Batchelder *et al.*[39]).

conformation with $\chi_1 = 180°$, $\chi_2 = 60°$ (Figure 7.9). This hopping takes place at various rates κ, where κ^{-1} is equal to the rotational correlation time. At temperatures above where bulk water on collagen freezes ($\approx 6°$ C), the interconversion of the Leu side chain is rapid. The small narrow water signal in Figure 7.8, spectra d–g, creates some difficulty in determining how much of the central intensity is due to water deuterons and how much is contributed by the deuterons in leucine. Indeed, the broader component at the base of this narrow signal is due to the $\eta = 1$ powder pattern of the leucine methyl deuterium atoms. The theoretical model used to interpret the observed data explains the transformation of the ^{2}H NMR spectra from axially symmetric ($\eta = 0$) powder pattern to one in which the asymmetry parameter, η, is very nearly 1 (Figure 7.8 f and g). However, the width of the calculated line shape is about 1.3 times greater than the width of the observed line shape. A possible method for reducing the calculated line width and also preserving the $\eta = 1$ line shape is to introduce a small fraction of other side chain conformations in the calculation. In fact, it turned out that in about 10% of the peptides examined, side chain conformations other than the above-mentioned predominant conformations are present. Accordingly, Batchelder *et al.*[39] calculated the principal components of the ^{2}H electric field tensor, assuming a four-site exchange model in which the methyl carbons occupy positions at the corners of a tetrahedron, allowing for different populations of the four possible orientations.

Instead of wide-line NMR approaches utilizing static ^{2}H, ^{13}C, or ^{15}N powder pattern spectra so far discussed, it is expected that a more detailed picture of dynamic feature with timescale of ms to ns, which is biologically very important as mentioned already is available from careful examination of isotropic NMR signals recorded by a single NMR spectra for complex proteins such as intact membrane proteins, if many of them were site-specifically assigned. This is obviously made possible if one examines suppressed peaks due to motional broadening in which motional frequency is interfered with frequency of proton decoupling or magic angle spinning as demonstrated in Figure 3.9. As an illustrative example, Figure 7.10 compares how backbone dynamics of bacterio-opsin (bO) in which retinal as a chromophore is removed from bR, is induced to result in site-specific suppression of peaks in ^{13}C DD-MAS and CP-MAS NMR spectra of [3-^{13}C]Ala-labeled bO (black traces) as compared with those of bR (gray traces), respectively.[18] The several ^{13}C NMR peaks including Ala 160 and 196 located in the E-F and F-G loops, respectively, and Ala 39, 53, 84, and 215 in the helices B, C, and G (see Figure 2.8 for sequence) are obviously suppressed due to acquisition of intermediate or slow motions, owing to removal of retinal, with correlation times in the order of 10^{-5} s. It was found that the M-like state of D85N or D85N/D96N at higher pH in which retinal–protein interaction is partly

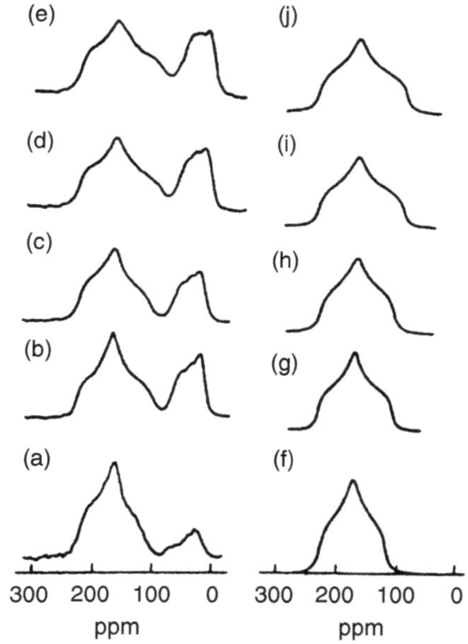

Figure 7.9. (a) Structure of leucine. (b and c) Projections of two of the nine possible rotamers of the leucine side chain. These are the two predominant configurations in crystals of leucine and small leucyl-containing peptides as revealed by X-ray crystallography (from Batchelder et al.[39]).

removed by neutralization of the negative charge at Asp 85 resulted in similar type of spectral changes.[40–42] Acquired motions with correlation times in the order of 10^{-4} s can be very conveniently monitored by selectively suppressed peaks from bR labeled with [1-^{13}C]Gly-, Ala-, Val-, etc., as far samples were spun as fast as 4 kHz.[40–42] It is emphasized that acquired such slow or intermediate motion with correlation times of ms to μs order is not necessarily limited to the cases of the above-mentioned bO or D85N mutants. Instead, ^{13}C NMR signals of membrane protein embedded in lipid bilayer as a monomer without forming 2D lattice are not necessarily fully visible, especially when [1-^{13}C]Gly, Ala, Val-labeled membrane proteins are examined[43–45] Undoubtedly, this type of consideration is the most important step to design experimental procedure by ^{13}C NMR.

7.3. Very Slow Motions with Frequency$<10^3$ Hz

Undoubtedly, molecular packing due to the protein–protein interaction is a very important factor to determine a dynamic feature of the aromatic side chains. So far, we demonstrated that line shape analysis of ^{2}H, ^{13}C, or ^{15}N powder patterns is a very convenient means to delineate rate constant for the backbone and side chain motions with intermediate or slow-frequency reorientation motion described above. It is of interest to examine rather low-rate constants for rotation of tyrosine and phenylalanine rings by two-fold flip–flop motions (10^3 to 1 Hz) in the solid state by means of 1D and 2D exchange spectroscopy for isotropic signals[46,47] as illustrated in Figure 7.1, although this problem has been extensively analyzed in globular proteins in solution state.[48–51] Naito *et al.*[45] showed that the rate constant for flip–flop motions of Tyr side chain in ^{13}C NMR spectra of Leu^6-enkephalin trihydrate can be estimated as 1.3×10^2 s^{-1} at ambient temperature based on a

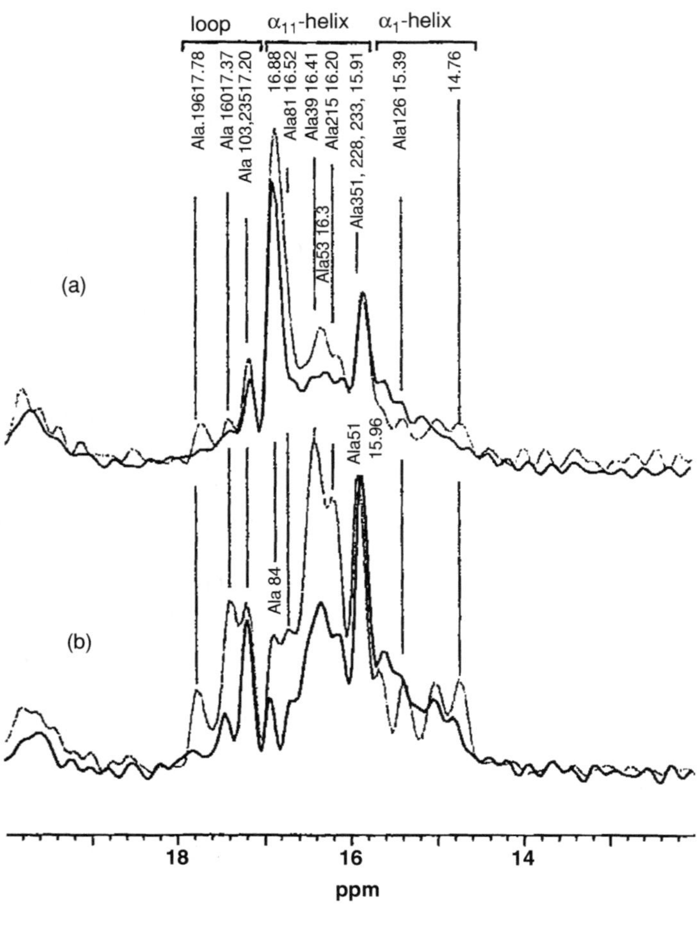

Figure 7.10. ^{13}C NMR spectra of bO (black traces) as compared with those of bR (gray traces), recorded by the DD-MAS (a) and CP-MAS (b) NMR methods, respectively.[18]

spectral simulation utilizing the two-site exchange model. They further recorded 1D or 2D exchange spectra in order to analyze the similar flip–flop motions whose rate constants are much smaller than the limiting value as estimated from the simple line shape analysis of the two-site exchange ($10\ s^{-1}$). In fact, the rate constant for the flip–flop motion of Ac-Tyr-NH_2 and the extended form of Met^5-enkephalin are found to be 1.94 and 1.45 s^{-1} at ambient temperature, respectively. They were also able to determine the rate constant of very slow flip–flop motions from an order of magnitude of 1 through $10^{-3}\ s^{-1}$ in the case of Tyr-OH or Tyr-NH_2. Accordingly, it is possible to utilize this approach to determine the very low fluctuation motions of aromatic side chains for a variety of small peptides in the solid.

References

[1] F. R. N. Gurd and T. M. Rothgeb, 1979, *Adv. Protein Chem.*, 33, 73–165.
[2] G. A. Petsko and D. Ringe, 1984, *Annu. Rev. Biophys. Bioeng.*, 13, 331–371.
[3] M. Karplus and J. A. McCammon, 1983, *Annu. Rev. Biochem.*, 53, 263–300.
[4] A. G. Palmer III, 1997, *Curr. Opin. Struct. Biol.*, 7, 732–737.
[5] A. G. Palmer III, C. D. Kroenke, and J. P. Loria, 2001, *Methods Enzymol.*, 339, 204–238.
[6] M. Akke, 2002, *Curr. Opin. Struct. Biol.*, 12, 642–647.
[7] J. K. Lanyi, 1993, *Biochim. Biophys. Acta*, 1183, 241–261.
[8] A. Maeda, H. Kandori, Y. Yamazaki, S. Nishimura, M. Hatanaka, Y.-S. Chon, J. Sasaki, R. Needleman, and J. K. Lanyi, 1997, *J. Biochem. (Tokyo)*, 121, 399–406.
[9] W. Stoekenius, 1999, *Protein Sci.*, 8, 447–459.
[10] M. Ferrand, A. J. Dianoux, W. Petry, and G. Zaccai, 1993, *Proc. Natl. Acad. Sci. USA*, 99, 9668–9672.
[11] R. A. Kinsey, A. Kintanar, M. -D. Tsai, R. L. Smith, N. Janes, and E. Oldfield, 1981, *J. Biol. Chem.*, 4146–4149.
[12] R. A. Kinsey, A. Kintanar, and E. Oldfield, 1981, *J. Biol. Chem.*, 256, 9028–9036.
[13] M. A. Keniry, H. S. Gutowsky, and E. Oldfield, 1984, *Nature*, 307, 383–386.
[14] R. Henderson, J. M. Baldwin, T. A. Ceska, F. Zemlin, E. Beckmann, and K. H. Downing, 1990, *J. Mol. Biol.*, 213, 899–929.
[15] N. Grigorieff, T. A. Ceska, K. H. Downing, J. M. Baldwin, and R. Henderson, 1996, *J. Mol. Biol.*, 259, 393–421.
[16] E. Pabay-Peyroula, G. Rummel, J. P. Rosenbusch, and E. M. Randau, 1997, *Science*, 277, 1676–1681.
[17] H. Luecke, H. -T. Richter, and J. K. Lanyi, 1998, *Science*, 280, 1934–1937.
[18] S. Yamaguchi, S. Tuzi, M. Tanio, A. Naito, J. K. Lanyis, R. Needleman, and H. Saitô, 2000, *J. Biochem. (Tokyo)*, 127, 861–869.
[19] M. Tanio, S. Inoue, K. Yokota, T. Seki, S. Tuzi, R. Needleman, J. K. Lanyi, A. Naito, and H. Saitô, 1999, *Biophys. J.*, 77, 431–442.
[20] H. Saitô, 1986, *Magn. Reson. Chem.*, 24, 835–852.

[21] H. Saitô and I. Ando, 1989, *Annu. Rep. NMR Spectrosc.*, 21, 208–290.

[22] S. Tuzi, S. Yamaguchi, A. Naito, R. Needleman, J. K. Lanyi, and H. Saitô, 1996, *Biochemistry*, 35, 7520–7527.

[23] H. Saitô, J. Mikami, S. Yamaguchi, M. Tanio, A. Kira, T. Arakawa, K. Yamamoto, and S. Tuzi, 2004, *Magn. Reson. Chem.*, 42, 218–230.

[24] S. Yamaguchi, S. Tuzi, K. Yonebayashi, A. Naito, R. Needleman, J. K. Lanyi, and H. Saitô, 2001, *J. Biochem.(Tokyo)*, 129, 373–382.

[25] S. K. Sarkar, C. E. Sullivan, and D. A. Torchia, 1983, *J. Biol. Chem.*, 258, 9762–9767.

[26] S. K. Sarkar, C. E. Sullivan, and D. A. Torchia, 1985, *Biochemistry*, 24, 2348–2354.

[27] K. Okuyama, K. Okuyama, S. Arnott, M. Takayanagi, and M. Kakudo, 1981, *J. Mol. Biol.*, 152, 427–443.

[28] D. M. Rice, R. J. Wittebort, R. G. Griffin, E. Meirovitch, E. R. Stimson, Y. C. Meinwald, J. H. Freed, and H. A. Scheraga, 1981, *J. Am. Chem. Soc.*, 103, 7707–7710.

[29] R. E. Stark, L. W. Jelinski, D. J. Ruben, D. A. Torchia, and R. G. Griffin, 1983, *J. Magn. Reson.*, 55, 266–273.

[30] T. G. Oas, C. J. Hartzell, T. J. McMahon, G. P. Drobny, and F. W. Dahlquist, 1987, *J. Am. Chem. Soc.*, 109, 5956–5962.

[31] C. J. Hartzell, M. Whitfield, T. G. Oas, and G. P. Drobny, 1987, *J. Am. Chem. Soc.*, 109, 5967–5969.

[32] G. S. Harbison, L. W. Jelinski, R. E. Stark, D. A. Torchia, J. Herzfeld, and R. G. Griffin, 1984, *J. Magn. Reson.*, 60, 79–82.

[33] T. G. Oas, C. J. Hartzell, F. W. Dahlquist, and G. P. Drobny, *J. Am. Chem. Soc.*, 109, 5962–5966.

[34] A. Naito, A. Fukutani, M. Uitdehaag, S. Tuzi, and H. Saitô, 1998, *J. Mol. Struct.*, 441, 231–241.

[35] A. Fukutani, A. Naito, S. Tuzi, and H. Saitô, 2002, *J. Mol. Struct.*, 602–603, 491–503.

[36] W. P. Rothwell and J. S. Waugh, 1981, *J. Chem. Phys.*, 75, 2721–2732.

[37] M. H. Frey, J. G. Hexem, G. C. Leo, P. Tsang, S. J. Opella, A. L. Rockwell, and L, M. Gierasch, 1983, in *Peptides Structure and Function: Proc. 8th American Peptide Symp.*, Pierce Chemical Co., Rockford, IL, p. 763.

[38] M. Kamihira, A. Naito, S. Tuzi, and H. Saitô, 1999, *J. Phys. Chem. A.*, 103, 3356–3363.

[39] L. S. Batchelder, C. E. Sullivan, L. W. Jelinski, and D. A. Torchia, 1982, *Proc. Nat. Acad. Sci. USA*, 79, 386–389.

[40] S. Yamaguchi, S. Tuzi, M. Tanio, A. Naito, J. K. Lanyi, R. Needleman, and H. Saitô, 2000, *J. Biochem. (Tokyo)*, 127, 861–869.

[41] Y. Kawase, M. Tanio, A. Kira, S. Yamaguchi, S. Tuzi, A. Naito, M. Kataoka, J. K. Lanyi, R. Needleman, and H. Saitô, 2000, *Biochemistry*, 39, 14472–14480.

[42] H. Saitô, S. Tuzi, S. Yamaguchi, M. Tanio, and A. Naito, 2000, *Biochim. Biophys. Acta*, 1460, 39–48.

[43] H. Saitô, S. Tuzi, M. Tanio, and A. Naito, 2002, *Annu. Rep. NMR Spectrosc.*, 47, 39–108.

[44] H. Saitô, T. Tsuchida, K. Ogawa, T. Arakawa, S. Yamaguchi, and S. Tuzi, 2002, *Biochim. Biophys. Acta*, 1565, 97–106.

[45] H. Saitô, K. Yamamoto, S. Tuzi, and S. Yamaguchi, 2003, *Biochim. Biophys. Acta*, 1616, 127–136.
[46] A. Naito, M. Kamihira, S. Tuzi, and H. Saitô, 1995, *J. Phys. Chem.*, 99, 12041–12046.
[47] S. Macura, Y. Huang, D. Suter, and R. R. Ernst, 1981, *J. Magn. Reson.*, 43, 259–281.
[48] C. Connor, A. Naito, K. Takegoshi, and C. A. McDowell, 1985, *Chem. Phys. Lett.*, 113, 123–128.
[49] M. Karplus and J. A. McCammon, 1983, *Annu. Rev. Biochem.*, 52, 263–300.
[50] G. Wagner, 1980, *FEBS Lett.*, 112, 280–284.
[51] B. T. Nall and E. H. Zuniga, 1990, *Biochemistry*, 29, 7576–7584.

Chapter 8

HYDROGEN-BONDED SYSTEMS

It is well known that hydrogen bonds play an important role in forming secondary structures of polypeptides and proteins,[1–3] and have been widely studied by various spectroscopic methods. High-resolution NMR has been used as one of the most powerful means for obtaining useful information about details of the hydrogen bond. Chemical shift is one of the most important parameters for providing information about molecular structures including hydrogen bonds. Accordingly, the chemical shifts for nuclei involved are sensitive to the spatial arrangement of the nuclei comprising hydrogen bonds, since the electronic structure around the amide carbonyl carbon and nitrogen in peptides and polypeptides is greatly affected by the nature of hydrogen bond.

However, it is difficult to estimate exactly the hydrogen bonding effect on the chemical shifts in solution, because the observed chemical shifts are often the averaged values between free and hydrogen bonded species. On the other hand, the chemical shifts in the solid state provide direct information about the hydrogen bond of the peptides and polypeptides with a fixed geometry. From such a viewpoint, systematic studies on the hydrogen bonding effect on the ^{1}H, ^{2}H, ^{13}C, ^{15}N, and ^{17}O chemical shifts of respective nuclei in the peptide unit in the solid state have been carried out to obtain detailed information about the hydrogen-bonded structure.[4–7] In this chapter, the experimental as well as theoretical account of hydrogen-bonded structure of peptides and polypeptides in the solid state will be described.

8.1. Hydrogen Bond Shifts

8.1.1. ^{13}C Chemical Shifts

8.1.1.1. N···O hydrogen bond length ($R_{N\cdots O}$) and ^{13}C chemical shift

The hydrogen bonding effect on the ^{13}C chemical shift of the carbonyl carbon in several amino acid residues has been investigated.[8–14] The observed isotropic chemical shifts (δ_{iso}) of C=O carbons in Gly, Ala, Val, D- and L-Leu, and Asp residues of oligopeptides were plotted against the N···O hydrogen bond length ($R_{N\cdots O}$), and are shown in Figure 8.1. There appears an approximately linear relationship between δ_{iso} and $R_{N\cdots O}$ of peptides considered here as

$$\delta_{iso} = a - bR_{N\cdots O}, \tag{8.1}$$

where a and b are constants. The expressions for these relationships as determined by the least-mean-squares method are given in ppm relative to tetramethylsilane (TMS).[4,12]

$$\delta_{iso}(\text{Gly}) = 206.0 - 12.4R_{N\cdots O} \tag{8.2a}$$

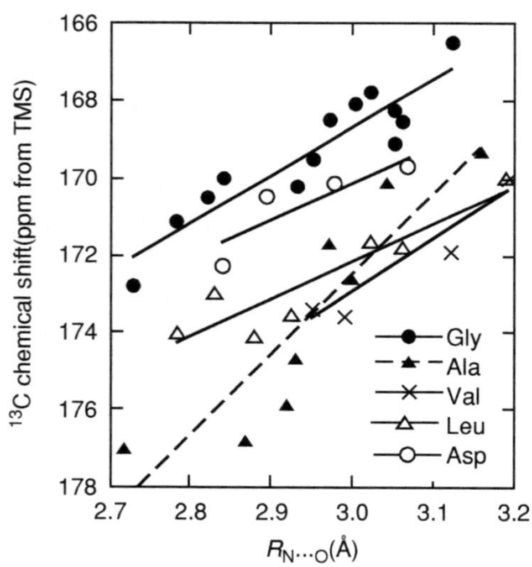

Figure 8.1. Correlation between the observed ^{13}C chemical shifts of the amide carbonyl carbons in Gly, Ala, Val, Leu, and Asp residues of oligopeptides in the solid state and the hydrogen bond length $R_{N\cdots O}$.[11]

$$\delta_{iso}(\text{Ala}) = 237.5 - 21.7R_{\text{N}\cdots\text{O}} \quad (8.2b)$$
$$\delta_{iso}(\text{Leu}) = 202.2 - 10.0R_{\text{N}\cdots\text{O}} \quad (8.2c)$$
$$\delta_{iso}(\text{Val}) = 215.4 - 14.2R_{\text{N}\cdots\text{O}} \quad (8.2d)$$
$$\delta_{iso}(\text{Asp}) = 199.0 - 9.6R_{\text{N}\cdots\text{O}} \quad (8.2e)$$
$$\delta_{22}(\text{Gly}) = 262.9 - 30.2R_{\text{N}\cdots\text{O}}, \quad (8.2f)$$

where δ_{iso} is expressed in ppm and $R_{\text{N}\cdots\text{O}}$ in Å. These relationships indicate that the hydrogen bond length can be determined through the observation of the ^{13}C chemical shift of the carbonyl carbon in the amino acid residues in peptides and polypeptides. The magnitude of the intercept *a* decreases in the order, Ala > Val > Gly > Leu = Asp. The magnitude of the slope *b* decreases in the same order as in the intercept *a*. From these results, it can be said that the values of *a* and *b* are characteristic for individual amino acid residues.

It is noted that the carbonyl ^{13}C chemical shifts, not only in oligopeptides (dimer or trimer) but also in the polypeptides ($(\text{Gly})_n$ and $(\text{Ala})_n$), give a similar dependence on hydrogen bond length. This suggests that the ^{13}C chemical shift of any carbonyl carbon, as a proton acceptor in the hydrogen bond between the amide >C=O and amide >N–H, is strongly influenced by the hydrogen bond length. Some glycine-containing peptides in the crystalline state have a hydrogen bond between the >C=O and $-\text{NH}^{3+}$, where the N-terminus is protonated as NH^{3+} and the C-terminus is unprotonated.[8] The carbonyl ^{13}C chemical shifts are linearly displaced upfield with decreasing $R_{\text{N}\cdots\text{O}}$ in contrast to the case of the >C=O...H–N< type hydrogen bond.

Next, we consider the ^{13}C chemical shift calculations on the amide carbonyl carbons of some amino acid residues by using the corresponding dipeptide hydrogen-bonded with two formamide molecules in calculations using the FPT INDO method as a function of $R_{\text{N}\cdots\text{O}}$.[8–11] Figure 8.2 shows the calculated isotropic shielding constant σ_{iso} of the amide carbonyl carbons of the model compounds for the amino acid residues in various kinds of peptides as a function of $R_{\text{N}\cdots\text{O}}$, where the dihedral angles (ϕ,ψ) are fixed in the typical β-sheet form. The calculated shielding constants (see Section 3.1) are all expressed in ppm with an opposite sign to that of the experimental chemical shifts. It is noted that the negative sign for the calculated shielding constant indicates deshielding, in contrast to the positive sign of the diamagnetic and paramagnetic terms. The relative change for the ^{13}C shielding constants is predominantly governed by the paramagnetic term. The isotropic ^{13}C shielding constants for all of the amino acid residues show approximately linear downfield shifts in the region of short $R_{\text{N}\cdots\text{O}}$ values. This experimental finding should be reasonably explained by the calculated

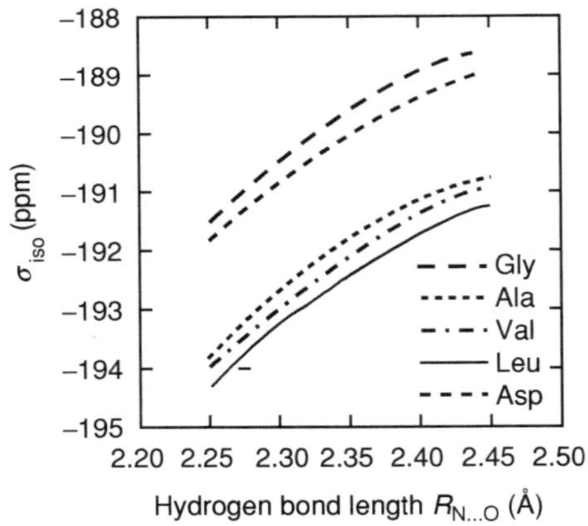

Figure 8.2. Plots of calculated isotropic ^{13}C chemical shielding constants (σ_{iso}) of the amide carbonyl carbon of the model peptides for the amino acid residues (Gly, Ala, Val, Leu, and Asp residues) against the hydrogen bond length $R_{N\cdots O}$.[11]

results in $R_{N\cdots O}$ region. The semi-empirical INDO MO approximation neglects some two-center electron integrals, and therefore the intermolecular interactions are considered to reproduce reasonably in the short $R_{N\cdots O}$ region. Further, ^{13}C chemical shielding constants of the carbonyl carbon in the hydrogen bond between the >C=O and $-NH^{3+}$ of the *N*-acetyl-*N′*-methylglycine amide are calculated as a function of the hydrogen bond length. The ^{13}C chemical shift is displaced upfield for decreasing values of $R_{N\cdots O}$, consistent with the experimental results. This trend is opposite to that found for peptides hydrogen-bonded between the amide >C=O and H–N< groups.

Similar relationship between the principal values of the ^{13}C chemical shift tensor and hydrogen bond length have been elucidated for the carbonyl carbons in Gly, Ala, Val, Leu, and Asp residues in the crystalline state by experiments and theoretical calculations.[3,11,12] It is already known that δ_{11} is in the amide sp^2 plane, and lies along a direction normal to the C=O bond, the δ_{22} component lies almost along the amide C=O bond, and the δ_{33} component is aligned perpendicular to the amide sp^2 plane (see Figure 6.12).[8,9,14,20] The exact tensor components are determined by the Herzfeld–Berger analysis[15] of spinning side bands of the ^{13}C CP-slow MAS NMR spectra of [1-^{13}C]Ala-containing peptides, as shown in the plots of the observed ^{13}C tensor components, δ_{11}, δ_{22} and δ_{33} against the N···O hydrogen bond length ($R_{N\cdots O}$) (Figure 8.3).[9] The experimental δ_{22} values are most sensitive to $R_{N\cdots O}$, and displaced upfield with $R_{N\cdots O}$, except for Ala-Pro-Gly • H_2O in which the covalent bond between the Ala and Pro residues

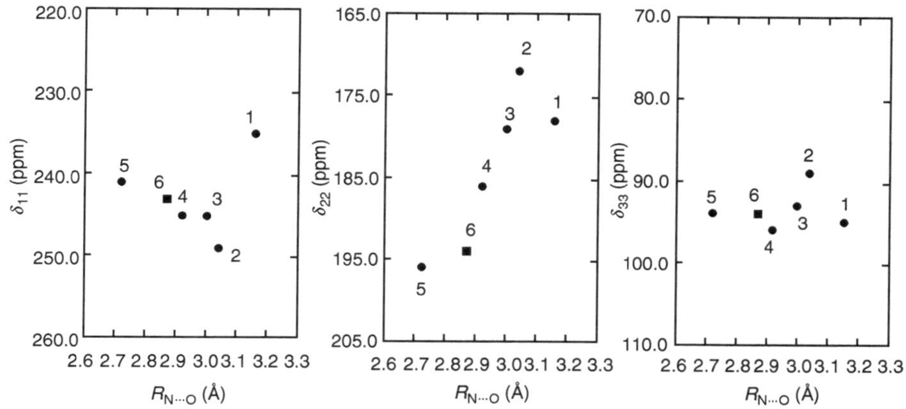

Figure 8.3. Plots of the observed ^{13}C tensor components, δ_{11} (left), δ_{22} (middle), δ_{33} (right), for the amide carbonyl carbon of the [1-^{13}C]Ala-containing peptides against the hydrogen bond length $R_{N\cdots O}$. (1) Ala-Pro-Gly•H_2O, (2) Ala-Ser, (3) Ala-Gly-Gly•H_2O, (4,5) Ac-Ala-NHMe, and (6) $(Ala)_n$.[9]

forms an imide bond rather than a peptide bond, but the experimental δ_{33} values, on the other hand, are almost independent of $R_{N\cdots O}$. Therefore, the observed large downfield shift in the isotropic ^{13}C chemical shifts, δ_{iso}, with a decrease of $R_{N\cdots O}$ comes from the behavior of the δ_{22} in overcoming the opposite changes in the δ_{11}. Similar results were also obtained on the amide carbonyl carbon of various amino acids.[22] Further, principal values for the chemical shift tensor of hydrogen-bonded Ala carbonyl carbons of gA were determined.[16]

It is found from the FPT INDO calculations on ^{13}C shielding tensors (Figure 8.4)[4,8,9] that the calculated σ_{22} expressed by the chemical shift scale is the most sensitive to a change of $R_{N\cdots O}$, and moves linearly downfield with a decrease in $R_{N\cdots O}$. At the same time, σ_{11} increases with a decrease in $R_{N\cdots O}$, whereas σ_{33} is insensitive to changes in $R_{N\cdots O}$, consistent with the experimental findings mentioned above.

The $R_{N\cdots O}$ values were further evaluated by the ^{13}C chemical shifts for $(Gly)_n$, $(Ala)_n$, $(Val)_n$, and $(Leu)_n$, in the solid, taking several conformations such as right-handed α_R-helix, β-sheet, 3_1-helix, and ω_L-helix by means of the C=O ^{13}C shifts.[12,13] The $R_{N\cdots O}$ values for the β-sheet form remained unchanged for these polypeptides in the range of 3.0–3.1 Å, while they are distributed in the range of 2.7–2.8 Å and 2.8 Å for the α_R-helix and 3_1-helix forms, respectively. In fact, the $R_{N\cdots O}$ values of 2.78 Å for the α_R-helix form in $(Val)_n$ and of 2.89 Å for the α_R-helix form in $(Ala)_n$ are not far from the $R_{N\cdots O}$ values of 2.89 and 2.87 Å, respectively, as determined by X-ray

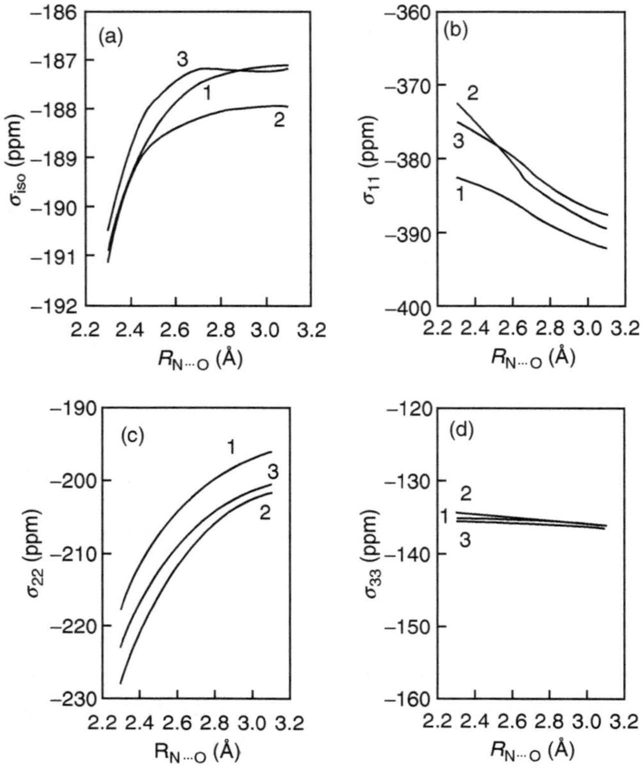

Figure 8.4. Variation of the calculated ^{13}C chemical shielding constants and its tensor components of the amide carbonyl carbon for the L-Ala residue in *N*-acetyl-*N'*-methyl-L-alanine amide forming two hydrogen bonds with two formamide molecules with the hydrogen bond length $R_{N\cdots O}$. (a) isotropic shielding constant σ_{iso}, (b) σ_{11}, (c) σ_{22}, and (d) σ_{33} for (1) α-helix, (2) β-sheet and (3) helix $((\phi, \psi) = (-88°, 155°))$.[9]

diffraction. The $R_{N\cdots O}$ values for the guest Gly residue incorporated into host polypeptides, (Leu, Gly*)$_n$ and (Ala, Gly*)$_n$ were also determined as 2.7 Å,[12,13] where the host Leu and Ala residues take the α_R-helix form. This is in agreement with the values of 2.7 and 2.8 Å for the host Leu and Ala residues. This finding indicates that the guest Gly residue is completely incorporated into host polypeptides with the α_R-helix form. Similar results are obtained in the case of (Val, Gly*)$_n$ and (Ala, Gly*)$_n$ with the β-sheet form. This indicates that the hydrogen bond length for the host residue in copolypeptides and proteins can be determined through observation of the δ value for the Gly residue.

8.1.1.2. Chemical shift tensor components and its direction: single crystalline approach

As demonstrated already on the basis of the calculated chemical shift data, δ_{11} moves somewhat upfield with a decrease in $R_{N\cdots O}$, which is roughly in the direction of the C(=O)–C_α bond, and δ_{33} is insensitive to a change of $R_{N\cdots O}$, which is perpendicular to the C_α–C(=O)–N plane.[5,7,14,20] Naturally, it is expected that the direction of the ^{13}C chemical shift tensor components of the amide carbonyl carbon depends on the hydrogen bond length and angle. The principal values of the ^{13}C chemical shift tensor are summarized in Table 8.1.[14] Then, ^{13}C NMR spectra were recorded for Gly[1] carbonyl carbon of ^{13}C-labeled Gly-Gly, Gly-Gly-HNO_3, and Gly-Gly-HCl • H_2O, to determine the magnitude and direction of these quantities described in Section 2.1.1,[14] in which crystal structures are known by X-ray diffraction.[17–19] In particular, ^{13}C NMR of the single crystal, for instance, [1-^{13}C, 10%]Gly-Gly, were recorded by changing the angle θ between the single

Table 8.1. Geometrical parameters determined by X-ray diffraction and neutron diffraction with the principal values of ^{13}C chemical shift tensor and orientation of the carbonyl carbons in Gly-containing peptides.[14]

	GlyGly	GlyGly • HNO_3	GlyGly • HCl • H_2O
Hydrogen bond length[a] (Å)			
N···O	2.97	3.12	–
O···O	–	–	2.73
H···O	–	2.38	2.36
Hydrogen bond angle[b] (degree)			
C=O···N	157	162	–
N=H···O	–	165	–
C=O···N	–	–	170.7
Dihedral angle (degree)			
ψ	152.4	148.9	162.1
ω	175.8	175.5	176.8
^{13}C chemical shift/ppm			
δ_{iso}	168.1	168.3	169.7
δ_{11}	242.3	248.1	242.1
δ_{22}	173.8	167.8	177.1
δ_{33}	88.2	89.1	87.9
Angle between δ_{22} and C=O	10.0°	5.0°	13.0°
Tilted angle from C=O–N	90.0°	85.0°	90.0°

[a]For GlyGly, GlyGly • HNO_3 and GlyGly • HCl • H_2O the experimental errors are ±0.007, 0.002, and 0.003 Å, respectively.[52,58,59]

[b]For GlyGly, GlyGly • HNO_3 and GlyGly • HCl • H_2O the experimental errors are ±0.4°, 0.1°, and 0.2°, respectively.[52,58,59]

crystal plane and an applied magnetic field direction with a home-made goniometer attached to rotor, and changing the rotational angle from 0° to 360° with an interval of 10°.[14] The determined ^{13}C chemical shifts of the [1-^{13}C,10%] Gly[1] carbonyl carbon of a Gly-Gly single crystal are plotted against the rotation angles (ψ) of the rotor with different prepared angles (θ) of 30°, 50°, and 90°, together with the simulated curve using the chemical shift tensor components (determined from the slow-speed MAS method given above) by the above-mentioned analysis with the least-squares method.

A typical static ^{13}C CP-NMR spectrum of the Gly[1] amide carbonyl carbon of a [1-^{13}C,10%]Gly-Gly single crystal is observed. It is expected that two carbonyl signals will appear in the spectrum, because there are only two Gly-Gly molecules with magnetically different environments in the unit cell. However, six peaks were resolved in the spectrum. They are caused by the fact that two individual signals are split into three peaks due to the quadrupolar coupling between the ^{13}C and ^{14}N nuclei. The extent of quadrupolar splitting depends on the angle between the magnetic field and the ^{13}C–^{14}N interatomic vector.[20] The splitting depends also on the strength of the magnetic field and decreases with the magnetic field strength.

In order to determine the position of the center peak of the split signals, ^{13}C NMR spectra of a Gly-Gly single crystal were measured at two different frequencies (67.8 and 125 MHz) using the same angle between the magnetic field and the single crystal. The ^{13}C signal was split into three peaks by the second order quadrupolar splittings, e^2qQ/h.[21] If $(e^2qQ/h) > 0$, the central peak is displaced by $2d$ at low field where d is the effect of quardrupolar interaction on the position of spectral lines, and the other peaks are displaced by d at low field, as compared with the case for $(e^2qQ/h) = 0$. If $(e^2qQ/h) < 0$, the central peak is displaced by $2d$ at high field, and the other peaks are displaced by d at high field.

From these results, the ^{13}C chemical shift tensor components and their directions are determined as shown in Table 8.1.[14] The angles between the direction of δ_{22} and the C=O bond for Gly-Gly, Gly-Gly • HNO_3, and Gly-Gly • HCl • H_2O are 10°, 5°, and 13°, respectively, and the angles between the direction of δ_{33} and the plane N–C=O are 90°, 85°, and 90°, respectively. The direction of δ_{11} is perpendicular to δ_{22} and δ_{33}. Therefore, the Euler angles between the PAS and the molecular axis system are (0°, 10°, 0°), (5°, 5°, 0°), and (0°, 13°, 0°) for Gly-Gly, Gly-Gly • HNO_3, and Gly-Gly • HCl • H_2O, respectively.

The hydrogen bond angles for Gly-Gly and Gly-Gly • HNO_3 are 157° and 162°, respectively (Table 8.1). The former deviates by 23° from a linear >C=O...H–N< hydrogen bond and the latter by 18°. Therefore, the deviation of the former is larger by 5° than that of the latter. Such a deviation of the hydrogen bond angle from the linearity of the hydrogen bond leads to a deviation of the symmetry of the electronic distribution around the >C=O bond. The behavior of ^{15}N-labeled the amide carbonyl carbon chemical shift tensor components is due to the hydrogen bonding and conformation. The chemical shift behavior of C=O bond is more largely due to changes of the hydrogen-bonded structure than by accompanying conformational changes. The amide-carbonyl chemical shifts of the hydrogen-bonded Gly^1 residue in Gly-Gly and Gly-Gly·HNO_3 peptides, and thus only with rotations in the angle ψ, the rotational angle around the $-CH_2-C(=O)$ bond in Gly^1 is employed because there is no angle ϕ in a Gly^1 residue. The values of the angle ψ for Gly-Gly and Gly-Gly·HNO_3 are very close to each other as shown in Table 8.1. Probably, the conformational effect on the amide-carbonyl chemical shift tensor components is not needed to be taken into account. As mentioned above, the angles between the direction of δ_{22} and the >C=O bond for Gly-Gly and Gly-Gly • HNO_3 are 10° and 5°, respectively. Therefore, the larger deviation of the direction of δ_{22} from the >C=O bond axis for Gly-Gly compared with that for Gly-Gly • HNO_3 comes from the larger deviation of the hydrogen bond angle from linearity in the >C=O···H–N< hydrogen bond. On the other hand, the hydrogen bond lengths of Gly-Gly and Gly-Gly • HNO_3 are 2.97 and 3.12 Å, respectively. This difference is very small namely 0.15 Å. The position of δ_{22} for Gly-Gly shifts much lower field than that for Gly-Gly • HNO_3. This agrees with the above-mentioned results[8,20] that the decrease in the hydrogen bond length leads to downfield shift.

8.1.2. ^{15}N Chemical Shifts

^{15}N NMR spectra were recorded to study polypeptides and proteins in the solid state.[23–26] The isotropic ^{15}N chemical shifts of a number of homopolypeptides in the solid state are significantly displaced as much as 1.2–10.0 ppm between the α-helix and β-sheet forms, although these amounts might be larger than those with respect to free state. Such a large chemical shift difference may come from changes in the hydrogen bond length and angle, and through a change in the dihedral angles (ϕ, ψ). Kuroki *et al.* have measured isotropic ^{15}N chemical shifts and the tensor components of Gly residue in a variety of peptides with a terminal Boc group in order to clarify the relationship between the ^{15}N chemical shift (relative to saturated $^{15}NH_4NO_3$ solution in water) and $R_{N\cdots O}$,[23,24] and have shown that there exists a clear relationship between the observed isotropic ^{15}N chemical shifts (δ_{iso}) of ^{15}N-labeled GlyNH in Boc-Gly, Boc-Gly-Ala, Boc-Gly-Phe, Boc-GlyAib, and Boc-Gly-Pro-OBzl in the solid state, and their $R_{N\cdots O}$ values determined by X-ray diffraction. The hydrogen bond lengths of peptides used are in the 2.95–3.08 Å range. On the other hand, the hydrogen bond angles (C=O···N) are in the 113°–155° range. The chemical shifts move to downfield with a decrease in $R_{N\cdots O}$. The δ_{11} and δ_{33} are more sensitive than δ_{22} to a change in $R_{N\cdots O}$. A change of 0.2 Å in $R_{N\cdots O}$ leads to a change of 20 ppm in δ_{11} and δ_{33}, but is a change of 5 ppm in δ_{22}. δ_{22} and δ_{33} moves linearly downfield with a decrease in $R_{N\cdots O}$, and δ_{11} moves downfield with $R_{N\cdots O}$ with large scatter. Such a behavior is governed not only by the hydrogen bond length, but also by the hydrogen bond angle. The direction of the Gly-NH ^{15}N chemical shielding tensor components has been determined by using a Boc-Gly-Gly-Gly-OBzl single crystal.[26] The δ_{11} component lies approximately along the N–H bond, the δ_{33} component lies approximately along the N–C_α bond, and the δ_{22} component is aligned in the direction perpendicular to the peptide plane (see Figure 7.5 for a=b=0°).

The FPT INDO calculation shows that the σ_{11} component lies approximately along the N–H bond, in agreement with the experimental results, but the σ_{22} component lies approximately along the N–C_α bond and the σ_{33} component is aligned in the direction perpendicular to the peptide plane, which is different from the experimental results. It is not easy to determine the direction of σ_{22} and σ_{33} with significant experimental accuracy, because their values are very close to each other. The calculated assignment of σ_{22} and σ_{33} seems to be acceptable, although it is difficult to distinguish them. The direction of σ_{11} can be easily determined, because its magnitudes are significantly different from the others. The reason why in the theoretical calculation the most shielded component σ_{33} is aligned in the direction perpendicular to the peptide plane is because the orbitals of the nitrogen lone-pair electrons are in this direction. The experimental results (that σ_{33} moves linearly downfield with a decrease in the hydrogen bond length, that σ_{33} is not related to hydrogen bond angle, and that σ_{11} and σ_{22} are related to hydrogen bond length and hydrogen bond angle) are reasonably explained by the calculation.

8.1.3. ^{17}O Chemical Shifts

Hydrogen-bonded structures and the ^{17}O NMR of Gly and Ala containing peptides and polypeptides were examined.[6,27–34] Solid state ^{17}O NMR of the amide carbonyl oxygen provides a deep insight into understanding the hydrogen-bonded structures of solid peptides and polypeptides, because this nucleus acts as a direct proton acceptor. As described already in Section 4.3, its natural abundance is very low (0.037%), and spectral interpretation is not always straightforward due to the quadrupolar interactions under the strength of currently available magnetic field. Static solid state ^{17}O NMR spectra of polypeptides are measured by CP method at several NMR frequencies (270, 400, and 500 MHz for ^1H).[27] NMR parameters such as chemical shift (δ), quadrupolar coupling constant (e^2qQ/h), and asymmetric parameter (η) can be determined by spectral simulation, but a combination of ultra high-field magnet (800 MHz for ^1H) and high-speed MAS (25 kHz) is also essential as discussed in Section 5.3. Indeed, such NMR parameters of $(Ala)_n$ in the solid with the α-helix and β-sheet forms were evaluated from the spectra already shown in Figure 4.21.[32,33] The ^{17}O chemical shifts are calibrated through external distilled water ($\delta = 0$ ppm). The sharp peaks at 375 and 140 ppm come from ZrO_2 oxygen contained in the NMR rotor and the spinning sideband, respectively. The ^{17}O NMR parameters thus obtained are summarized in Table 8.2[32] together with the hydrogen bond length determined by X-ray crystallographic studies. The δ_{iso} values for $(Ala)_n$ with the α-helix form and β-sheet form are 319 and 286 ppm, respectively, and the corresponding e^2qQ/h values are 8.59 and 8.04 MHz, respectively. The ^{17}O chemical shift values move to high field, and the e^2qQ/h values decrease with a decrease in $R_{N\cdots O}$. The ^{17}O NMR parameters of solid 10% ^{17}O-labeled $(Gly)_n$ with the 3_1-helix (PGII) (Figure 8.5) and β-sheet (PGI) forms determined by the spectral simulation are shown in Table 8.3 together with the hydrogen bond length determined by X-ray crystallographic studies. The δ_{iso} values for PGI and PGII are 304 and 293 ppm, respectively, and the corresponding e^2qQ/h values are 8.36 and 8.21 MHz, respectively. The amide carbonyl ^{17}O chemical shifts in polyglycines move downfield and that the e^2qQ/h decreases by a decrease in the hydrogen bond length $R_{N\cdots O}$. Further, the asymmetric parameters (η) for PGI and PGII are 0.30 and 0.33, respectively.

As mentioned above, solid state ^{17}O NMR spectral pattern is greatly influenced by the NMR frequency. Kuroki *et al.* observed solid state ^{17}O CP NMR spectra of Gly*-Gly (6% ^{17}O-labeled) at 36.6 MHz (270 MHz for ^1H), 54.2 MHz (400 MHz for ^1H), and 67.8 MHz (500 MHz for ^1H) (see

Table 8.2. The ^{17}O NMR parameters of solid $(Ala)_n$ as determined from high-speed MAS-NMR measurements with hydrogen bond lengths ($R_{N\cdots O}$).

	α-helix	β-sheet
e^2qQ/h(MHz)	8.59	8.04
η	0.28	0.28
δ_{iso} (ppm)	319	286
$R_{N\cdots O}$ (Å)	2.87	2.83

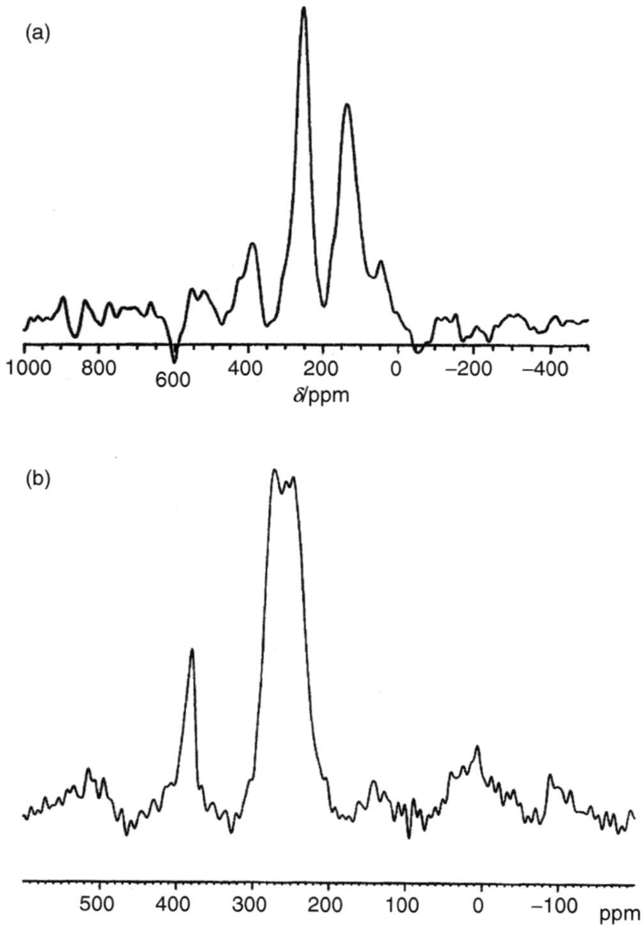

Figure 8.5. Observed ^{17}O CP/MAS NMR spectrum of polyglycine II at a MAS rate of 8 kHz at 67.8 MHz (a) and that at a MAS rate of 30 kHz with proton decoupling at 108 MHz (b).[33]

Figure 4.20) to determine ^{17}O NMR parameters. The ^{17}O spectrum at 36.6 MHz consists of two splitting signal groups. Their separation is decreased with an increase in NMR frequency, and at 67.8 MHz the two splitting signals coalesce. This means that such a splitting comes from the quadrupolar interaction and if one wants to obtain a sharp ^{17}O signal, high-frequency NMR is needed.

^{17}O chemical shielding tensor and quadrupolar coupling constant were calculated by MO method.[29] The direction of the principal axes of the chemical shielding of the carbonyl oxygen in *N*-acetyl-*N*′-methylglycine

Table 8.3. Determined ^{17}O NMR parameters of solid polyglycines with β-sheet and 3_1-helix forms together with hydrogen bond lengths ($R_{\mathrm{N\cdots O}}$).

	PGI (β-sheet form)	PGII (3_1-helix form)
MAS experiments by 108.6 MHz		
e^2qQ/h (MHz)	8.36	8.21
η	0.30	0.33
δ_{iso} (ppm)	304	293
Static experiments by 67.8 MHz[a]		
e^2qQ/h (MHz)	8.55	8.30
η	0.26	0.29
δ_{iso} (ppm)	299	288
$R_{N\cdots O}$ (Å)	2.95[b]	2.73[c]

[a]Ref. [27].
[b]Ref. [60].
[c]Ref. [53].

amide hydrogen-bonded with two formamide molecules has been determined by FPT-MNDO-PM3 method. The σ_{22} component lies approximately along the C=O bond, the σ_{11} component is aligned in the direction perpendicular to the C=O bond on the peptide plane, and the σ_{33}, which is the most shielded component is aligned in the direction perpendicular to the peptide plane. It is very interesting result that the most shielded component σ_{33} is not aligned along the direction of the C=O bond or the direction of lone pair electron that is aligned 120° or −120° from the C=O bond direction on the peptide plane. The sp^2 hybrid property of the carbonyl bond is lost due to the double bonding property of the peptide bond. The δ_{iso} values in both peptides and polypeptides move upfield with decrease in the hydrogen bond length ($R_{\mathrm{N\cdots O}}$). The plots of the observed principal values of ^{17}O chemical shifts against the hydrogen bond length show that every principal value in both peptides and polypeptides moves upfield with decrease in the hydrogen bond length. It turns out that all of the principal values of the ^{17}O chemical shift of Gly carbonyl carbon in the model system move largely upfield with an increase in $R_{\mathrm{N\cdots O}}$.

The plots of the observed e^2qQ/h values against the hydrogen bond length show that the e^2qQ/h values decrease linearly with a decrease of the hydrogen bond length ($R_{\mathrm{N\cdots O}}$) between the amide nitrogen and oxygen atoms. This relationship is expressed by

$$e^2qQ/h\,(\mathrm{MHz}) = 5.15 + 1.16R_{\mathrm{N\cdots O}}(\text{Å}). \tag{8.3}$$

This change comes from change of the q values, which are the largest component of electric field gradient tensor (eq). This experimental result shows that the decrease in the hydrogen bond length leads to the decrease of electric field gradient. The e^2qQ/h value is very sensitive to the change of hydrogen-bonding length.

Wu and Dong measured 2D MQMAS-NMR spectra of [$^{17}O_2$]-D-alanine and [$^{17}O_2$]-D,L-glutamic acid • HCl in the crystalline state.[34] They showed that two oxygen atoms of the alanine molecule experience different chemical environments. A combined analysis of the 2D MQMAS and 1D MAS spectra gave the NMR parameters such as O1, $\delta_{iso} = 275 \pm 5$ ppm, quadrupolar coupling constant $e^2qQ/h = 7.60 \pm 0.02$ MHz, $\eta = 0.60 \pm 0.01$; O2, $\delta_{iso} = 262 \pm 5$ ppm, $e^2qQ/h = 6.40 \pm 0.02$ MHz, $\eta = 0.65 \pm 0.01$. The L-alanine exists in its zwitterionic form and the two oxygen atoms have quite different hydrogen bonding environments: the stronger hydrogen bonding environment at O2 results in a small but significant lengthening compared with O1.

8.1.4. ^{1}H Chemical Shifts

It is also interesting to clarify whether there is a relationship between the hydrogen bond length between amide nitrogen and oxygen atoms ($R_{N\cdots O}$) and solid state ^{1}H NMR chemical shift. For this purpose, it is necessary to remove larger dipolar interactions from nearby proton nuclei in order to obtain a small line width. The combined rotational and multipulse spectroscopy (CRAMPS) method is one of the solutions to obtain ^{1}H NMR spectrum with reasonable resolution.[35] For obtaining high-resolution ^{1}H NMR spectrum for peptides and proteins in the crystalline state, there are some reports in which a combination of MAS and CRAMPS techniques is used.[36–38] However, the CRAMPS needs slow-speed MAS for sampling the data points during acquisition (e.g., the BR24 multipulse sequence needs less than about 3 kHz for sampling). The spinning rate of 3 kHz is not enough to give a high-resolution ^{1}H NMR spectrum for analyzing amide proton chemical shift in peptides and polypeptides because the amide proton is directly bonded to the quadrupolar ^{14}N nucleus. As reported by McDermott,[36] amino acids take up NH_3^+ forms, and so the quadrupolar effect on proton line width by the ^{14}N becomes very small because of its high symmetry. Therefore, the NH_3^+ proton line width in the amino acids becomes much sharper than that of amide protons in peptides and polypeptides considered here.

Let us consider the relationship between the hydrogen bond length between amide nitrogen and oxygen atoms ($R_{N\cdots O}$) of hydrogen-bonded Gly-containing peptides and polypeptides in the solid state and data obtained from the observation of high-resolution ^{1}H NMR spectrum at high speed

MAS of 30 kHz and high frequency of 800 MHz for removal of the dipolar coupling and quadrupole coupling with amide ^{14}N.[39]

Figure 8. 6 shows ^1H MAS NMR spectra of Gly-containing peptides and $(Gly)_n$ obtained by a single-pulse method with a MAS rate of 30 kHz at 800 MHz.[39] The amide proton, α-proton, and the side-chain protons are straightforwardly assigned because their peaks are clearly resolved from each other. The chemical shifts of the other functional groups are approximately the same as their values in the corresponding amino acids in aqueous solution. The amide proton chemical shift must be assigned carefully. Thus, if the corresponding peak is overlapped with the other peaks, it must be decomposed by using computer-fitting so that the chemical shift can be determined.

A new NMR technique of measuring high-resolution solid state ^1H NMR is designed by using the FSLG-2 homonuclear dipolar decoupling method[41,42] combined with high-speed MAS. High-resolution solid state ^1H NMR spectra of polypeptides are thus successfully observed, and gave more reasonable resolution for the amide proton signals of which the chemical shifts have information about the hydrogen-bonded structure, as compared with other high-resolution solid state ^1H NMR methods.[40] Table 8.4 shows the determined Gly amide proton chemical shift values of peptides and polypeptides in the solid state together with the hydrogen bond length between amide nitrogen and oxygen atoms ($R_{N\cdots O}$), as determined from X-ray diffraction. Here, it can be said that the reduction of $R_{N\cdots O}$ may lead to a decrease in the hydrogen bond length between the amide proton and oxygen atom.

Figure 8.7 shows the plots of the determined ^1H chemical shift values (δ) of hydrogen-bonded Gly amide protons of Gly-containing peptides and polyglycines in the solid state against the hydrogen bond lengths between amide nitrogen and oxygen atoms ($R_{N\cdots O}$), as determined from X-ray diffraction. The bars indicate the experimental errors in the spectra. As $R_{N\cdots O}$ is decreased from 3.12 to 2.72 Å, the amide ^1H chemical shift moves to downfield by 1.28 ppm from 7.76 to 9.04 ppm. Thus, it can be said that the amide ^1H chemical shift moves downfield with decreasing hydrogen bond lengths between amide nitrogen and oxygen atoms ($R_{N\cdots O}$). This means that the hydrogen bond length $R_{N\cdots O}$ can be estimated through the observation of the amide ^1H chemical shift.

The ^1H chemical shielding calculations of hydrogen-bonded Gly amide protons of two hydrogen-bonded Gly-Gly molecules have been made by using the Gaussian 96 program with *ab initio* 6-31G** basis set by changes of $R_{N\cdots O}$ from 3.5 to 2.6 Å, as referred to its crystal structure determined by X-ray diffraction. The calculated chemical shifts move to downfield by 2.5 ppm from 6.9 to 9.4 ppm as $R_{N\cdots O}$ is decreased from 3.30 to 2.72 Å.

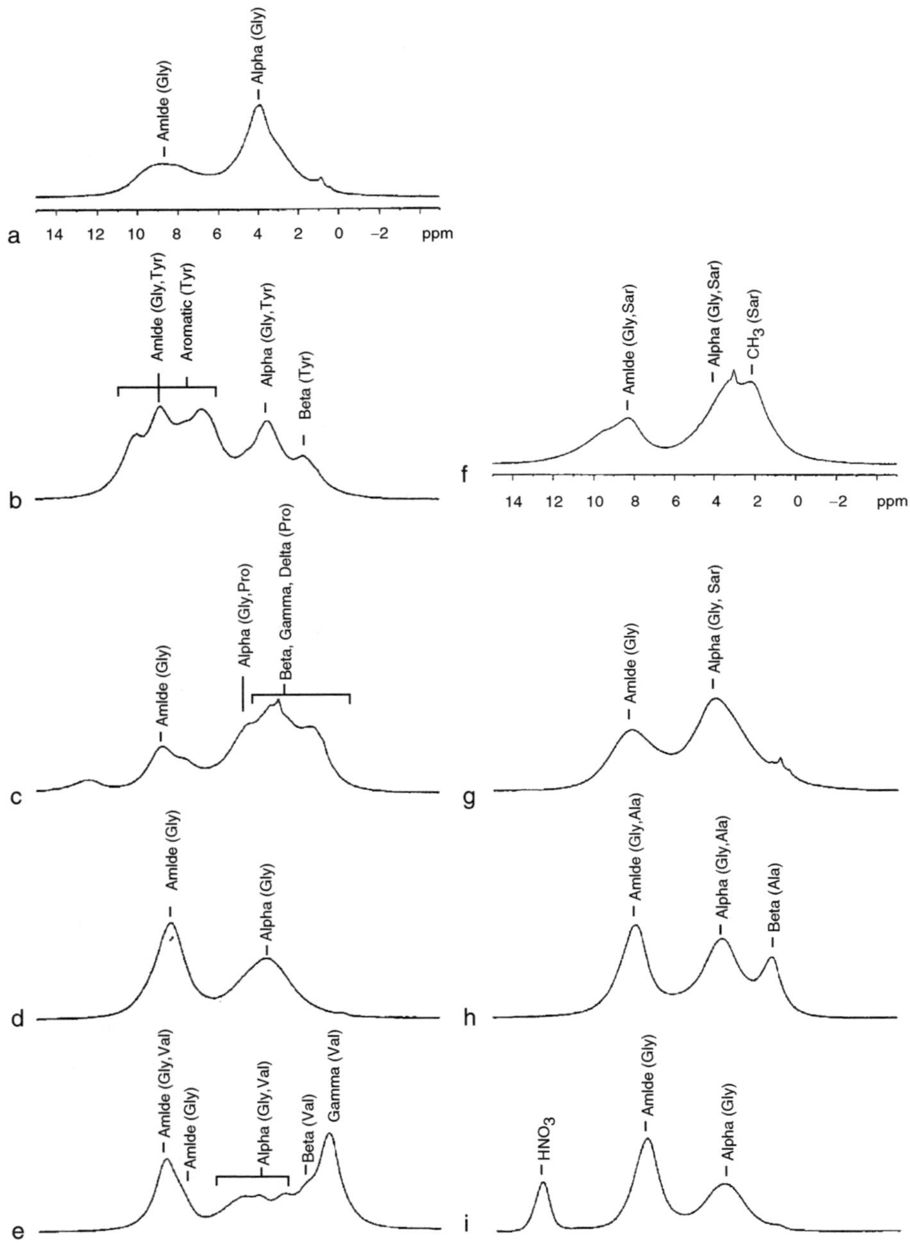

Figure 8.6. ^{1}H MAS-NMR spectra of Gly-containing peptides and $(Gly)_n$ obtained by single-pulse method with a MAS rate of 30 kHz at 800 MHz: (a) $(Gly)_n$ (form II), (b) Tyr-Gly-Gly, (c) Pro-Gly-Gly, (d) Gly-Gly, (e) Val-Gly-Gly, (f) Sar-Gly-Gly, (g) $(Gly)_n$ (form I), (h) Ala-Gly-Gly, and (i) Gly-Gly•HNO_3.[39]

Table 8.4. ^{1}H NMR chemical shifts of hydrogen-bonded glycine residue amide proton for peptides as determined by FSLG-2 proton MAS-NMR and their geometrical parameters (hydrogen bond lengths ($R_{N \cdots O}$) as referred by X-ray diffraction data.[a]

Sample	Hydrogen-bonded glycine amide proton chemical shift δ (ppm)	Hydrogen bond length $R_{N \cdots O}$ (Å)
Polyglycine (form II)	9.84	2.73
Tyr-Gly-Gly	9.74	2.88
Pro-Gly-Gly	9.26	2.89
Gly-Gly	9.06	2.94
Polyglycine (form I)	8.90	2.95
Sar-Gly-Gly	8.72	3.06
Val-Gly-Gly	8.68	3.05
Ala-Gly-Gly	8.38	3.00
Gly-Gly-HNO_3	7.38	3.12

[a]Refs. [52,53,57–59].

This shows that the calculation explains qualitatively the experimental results. However, quantitative agreement is not obtained. This may be because strictly speaking the position of the amide proton in the >N–H$\cdots$O=C< hydrogen bond depends on $R_{N \cdots O}$,[43] but in this calculation the N–H bond length is fixed at 1.0 Å.

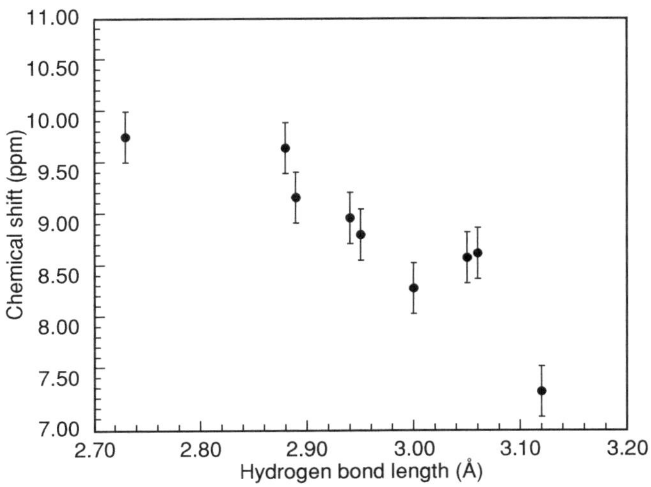

Figure 8.7. Plots of the observed ^{1}H chemical shift values (δ) of hydrogen-bonded Gly amide protons of Gly-containing peptides and polyglycines in the solid state against the hydrogen bond lengths between amide nitrogen and oxygen atoms ($R_{N \cdots O}$) as determined from X-ray diffraction. The bars indicate the experimental errors in the spectra.[39]

Kimura *et al.* determined the hydrogen bonded amide N–H bond lengths of $(Ala)_n$ with the α-helix and β-sheet forms by the observation of the ^1H CRAMPS NMR spectra of fully ^{15}N-labeled $(Ala)_n$s in the solid state.[44] The N–H dipolar spinning sideband pattern of α-helical $(Ala)_n$ is different from that of β-sheet $(Ala)_n$. These sideband patterns are sensitive to N–H bond length. From the sideband pattern analysis, the N–H bond lengths for α-helical and β-sheet $(Ala)_n$s are determined to be 1.09 and 1.12 Å, respectively. Thus, the N–H bond length for the former is less shorter than that for the latter.

8.2. ^2H Quadrupolar Coupling Constant

From a comparison of the observed static solid state ^2H NMR spectra of amide ^2H-labeled PGI(a), ^2H-labeled PGII(b), and ^2H-labeled PLA(c) and the theoretically simulated spectra, the quadrupolar coupling constant (e^2qQ/h) and asymmetric parameter of field gradient tensor (η) have been determined. The electric field gradient (eq) is a sensitive measure of the electronic charge distributions in the vicinity of the nucleus. Therefore, static solid state ^2H NMR spectra of amide ^2H-labeled peptides and polypeptides yield e^2qQ/h and η values based on spectral simulation as shown in Table 8.5, together with the hydrogen bond lengths ($R_{N\cdots O}$) as determined by X-ray or neutron diffraction.[49–55] The η values of peptides and polypeptides were a common value of 0.05.

In Figure 8.8, the plots of the observed e^2qQ/h values against the hydrogen bond length ($R_{N\cdots O}$) were shown. The e^2qQ/h value decreases with a decrease in $R_{N\cdots O}$. The experimental result shows that the reduction of the hydrogen

Table 8.5. Nuclear quadrupolar coupling constant (e^2qQ/h) and asymmetry parameter (η) of amide deuterium of peptides and polypeptides as determined by static ^2H NMR with their hydrogen bond lengths ($R_{N\cdots O}$).[a]

Sample	NMR parameters		Hydrogen bond length $R_{N\cdots O}$ (Å)
	e^2qQ/h (kHz)[a]	η	
GlyGly	174	0.1	2.94
AlaGly	169	0.1	2.85
PGI	175	0.1	2.95
PGII	165	0.1	2.73
PLA	173	0.1	2.87

[a]Refs. [50,51,53,54,60].

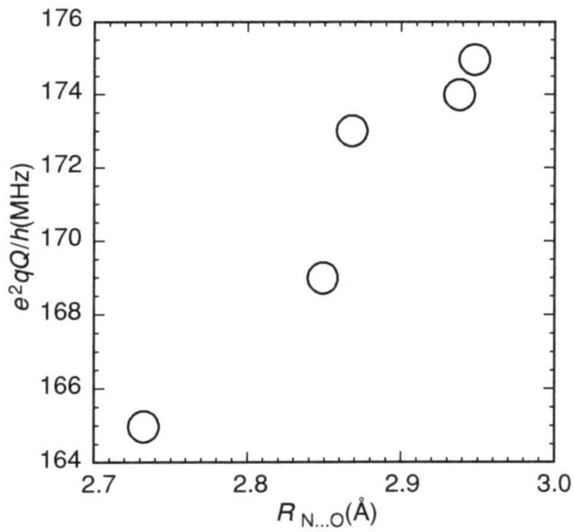

Figure 8.8. The plots of the experimental e^2qQ/h values of N^2H-GlyGly, N^2H-AlaGly, N^2H-PGI, N^2H-PGII, and N^2H-PLA against the hydrogen bond length ($R_{N\cdots O}$).[45]

bond length leads to a linear decrease in electric gradient (*eq*). The *eq* value is very sensitive to a change in hydrogen bond length. This experimental finding is consistent with the experimental results of hydrogen-bonded amide ^{17}O and ^{2}H nuclei of polypeptides.[23–34,45–48] From this relation, useful information about the hydrogen bond length ($R_{N\cdots O}$) in peptides and polypeptides can be obtained by observation of the e^2qQ/h value. It is worthwhile to point out here that the correlation between the deuterium quadrupole coupling constants and hydrogen bond lengths for a variety of crystals including amino acids were reported some 40 years ago.[54–56]

References

[1] G. A. Jeffrey, 1997, *An Introduction to Hydrogen Bonding*, Oxford University Press, New York.

[2] G. C. Pimentel and A. L. McClellan, 1960, *The Hydrogen Bond*, Freeman, San Francisco.

[3] S. Scheiner, 1997, *Hydrogen Bonding: A Theoretical Perspective*, Oxford University Press, New York.

[4] A. Shoji, S. Ando, S. Kuroki, I. Ando, and G. A. Webb, 1993, *Annu. Rep. NMR Spectrosc.*, 26, 55–98.

[5] N. Asakawa, T. Kameda, S. Kuroki, H. Kurosu, S. Ando, I. Ando, and A. Shoji, 1998, *Annu. Rep. NMR Spectrosc.*, 35, 55–137.

[6] S. Kuroki, K. Yamauchi, I. Ando, A. Shoji, and T. Ozaki, 2001, *Curr. Org. Chem.*, 5, 1001–1016.
[7] I. Ando, S. Kuroki, H. Kurosu, and T. Yamanobe, 2001, *Prog. Nucl. Magn. Reson. Spectrosc.*, 39, 79–133.
[8] S. Ando, T. Yamanobe, I. Ando, A. Shoji, T. Ozaki, R. Tabeta, and H. Saitô, 1988, *J. Am. Chem. Soc.*, 110, 3380–3386.
[9] N. Asakawa, S. Kuroki, H. Kurosu, I. Ando, A. Shoji, and T. Ozaki, 1992, *J. Am. Chem. Soc.*, 114, 3261–3265.
[10] N. Asakawa, H. Kurosu, I. Ando, A. Shoji, and T. Ozaki, 1994, *J. Mol. Struct.*, 317, 119–129.
[11] K. Tsuchiya, A. Takahashi, N. Takeda, N. Asakawa, S. Kuroki, I. Ando, A. Shoji, and T. Ozaki, 1995, *J. Mol. Struct.*, 350, 233–240.
[12] T. Kameda, N. Takeda, S. Kuroki, S. Ando, I. Ando, A. Shoji, and T. Ozaki, 1996, *J. Mol. Struct.*, 384, 17–23.
[13] T. Kameda and I. Ando, 1997, *J. Mol. Struct.*, 412, 197–203.
[14] N. Takeda, S. Kuroki, H. Kurosu, and I. Ando, 1999, *Biopolymers*, 50, 61–69.
[15] J. Herzfeld and A. E. Berger, 1980, *J. Chem. Phys.*, 73, 6021–6030.
[16] Q. Teng, M. Igbal, and T. A. Cross, 1992, *J. Am. Chem. Soc.*, 114, 5312–5321.
[17] A. B. Biswast, E. W. Hughes, B. D. Sharma, and J. N. Wilson, 1968, *Acta Crystallogr.*, B24, 40–50.
[18] S. N. Rao and R. Parthasarathy, 1973, *Acta Crystallogr.*, B29, 2379–2388.
[19] T. F. Koetzle, W. C. Hamilton, and R. Parthasarathy, 1972, *Acta Crystallogr.*, B28, 2083–2090.
[20] A. Naito, S. Ganapathy, K. Akasaka, and C. A. McDowell, 1981, *J. Chem. Phys.*, 74, 3190–3197.
[21] M. Mehring, 1983, *Principles of High Resolution NMR in Solids*, Springer-Verlag, Berlin.
[22] Z. Gu and A. McDermott, 1993, *J. Am. Chem. Soc.*, 115, 4282–4285.
[23] S. Kuroki, S. Ando, I. Ando, A. Shoji, T. Ozaki, and G. A. Webb, 1990, *J. Mol. Struct.*, 240, 19–29.
[24] S. Kuroki, N. Asakawa, S. Ando, I. Ando, A. Shoji, and T. Ozaki, 1991, *J. Mol. Struct.*, 245, 69–80.
[25] R. E. Stark, L. W. Jelinski, D. J. Ruben, D. A. Torchia, and R. G. Griffin, 1983, *J. Magn. Reson.*, 55, 266–273.
[26] Y. Hiyama, C. H. Niu, J. V. Silverton, A. Bavoso, and D. A. Torchia, 1988, *J. Am. Chem. Soc.*, 110, 2378–2383.
[27] S. Kuroki, A. Takahashi, I. Ando, A. Shoji, and T. Ozaki, 1994, *J. Mol. Struct.*, 323, 197–208.
[28] S. Kuroki, I. Ando, A. Shoji, and T. Ozaki, 1992, *J. Chem. Soc., Chem. Commun.*, 433–434.
[29] S. Kuroki, S. Ando, and I. Ando, 1995, *Chem. Phys.*, 195, 107–116.
[30] A. Takahashi, S. Kuroki, I. Ando, T. Ozaki, and A. Shoji, 1998, *J. Mol. Struct.*, 442, 195–199.
[31] S. Kuroki, K. Yamauchi, H. Kurosu, S. Ando, I. Ando, A. Shoji, and T. Ozaki, 1999, *Modeling NMR Chemical Shifts, ACS Symp. Ser.*, 732, 126–137.
[32] K. Yamauchi, S. Kuroki, I. Ando, A. Shoji, and T. Ozaki, 1999, *Chem. Phys. Lett.*, 302, 331–336.

[33] K. Yamauchi, S. Kuroki, and I. Ando, 2002, *J. Mol. Struct.*, 602/603, 171–175.

[34] G. Wu and S. Dong, 2001, *J. Am. Chem. Soc.*, 123, 9119–9125.

[35] B. C. Gerstein, R. G. Pembleton, R. C. Wilson, and L. M. Wyan, 1977, *J. Chem. Phys.*, 66, 361–362.

[36] a: A. McDermott and C. F. Ridenour, 1996, *Encyclopedia of NMR*, D. M. Grant and R. K. Harris, Eds., John Wiley, New York, p. 3820, b: Y. Wei and A. E. McDermott, 1999, *Modeling NMR Chemical Shifts, ACS Symp. Ser.*, 732, 177–193.

[37] A. Shoji, H. Kimura, T. Ozaki, H. Sugisawa, and K. Deguchi, 1996, *J. Am. Chem. Soc.*, 118, 7604–7607.

[38] A. Shoji, H. Kimura, and H. Sugisawa, 2002, *Annu. Rep. NMR Spectrosc.*, 45, 69–150.

[39] K. Yamauchi, S. Kuroki, K. Fujii, and I. Ando, 2000, *Chem. Phys. Lett.*, 324, 435–439.

[40] K. Yamauchi, S. Kuroki, and I. Ando, 2002, *J. Mol. Struct.*, 602/603, 9–16.

[41] A. Bielecki, A. C. Kolbert, and M. H. Levitt, 1989, *Chem. Phys. Lett.*, 155, 341–346.

[42] (a) M. H. Levitt, A. C. Kolbert, A. Bielecki, and D. J. Ruben, 1993, *Solid State Nucl. Magn. Reson.*, 2, 151–163, (b) S. Hafner and H. W. Spiess, 1996, *J. Magn. Reson., Ser. A*, 121, 160–166, (c) S. Hafner and H. W. Spiess, 1997, *Solid State Nucl. Magn. Reson.*, 8, 17–24.

[43] S. Hori, K. Yamauchi, S. Kuroki, and I. Ando, 2002, *Int. J. Mol. Sci.*, 3, 907–913.

[44] H. Kimura, A. Shoji, H. Sugisawa, K. Deguchi, A. Naito, and H. Saitô, 2000, *Macromolecules*, 33, 6627–6629.

[45] S. Ono, S. Kuroki, I. Ando, H. Kimura, and K. Yamauchi, 2002, *J. Mol. Struct.*, 602/603, 49–58.

[46] Y. Hiyama, J. V. Silverton, D. A. Torchia, J. T. Gerig, and S. J. Hammond, 1986, *J. Am. Chem. Soc.*, 108, 2715–2723.

[47] C. M. Gall, J. A. Diverdi, and S. J. Opella, 1981, *J. Am. Chem. Soc.*, 103, 5039–5043.

[48] M. H. Frey, J. A. DiVerdi, and S. J. Opella, 1985, *J. Am. Chem. Soc.*, 107, 7311–7315.

[49] P.-G. Jönsson and Å. Kvick, 1972, *Acta Crystallogr.*, B28, 1827–1833.

[50] Å. Kvick, A. R. Al-Karaghouli, and T. F. Koetzle, 1977, *Acta Crystallogr.*, B33, 3796–3801.

[51] P. Michel, H. J. Kochet, and G. Germain, 1970, *Acta Crystallogr.*, B26, 410–417.

[52] F. H. C. Crick and A. Rich, 1955, *Nature (London)*, 176, 780–781.

[53] S. Arnott and S. D. Dover, 1967, *J. Mol. Biol.*, 30, 209–212.

[54] T. Chiba, 1964, *J. Chem. Phys.*, 41, 1352–1358.

[55] R. Blinc and D. Hadzi, 1966, *Nature*, 212, 1307–1309.

[56] M. J. Hunt and A. L. MaCkay, 1974, *J. Magn. Reson.*, 15, 402–414.

[57] W. T. Astbury, C. E. Dalgleish, S. E. Darmon, and G. B. B. M. Sutherland, 1948, *Nature (London)*, 69, 596–600.

[58] W. M. Carson and M. L. Hackert, 1978, *Acta Crystallogr.*, B34, 1275–1280.

[59] V. Lalitha, E. Subramanian, and R. Parthasarathy, 1986, *Ind. J. Pept. Protein Res.*, 27, 223–228.

[60] J. P. Glusker, H. L. Carrel, M. Berman, B. Gallen, and R. M. Peck, 1977, *J. Am. Chem. Soc.*, 99, 595–601.

[61] V. Lalitha, R. Murali, and E. Subramanian, 1986, *Ind. J. Pept. Protein Res.*, 27, 472–477.

[62] V. Lalitha and E. Subramanian, 1985, *Ind. J. Pure Appl. Phys.*, 23, 506–508.

Chapter 9

FIBROUS PROTEINS

Fibrous proteins such as collagen, silk fibroin, keratin are elongated and often play structural roles because of their rigid or elastic properties, as in polysaccharides as their counterpart in plant kingdom. They take regular secondary structures such as collagen triple helix, β-sheet, and α-helix, depending upon their primary structure and sample history. Conformation and dynamics of these fibrous proteins have been extensively analyzed with preparations from natural abundance source by solid state NMR since earlier days for development of solid state NMR as a complementary means to X-ray fiber diffraction studies. Their spectral features are usually simple compared with those of the globular and membrane proteins, because single regular secondary structures consisting of a limited number of amino acid residues are present. In such studies, solid state NMR, using appropriate synthetic polypeptides as structural models, turned out to be very useful as an alternative means to fiber X-ray diffraction. The major advantage of solid state NMR approach is that amorphous preparations that are not tractable by X-ray diffraction can be equally examined together with their crystalline preparations. Therefore, conformational changes associated with crystallization can be very conveniently monitored by solid state NMR.

9.1. Collagen Fibrils

Collagen is the main constituent of higher animal frameworks such as the bones, tendons, skin, ligaments, blood vessels, and supporting membraneous tissues. In spite of this great diversity of role, there are only about a dozen distinct, but closely related types of polypeptide chains. In particular, collagen has a rodlike shape, with a dimension of about 3000 × 15 Å and is composed of a triple-stranded helix, which is assembled into cylindrical fibrils having a diameter of 50–2000 Å.[1] The individual triple helical chains are composed of repeating pattern (Gly-X-Y)$_n$ where X and Y are frequently occupied by prolyl and 4-hydroxyprolyl residues, respectively. The most characteristic feature of amino acid composition for vertebrate collagen is the constancy of the content of Gly (33 ± 1.3%), Ala (10.8 ± 0.9%), Pro (11.8 ± 0.9%), and Hyp (9.1 ± 1.3%).[2] Therefore, these amino (imino) acid residues constitute 65% of the total residues. Further, 72% of the total ^{13}C peaks for collagen could be ascribed to those of Gly, Ala, Pro, Hyp, and Glu, when Glu (7.4 ± 10%) is taken into account, although the rest is an assembly of minor amino acid residues. In region where the tripeptide repeating pattern exists, the polypeptide chains have a conformation related to the form II (3_1-helix) of (Gly)$_n$ and (Pro)$_n$ but three chains are twisted around each other in a supercoiled conformation to result in the triple helix.[2,3] A number of synthetic polytripeptides (Gly-X-Y)$_n$ taking the triple-helix conformation similar to that of collagen as proved by X-ray diffraction studies have been extensively utilized,[2,3] as model systems for conformational characterization of collagen fibrils.

Accordingly, conformational characterization of collagen fibril is feasible based on the isotropic ^{13}C NMR signals of major amino acid residues, with reference to the conformation-dependent displacement of ^{13}C chemical shifts from collagen and these model polypeptides (see Table 6.2).

9.1.1. Conformational Characterization of Collagen Fibrils: High-Resolution NMR Studies

As illustrated in Figure 9.1, the 75.46 MHz ^{13}C CP-MAS NMR spectra of collagen fibrils from bovine tendon as well as those of model peptides taking either the collagen-like triple helix or 3_1-helix give rise to well-resolved ^{13}C NMR peaks, when they are recorded at higher magnetic field, either 7.0 or 9.3 T.[4,5] This is in contrast to the ^{13}C NMR spectra taken at a lower magnetic field (1.4 T)[6] in which the observed ^{13}C NMR signals were not well resolved, because individual C_α and carbonyl peaks of respective residues are split into asymmetric doublet patterns owing to the presence of the second order ^{14}N quadrupole effect on the ^{14}N–^{13}C dipolar interaction at such lower field.[7,8] The assignment of the well-resolved ^{13}C peaks is straightforward with reference to the data of the conformation-dependent displacements of ^{13}C chemical shifts so far accumulated (see Tables 6.1 and 6.2). Naturally, the ^{13}C chemical shifts of Pro, Ala, Gly, and Hyp residues from collagen fibrils are very close to those of model systems, (Pro-Ala-Gly)$_n$ and (Pro-Pro-Gly)$_n$, and (Hyp)$_n$ taking the similar triple helix and 3_1-helix,

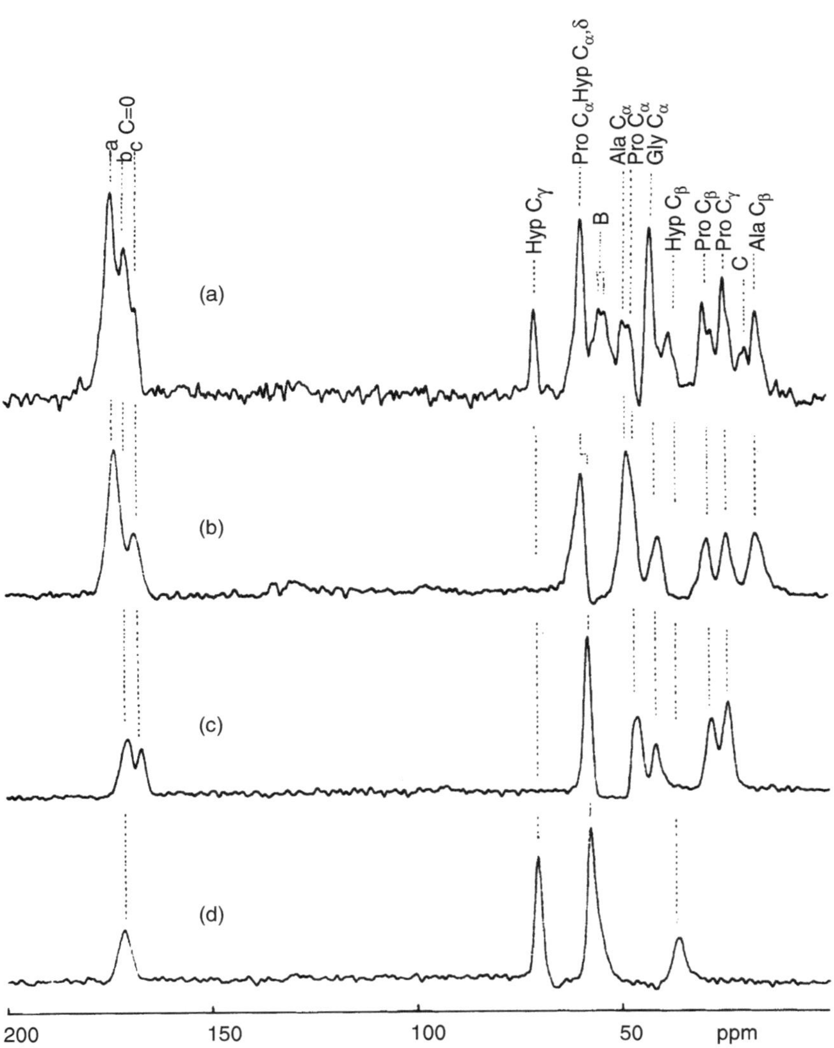

Figure 9.1. 75.46 MHz ^{13}C CP-MAS NMR spectra of collagen from bovine Achilles tendon (a) and of model polypeptides taking collagen-like triple helix [b, $(Pro\text{-}Ala\text{-}Gly)_n$] and [c, $(Pro\text{-}Pro\text{-}Gly)_{10}$], or 3_1-helix [d, $(Hyp)_n$].

respectively. The forms I and II of Pro residue can be readily distinguished by means of the difference in the ^{13}C chemical shifts between the C_β and C_γ ($\Delta_{\beta,\gamma}$) carbons: 9.3 and 2.4 ppm, respectively, which are within the range of values for *cis*- and *trans*-X-Pro in solution.[9,10] The peaks B and C, however, were ascribed to the minor amino acid residues which amount to ca. 35%. It is difficult, however, to locate signals of the corresponding $C_β$ and other side-chain signals from such minor components. This is because such $C_β$ ^{13}C NMR signals could be significantly suppressed for a variety of peptides and membrane proteins even in the solid state,[11–16] as a result of interference of fluctuation frequency with frequency of the proton decoupling, leading to failure of attempted peak-narrowing (see Section 3.3). With recent introduction of high-field magnet (750 MHz for ^{1}H), it is possible to study collagen in native tissues such as cartilage.[16a]

^{13}C spin–lattice relaxation times of collagen fibrils and model polypeptides are substantially different among carbons of a variety of amino acid residues at different positions, as summarized in Table 9.1. The very short ^{13}C T_1 values of the methyl carbons are mainly caused by dipolar interactions with the methyl protons undergoing rapid C_3 rotation in a timescale of 10^{-8} s.[17] The obvious gradient in the T_1 values from the terminal methyl to the backbone carbon, as observed for Leu residue in Pro-Leu-Gly-NH_2 (Table 9.2) and Ala residues in the others can be similarly explained. Further, the ^{13}C spin–lattice relaxation times of both the $C_β$ and $C_γ$ carbons of Pro and Hyp in fibrils are substantially reduced (1–5 s) as compared with those of some crystalline oligopeptides (20–30 s). Interestingly, another type of T_1 gradient is also seen for Pro and Hyp residues in collagen, collagen-like peptides, and oligopeptides: $NT_1^{\gamma} < NT_1^{\beta} < NR_1^{\delta} < NT_1^{\alpha}$, where NT_1^N stands for the T_1 value at the C^N position. The presence of interconversion between two half-chair forms is responsible for the relaxation process, with lifetime of 10^{-11}–10^{-12} s for the two states and a typical range of motion for the C–H vector of 50–70°, in contrast to the data in aqueous solution by London.[18] It was also shown that imino residues in ^{2}H-labeled collagen, DL-proline, and DL-proline hydrochloride have flexible rings with root-mean-square angular fluctuations in the 11–30° range at 22°C,[19,20] as revealed by ^{2}H NMR line shape analysis.

9.1.2. Dynamics Studies of Collagen Based on ^{2}H or ^{13}C Powder Patterns

Dynamics of collagen fibrils can also be studied by analysis of ^{2}H quadrupole interaction or ^{13}C CSA available from their respective powder patterns sensitive to motions with frequency of 10^6 and 10^4 Hz, respectively. The T_1 and NOE values obtained for collagen fibrils enriched with [2-^{13}C]- or [1-^{13}C]Gly are similar to the values obtained in solution as shown in Table 9.3, suggesting that the molecule reorients in the fibril.[21–23] Nevertheless, free rotation about the long axis can be excluded in view of

Table 9.1. ^{13}C spin–lattice relaxation times of collagen and its model polypeptides (s).[4]

	3_1-Helix				Triple helix		Collagen
	$(Gly)_n$	$(Pro)_n$	$(Hyp)_n$	$(Ala\text{-}Gly\text{-}Gly)_n$	$(Pro\text{-}Ala\text{-}Gly)_n$	$(Pro\text{-}Gly\text{-}Pro)_n$	Skin
Gly C_α	6.7			9.8	15	6.6	4.8
C=O	12			21	*	21	17
Ala C_α				22	15*		7.1
C_β				0.80	0.68		0.64
C=O			23		34*		22
Pro C_α		11			28	11	8.7
C_β		2.8			5.2	2.8	1.4
C_γ		2.3			5.2	2.4	2.4
C_δ		4.8			15*	6.8	5.5
C=O		14			34*	17	22
Hyp C_α			16*				7.0
C_β			1.2				2.0
C_γ			1.3				4.9
C_δ			5.5				7.0
C=O			16				

*Overlapped signals.

Table 9.2. ^{13}C Spin–lattice relaxation times of some tripeptides (s).[5]

	Gly-Pro-Ala	Ala-Gly-Gly	Ala-Pro-Gly	Pro-Leu-Gly-NH_2
Ala C_α	6.8	5.0	6.0	
C_β	1.7	1.1	1.5	
C=O	20	17	29	
Pro C_α	31		32	750
C_β	23		2.4	28
C_γ	20		2.0	17
C_δ	30		–	63
C=O	25		31	–
Leu C_α				750
C_β				35
C_γ				13
C_δ				1.0
Gly C_α	29	29, 17 (C-terminal)	34	130
C=O	33	25, 11 (C-terminal)	19	

anisotropic ^{13}C and ^{2}H NMR powder patterns illustrated in Figures 7.7 and 7.8. The simplest model for reorientation that is compatible with the NMR data is a two-site model in which the molecule is assumed to jump between

Table 9.3. T_1 and nuclear Overhauser enhancement (NOE) values measured for collagen labeled with [2-^{13}C]- and [1-^{13}C]glycine.[23]

	T_1(s)		NOE	
	Solution	Fibrils	Solution	Fibrils
[2-^{13}C]Gly-collagen	0.03	0.13	1.44	1.74
[1-^{13}C]Gly-collagen	0.54	2.0	1.52	1.60

two azimuthal orientations separated by an angle $\Delta\phi$. For simplicity, the residence time in each orientation is assumed to be the same, and that the time required to jump from one orientation to the other is negligible compared to the residence time: $\Delta\phi \simeq 30°$ if $R_j \simeq 10^8\ \mathrm{s}^{-1}$ (R_j^{-1} is the residence time). Consistent with the relaxation data, the carbonyl signal of fibril has the line shape of axially asymmetric chemical shift tensor with a width of 100 ppm as compared with that of 150 ppm of polycrystalline Gly-Gly as static reference. It is also possible to obtain $\Delta\phi$ from analysis of ^{2}H NMR line shape of [3,3,3-$^2H_3$]Ala-labeled collagen fibril,[24] because the deuterium field gradient tensor is axially symmetric and that the symmetric axis is along with the C–^{2}H bond axis. The molecule in solution can undergo free rotation about its long axis, ($R_1 \simeq 10^7\ \mathrm{s}^{-1}$), whereas end-over-end rotation is slow ($R_2 \simeq 10^2\ \mathrm{s}^{-1}$). The $\approx$37 kHz quadrupolar splitting observed for the frozen fibrils would be expected to collapse by a factor of $(1–3\cos^2\theta)/2$ in the presence of free rotation about the long axis of the collagen molecule (see Section 4.4). Here, θ is the angle formed between the C_α–C_β bond axis and the long axis of the molecule. The observed quadrupolar splitting for labeled collagen in solution is $\approx$10 kHz, which requires the θ to be $\approx 70°$. Further, the best fit of the data indicates that the collagen molecule in the fibril undergoes reorientation over a $\approx$ 30–40° range in azimuthal angle (that is $2\delta \approx$ 30–40°). This angle is in good agreement with an estimate based on the above-mentioned ^{13}C T_1 data, $2\delta \approx 30°$. The backbone[21–24] as well as side-chain motions[25] with different timescale in the collagen fibrils were also discussed in Chapter 7 (see Figures 7.4, 7.7, and 7.8).

9.2. Elastin

Elastin is a major protein constituent of connective tissue such as skin, lung, vessel, and ligament, conferring upon these tissues macroscopic rubber-like properties of high extensibility and small elastic modulus.[26,27] Elastin is an extremely insoluble molecule due to the extensive cross-linking by desmosine and isodesmosine at Lys residue to form an insoluble three-dimensional elastic network. Its soluble precursor, tropoelastin as a long single polypeptide chain containing about 850 residues, with $M_r \approx 70{,}000$, is deposited into the extracellular space and rapidly forms elastic fiber with the help of several microfibril proteins. Two major types of alternating domains are found in tropoelastin: (1) hydrophilic cross-linking domains rich in Lys and Ala, and (2) hydrophobic domains rich in Val, Pro, Ala, and Gly, which often occur in repeats of VPGVG or VGGVG.[27] In this connection, Urry and coworkers[28,29] proposed that the VPGVD unit adopts a type II β-turn structure around the Pro-Gly pair, repetition of which gives rise to a right-handed helix termed β-spiral. As to the network structure, two types of models, the single-phase model[30] and two-phase model,[31,32] have been proposed to explain the behavior of elastin. The single-phase model is known as a random chain model, which considers elastin to be like a typical rubber and any diluent present in the system is randomly distributed throughout the swollen material.[30] This model, however, does not agree with the reversible structural changes in elastin as indicated by fluorescence probe when it is stretched.[33] The latter model assumes that elastin swollen in water is a two-phase system composed of globular domains of proteins (connected by cross-links) and the aqueous diluent consigned to the spaces between the domains.

Based on characterization of the most common motifs in the structure and sequence of insoluble elastin, a number of peptides have been proposed as a model for elastin. The proposed β-spiral form[28] for $(\mathrm{VPGVG})_n$ is not consistent with the random coil structure of the single-phase model as mentioned above, however. It was also shown that short $(\mathrm{VPGVG})_n$, where n varies between 1 and 5, undergoes an extended ↔ β-turn transition with increasing temperature, suggesting that the induction of the β-spiral occurs at the level of single pentameric units.[34]

Nevertheless, both experimental[35] and simulation studies[36] suggest that a rigid, well-ordered β-spiral model is not a good description of elastin in water, although local and fluctuating β-spiral structure may be present to a certain degree.

9.2.1. Swollen Elastin

It turned out that such chemically cross-linked elastin fiber exhibits elasticity when it is swollen and yields well-resolved ^{13}C NMR signals by solution NMR technique,[37–39] although unswollen elastin is brittle and has dipolar broadened line widths that cannot be observed by conventional solution NMR spectrometer. Scalar decoupled ^{13}C NMR signals of elastin from calf ligamentum nuchae vary markedly dependent upon solvent (diluent) polarity.[37] An elastin spectrum is not obtained in the solvents such as benzene and ethanol of lowest polarity because of line widths broader than 10^3 Hz. In contrast, line widths for backbone carbons are of the order of 50 Hz in the highly polar solvents, DMSO, 0.15 M NaCl and formamide, as a result of swelling by these diluents. The resulting correlation times for elastin in 0.15 M NaCl from measured T_1, line width, and NOE values were 55, 500, and 6 ns, respectively, using a single correlation time model. This is because the backbone segmental motion is highly heterogeneous: the T_1 and NOE values are sensitive to the fast motions, while the line width is rather sensitive to slow motions. Therefore, Lyerla and Torchia[38] used a log χ^2 distribution of correlation times introduced by Schaefer,[40,41] defined by

$$G_p(\mathrm{s}) = (\mathrm{ps})^{p-1}\mathrm{e}^{-p\mathrm{s}}\mathrm{p}/\Gamma(p) \tag{9.1}$$

$$s = \log_b[1 + (b-1)\tau/\tau], \tag{9.2}$$

in order to obtain consistent correlation times, $\bar{\tau}$, based on T_1, line width (Δ), and NOE values. Here, $G_p(\mathrm{s})$ is the probability of finding a correlation time t corresponding to $s = s(t)$. The width of the distribution is governed by the variables, p and b, and the use of a log $(\tau/\bar{\tau})$ argument in defining G allows very broad distributions to be employed. A comparison of the calculated $\bar{\tau}$ values for elastin, assuming a distribution function having $p = 14$, $b = 1000$ employed for *cis*-polyisoprene leads to several interesting conclusions about elastin chain motion (Table 9.4). Comparable $\bar{\tau}$ values were obtained for *cis*-polyisoprene (0.4 ns)[37] and elastin swollen by solvents containing a polar organic component (1.8–2.7 ns) in agreement with a variety of studies that have shown that elastin behavior in these solvents approximates that of an ideal rubber. Thus, the NMR data indicate that a viscoelastic rubber-like

Table 9.4. Comparison of $\bar{\tau}$ values (ns) obtained for elastin[27] and *cis*-polyisoprene[29] using a log χ^2 distribution of correlation times having $p = 14$, $b = 1000$.

Sample	Solvent	$\bar{\tau}$
cis-Polyisoprene	None	0.4
Elastin	0.15 M NaCl-formamide	1.8
Elastin	0.15 M NaCl-ethanol	2.5
Elastin	DMSO	2.7
Elastin	0.15 M NaCl	80

network is an appropriate model for the mobile elastin component observed in the scalar decoupled spectra.

9.2.2. Unswollen Elastin and Elastin-Mimetic Polypeptides

Naturally, CP-MAS NMR approach is the most appropriate means to record ^{13}C NMR signals from brittle samples of unswollen elastin. Kumashiro and coworkers recorded solid state NMR to characterize lyophilized α-elastin, an acid-soluble form of elastin prepared by treatment with oxalic acid, from normal and undercross-linked pig aortas.[42] Chemical shifts of various peaks indicated that the overall structures of normal and undercross-linked elastin were similar but that there was more mobility in the normal protein as manifested from obviously shortened ^{13}C spin–lattice relaxation times in the rotating frame, ^{13}C $T_{1\rho}$, although ^{13}C T_1 and ^{1}H $T_{1\rho}$ values are not significantly varied. The data suggest that loss of function in the elastic fiber appears to be correlated with changes in the motions or fluctuation on the kHz timescale.

To clarify a pivotal role of water for structure–function relationship of elastin, ^{13}C CP-MAS NMR spectra of lyophilized and hydrated elastin from bovine nuchal ligament were examined for samples with different degrees of hydration (0–100%) calculated on the basis of 100% hydration being equal to 60% water by mass.[43] There is a marked spectral change at 30% hydration between fully hydrated and dehydrated states recorded by the CP-MAS NMR (Figure 9.2). In the hydrated elastin, the carbonyl peak at 173 ppm, C_α carbons at 60, 57, and 43 ppm (Gly C_α) and aliphatic side chain at 20–30 ppm were substantially suppressed as compared with those of lower hydration less than 30%. The peak-intensities of the former three peaks, however, were recovered when the corresponding ^{13}C NMR spectra were recorded by DD-MAS NMR because of significant mobility in fully hydrated elastin. In fact, the aliphatic carbons of hydrated elastin have very short T_1 values (<1 s) as compared with those of lyophilized preparation at 37 °C (12 and 2 s, respectively), as a result of motions of the protein with the timescale of 10^{-8} s. In addition, the hydrated sample at 37 °C has ≈1 ms of ^{1}H $T_{1\rho}$ value, in contrast to the case of lyophilized preparation as 8–9 ms. This is the consequence of the presence of lower frequency motions in the order of kHz in addition to the motions of 10^{-8} s. [1-^{13}C]Gly-, [2-^{13}C]Gly-, or [^{15}N]Gly-labeled elastin samples were prepared with the use of primary cultures of neonatal rat smooth muscle cells (NRSMC) to provide abundant quantities of insoluble elastin and elastic fibers.[44] ^{13}C CP-MAS NMR spectrum of hydrated [1-^{13}C]Gly-elastin exhibits the peak at 171.9 ppm with a width of ca. 200 Hz. The ^{13}C T_1 of this peak is 5.1 (±0.5) s, consistent with that of preparation of natural abundance in hydrated state.[43] The signal-to-noise ratio of the carbonyl signal recorded by DD-MAS experiment is actually much greater than that observed by CP-MAS experiment, because several portions of the backbone were not observable with CP because of motions involved.

Solid state NMR methods were used[45] for examination of polypeptide and hydrate ordering in both elastic (hydrated) and brittle (dry) elastin fibers: (1) tightly bound waters are absent in both dry and hydrated elastin and (2) that the backbone in the hydrated protein is highly disordered with large amplitude motions. The hydrate was studied by ^{2}H and ^{17}O NMR and no evidence of solid-like ^{2}H and ^{17}O signals was available from fully hydrated elastin. An upper limit of the order parameter, $S = \langle\Delta\sigma\rangle/\Delta\sigma < 0.1$, was determined for the backbone carbonyl group in hydrated elastin, although $S \approx 0.9$ is obtained in most proteins. Here $\langle\Delta\sigma\rangle$ and $\Delta\sigma$ denote the width of the ^{13}C CSA for the hydrated and dry elastin. A useful, albeit approximate model frequently used for interpreting backbone order parameters is based on the picture of a vector rotationally diffusing within a cone of semi-angle θ_0. In this model, $S = \cos\theta_0(1 + \cos\theta_0)/2$ [46] and the upper limit for S established here corresponds to $\theta_0 > 80°$.

^{13}C CP-MAS NMR spectra of elastin-mimetic polypeptide, (LGGVG)$_n$, showed that the peptide does not adopt a single conformation in the solid, providing further support to models for elastin that involve significant conformational heterogeneity.[47] A recombinant, elastin-mimetic protein, poly(Lys-25) having a repeat sequence of [(VPGVG)$_4$(VPGKG)]$_{39}$ with molecular weight of 81 kDa was prepared by using microbial protein expression, and its cross-linked polymer at the ε-amino group of lysine by means of *N*-hydroxysuccimide.[48,49] The resulting hydrogel exhibited reversible, temperature-dependent expansion and contraction.

Secondary structure and backbone dynamics of an elastin-mimetic polypeptide "Poly(Lys-25)" were examined:[50–52] type II β-turn structure was dominant at the Pro-Gly pair but a type I β-turn was rejected based on the Pro ^{15}N, ^{13}C$_\alpha$, and ^{13}C$_\beta$ isotropic shifts and the Gly 3 C$_\alpha$ isotropic and anisotropic chemical shifts. Backbone motion increases only slightly below 20% hydration level, whereas both the backbone and the side chains undergo large amplitude motions above 30% hydration,[52] consistent with the data of native elastin (see Figure 9.2).[44] The root-mean-square fluctuation angles as obtained from ^{13}C–^{1}H and ^{1}H–^{1}H dipolar coupling using 2D isotropic–anisotropic correlation experiments are found to be 11–18° in the dry protein and 16–21° in the 20% hydration. The reduced peak intensities by CP-MAS experiment at higher hydration are not always caused by reduced CP-efficiency alone, because the peak-intensities in the DD-MAS NMR are also reduced in the presence of motions of intermediate frequency in the order of 10^4–10^5 Hz. Decreased ^{1}H $T_{1\rho}$ experiment at 30% hydration indicated that

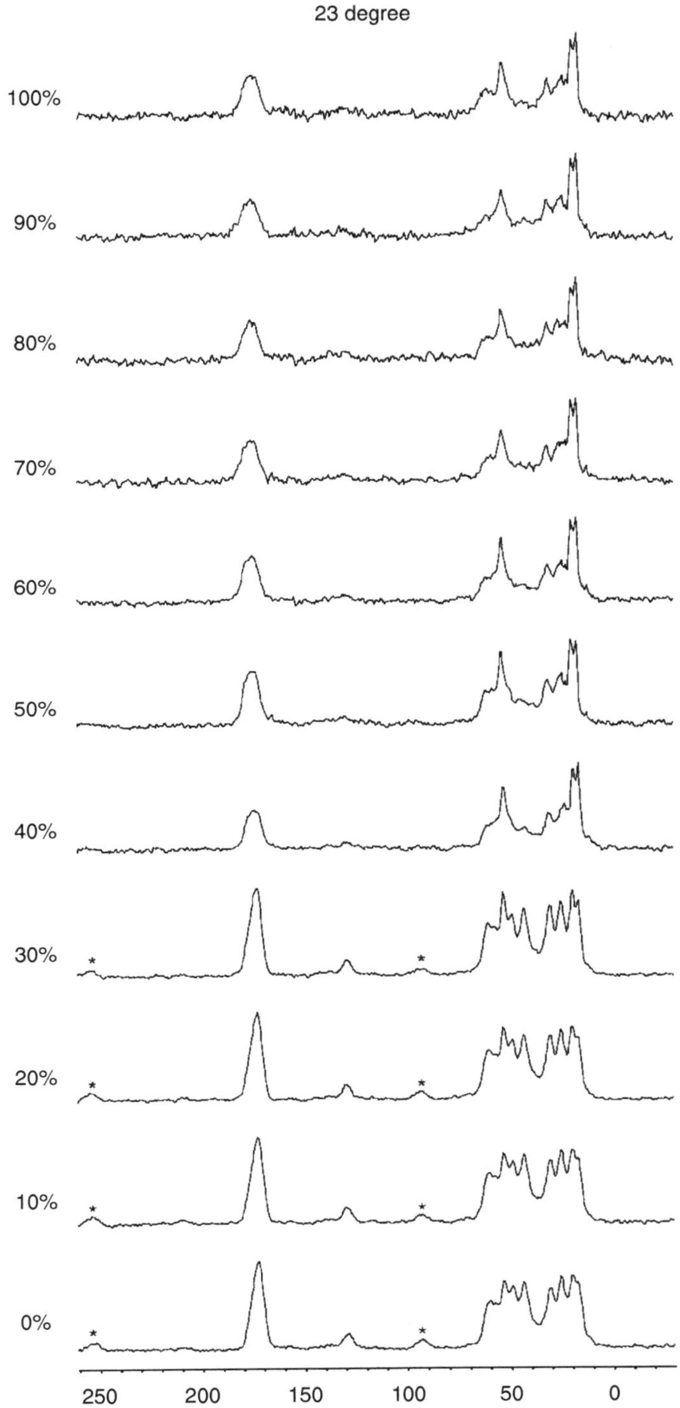

Figure 9.2. ^{13}C CP-MAS NMR spectra of elastin with hydration levels of 0–100% (from Perry *et al.*[43]).

significant motions occur on the microsecond timescale. The other elastin-mimetic peptide, the central part pentameric unit of $(VPGVG)_3$, was analyzed by the determination of a number of distance and torsion angles.[53]

9.3. Cereal Proteins

The prolamin storage proteins account for nearly half of the total proteins present in most mature seeds.[54–56] Most of them are insoluble in water and in salt solutions but some are soluble in alcohol–water mixture. These proteins are very rich in glutamine and proline and are characterized by unusual primary structures consisting of the repetition of well-defined peptide sequence. Three main groups of prolamins are defined according to their average molecular weight as well as to their content of amino acid residues containing sulfur and similar fractions available from different cereals (Table 9.5). The properties of these proteins and particularly their response to hydration are believed to account for the characteristic viscosity and elasticity of these systems.[54–56] The alcohol-insoluble fraction of wheat gluten (glutenin) is believed to be responsible for its elasticity whereas the alcohol-soluble fraction (gliadins) seems to account for viscosity properties. Such properties should also determine the behavior of dough as well as some of the final properties of baked products.[54,56]

The approximate molecular weight of barley C-hordein is 40,000 and its primary structure consists of short N- and C-terminal domains with 12 and 6 residues, respectively, and a large central repetitive domain based on the repeating peptide sequence (consensus: PQQPFPQQ).[57] Gluten is a complex mixture of many protein fractions with different molecular weights and structural properties. Two types of subunits of HMW glutenins are recognized, which are called *x*- and *y*-types, and all comprise three distinct

Table 9.5. Prolamin protein fractions found in wheat, barley, and rye.

Type of fraction	Wheat	Barley	Rye
HMW (high molecular units)	HMW subunits (glutenins)	D-Hordein	HMW secalins
S-Poor	ω-Gliadins	C-Hordein	ω-Secalins
S-Rich	γ-Gliadins	γ-Hordein	γ-Secalins 40 K
	α-Gliadins	–	–
	LMW subunits (Glutenins)	B-Hordein	–
			γ-Secalins 75 K

domains: two nonrepetitive domains at the N- and C-terminus, flanking a central repetitive domain based on three different motifs: hexapeptides (PGQGQQ), nonapeptides (consensus GYYPTSP/LQQ), and, in *x*-type subunit only, tripeptides (consensus GQQ).[58]

Water addition to C-hordein caused more efficient ^{1}H spin–lattice relaxation and that the temperature at which the maximum of spectral density function is moved to lower temperature.[56] This can be accounted for by plasticization of protein enhancing rapid motions (of the order of ω_0, 100 MHz) and chemical exchange. Further, it was shown that ^{13}C NMR peak-intensities of ω-gliadins recorded by CP-MAS NMR were substantially suppressed by hydration with 48% water together with reduced ^{1}H T_1 data. These findings were interpreted in terms of proposed model for network. The ^{13}C CP-MAS and DD-MAS NMR spectra of dry and hydrated barley storage protein, C-hordein, were recorded as a model for wheat *S*-poor prolamins (see Table 9.5), together with those of model synthetic peptides $(Pro)_2(Gln)_6$ (I) and $(Pro\text{-}Gln\text{-}Gln\text{-}Pro\text{-}Phe\text{-}Pro\text{-}Gln\text{-}Gln)_3$ (II) under dry or hydrated conditions.[59] The spectral features of C-hordein as well as these peptides were appreciably different from each other depending upon the extent of hydration, reflecting different domains that adopt different types of conformations as well as dynamics. In particular, considerable proportions of the peak intensities were lost in the CP-MAS spectra, and well-resolved ^{13}C NMR signals emerged in DD-MAS NMR spectra owing to acquisition of molecular motions by swelling. Further, ^{13}C spin–lattice relaxation times of C-hordein and peptide-II were reduced by more than one order of magnitude by hydration, reflecting the presence of well-swollen molecular chains. In contrast, the T_1 values of peptide-I upon hydration remained one-third of those in the dry state. In addition, ^{13}C NMR signals of the aromatic side chain of Phe residues disappeared on hydration owing to interference between the frequency of the acquired flip-flop motion and the proton decoupling frequency. This information gives a new insight into establishing the structural properties of the studied protein system. A model may be put forward for a gel-type structure in which the more rigid part of the system involves intermolecular hydrogen-bonded Gln side chains as well as some hydrophobic ''pocket'' involving Pro and Phe residues.

In order to systematically examine the role of (1) disulfide bonds, (2) irregular chain ends, (3) length of repetitive chain, and (4) heterogeneity on the structure of the hydrated systems, ^{13}C NMR spectra were compared among several kinds of subunits of whole HMW fractions, 1Dx5 SS (single 88 kDa protein with SS bonds), 1Dx5 (single 88 kDa protein without SS bond), 58 kDa peptide (central 1Dx5 protein), and 21 mer (as 1Dx5 repeat unit).[60] For all the systems investigated, a network is formed upon hydration, even for short peptide chains with only 21 monomers. In all cases, the network formed seems to be held together by junction zones (or train sections) involving hydrogen-bonded glutamine side chains, close to hydrophobic interactions established between the side chains of hydrophobic residues. ^{13}C and ^{1}H solid state NMR of HMW glutenin subunit 1Dx 5 were compared with those of the form alkylated to block cysteine residues, *alk* 1Dx5, in order to examine the effect of disulfide bonds on the hydration behavior. As illustrated in Figure 9.3,[61] ^{13}C NMR spectra of both forms are very similar in the dry state but some differences may be found upon hydration. Hydration results in a marked decrease in the peak intensities, mainly because of the increasing mobilization of the protein molecules in the 10^{-8} s frequency range. However, a decrease in the CP-MAS intensity may also be observed if the frequency of internal motions interferes with the proton decoupling frequency (10^{-5} s)[11] or with the MAS frequency (10^{-4} s). The molecular motion of the two-protein dynamics populations was further characterized by ^{13}C T_1 and ^{1}H $T_{1\rho}$, T_2, and T_1 relaxation times. The results suggest that hydration leads to the formation of a network held by a cooperative action of hydrogen-bonded glutamines and some hydrophobic interactions. The presence of disulfide bonds was observed to promote easier plasticization of the protein and the formation of a more mobile network, probably involving a higher number of loops and/or larger loops.

The role of nonrepetitive terminal domain present and the length of the central repetitive domain in the hydration of 1Dx5 were studied by ^{13}C and ^{1}H NMR measurements of an alkylated 1Dx5 subunit (*alk* 1Dx5), a recombinant 58 kDa peptide corresponding to the central repetitive domain of 1Dx5, and two synthetic peptides (with 6 and 21 amino acid residues) based on the consensus repeat motifs of the central domain.[62] These results are consistent with the proposal of a hydrated network held by hydrogen-bonded glutamines and possibly hydrophobic interactions. The nonrepetitive terminal domains were found to induce water insolubility and a generally higher network hindrance. ^{13}C CP-MAS and DD-MAS NMR signals of 21 mer were substantially suppressed at low water content due to the presence of low-frequency fluctuation motions in the megahertz and kilohertz frequencies. This means that network structure upon hydration was formed even for the 21 mer but 6 mer peptide was too short to allow this structure.

9.4. Silk Fibroin

Silk fibroins are the structural proteins synthesized by silkworms for their cocoon, by spiders for their webs, etc. The cocoon silk fibroin has long been the main subject of scientific as well as technological interest, as

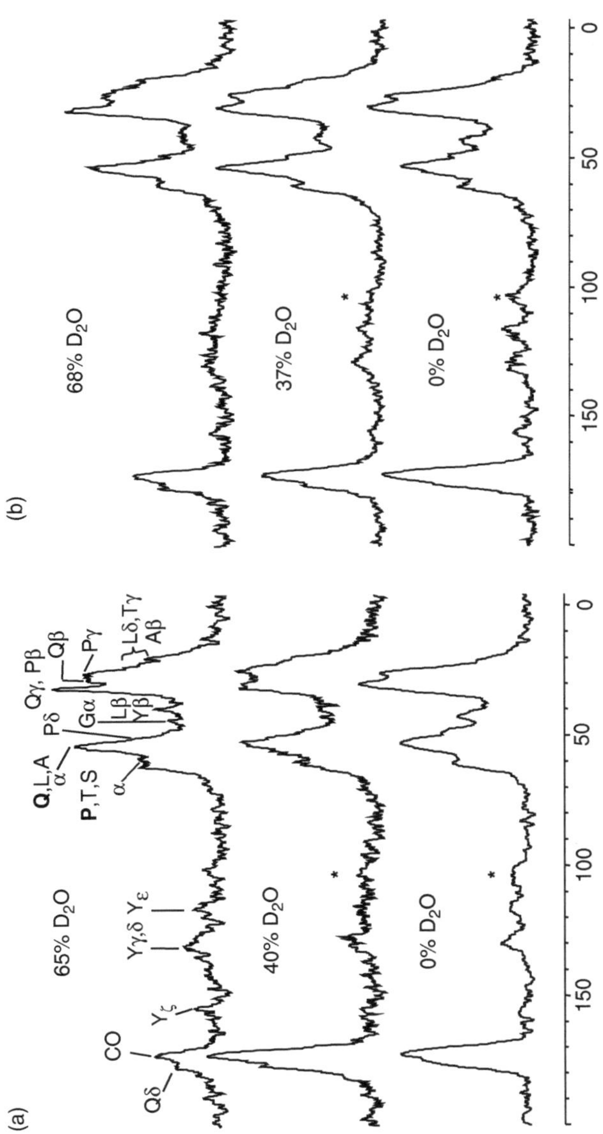

Figure 9.3. ^{13}C CP-MAS NMR spectra of (a) 1Dx5 and (b) *alk* 1Dx5 at different hydration levels (from Alberti *et al.*[61]).

an excellent material for fiber production. The most common fibroin from *Bombyx mori* contains amino acids, Gly (42.9%), Ala (30.0%), Ser (12.2%), and Tyr (4.8%), consisting largely of a repeat sequence of six amino acid residues such as (Gly-Ala-Gly-Ala-Gly-Ser)$_n$.[63] The amino acid composition of cocoon fibroin from *Samia cynthia ricini*, however, contains Ala (48.4%), Gly (33.2%), Ser (5.5%), and Tyr (4.5%). It is known that these silk fibroins take several kinds of conformations, depending on species of silkworms and conditions of sample preparations:[63–66] they take two distinct structures, more stable silk II (antiparallel β-sheet) present in *fiber*s and less stable silk I,[66–68] or α-helical forms in *film* which was dried gently from liquid silk removed from silk gland, for *B. mori* or *S. c. ricini*, respectively.

In recent years, much interest has been paid to spider dragline silk because of high tensile strength that is comparable to steel and is only slightly inferior to Kevlar, and elasticity that is comparable to rubber.[69] Dragline silk is thought to be composed of mainly two proteins, Spidroin I and II.[70] Consensus repeats from the various silk proteins reveals four types of shared amino acid motifs: (1) GPGGX/GPGQQ, (2) GGX, (3) poly-Ala/poly-Gly-Ala, and (4) a ''spacer'' sequence with amino acids that do not conform to the typical amino acid composition of spider silks. Each module is then associated with its impact on the mechanical properties of a silk fiber. In particular, correlations are proposed between an alanine-rich ''crystalline module'' and tensile strength and between a proline-containing ''elasticity module'' and extensibility.

Silk fibroin has proved to be one of the most appropriate objective to be characterized by a variety of solid state NMR techniques so far developed.[71,72]

9.4.1. Cocoon Fibroins

Dimorphic structures of cocoon fibroins are readily distinguished by ^{13}C chemical shifts of individual amino acid residues with reference to the data of their conformation-dependent displacements of ^{13}C chemical shifts (Tables 6.1 and 6.2). The ^{13}C NMR peaks of Ala, Gly, and Ser residues of *B. mori* fibroin are well separated in the dimorphs as illustrated in Figure 9.4,[73] because the amino acid composition is limited to the following four kinds of residues: Gly, Ala, Ser, and Tyr, as mentioned above. A later study[74] showed that the Tyr C_α peak is superimposed on the Ser C_α peak and Tyr C_β peak is suppressed, while the side-chain peaks (C_γ, C_δ, C_ε, and C_ζ) are well resolved. The improved spectral resolution to give the three peaks at the carbonyl signals was achieved when highly crystalline preparation was

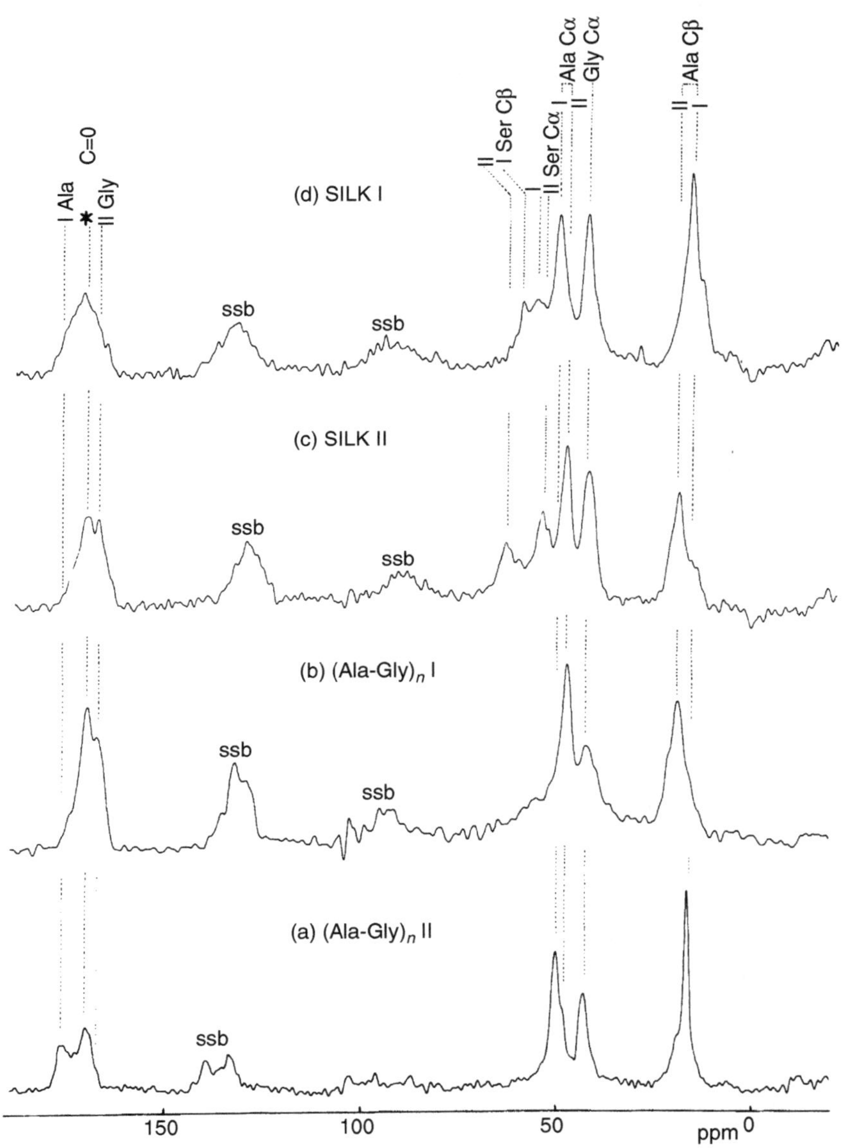

Figure 9.4. ^{13}C CP-MAS NMR spectra of the crystalline fraction from *B. mori* silk fibroin and (Ala-Gly)$_n$ in the solid state. I and II stands for the peaks from silk I and II, respectively.[75]

used.[75] The suppressed C_β signals, as encountered for Tyr residue, were also noted for a variety of minor amino acid residues in collagen fibrils[5] as a result of interference of frequencies of fluctuation motions with decoupling frequency.[11–16] Consistent with expectations based on earlier X-ray diffraction and IR data, the ^{13}C chemical shifts of Ala and Gly residues in silk I and silk II are the same as those of (Ala-Gly)$_n$ II and I, respectively. In a similar manner, the ^{13}C NMR signals are well distinguished between two kinds of preparations with the α-helix and β-sheet forms for *S. c. ricini* fibroin as major forms (Figure 9.5). The former and latter samples were obtained by casting liquid silk (film) directly taken from the posterior silk gland of the mature larva on a PMMA plate and precipitate of *S. c. ricini* fibroin digested by chymotrypsin, respectively. The carbonyl signals at 176.2 and 172.3 ppm are readily ascribed to Ala and Gly residues taking the α-helix conformation [176.4 and 171.6 ppm from (Ala)$_n$ and (Ala, Gly*)$_n$, random copolymer, 5% Gly* ([1-^{13}C] Gly), respectively] (see Tables 6.1, 6.2 and 9.1). In addition, the C_α and C_β signals of Ala residue (52.5 and 15.7 ppm, respectively) are consistent with those of (Ala)$_n$ (52.4 and 14.9 ppm, respectively).

The primary structure of *B. mori* fibroin is mainly Ala-Gly alternative copolypeptide, but Gly-Ala-Ala-Ser units appear frequently and periodically. The role of such Gly-Ala-Ala-Ser units was examined for the secondary structure of sequential model peptides containing Gly-Ala-Ala-Ser units selected from the primary sequence of *B. mori* fibroin by ^{13}C CP-MAS NMR.[76] It was shown that the presence of the Ala-Ala units in *B. mori* fibroin chain will act as one of the inducing factors of the structural transition for silk fiber formation. The structural role of tyrosine in *B. mori* fibroin was also studied by the introduction of one or more Tyr into (AG)$_{15}$.[77] Silk II is able to accommodate a single Tyr residue as viewed from ^{13}C NMR spectra. Interestingly, silk I remains stable when Tyr is positioned near the chain terminus (AG)$_{12}$YG(AG)$_2$, but the conformation is driven towards silk II when Tyr is located in the central region of (AG)$_7$YG(AG)$_7$. Closer examination of ^{13}C NMR spectra of silk II structure of *B. mori* fibroin and some model peptides including (AG)$_{15}$ revealed a broad and asymmetric peak, yielding three isotropic chemical shifts at 22.2, 19.6, and 16.7 ppm for Ala C_β signals with relative fractions of 27%, 46%, and 27%.[78] The former two peaks can be ascribed to an antiparallel β-sheet conformation based on recent X-ray diffraction results of the silk II structure of *B. mori* fibroin.[79] The last peak is naturally ascribed to the remaining silk I, although they claimed as a ''distorted β-turn'' structure. They showed that such a structural change from silk I to silk II forms can be observed for (AG)$_n$ ($n > 9$) and stretching of fibroin fiber yielded such heterogeneous structure depending upon its ratio.

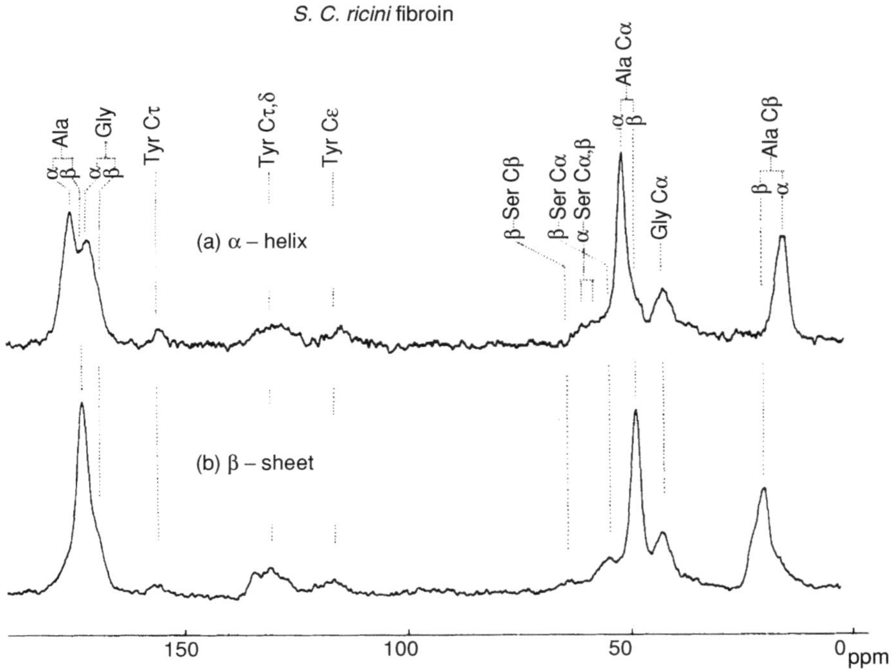

Figure 9.5. ^{13}C CP-MAS NMR spectra of α-helix (a) and β-sheet form (b) from *S. c. ricini* silk fibroin. The Greek letters α and β stand for the α-helix and β-sheet forms, respectively.[75]

Among silkworms, the fiber properties are considerably different among *B. mori* silk and wild silkworm silk such as *A. pernyi*, *A. yamanai, and S. c. ricini*. The primary structure of the latter fibroins is composed of alternative blocks of polyalanine (PLA) and glycine-rich regions.[80] The liquid silk from the silk gland of *S. c. ricini* takes a mobile α-helix structure as viewed from Ala ^{13}C C_β signal by ^{13}C DD-MAS NMR (up to 96 h from start), and finally aggregated to assume a β-sheet structure as viewed from CP-MAS NMR (96–101 h).[81] In contrast, ^{13}C NMR peak of [1-^{13}C]Gly adjacent to the N-terminal Ala residue of PLA in *S. c. ricini* [Gly-*Gly*-$(Ala)_{12-13}$ sequence] was assigned.[82] Subsequently slow exchange was observed in this region between helix and coil forms in the NMR timescale by raising temperature up to 45 °C. Further, ^{13}C CP-MAS NMR spectra of [2-^{13}C]Ser-, [3-^{13}C]Ser-, [3-^{13}C]Tyr-labeled *A. pernyi* silk fibroin was recorded after preparation by oral administration to fifth instar larvae from 3 to 6 old days for 4 days.[83] It was found that 65% of Ser residues are in the α-helical state as viewed from their ^{13}C chemical shifts of C_α and C_β carbons and ascribed to those located at the N-terminal of PLA and are considered to be incorporated into the α-helix of PLA. The Tyr and other Ser residues take the random coil form.

As described in Section 5.1, secondary structure with atomic resolution of polypeptides and fibrous proteins can be determined, if a variety of orientational constraints such as ^{13}C and ^{15}N CSAs, ^{13}C–^{15}N dipolar interaction, etc. are available from samples of mechanically oriented systems with respect to the applied magnetic field. To determine such structure with the ϕ, ψ torsion angles of *B. mori* fibroin, static ^{13}C and ^{15}N spectra of [1-^{13}C]Gly, [1-^{13}C]Ala, [^{15}N]Gly, and [^{15}N]Ala-labeled preparations were revealed[84] to reveal the constraints, angles θ_{CO} and θ_{NH}, and θ_{CN} and θ_{NC} from the ^{13}C and ^{15}N CSAs and ^{13}C–^{15}N dipolar interaction, respectively, with respect to the fiber axis Z_{FAS}, as illustrated in Figure 9.6 (right). For this purpose, the silk fibers were wound onto a form, which produces sheets of highly oriented fibers. These sheets were fixed with quick-setting epoxy and then cut into 4.5 mm × 10 mm pieces, stacked together, and fixed with the epoxy to form a block of 4.5 mm × 4.5 mm × 10 mm that fit within the radio frequency coil of the NMR probe. The resulting ^{13}C NMR spectra of an oriented block of [1-^{13}C]Gly-labeled *B. mori* fibroin thus obtained exhibit the orientation-dependent spectral change as a function of the β_L, the angle between the fiber axis and the applied magnetic field. Then, unique ϕ, ψ torsion angles were determined by the combination of the calculated orientations, which satisfy the observed bond orientations, as illustrated for Ala and Gly residues, in Figure 9.6. The best-fit solution based on the minimum bond orientation difference for Ala is $\phi = -140°$, $\psi = 142°$. Similarly, for Gly the best-fit values are $\phi = -139°$, $\psi = 135°$. The torsion angles by X-ray diffraction study are $\phi = -139°$, $\psi = 140°$.[85]

Instead of requiring macroscopically oriented samples, secondary structure of silk fibroin with atomic resolution from [1-^{13}C]Ala-labeled *S. c. ricini* in fiber and film was determined[86] by DOQSY experiments[87] to correlate two carbon orientations. The C=O distances between carbonyl carbons in amino acids adjacent in the primary structures in proteins are, with few exceptions, shorter than other C=O contacts. The former is found to lie between 2.7 and 3.7 Å depending on (ϕ, ψ), the latter usually exceed 4.2 Å. For short excitation/reconversion times (initial rate), the DOQSY excitation depends on the inverse of the sixth power of the distance and the spectra can be approximately described by Ala C=O pairs adjacent in the primary structure. However, signals arising from the nonadjacent contacts may be up to 25% of the total signal intensity. The most probable conformation for the fiber thus obtained is (ϕ, ψ) = (−135°, 150°), and 70% of the probability function $P(\phi, \psi)$ falls within the β-sheet region. For the film, the bimodal distribution of $P(\phi, \psi)$ between the right- and left-handed α-helices cannot be reduced to a singly peaked distribution because the C_α chemical

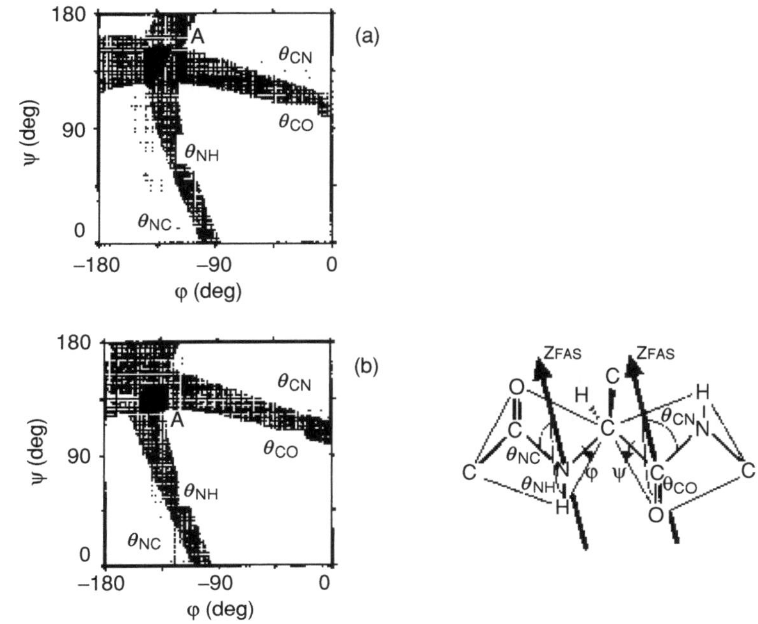

Figure 9.6. Variation in bond orientations constrained by the $C_{\alpha(i-1)}$–$C_{\alpha(i+1)}$ axis being parallel to the FAS and by the NMR orientational constraints for the (a) Ala and (b) Gly residues. The conformational space is restricted to an experimental error of $\pm 5°$ (from Demura et al[84]).

shift difference between the two conformations is only minor (see Table 6.2). The film shows a maximum at ($\pm 60°$, $\pm 45°$), corresponding to 60% of $P(\phi, \psi)$. Within the 15° grid resolution of their analysis, these regions of the Ramachandran plot coincide with the standard conformation for the antiparallel β-sheet (−140°, 135°) and for the α-helix (−57°, −47°). 2D spin-diffusion spectra of several ^{13}C- and ^{15}N-labeled (Ala-Gly)$_{15}$ were recorded[88] as a model for silk I form of *B. mori* fibroin under off magic angles spinning (OMAS).[89,90] Silk I form of this peptide was prepared after dissolving the peptide in 9 M LiBr, followed by dialysis against water. The conformation of (Ala-Gly)$_{15}$ with silk I structure thus obtained is a repeated β-turn type II-like structure, with the torsion angles as (− 60 ($\pm$ 5)°,130 ($\pm$ 5)°) and (70 ($\pm$ 5)°, 30 ($\pm$ 5)°) for Ala and Gly residues, respectively. The β-turn type II-like form was also confirmed by measurement of the distance in the intramolecular hydrogen bond between ^{13}C=O carbon of the 14th Gly residue and ^{15}N of the 17th Ala residue as 4.0 Å by REDOR. 2D-diffusion and REDOR spectra of several ^{13}C- and ^{15}N-labeled sequential peptides for repeated helical region in *S. c. ricini* fibroin were analyzed to reveal the torsion angles of the α-helical region.[91,92] They demonstrated that the local structure of N- and C-terminal residues, and also the neighboring residues of α-helical poly-Ala chain in the model peptides is more strongly wound structure than found in typical α-helix structure because of the presence of the observed intramolecular N–H···O=C distance of 4.8–5.0 Å, irrespective of 4.1 Å as expected value for the normal α-helix form. It is noted that such intramolecular N–H ···O=C distance between i and i+4 residue in some α-helical polypeptides in the solid and lipid bilayer turned out to be 4.5 ± 0.1 Å.[93,94] It is also pointed out that the interatomic distance from the REDOR experiment $\langle 1/r^3 \rangle^{-1/3}$ is not always the same as that determined by diffraction method, $\langle r \rangle$, and tends to give rise to prolonged value if molecular motions including local small amplitude motion of particular residues or rotation about the helical axis are present, where r stands for the interatomic distance.[95,96]

9.4.2. Spider Dragline Fibroins

The dragline silk of *Nephila clavipes* is produced by the ampullate glands and is drawn out of the spider from the spinnerets. It is conceivable that a unique combination of tensile strength and elasticity of such dragline silk could be ascribed to its complex and heterogeneous microstructure as revealed by solid state NMR or X-ray diffraction, etc. This means that the organization of the dragline fibroin is fundamentally different from that of the cocoon fibroin such as *B. mori* silk, although the average amino acid composition of the former is roughly similar to that of the latter: Gly (42%), Ala (25%), Gln (10%), Leu (4%), Arg (4%), Tyr (3%), and Ser (3%) for *N. clavipes*[97–99] while Gly (43%), Ala (30%), Ser (12%), Tyr (5%) for *B. mori*.[63] ^{13}C NMR spectral feature of *N. clavipes*[100] is very similar to that of silk II from *B. mori* fibroin,[73,74] except for the differential proportion of Ser and Gln residues. In particular, ^{13}C NMR signals both from the crystalline and amorphous regions, arising from Ala C_α, C_β, Gly C_α, Gln $C_{\beta,\gamma}$, Tyr $C_{\gamma,\delta}$, C_ε, C_ζ, and C=O carbons, can be detected. Accordingly, it is observed that *N. clavipes* fibroin takes the β-sheet form as manifested from the conformation-dependent ^{13}C chemical shifts of Ala C_α, C_β, and Ala, Gly, and Gln C=O peaks. An excellent fit to the intensities from the Ala C_β was obtained by a two-component fit in which 40% of the carbons have a carbon spin–lattice relaxation times (T_1) of 0.48 s and 60% relax more slowly, with a T_1 of 2 s. This observation indicates that Ala in dragline silk is present in two different motional environments.

To gain insight into the two-component nature, ^2H NMR spectra of [3,3,3-$^2H_3$]Ala-labeled dry *N. clavipes* fibroin and its wet supercontracted preparation were recorded.[101] The axially symmetric powder pattern, narrowed by rapid methyl group reorientation, contributes 90% of the intensity and a static component, with perpendicular singularities at ± 60 kHz, which contributes 10% of the intensity. The ^2H NMR spectrum of supercontracted silk, prepared by soaking fibers in deuterium-depleted H_2O for 12 h, then blotting them with filter paper, shows that the methyl component is unchanged upon wetting. However, the static signal is replaced by a peak centered at zero frequency, which is typical for averaging the quadrupole coupling by fast isotropic reorientation. It was shown that about 40% of Ala methyl groups are present in β-sheet that are highly oriented parallel to the fiber axis. The other 60% are poorly oriented and have a larger volume available for unimpeded methyl reorientation. The existence of two motional and structural environments for the methyl groups is also supported by the ^{13}C T_1 data. A possible model for dragline silk is that some of the residues are present in a classical crystalline phase while the remainder is in protocrystals possibly preformed β-sheet.

Local structure of spider dragline silk from [1-^{13}C] Ala- or Gly-labeled *Nephila madagascariensis* was analyzed by 2D spin-diffusion NMR. The proton-driven 2D diffusion spectra of [1-^{13}C]Gly- and Ala-labeled *N. madagascariensis* was examined[102] at $T = 150$ K and 10 s mixing time are given in the left and right traces, respectively, in Figure 9.7. The spectrum of [1-^{13}C]Gly-labeled silk (left) shows a rather broad exchange pattern, while virtually no off-diagonal intensity is observed from the Ala carboxyl carbon. The absence of cross-peaks in Ala indicates $(Ala)_n$ segments adopts a highly ordered β-sheet structure. The microstructure of the glycine-rich domains is found to be ordered. The simplest model that explains the experimental finding is a 3_1-helical structure, present in form II crystal of $(Gly)_n$ and $(Ala\text{-}Gly\text{-}Gly)_n$. This observation may help to explain the extraordinary mechanical properties of this silk, because 3_1-helices can form interhelix hydrogen bonds. It is also demonstrated that the presence of the 3_1-helix in the glycine-rich domain can be readily identified by 1D experiments utilizing the conformation-dependent ^{13}C chemical shifts of their Gly C_α and C=O peaks (43.3 and 172.5 ppm, respectively) with reference to those of $(Gly)_n$ II and $(Ala\text{-}Gly\text{-}Gly)_n$ II (43.5–43.0 and 172.2–172.1 ppm, respectively).[4]

In relation to the design principles of the noncrystalline glycine-rich domains of spider dragline silk, 2D solid state NMR was applied[103] to determine the distribution of the backbone torsion angles (ϕ, ψ) as well as the orientation of the polypeptide backbone toward the fiber at both the Gly and Ala residues. The DOQSY spectrum of [1-^{13}C]Ala-labeled spider

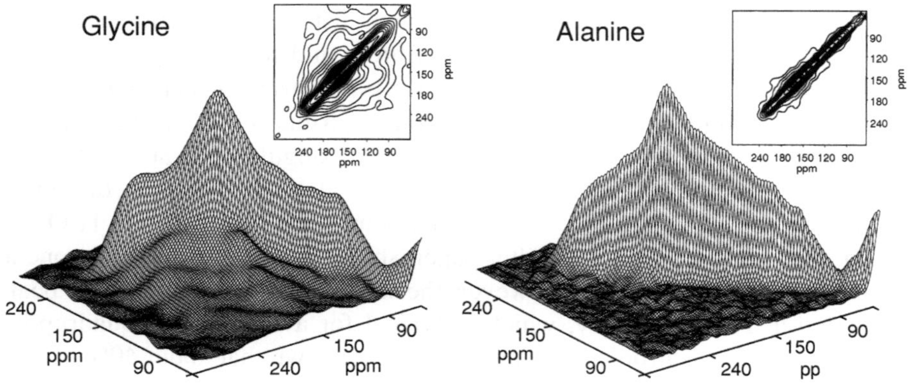

Figure 9.7. Proton-driven spin-diffusion spectra of [1-^{13}C]Gly- (left) and [1-^{13}C]Ala-labeled (right) *N. madagascariensis* dragline silk at $T = 150$ K. A mixing time of 10 s was used. The spectrum was acquired with 128 transients per data point in t_1; 96 spectra have been recorded in the F_1 domain. The data matrix of 96 × 128 points was zero-filled to 256 × 256. As inset, the contour plots of the same data (from Kümmerlen *et al.*[102]).

silk from *Nephila edulis* sample showed that Ala residue takes a clear preference for the β-sheet region with the most abundant conformation (ϕ, ψ) = (−135°, 150°), which within the grid resolution of 15° used corresponds well with known β-sheet structure. The isotropic chemical shift of the C_α carbon of Ala residue was used as a further constraint to distinguish between (ϕ, ψ) = ($-\phi$, $-\psi$). Alternatively, conformation of Gly residue was interpreted in terms of a 3_1-helix and an extended β-sheet forms, on the basis of the DOQSY spectrum of [1-^{13}C]Gly-labeled *N. edulis* silk. Complementary to the local structure, the orientation of the peptide chain toward the fiber axis (α_F, β_F) was determined on the basis of direction exchange with correlation for orientation-distribution evaluation and reconstruction (DECODER) spectra.

REDOR NMR on strategically ^{13}C and ^{15}N labeled samples was used to study the conformation of the LGXQ (X = S, G, or N) motif in the major ampullate gland dragline silk from the spider *N. clavipes*.[104] This motif is remarkably well conserved between repeats in the amino acid sequence and that G–S residues are often found in bends. While there is likely some heterogeneity in the structures formed by the LGXQ sequences, the data indicate that they all form compact turn-like structure. Indeed, the best fit among the standard structures is that of type I β-turn, which is often found connecting two strands of antiparallel β-strands.

9.4.3. Supercontracted Dragline Silk

It is known that water causes dragline silk unconstrained at one end to supercontract or shrink to about half of its original length.[105,106] It is noted that water is the only substance that produces an appreciable amount of contraction among water and various alcohols.[107] The major ampullate dragline silk is the only one that contracts to an appreciable extent, and the other fibroin from minor ampullate of *N. clavipes* or *B. mori* fibroin do not show such property.[107,108] Supercontraction is characterized by randomization of the orientation of the β-sheet, as viewed from diffraction patterns.[108] Despite the huge impact of the supercontraction process on the outside appearance of the dragline silk, no major structural changes are detected in the local structures as seen by spin-diffusion NMR upon supercontraction induced by urea.[109] This is obviously because spin-diffusion NMR approach is not always suitable to analyze the problem when protein dynamics plays an important role. Significant differences in the dynamics of the polypeptide chain upon supercontraction are detected by means of the second moment analysis of 1D ^{13}C CP powder patterns taken at room temperature.[109]

Solid state ^{13}C and ^{2}H NMR were recorded to study *N. clavipes* silk fibers, so as to address the molecular origins of supercontraction in the wet silk.[110] A substantial fraction of the ^{13}C CP-MAS NMR peaks, including Gly, Gln, Tyr, Ser, and Leu residues in the protein backbone, was suppressed in the water-wetted state (solid trace) as compared with those of dry state (dashed trace), as illustrated in Figure 9.8 (left). In contrast, ^{13}C NMR peak-intensities of wetted state is increased as recorded by DD-MAS NMR with a delay time of 1 s (Figure 9.8, right) as compared with those of dry state. This is obviously caused by acquisition of large-amplitude rapid, reorientational motions as indicated also by shortened ^{13}C spin–lattice relaxation times (frequency in the order of 10^8 Hz). Static ^{13}C spin-echo spectrum of [1-^{13}C]Gly-labeled dragline silk exhibits two Gly carbonyl populations, one nearly isotropically averaged and the other essentially static. They estimated that $40 \pm 5\%$ of Gly residues are found in the broad powder pattern, while the remainder ($60 \pm 5\%$) reorients at rates faster than 20 kHz (motional frequency in the order of 10^4 Hz). The observation of the peak-suppression caused by this kind of slow motions is consistent with that of many hydrated fibrous proteins so far discussed such as elastin, cereal proteins and their model polypeptides, as described earlier. ^{2}H NMR of silk samples that incorporated one terminal methyl group provides probe for dynamics at specific side changes along the fiber. The wet Leu-labeled dragline silk

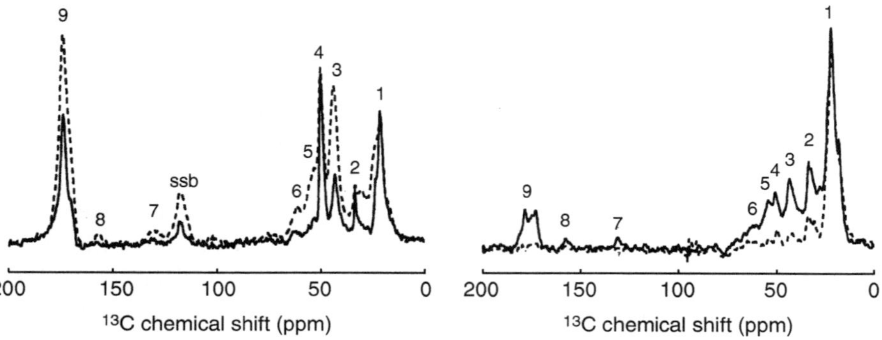

Figure 9.8. ^{13}C CP-MAS NMR (left) and DD-MAS NMR (right) spectra of *N. clavipes* dragline silk in the dry and the water-wetted state. Dashed line (. . .) is the spectra of the dry silk; solid line (-) is the spectra of the silk immersed in water. Peak assignments include the followings: (1) Ala C_β; (2) Gln C_γ; (3) Gly C_α; (4) Ala C_α; (5) Gln C_α, Leu C_α, Ser C_α; (6) Ser C_β; (7) Tyr C_δ; (8) Tyr C_ζ; (9) all C=O sites (from Yang et al.[110]).

takes two distinct populations of Leu with dramatically different properties, including narrowed component of Lorentzian line ascribed to Leu residues whose side chains reorient both rapidly and nearly isotropically. This is consistent with the observation that when wet silk supercontracts, the observed shrinkage to about half its original length is accompanied by a more than twofold increase in the fiber's cross-sectional area.[111]

Regenerated spider silk was examined by scanning electron micrographs, mechanical properties, wide-angle X-ray diffraction (WAXD), and structural change of fiber by postspinning drawing and water treatment by ^{13}C CP-MAS NMR.[112] The fraction of Ala residues in β-sheet structures increases linearly with the extent of drawing applied to the specimen. Thus, the drawing-induced improvement of the mechanical properties of the regenerated silk appears to be, at least in part, due to a reinforcement of the fibers by very strong and stiff poly-Ala microcrystallites. Soaking a single-drawn material in water for 4 h, with the fiber ends unconstrained, results in a large increase in the fraction of Ala residues adopting the β-sheet conformation. Solid state NMR techniques were used to study two different types of spider silk from two Australian orb-web spider species, *N. edulis* and *Argiope keyserlingi*.[113] ^{13}C- and ^{1}H-spin–lattice relaxation times (T_1s) indicated that different types of silks have different molecular motion on both slower (kHz) and faster (MHz) timescales. Shorter T_1 times usually indicate more mobile, flexible regions on the faster timescale, and longer T_1 times suggest crystalline, more durable, but less flexible areas. The generally shorter relaxation times of Gly compared to Ala groups support the model of silk in which Gly-rich blocks in silk constitute the amorphous, elastic regions and Ala-rich blocks dominate the more crystalline, structured areas. The increased relaxation rate due to hydration as viewed from both Zeeman (T_1) and rotational frame of reference ($T_{1\rho}$) of ^{13}C and ^{1}H reflects increases in protein mobility on the kHz and MHz timescale, but does not necessarily suggest major structural changes in hydrated silk.

2D wide-line separation (WISE) NMR[114] is implemented to correlate ^{13}C chemical shifts with mobility by observing the corresponding ^{1}H line widths and line shapes in water-saturated spider dragline silk.[115] The WISE NMR spectrum of native *N. clavipes* spider dragline silk exhibits ^{1}H line widths that are $\approx$40 kHz for all carbon environments characteristic of a rigid system. In contrast, the water-saturated case displays a component of the ^{1}H line that is narrowed to $\approx$5 kHz for the Gly C_α resonated at 42.5 ppm and newly resolved Ala helical environment while the Ala C_β corresponding to the β-sheet conformation remains broad (Figure 9.9). These results indicate that water permeates the amorphous, Gly-rich matrix and not the crystalline, poly-Ala β-sheets. The appearance of this mobile Ala helical environment in the WISE spectrum of the wet silk could explain the increase in elasticity and decrease in stiffness observed for wet, supercontracted dragline silk. Spin diffusion was monitored between the mobile regions and the poly-Ala β-sheet crystallites by including a mixing period in the WISE NMR pulse sequence. The ^{1}H line width of the Ala C_β representing the β-sheet conformation (20.3 ppm) was monitored as a function of τ_m. The narrowing of the ^{1}H line widths with increasing mixing time is due to ^{1}H magnetization from the mobile region participating in the CP process as a result of spin diffusion. The τ_m where this process comes to spatial equilibrium is related to the length scale of the poly-Ala crystalline domains. The average poly-Ala crystalline domains based on the spin-diffusion experiment completed at 50 ms are 6 ± 2 nm, in excellent agreement with WAXD measurements that measured crystallites having dimension of $2 \times 5 \times 7$ nm.[116]

9.5. Keratins

Since keratins, contained in wools, hairs, feathers, etc. generally have periodic amino acid sequences and higher order structure, clarification of their fine structure is very important, not only understanding their physical and chemical properties and functions but also obtaining information about the molecular design of synthetic polypeptides.[117–119] The native wool fiber consists of intermediate filaments (‘‘microfibrils’’) composed of low-sulfur proteins which are embedded in a nonfilamentous matrix. The nonfilamen-

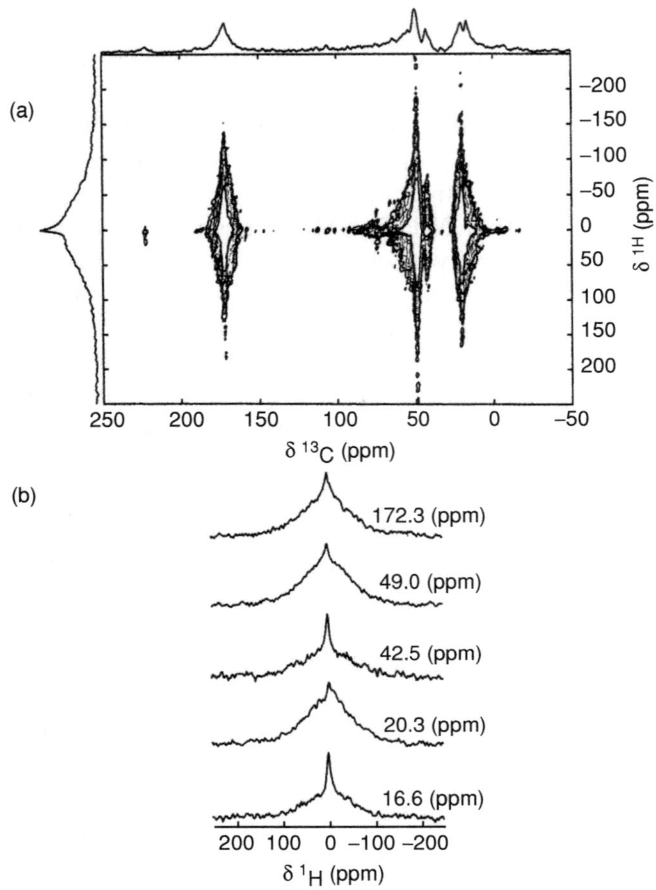

Figure 9.9. WISE NMR spectrum of wet, supercontracted *N. clavipes* spider dragline silk (a) and ^{1}H slices at specified ^{13}C chemical shifts (b) (from Holland *et al.*[115]).

tous matrix usually contains two classes of proteins: one is high-sulfur protein (SCMKB) and the other is protein containing Gly and Tyr residues (high-Gly-Tyr protein:HGT).[117–119] Wool keratin can be divided into three main fractions after reducing the disulfide bonds and protecting the thiol groups with iodoacetic acid to form *S*-(carboxyl-methyl)keratin (SCMK)[120,121] such as low-sulfur proteins SCKMA, SCMKB, and HGT. In addition, helix-rich fragments are obtained from SCMKA (SCMKA-hf) after partial hydrolysis with α-chymotrypsin.[122] The α-helix content of SCMKA in an aqueous solution was determined to be about 50% by means of ORD[123] and CD.[124] The sequence of several amino acid residues, $(a\text{-}b\text{-}c\text{-}d\text{-}e\text{-}f\text{-}g)_n$, in which the hydrophobic and hydrophilic residues are located at *a* and *d* positions and other positions, respectively, has been determined in SCMKA as well as tropomyosin, myosin, and paramyosin, which assume a coiled-coil α-helix rope.[117–119,123]

The amino acid compositions of wool (Merino 64), SCMKA, SCMKA-hf, SCMKB, and HGT proteins are different from each other.[124] The contents of helix-forming residues such as the Thr, Ser, Pro, Gly, and Cys residues as the major amino acid residues decrease in the order of wool, SCMKA, and SCMKA-hf. On the other hand, SCMKB contains Thr, Ser, Pro, and Cys residues, and HGT contains Ser, Gly, Tyr, and Phe residues. In this section, we are mainly concerned with wool keratins to be one of the typical keratin proteins and are briefly compared with other keratins.

9.5.1. Wool Keratins

9.5.1.1. Conformational characterization

^{13}C CP-MAS NMR spectra of wool (Merino 64), SCMKA, SCMKA-hf, SCMKB, and HGT in the solid state are shown in Figure 9.10, together with ^{13}C CP-MAS DDph[123a] NMR spectra (right traces) for their aliphatic carbons.[124] The observed ^{13}C chemical shifts are summarized in Table 9.6, together with their assignment of peaks with reference to homopolypeptides in the solid state.[4,73,125–138]

Conformational characterization of wool keratins is readily feasible based on the conformation-dependent ^{13}C chemical shifts[130,131] of the main-chain carbonyl carbons which are resolved to 175.8 ± 0.8 (α_R-helix) and 170.9 ± 1.2 ppm (β-sheet form), as shown in the expanded spectra (Figure 9.10).[124] The minor shoulder peak at about 180 ppm arises from the side-chain carbonyl carbons of Asp C_γ, Glu C_δ, and carboxymethyl Cys C_ε, while the other at about 166 ppm comes from the NMR rotor made of polyimide. The peak at 172 ppm is present as a single peak for SCMKB and HGT, because they are rich in the β-sheet form. The relative intensity of this peak decreases in the order of wool, SCMKA, and SCKMA-hf, however. The relative peak intensities were evaluated as 5.0%, 52.0%, 37.4%, and 5.6% for the peaks at 176, 172, 180, and 160 ppm, respectively, based on convoluted spectra. The relative peak intensity of the side-chain carbonyl carbons (ca. 180 ppm) were determined to be 11%, 12%, 11%, and 1.6% for SCMKA, SCMKA-hf, SCMKB, and HGT, respectively, consistent with their amino acid compositions.[124] The proportion of the α-helix component increases in the order of wool, SCMKA, and SCKMA-hf.

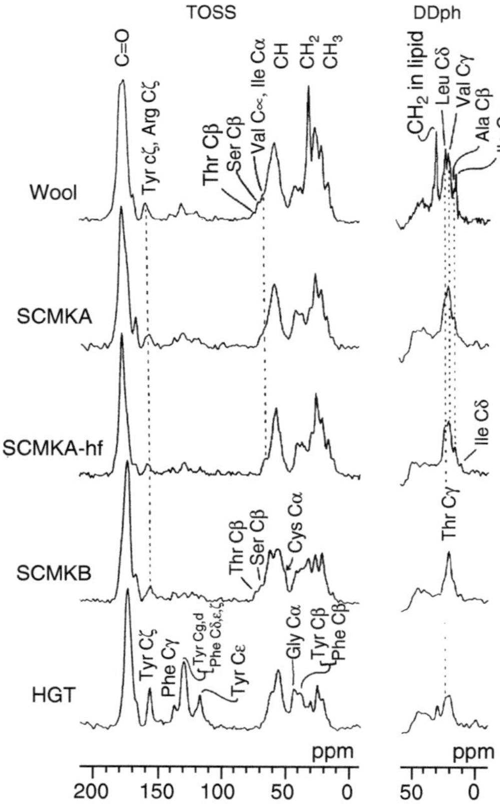

Figure 9.10. ^{13}C CP-MAS TOSS NMR spectra of wool, SCMKA, SCMKA-hf, SCMKB, and HGT in the solid state and ^{13}C CP-MAS DDph NMR spectra in the aliphatic carbons region.[124]

SCMKB and HGT are believed to originate chiefly from a nonfilamentous matrix between the microfibrils of the wool fiber, but no evidence has yet been obtained for the existence of any ordered structure in solution or in the solid state.[117–119] The β-sheet form are present in both SCMKB and HGT film cast from formic acid solution as identified by X-ray diffraction and CD.[139,140] Undoubtedly, the ^{13}C NMR data are consistent with this view, as summarized in Table 9.6 and the existence of the β-sheet can be readily confirmed by means of the Thr and Ser C_β carbons at 71.8 and 67.8 ppm (Table 9.6).

No significant peak at 65–68 ppm is available from HGT, although the proportion of the Ser residue in HGT is almost the same as that of SCMKB. This implies that the proportion of Ser residues in HGT in the

Table 9.6. Observed ^{13}C NMR chemical shifts for wool, SCMKA, SCMKA-hf, SCMKB, and HGT in the solid state.[124]

Chemical shifts[a]/ppm					Assignment[b]
Wool	SCMKA	SCMKA-hf	SCMKB	HGT	
175.3	176.0	176.3			C=O (α-helix)
173.2			172.8	172.3	C=O (β-sheet)
156.2	155.1	157.2	154.8	156.0 (156.4)	Tyr C_ζ and Arg C_ζ
				136.9 (137.3)	Phe C_γ
128.5	129.1	128.2		128.9 (128.9)	Phe $C_{\delta,\varepsilon,\zeta}$ and Tyr C_γ
				116.3	Tyr C_ε
72.2			71.8		Thr C_β (β-sheet)
68.6			67.8		Ser C_β (β-sheet)
65.2	c	64.4			Val C_α and Ile C_α (α-helix)
			60.6		Thr C_α (β-sheet) and Pro C_α
				60.2	d
56.6	56.9	56.4	54.8	54.6	C_α methine
				42.5	Gly C_α
40.5	40.3	40.3	40.1		d
				38.4	Tyr C_β and Phe C_β
36.3	35.6	36.7	36.6		d
30.1 (30.0)					Mainly CH_2 in lipid
	28.8	28.9	30.8	30.2	d
25.4	25.0	25.2	25.4	24.6	d
(23.1)	c	(23.7)		(24.6)	Leu C_δ
20.3 (20.3)	20.7 (20.9)	21.0 (21.1)		20.6 (20.8)	Mainly Val C_γ
			20.3 (20.7)		Mainly Thr C_γ (β-sheet)
(16.5)[c]	16.2 (16.5)	16.1 (16.5)	c		Mainly Ala C_β (α-helix)
14.9 (14.8)					Mainly Ile $C_{\gamma'}$
11.9 (11.7)	12.1	12.4 (11.9)	12.1		Ile C_δ

[a]The numbers in the parentheses are chemical shifts of the peaks observed in ^{13}C CP-MAS DDph spectra.
[b]The assignment was made by reference data of homopolypeptides in the solid state. See Section 6.4.
[c]Observed as the shoulder peak.
[d]Unassigned at this stage.

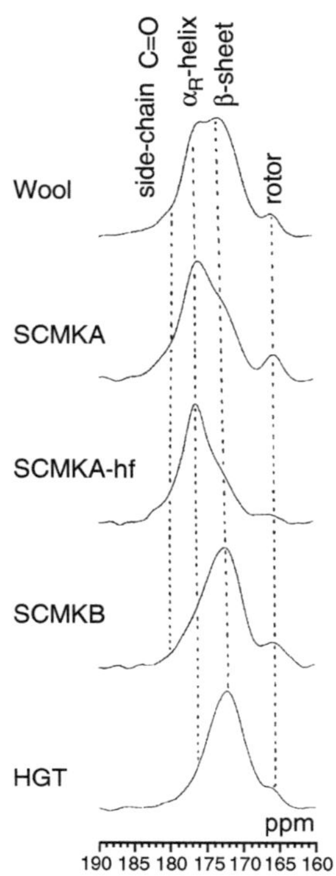

Figure 9.11. Expanded ^{13}C CP-MAS TOSS NMR spectra for the carbonyl carbon region in wool, SCMKA, SCMKA-hf, SCMKB, and HGT.[124]

β-sheet form is less than that of SCKMB in the β-sheet form. It has been reported that ^{13}C peaks of the Ser C_β carbons of silk fibroin samples in the silk I form (loose helix[141]) appear at 59.0–61.5 ppm.[73] The peak at 60.2 ppm in the ^{13}C CP-MAS NMR spectrum of HGT may come from the Ser C_β carbons in a silk I like form. Similarly, the ^{13}C chemical shift (38.4 ppm) of the peak assignable to the Tyr and Phe C_β carbons is in between the values of the α_R-helix and β-sheet forms. This implies that the Tyr and Phe residues of HGT reside in conformations other than the α_R-helix and β-sheet forms.

SCMKA was considered to originate chiefly from the microfibrils,[117–119] and the proportion of the α-helix form aqueous solution was determined to be about 50% by means of ORD[142] and CD.[143] Also, the proportion of the α-helix form of SCMKA-hf in aqueous solution was determined to be about 84% and 90% by means of ORD[144] and CD,[145] respectively. The relative proportion of the α-helix form as evaluated from the solid state NMR is 75%

and 35% for SCMKA-hf and SCMKA, respectively. Indeed, the proportion of the α-helix form in SCMKA-hf is higher than that in SCMKA. Moreover, ^{13}C chemical shifts of Val and Ile C_α and Ala C_β carbons in SCMKA and SCMKA-hf suggest that they are rich in the α_R-helix structure (Figure 9.12 and Table 9.6).

The relative proportion of the α-helix and β-sheet forms in wool was evaluated as 42% and 58%, respectively from the main Chain carbonyl carbons. The ^{13}C NMR peaks assignable to the Val and Ile C_α and Ala C_β carbons in the α_R-helix form are resonated at 65.2 and 16.5 ppm, respectively. The peak at 30 ppm of wool is very strong, although no such peak was identified for the other samples (Figure 9.10). In general, wool contains lipids of 2.5 wt.% in the wool cell membrane. In fact, lipid components are removed from SCKM samples and so its peak is not identified.

The microfibrils of wool are composed of protofibrils, which consist of coiled-coil α-helix ropes.[117–119] The sequence for seven amino acid residues which take the coiled-coil α-helix is found in SCMKA (especially in SCKMA-hf), as well as in tropomyosin, myosin, and paramyosin. The Ala C_β peak of wool is ascribed to 16.5 ppm in wool, SCKMA, and SCKMA-hf (Table 9.6). Further, the two peaks at 15.8 and 16.7 ppm for tropomyosin in the solid state were ascribed to Ala C_β carbons in the external and internal sites of the coiled-coil structure, respectively.[146,147] Therefore, the ^{13}C chemical shifts of Ala C_β carbon peaks observed in wool, SCKMA, and SCKMA-hf suggest that a number of the Ala residues in these samples are located in the internal site of the coiled-coil structure.

9.5.1.2. *Conformational changes by stretching and heating*

It has been reported that the β-sheet form appears in SCKMA films upon stretching and the α-helix form disappears upon heating, as studied by X-ray diffraction[148] and IR.[149] As a result, ^{13}C CP-MAS TOSS NMR are varied under the condition of unstretched, stretched, and heated SCMKA films as shown in Figure 9.12.[150] The relative intensities of the unstretched film agree with those of the lyophilized product. Obviously, it is evident that the relative intensity of the β-sheet form is increased with the stretching ratio at the expense of the peak-intensity of the α-helix content.

Further, the C_α ^{13}C peaks in SCMKA films are displaced upfield upon stretching concomitant with the spectral change as observed from the C=O ^{13}C peaks. To visualize this situation, Yoshimizu *et al.*[151] performed a spectral simulation by assuming the Gaussian function for a peak with

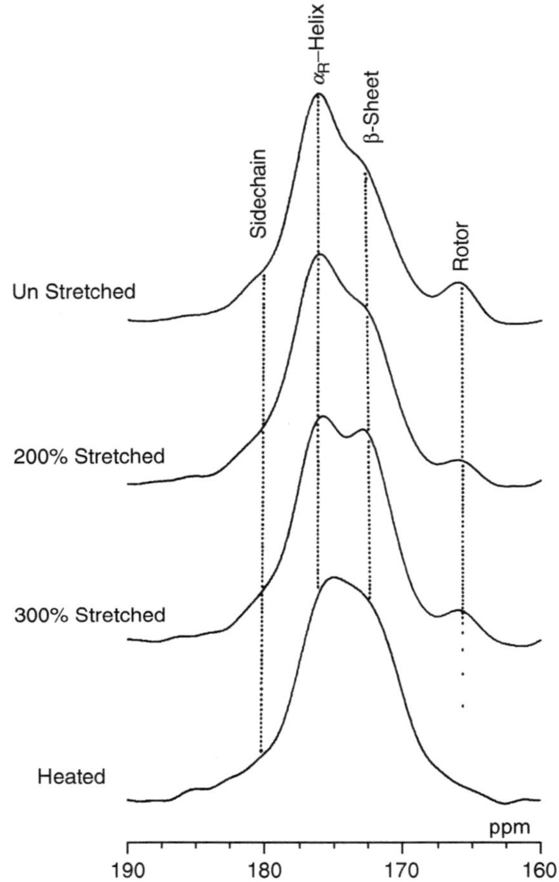

Figure 9.12. Expanded ^{13}C CP-MAS TOSS NMR spectra for the carbonyl carbon region in SCMKA films cast from aqueous solution.[124]

half-width 3.5 ppm. The observed and simulated spectra in the C_α carbon region agree with each other. Alternatively, similar results were obtained when the simulation was performed under the assumption that the N- and C-terminal domains are in the β-sheet and only the α_R-helix content of the rod domain changes based on the amino acid sequence data of the component. Therefore, the increase in the content of the β-sheet form by stretching was also justified from the analysis of the C_α ^{13}C signals.

It is also likely that the random-coil form may appear by heating at 200 °C for 3 h as judged from the broadened line widths. In fact, Sakabe *et al.*[152] reported that X-ray diffraction of the α-helix form completely vanishes in the SCMKA film heated at 200 °C for 3 h in vacuum. The relative proportion of the α-helix content was 25% and 8% for SCMB and HGT, respectively, as judged by the ^{13}C NMR peak-intensities, although their X-ray diffraction patterns show only the existence of β-sheet form.[139,140] The same result was obtained on the nonhelical fragment of SCMKA after partial hydrolysis with α-chymotrypsin:[153] it has been reported that the X-ray diffraction pattern of this fragment shows no reflections.[152] In this connection, NMR approach is more sensitive than X-ray diffraction to identify this sort of α-helix. This is because X-ray diffraction is sensitive to locate the well-ordered secondary structure, while NMR approach is also sensitive to less-ordered secondary structure.

9.5.2. Human Hair Keratin

Human hair keratin is one of several mammalian structural components formed from α-keratin like wool, nail, and horn.[154] The histological structure of the hair fiber consists of two components, cortex and cuticle. The cortex, comprising 85–90% of hair consists of spindle-shaped macrofibril, which have two main structures, microfibril and matrix.[155] The keratin intermediate filament associated proteins (KIFap) to be the largest fraction of the complex contains a significant amount of disulfide bonds which can form the cross-linked structure through intra- and/or intermolecular covalent bond form.[118–120] The KIFap is basically an amorphous structure, and only limited studies on the structure have been reported. The presence of a well-conserved common pentapeptide repeat of (Cys-X-Pro-Y-Cys)$_n$ has been pointed out by the amino acid sequence analysis.[118–120] The conformation is considered to be β-bend, which is stabilized by the disulfide bond between the side chains of Cys residues.

^{13}C CP-MAS NMR spectrum of ^{13}C-enriched human hair gave a strong peak at 12–20 ppm, which is assigned to Cys S-$^{13}CH_3$ carbon.[147] The signal patterns for the carbonyl carbons of untreated and 20 h reduction-treated hairs appearing at 172–180 ppm are almost the same and thus conformation of keratin protein did not change by the reduction treatment. The Cys S-$^{13}CH_3$ signal of the reduction-treated hair consists of two peaks. The upfield peak appeared as a shoulder can be assigned to the Cys S-$^{13}CH_3$ carbons in the amorphous domain and the downfield peak can be assigned to that in the ordered structure domain.

9.5.3. Hoof Wall Keratin

Hoof wall is analogous fingernail and can be considered as a mechanically stable, multidirectional, fiber-reinforced composite consisting of two main structures like wool, human hair, etc. ^{13}C CP-MAS spectrum of normally hydrated α-keratin (a),[145] from normally hydrated quine hoof, and dehydrated α-keratin (b), shows an asymmetric signal at about 175 ppm by the carbonyl carbon, a broad signal at about 54 ppm by the C_α carbon and a complex line shape in the 10–35 ppm region by the alkyl components of the side chains. These peak assignments are very similar to those in wool keratin. The carbonyl signal consists of two peaks. When the keratin sample is dehydrated, the intensity of the downfield peak at about 175.6 ppm diminishes relative to the upfield peak at about 172.7 ppm. 2D WISE[156] spectra were recorded for the normally hydrated keratin sample (a) and the dehydrated keratin sample (b). The hydrated sample shows the well-resolved peaks for the alkyl side-chain regions, indicating a degree of molecular motion. There is particular narrowing of the ^{1}H line shapes in the ^{13}C chemical shift range at about 17 and 36 ppm. These regions of the ^{13}C spectrum correspond to the C_β carbons in Ala and Asp/Glu/Gln residues, respectively, suggesting that these are highly mobile sites.

9.6. Bacteriophage Coat Proteins

The filamentous bacteriophage, fd, Pf1, etc., are nucleoprotein complexes of apparently simple design.[157] They are all filaments with a single-stranded circle of DNA surrounded by several thousand copies of a small protein. The virus particle contains no membrane components and the coat protein act as a structural protein. The coat protein, however, exists as a membrane-bound protein during the viral life cycle after infection and prior to assembly.[157–159] In all models of the coat proteins in these viruses, nearly all of the α-helix is extended approximately parallel to the filament axis. This result in the peptide carbonyl groups with their substantial local diamagnetic susceptibility being aligned parallel to each other and the filament axis. The resulting large net diamagnetic anisotropy and the liquid-crystalline character of the viral solutions are combined to align the viral particles parallel to the applied magnetic field.[160]

9.6.1. Unoriented fd Coat Protein

The backbone motions of the peptide units are available from line shape analysis of powder patterns, arising from their respective CSAs such as ^{13}C C_α and C=O carbons, ^{15}N, etc. Gall *et al.*[161] recorded ^{2}H, ^{13}C, and ^{15}N NMR spectra of fd bacteriophage coat protein to analyze the motions of their aromatic amino acids. The presence of background signals from natural abundant nuclei in the ^{13}C-labeled sample, however, represents a serious obstacle to line shape analysis, as illustrated for ^{13}C NMR spectra of [$^{13}C_\varepsilon$]Tyr-labeled fd. The slow magic angle spinning spectrum (0.38 kHz), arising from the narrow sidebands from Tyr residues, was recorded to distinguish the powder pattern from the natural abundance background. The calculated powder pattern from the sideband intensities is consistent with the difference spectrum between the labeled and unlabeled stationary samples and indicates that the Phe and Tyr rings undergo 180° flips about the C_β–C_γ bond axis more often than 10^6 Hz as well as small amplitude rapid motions in other directions.

The dynamics of the coat protein in fd bacteriophage as revealed by ^{15}N and ^{2}H NMR experiments[162] showed that the virus particles and the coat protein subunits are immobile on the timescales of the ^{15}N CSA (10^3 Hz) and ^{2}H quadrupole (10^6 Hz) interactions. Like the structural form of the protein, the membrane-bound form has four mobile residues at the C-terminus.[163] The membrane-bound form of the coat protein in lipid bilayer differs from the structural form in having several mobile residues at the C-terminus. Many of the side chains of residues with immobile backbone sites undergo large amplitude jump motions. The dynamics are generally similar in both the structural and membrane-bound forms of the protein.

9.6.2. Oriented Coat Protein

^{13}C and ^{31}P NMR spectra of the protein and DNA backbone structures, respectively, are compared for fd and Pf1 bacteriophages of magnetically oriented solution by their large net diamagnetic anisotropy.[164] Striking differences are observed between fd and Pf1 in both their protein and DNA structures. The ^{15}N NMR spectra of fd in Figure 9.13 are dominated by the resonance band from the many similar peptide amide backbone sites. The amino groups from the five Lys and the N-terminal Ala contribute to the narrow peak near 10 ppm (Figure 9.13(b) and (c)). Intensity from DNA sites is lost in the base line at this level of presentation because the phages

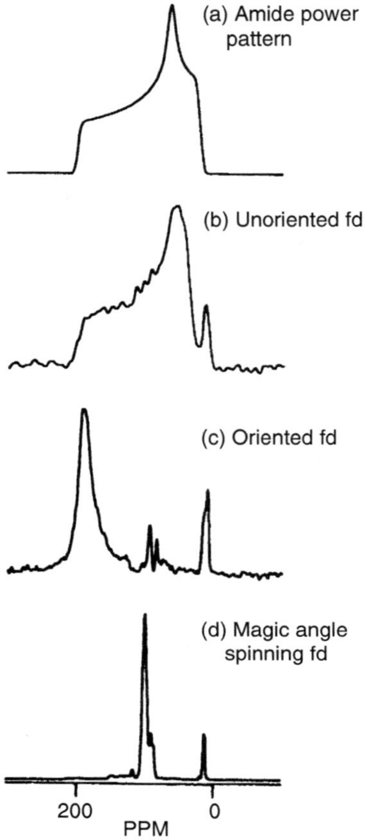

Figure 9.13. ^{15}N NMR spectra of uniformly ^{15}N-labeled fd at 25.3 MHz. (a) Calculated powder pattern based on the spectrum of acetyl [^{15}N]glycine (δ_{33} =192, δ_{22} =51, and δ_{11} =10 ppm), (b) unoriented fd solution, (c) magnetic field-oriented fd solution, 45 mg/mL, pH 8; (d) magic angle spinning of fd gel at 15.2 MHz, ≈200 mg/mL, pH 6 (from Cross et al.[164]).

are >90% by weight protein. There are a number of remarkable features about the ^{15}N NMR spectrum of oriented fd in Figure 9.13(c). Chief among them is the absence of an equal distribution of intensity about the isotropic chemical shifts given in Figure 9.13(d). The observation of nearly all of the ^{15}N amide signals at 185 ppm, which is approximately equal to δ_{11} of the amide chemical shift tensor, clearly demonstrates that the sample is oriented in the magnetic field. This principal element of the chemical shift tensor has been assigned to the N–H bond direction. A limited range of angles for the N–H bonds relative to the filament axis reflects the spreads in individual resonance frequencies. The ^{15}N chemical shift resonance band at δ_{11} has a dipolar splitting of 14 kHz, which is close to the maximum calculated splitting corresponding to the N–H bond parallel to the applied magnetic field, as demonstrated for 2D chemical shift and heteronuclear dipolar spectrum of fd. The other distinguishable nitrogen sites have very small or no dipolar splittings, due to motional averaging (Lys, Ala-1), lack of a bonded hydrogen (Pro-6), or the orientation of the N–H bond being very near the magic angle ($\approx 55°$) with respect to the applied magnetic field (Trp-26). The spectra for Pf1 are clearly similar to those observed for fd. However, there are important differences, especially in that there is a pronounced shoulder on the upfield side of the main amide resonance peak in the Pf1 chemical shift spectrum. This means that in Pf1 coat protein, two distinct sections of a helix are present, the smaller of which is tilted with respect to the filament axis by about 20°. The DNA backbone structure of fd is completely disordered, while the DNA backbone of Pf1 is uniformly oriented such that all of the phosphodiester groups have the O–P–O plane of the nonesterified oxygens approximately perpendicular to the filament axis. The major coat protein (pVIII) subunits are essentially continuous α-helices in the capsid with their axes roughly aligned with the long axis of the virion, as shown by X-ray fiber diffraction.[165] Solid state NMR spectroscopy was used to analyze the conformational heterogeneity of the major coat protein (pVIII).[166] Both one- and two-dimensional solid state NMR spectra of magnetically aligned samples of fd bacteriophage reveal that an increase in temperature and a single-site substitution (Tyr-21 to Met, Y21M) reduced the conformational heterogeneity observed throughout wild-type pVIII. The NMR results are consistent with previous studies, indicating that conformational flexibility in the hinge-bend segment that links the amphipathic and hydrophobic helices in the membrane-bound form of the protein plays an essential role during phage assembly, which involved a major change in the tertiary, but not secondary, structure of the coat protein.

9.6.3. Magnetically Aligned Fibrous Proteins

2D $^{1}H/^{15}N$ PISEMA spectra of uniformly ^{15}N-labeled Y21M fd coat protein of magnetically aligned filamentous bacteriophage were recorded to determine the atomic resolution structure (Figure 9.14).[167] The samples for the solid state NMR experiments were 5-mm o.d. thin-wall glass tubes containing 180 μL of 60 mg/mL solutions of bacteriophage in 5 mM sodium borate buffer, pH 8.0, and 0.1 mM sodium azide. The temperature of the sample

Figure 9.14. Models of a section of the Y21M fd filamentous bacteriophage capsid built from the coat protein subunit structure, which was determined by solid state NMR spectroscopy. The symmetry was derived from the fiber diffraction studies. (a) and (b): Representations of the electrostatic potential on the molecular surface of the virus. (a) is a bottom view and (b) is a side view along the virus axis obtained by using the program GRASP. (c) and (d): View of the capsid structure showing the arrangement of the coat proteins in pentamers and further assembly of the two-fold helical structure obtained by using the program INSIGHT II Accelrys (from Zeri et al.[177]).

was controlled at 65 °C. The resulting 2D PISEMA spectrum contains resonances from all the structured amide sites of the coat protein. Each resonance is characterized by orientationally dependent frequencies from the ^{1}H–^{15}N heteronuclear dipole–dipole coupling and ^{15}N-chemical shift interactions, which, in combination with the magnitudes and orientations of the spin-interaction tensors in the molecular frame of the peptide bond,[168] provide input for structure determination.[169,170] The 2D PISEMA spectrum has a distinct "wheel-like" pattern of resonances because the helix axes are aligned approximately parallel to the direction of the applied magnetic field. The resonances were assigned by using a modified version of the shotgun NMR approach.[171] The 2D PISEMA spectra from one uniformly ^{15}N-labeled and several selectively ^{15}N-labeled samples were analyzed with a combination of polarity index slant angle (PISA) wheels[172,173] and dipolar waves.[174] The atomic resolution structure was then calculated through structural fitting of the ^{15}N–^{1}H dipolar coupling and ^{15}N CSA (see Section 5.1).[175] The molecular surface of the capsid is visualized[177] according to the electrostatic potential calculated by the program GRASP[176] and also by the program INSIGHT II. The structure of the coat protein determined by X-ray fiber diffraction has been described as a single gently curving α-helix. This work, however, gives very accurate values for the tilt of the helical regions, with residues 8–38 having a 23° tilt and the C-terminal region starting at residues 40, having a somewhat smaller tilt of 18° with respect to the filament.

References

[1] W. Traub and K. A. Piez, 1971, *Adv. Protein Chem.*, 25, 243–352.
[2] J. P. Carver and E. R. Blout, 1967, in *Treatise on Collagen*, vol. 1, G. N. Ramachandran, Ed., Academic Press, New York, pp. 441–526.
[3] G. N. Ramachandran, 1967, in *Treatise on Collagen*, vol. 1, G. N. Ramachandran, Ed., Academic Press, New York, pp. 103–183.
[4] H. Saitô, R. Tabeta, A. Shoji, T. Ozaki, I. Ando, and T. Miyata, 1984, *Biopolymers*, 23, 2279–2297.
[5] H. Saitô and M. Yokoi, 1992, *J. Biochem. (Tokyo)*, 111, 376–382.
[6] J. Schaefer, E. O. Stejskal, C. F. Brewer, H. D. Kaiser, and H. Sternlicht, 1978, *Arch. Biochem. Biophys.*, 190, 657–661.
[7] J. G. Hexem, M. H. Frey, and S. J. Opella, 1981, *J. Am. Chem. Soc.*, 103, 224–226.
[8] A. Naito, S. Ganapathy, and C. A. McDowell, 1981, *J. Chem. Phys.*, 74, 5393–5406.
[9] I. Z. Siemion, T. Wieland, and K. H. Pook, 1975, *Angew. Chem.*, 87, 712–714.
[10] R. Deslauriers and I. C. P. Smith, 1976, in *Topics in Carbon-13 NMR Spectroscopy*, vol. 2, G. C. Levy, Ed., John Wiley, New York, pp. 1–80.
[11] W. P. Rothwell and J. S. Waugh, 1981, *J. Chem. Phys.*, 74, 2721–2732.
[12] H. Saitô, R. Tabeta, F. Formaggio, M. Crisma, and C. Toniolo, 1988, *Biopolymers*, 27, 1607–1617.
[13] M. Kamihira, A. Naito, K. Nishimura, S. Tuzi, and H. Saitô, 1998, *J. Phys. Chem. B*, 102, 2826–2834.
[14] H. Saitô, S. Tuzi, S. Yamaguchi, M. Tanio, and A. Naito, 2000, *Biochim. Biophys. Acta*, 1460, 39–48.
[15] H. Saitô, S. Tuzi, M. Tanio, and A. Naito, 2002, *Annu. Rep. NMR Spectrosc.*, 47, 39–108.
[16] H. Saitô, 2004, *Chem. Phys. Lipids*, 132, 101–121.
[16a] D. Huster, J. Schiller, and K. Arnold, 2002, *Magn. Reson. Med.*, 48, 624–632.
[17] A. Naito, S. Ganapathy, K. Akasaka, and C. A. McDowell, 1983, *J. Magn. Reson.*, 54, 226–235.
[18] R. E. London, 1978, *J. Am. Chem. Soc.*, 100, 2678–2685.

[19] S. K. Sarkar, P. E. Young, and D. A. Torchia, 1986, *J. Am. Chem. Soc.*, 108, 6459–6464.
[20] S. K. Sarkar, Y. Hiyama, C. H. Niu, P. E. Young, J. T. Gerig, and D. A. Torchia, 1987, *Biochemistry*, 26, 6793–6800.
[21] D. A. Torchia and D. L. VanderHart, 1976, *J. Mol. Biol.*, 104, 315–321.
[22] L. W. Jelinski and D. A. Torchia, 1979, *J. Mol. Biol.*, 133, 45–65.
[23] D. A. Torchia, 1982, *Methods Enzymol.*, 82, 174–186.
[24] L. W. Jelinski, C. E. Sullivan, and D. A. Torchia, 1980, *Nature*, 284, 531–534.
[25] L. W. Jelinski and D. A. Torchia, 1980, *J. Mol. Biol.*, 138, 255–272.
[26] L. B. Sandberg, 1976, in *International Review of Connective Research, vol.* 7, D. A. Hall and D. S. Jackson, Eds., Academic Press, New York, p. 160.
[27] B. Lin and V. Daggett, 2002, *J. Muscle Res. Cell Motil.*, 23, 561–573.
[28] C. M. Venkatachalam and D. W. Urry, 1981, *Macromolecules*, 14, 1225–1229.
[29] D. W. Urry, D. K. Chang, N. R. Krishna, D. H. Huang, T. L. Trapane, and K. U. Prasad, 1989, *Biopolymers*, 28, 819–833.
[30] C. A. Hoeve and P. J. Flory, 1974, *Biopolymers*, 13, 677–686.
[31] T. Weis-Fogh and S. O. Anderson, 1970, *Nature*, 227, 718–721.
[32] W. R. Gray, L. B. Sandberg, and J. A. Foster, 1973, *Nature*, 246, 461–466.
[33] J. M. Goslin, F. F. Yew, and F. Weis-Fogh, 1975, *Biopolymers*, 14, 1811–1825.
[34] H. Reiersen, A. R. Clarke, and A. R. Rees, 1998, *J. Mol. Biol.*, 283, 255–264.
[35] J. M. Gosline, 1978, *Biopolymers*, 17, 697–707.
[36] B. Li, D. O. V. Alonso, and V. Daggett, 2001, *J. Mol. Biol.*, 305, 581–592.
[37] D. A. Torchia and K. A. Piez, 1973, *J. Mol. Biol.*, 76, 419–424.
[38] J. R. Lyerla, Jr. and D. A. Torchia, 1975, *Biochemistry*, 14, 5175–5183.
[39] D. A. Torchia and D. L. VanderHart, 1979, in *Topics in Carbon-13 NMR Spectroscopy*, G. C. Levy, Ed., John Wiley, New York, pp. 325–360.
[40] J. Schaefer, 1973, *Macromolecules*, 6, 882–888.
[41] J. Schaefer, 1974, in *Topics in Carbon-13 NMR Spectroscopy*, vol. 1, G. C. Levy, Ed., John Wiley, New York, pp. 149–208.
[42] K. Kumashiro, M. S. Kim, S. E. Kaczmarek, L. B. Sandberg, and C. D. Boyd, 2001, *Biopolymers*, 59, 266–275.
[43] A. Perry, M. P. Stypa, B. K. Tenn, and K. K. Kumashiro, 2002, *Biophys. J.*, 82, 1086–1095.
[44] A. Perry, M. P. Stypa, J. A. Foster, and K. K. Kumashiro, 2002, *J. Am. Chem. Soc.*, 124, 6832–6833.
[45] M. S. Pometun, E. Y. Chekmenev, and R. J. Wittebort, 2004, *J. Biol. Chem.*, 279, 7982–7987.
[46] G. Lipari and A. Szabo, 1980, *Biophys. J.*, 30, 489–506.
[47] K. K. Kumashiro, T. L. Kurano, W. P. Niemczura, M. Martino, and A. M. Tamburro, 2003, *Biopolymers*, 70, 221–226.
[48] R. A. McMillan, T. A. T. Lee, and V. P. Conticello, 1999. *Macromolecules*, 32, 3643–3648.
[49] R. A. McMillan and V. P. Conticello, 2000, *Macromolecules*, 33, 4809–4821.
[50] M. Hong, R. A. McMillan, and V. P. Conticello, 2002, *J. Biomol. NMR*, 22, 175–179.
[51] M. Hong, D. Isailovic, R. A. McMillan, and V. P. Conticello, 2003, *Biopolymers*, 70, 158–168.
[52] X. L. Yao, V. P. Conticello, and M. Hong, 2004, *Magn. Reson. Chem.*, 42, 267–275.

[53] X. L. Yao and M. Hong, 2004, *J. Am. Chem. Soc.*, 126, 4199–4210.
[54] A. S. Tatham, P. R. Shewry, and P. S. Belton, 1991, in *Advances in Cereal Science and Technology*, vol. 10, Y. Pomeranz, Ed., American Association of Cereal Chemistry, St Paul, p. 1.
[55] P. R. Shewry and A. S. Tatham, 1990, *Biochem. J.*, 267, 1–12.
[56] A. M. Gil, 1995, in *Magnetic Resonance in Food Science*, P. S. Belton, I. Delgadillo, A. M. Gil, and G. A. Webb, Eds., Royal Society of Chemistry, London, pp. 272–286.
[57] S. K. Rasmussen and A. Brandt, 1986, *Carlsberg Res. Commun.*, 51, 371–379.
[58] P. R. Shewry, N. G. Halford, and A. S. Tatham, 1992, *J. Cereal Sci.*, 15, 105–120.
[59] A. M. Gil, K. Masui, A. Naito, A. S. Tatham, P. S. Belton, and H. Saitô, 1996, *Biopolymers*, 41, 289–300.
[60] A. M. Gil, E. Alberti, A. Naito, K. Okuda, H. Saitô, A. S. Tatham, and S. Gilbert, 1999, in *Advances in Magnetic Resonance in Food Sciences*, Royal Society of Chemistry, London, pp. 126–134.
[61] E. Alberti, S. M.-Gilbert, A. S. Tatham, P. R. Shewry, and A. M. Gil, 2002, *Biopolymers*, 67, 487–498.
[62] E. Alberti, S. M. Gilbert, A. S. Tatham, P. R. Shewry, A. Naito, K. Okuda, H. Saitô, and A. M. Gil, 2002, *Biopolymers*, 65, 158–168.
[63] N. Hojo, 1980, *Zoku Kenshi no Kozo* (*Structure of Silk Fibers*), Shinshu University, Ueda, Japan.
[64] F. Lucas, and K. M. Rudall, 1968, in *Comprehensive Biochemistry*, Chapter 7, vol. 26B, M. Florkin and E. Scotz, Eds., Elsevier, Amsterdam.
[65] R. D. B. Fraser and T. P. MacRae, 1973, in *Conformation of Fibrous Proteins and Related Synthetic Polypeptides*, Chapter 13, Academic Press, New York.
[66] B. Lotz and F. C. Cesari, 1979, *Biochimie*, 61, 205–214.
[67] M. Shimizu, 1941, *Bull. Imp. Sericult. Expt. Sta. Jpn.*, 10, 475–494.
[68] O. Kratky, E. Schauenstein, and A. Sekora, 1950, *Nature*, 165, 319–320.
[69] E. Oroudjev, J. Soares, S. Arcdiacono, J. B. Thompson, S. A. Fossey, and H. G. Hansma, 2002, *Proc. Natl. Acad. Sci. USA*, 99, 6460–6465.
[70] C. Y. Hayashi, N. H. Shipley, and R. V. Lewis, 1999, *Int. J. Biol. Macromol.*, 24, 271–275.
[71] C. Zhao and T. Asakura, 2001, *Prog. Nucl. Magn. Reson. Spectrosc.*, 39, 301–352.
[72] T. Kameda and T. Asakura, 2002, *Annu. Rep. NMR Spectrosc.*, 46, 101–149.
[73] H. Saitô, R. Tabeta, T. Asakura, Y. Iwanaga, A. Shoji, T. Ozaki, and I. Ando, 1984, *Macromolecules*, 17, 1405–1412.
[74] H. Saitô, M. Ishida, M. Yokoi, and T. Asakura, 1990, *Macromolecules*, 23, 83–88.
[75] M. Ishida, T. Asakura, M. Yokoi, and H. Saitô, 1990, *Macromolecules*, 23, 88–94.
[76] T. Asakura, R. Sugino, T. Okumura, and Y. Nakazawa, 2002, *Protein Sci.*, 11, 1873–1877.
[77] T. Asakura, K. Suita, T. Kameda, S. Afonin, and A. S. Ulrich, 2004, *Magn. Reson. Chem.*, 42, 258–266.
[78] T. Asakura and J. Yao, 2002, *Protein Sci.*, 11, 2706–2713.
[79] Y. Takahashi, M. Gehoh, and K. Yuzuriha, 1999, *Int. J. Biol. Macromol.*, 24, 127–138.
[80] T. Asakura, T. Ito, M. Okudaira, and T. Kameda, 1999, *Macromolecules*, 32, 4940–4946.
[81] Y. Nakazawa, T. Nakai, T. Kameda, and T. Asakura, 1999, *Chem. Phys. Lett.*, 311, 362–366.

[82] Y. Nakazawa and T. Asakura, 2002, *FEBS Lett.*, 529, 188–192.
[83] Y. Nakazawa and T. Asakura, 2002, *Macromolecules*, 35, 2393–2400.
[84] M. Demura, M. Minami, T. Asakura, and T. A. Cross, 1998, *J. Am. Chem. Soc.*, 120, 1300–1308.
[85] R. E. Marsh, R. B. Corey, and L. Pauling, 1955, *Biochim. Biophys. Acta*, 16, 1–34.
[86] J. D. van Beek, L. Beaulieu, H. Schäfer, M. Demura, T. Asakura, and B. H. Meier, 2000, *Nature*, 405, 1077–1079.
[87] K. Schmidt-Rohr, 1996, *Macromolecules*, 29, 3975–3981.
[88] T. Asakura, J. Ashida, T. Yamane, T. Kameda, Y. Nakazawa, K. Ohgo, and K. Komatsu, 2001, *J. Mol. Biol.*, 306, 291–305.
[89] T. Nakai, J. Ashida, and T. Terao, 1988, *J. Chem. Phys.*, 88, 6049–6058.
[90] B. Blümich and A. Hagemeyer, 1989, *Chem. Phys. Lett.*, 161, 55–59.
[91] Y. Nakazawa and T. Asakura, 2003, *J. Am. Chem. Soc.*, 125, 7230–7237.
[92] Y. Nakazawa, M. Bamba, S. Nishio, and T. Asakura, 2003, *Protein Sci.*, 12, 666–671.
[93] S. Kimura, A. Naito, H. Saitô, K. Ogawa, and A. Shoji, 2001, *J. Mol. Struct.*, 562, 197–203.
[94] S. Kimura, A. Naito, S. Tuzi, and H. Saitô, 2002, *J. Mol. Struct.*, 602–603, 125–131.
[95] A. Naito, K. Nishimura, S. Kimura, M. Aida, N. Yasuoka, S. Tuzi, and H. Saitô, 1996, *J. Phys. Chem.*, 100, 14995–15004.
[96] A. Fukutani, A. Naito, S. Tuzi, and H. Saitô, 2002, *J. Mol. Struct.*, 602–603, 491–503.
[97] E. K. Tillinghast and T. Christenson, 1984, *J. Arachnol.*, 12, 69–74.
[98] R. W. Work and C. T. Young, 1987, *J. Arachnol.*, 15, 65–80.
[99] S. J. Lombardi and D. L. Kaplan, *J. Arachnol.*, 1990, 18, 297–306.
[100] A. Simmons, E. Ray, and L. W. Jelinski, 1994, *Macromolecules*, 27, 5235–5237.
[101] A. H. Simmons, C. A. Michal, and L. W. Jelinski, 1996, *Science*, 271, 84–87.
[102] J. Kümmerlen, J. D. van Beek, F. Vollrath, and B. H. Meier, 1996, *Macromolecules*, 29, 2920–2928.
[103] J. D. van Beek, S. Hess, F. Vollrath, and B. H. Meier, 2002, *Proc. Natl. Acad. Sci. USA*, 99, 10266–10271.
[104] C. A. Michal and L. W. Jelinski, 1998, *J. Biomol. NMR*, 12, 231–241.
[105] R. W. Work and N. Morosoff, 1982, *Text. Res. J.*, 52, 349–356.
[106] R. W. Work, 1985, *J. Exp. Biol.*, 118, 379–404.
[107] L. W. Jelinski, A. Blye, O. Liivak, C. Michal, G. LaVerde, A. Seidel, N. Shah, and Z. Yang, 1999, *Int. J. Biol. Macromol.*, 24, 197–201.
[108] A. D. Parkhe, S. K. Seeley, K. Gardner, L. Thompson, and R. V. Lewis, 1997, *J. Mol. Recogn.*, 10, 1–6.
[109] J. D. van Beek, J. Kümmerlen, F. Vollrath, and B. H. Meier, 1999, *Int. J. Biol. Macromol.*, 24, 173–178.
[110] Z. Yang, O. Liivak, A. Seidel, G. LaVerde, D. B. Zax, and L. W. Jelinski, 2000, *J. Am. Chem. Soc.*, 122, 9019–9025.
[111] R. W. Work, 1977, *Text. Res. J.*, 47, 650–662.
[112] A. Seidel, O. Liivak, S. Calve, J. Adaska, G. Ji, Z. Yang, D. Grubb, D. B. Zax, and L. W. Jelinski, 2000, *Macromolecules*, 33, 775–780.

[113] A. I. Kishore, M. E. Herberstein, C. L. Craig, and F. Separovic, 2002, *Biopolymers*, 61, 287–297.
[114] K. Schmidt-Rohr and H. W. Spiess, 1994, *Multidimensional Solid-state NMR and Polymers*, Academic Press, London.
[115] G. P. Holland, R. V. Lewis, and J. L. Yager, 2004, *J. Am. Chem. Soc.*, 126, 5867–5872.
[116] D. T. Grubb and L. W. Jelinski, 1997, *Macromolecules*, 30, 2860–2867.
[117] R. D. Fraser, T. P. MacRae, and G. E. Rogers, 1972, *Keratins: Their Composition, Structure and Biosynthesis*, C. C. Thomas, Springfield, IL.
[118] J. H. Bradbury, 1973, *Adv. Protein Chem.*, 27, 111–211.
[119] R. D. Fraser and T. P. MacRae, 1973, *Conformation in Fibrous Proteins*, Chapter 16, Academic Press, New York.
[120] I. J. O'Donnell and E. O. P. Thompson, 1964, *Aust. J. Biol. Sci.*, 17, 973–989.
[121] L. M. Dowling and W. G. Crewther, 1974, *Pre. Biochem.*, 4, 203–226.
[122] W. G. Crewther and L. M. Dowling, 1971, *J. Appl. Polym. Sci. Appl. Polym. Symp.*, 18, 120.
[123] W. G. Crewther, L. M. Dowling, and A. S. Inglis, 1986, *Biochem. J.*, 236, 695–703.
[123a] S. J. Opella and M. H. Frey, 1979, *J. Am. Chem. Soc.*, 101, 5854–5856.
[124] H. Yoshimizu and I. Ando, 1990, *Macromolecules*, 23, 2908–2912.
[125] T. Taki, S. Yamashita, M. Satoh, M. Shibata, T. Yamashita, R. Tabeta, and H. Saitô, 1981, *Chem. Lett.*, 1803–1806.
[126] H. Saitô, R. Tabeta, A. Shoji, T. Ozaki, and I. Ando, 1983, *Macromolecules*, 16, 1050–1057.
[127] H. Saitô, R. Tabeta, I. Ando, T. Ozaki, and A. Shoji, 1983, *Chem. Lett.*, 1437–1440.
[128] I. Ando, H. Saitô, R. Tabeta, A. Shoji, and T. Ozaki, 1984, *Macromolecules*, 17, 457–461.
[129] A. Shoji, T. Ozaki, H. Saitô, R. Tabeta, and I. Ando, 1984, *Macromolecules*, 17, 1472–1479.
[130] H. Saitô, R. Tabeta, A. Shoji, T. Ozaki, I. Ando, and T. Asakura, 1985, in *Magnetic Resonance in Biology and Medicine*, G. Govil, C. Khetrapal, and A. Saran, Eds., Tata McGraw-Hill, New Delhi, pp. 195–215.
[131] H. Saitô and I. Ando, 1989, *Annu. Rep. NMR Spectrosc.*, 21, 209–290.
[132] I. Ando and S. Kuroki, 1996, *Encyclopedia of NMR*, John Wiley, New York, pp. 4458–4468.
[133] S. Kuroki, K. Yamauchi, I. Ando, A. Shoji, and T. Ozaki, 2001, *Curr. Org. Chem.*, 5, 1001–1015.
[134] I. Ando, S. Kuroki, H. Kurosu, and T. Yamanobe, 2001, *Prog. Nucl. Magn. Reson. Spectrosc.*, 39, 79–133.
[135] D. Mueller and H. R. Kricheldorf, 1981, *Polym. Bull.*, 6, 101–108.
[136] H. R. Kricheldorf, H. R. Mueller, and K. Ziegler, 1984, *Polym. Bull.*, 9, 284–291.
[137] H. R. Kricheldorf, M. Mutter, F. Mazer, D. Mueller, and D. Forster, 1983, *Biopolymers*, 22, 1357–1372.
[138] H. R. Kricheldorf and H. R. Mueller, 1983, *Macromolecules*, 16, 615–623.
[139] T. Amiya, T. Miyamoto, and H. Inagaki, 1980, *Biopolymers*, 19, 1093–1097.
[140] T. Amiya, A. Kawaguchi, T. Miyamoto, and H. Inagaki, 1990, *J. Soc. Fiber Sci. Technol. Jpn.*, 36, 479–483.

[141] T. Asakura and T. Yamaguchi, 1987, *Nippon Sanshigaku Zasshi*, 56, 300–304.
[142] B. S. Harrap and J. M. Gillespie, 1963, *Aust. J. Biol. Sci.*, 16, 542–563.
[143] T. Amiya, K. Kajiwara, T. Miyamoto, and H. Inagaki, 1982, *Int. J. Biol. Macromol.*, 4, 165–172.
[144] M. J. Duer, N. McDougal, and R. C. Murray, 2003, *Phys. Chem. Chem. Phys.*, 5, 2894–2899.
[145] N. Nishikawa, Y. Horiguchi, T. Asakura, and I. Ando, 1999, *Polymer*, 40, 2139–2144.
[146] S. Tuzi, S. Sakamaki, and I. Ando, 1990, *J. Mol. Struct.*, 221, 289–297.
[147] S. Tuzi and I. Ando, 1989, *J. Mol. Struct.*, 196, 317–325.
[148] H. Sakabe, T. Miyamoto, and H. Inagaki, 1981, *Sen-I Gakkaishi*, 37, T273–T278.
[149] J. Koga, K. Kawaguchi, E. Nishino, K. Joko, N. Ikuta, I. Abe, and T. Hirashima, 1989, *J. Appl. Polym. Sci.*, 37, 2131–2140.
[150] H. Yoshimizu, H. Mimura, and I. Ando, 1991, *Macromolecules*, 24, 862–866.
[151] H. Yoshimizu, H. Mimura, and I. Ando, 1991, *J. Mol. Struct.*, 246, 367–379.
[152] H. Sakabe, T. Miyamoto, and H. Inagaki, 1982, *Sen-I Gakkaishi*, 38, T517–T522.
[153] H. Yoshimizu and I. Ando, unpublished data.
[154] J. M. Gillespie, 1990, *Cellular and Molecular Biology of Intermediate Filaments*, R. A. Goldman and P. M. Steinert, Eds., Plenum Press, New York, pp. 95–128.
[155] C. R. Robbins, 1994, in *Chemical and Physical Behavior of Human Hair*, Springer-Verlag, New York, pp. 1–92.
[156] K. Schmidt-Rohr, J. Clauss, and H. W. Spiess, 1992, *Macromolecules*, 25, 3273–3277.
[157] D. A. Marvin and B. Hohn, 1969, *Bacteriol. Rev.*, 33, 172–209.
[158] L. Makowski, 1984, in *Biological Macromolecules and Assembly*, A. McPherson, Ed., John Wiley, New York, pp. 203–253.
[159] D. A. Marvin, 1998, *Curr. Opin. Struct. Biol.*, 8, 150–158.
[160] J. Torbet and G. Maret, 1979, *J. Mol. Biol.*, 134, 843–845.
[161] C. M. Gall, T. A. Cross, J. A. DiVerdi, and S. J. Opella, 1982, *Proc. Natl. Acad. Soc. USA*, 79, 101–105.
[162] L. A. Colnago, K. G. Valentine, and S. J. Opella, 1987, *Biochemistry*, 26, 847–854.
[163] G. Leo, L. A. Colnago, K. G. Valentine, and S. J. Opella, 1987, *Biochemistry*, 26, 854–862
[164] T. A. Cross, P. Tsang, and S. J. Opella, 1983, *Biochemistry*, 22, 721–726.
[165] D. A. Marvin, R. D. Hale, C. Nave, and M. H. Citterich, 1994, *J. Mol. Biol.*, 235, 260–286.
[166] W. M. Tan, R. Jelinek, S. J. Opella, P. Malik, T. D. Terry, and R. N. Perham, 1999, *J. Mol. Biol.*, 286, 787–796.
[167] A. C. Zeri, M. F. Mesleh, A. A. Nevzorov, and S. J. Opella, 2003, *Proc. Natl. Acad. Sci. USA*, 100, 6458–6463.
[168] C. H. Wu, A. Ramamoorthy, and S. J. Opella, 1994, *J. Magn. Reson.*, A109, 270–274.
[169] T. G. Oas, C. J. Hartzell, F. W. Dahlquist, and G. P. Drobny, 1987, *J. Am. Chem. Soc.*, 109, 5962–5966.
[170] S. J. Opella, P. L. Stewart, and K. G. Valentine, 1987, *Q. Rev. Biophys.*, 19, 7–49.
[171] F. M. Marassi and S. J. Opella, 2003, *Protein Sci.*, 12, 403–411.

[172] F. M. Marassi and S. J. Opella, 2000, *J. Magn. Reson.*, 144, 150–155.
[173] J. Wang, J. Denny, C. Tian, S. Kim, Y. Mo, F. Kovacs, Z. Song, K. Nishimura, Z. Gan, R. Fu, J. R. Quine, and T. A. Cross, 2000, *J. Magn. Reson.*, 144, 162–167.
[174] M. F. Mesleh, G. Veglia, T. M. DeSilva, F. M. Marassi, and S. J. Opella, 2002, *J. Am. Chem. Soc.*, 124, 4206–4207.
[175] A. A. Nevzorov and S. J. Opella, 2003, *J. Magn. Reson.*, 160, 33–39.
[176] A. Nicholls, K. A. Sharp, and B. Honig, 1991, *Proteins*, 11, 281–296.
[177] A. C. Zeri, M. F. Mesleh, A. A. Nevzorov, and S. J. Opella, 2003, Proc. Natl. Acad. Sci. USA, 100, 6458–6463.

Chapter 10

POLYSACCHARIDES

The physical properties of a polysaccharide as well as its biological response may be strongly related to its secondary structures in the solid, gel, or solution states. A variety of monomers in polysaccharides such as glucose, galactose, glucosamine, etc. and linkage positions such as $1 \rightarrow 2$, $1 \rightarrow 3$, $1 \rightarrow 4$, etc. anomeric forms such as α and β, together with a degree of branching lead to an extraordinary variety of primary structures. The secondary structure of an individual polysaccharide is generally defined by a set of torsion angles (ϕ, ψ) about the glycosidic linkages. On the basis of the conformation-dependent ^{13}C chemical shifts, it is expected that ^{13}C NMR spectra can be utilized to distinguish one of the crystalline polymorphs in these polysaccharides from the others. In fact, it is expected that their ^{13}C chemical shifts of carbons located at the glycosidic linkages are displaced in line with their particular conformations. Saitô *et al.*[1] suggested a correlation of the ^{13}C chemical shifts of the C-1 and C-4 carbons of (1 → 4)-α-D-glucans with their torsions, ϕ and ψ, respectively, although some variations for such relations were later proposed.[2,3] This means that the conformation-dependent ^{13}C chemical shifts, if any, are not always simple for polysaccharides in contrast to those for peptides and proteins already described above. Nevertheless, it is also emphasized that distinction of the polymorphic structures of any given polysaccharide, including a possibility of forming either single or multiple helices, is made possible by a careful examination of their ^{13}C chemical shifts at carbons not always close to the glycosidic linkages, as encountered for (1 → 3)-β-D-glucan.[4]

In practice, it is not always straightforward to determine such torsion angles at the glycosidic linkages, because the experimentally available data from a current fiber diffraction study may not be sufficient to arrive at the final structure. In fact, three-dimensional structures of polysaccharides have been revised several times as in the case of B-amylose.[5–7] In particular, distinguishing between multiple-stranded helices and nested single helices is often one of the most difficult problems found in interpreting fiber diffraction.[8] It is thus a major advantage to record ^{13}C NMR spectra of these polysaccharides to reveal the secondary structure as a complementary means, because structural information by solid state NMR is equally available from the domain of noncrystalline portion.

10.1. Distinction of Polymorphs

10.1.1. (1 → 3)-β-D-Glucan[9–13] and -Xylan[14]

A variety of (1 → 3)-β-D-glucans have been isolated from different sources: a low molecular weight glucan from sea weed (laminaran) is freely soluble in aqueous solution, while high molecular weight glucans from bacteria (curdlan from *Alcaligenes faecalis* var. *myxogenes*) or fungi (lentinan from *Lentinus edodes*, schizophyllan from *Schizophyllum commune*, HA-β-glucan from *Pleurotus ostreatus*, etc.) are insoluble in a neutral aqueous solution.[9,10] It has been demonstrated that gel-forming and antitumor activity are characteristic of these high molecular weight glucans.[15] It appears, therefore, that elucidation of the secondary structures of these glucans by means of ^{13}C NMR spectra is essential to gain insight into a relationship between the secondary structure and their biological and physical properties.

A linear high molecular weight glucan, curdlan, is not soluble in a neutral aqueous solution, and is capable of forming an elastic gel by heating its aqueous suspension at a temperature above 60 °C, followed by cooling.[9,10] Three clearly different ^{13}C NMR spectra were available from curdlan in the solid state by changing the manner of sample preparation (Figure 10.1). The "anhydrous" sample (a) as received from a commercial source (Wako Pure Chemicals, Osaka) or lyophilized from DMSO solution was converted to the "hydrated" form by hydration (b) by placing it in a desiccator at 96% RH overnight. The "annealed" sample (c) was prepared by heating an aqueous suspension at a temperature above 150 °C, followed by slow cooling.[4,16] The annealed sample (c) has been identified as the triple helix form, on the basis of powder X-ray diffraction as compared with the data from a fiber diffraction study.[17] The hydrated sample (b) is considered to be somewhat flexible single-helix form, because the ^{13}C chemical shifts turn out to be identical to those observed in an elastic gel by high-resolution NMR measurements on liquid and solid components.[16] The anhydrous sample (a) is thought to be a single-chain form, because this is a dehydrated form of the sample (b), and can be readily converted to the hydrated form (b) by exposure to high RH. The C-3 ^{13}C chemical shift of the single-chain form (89.8 ppm) is displaced downfield by 2.5–3.3 ppm from that of the single helix

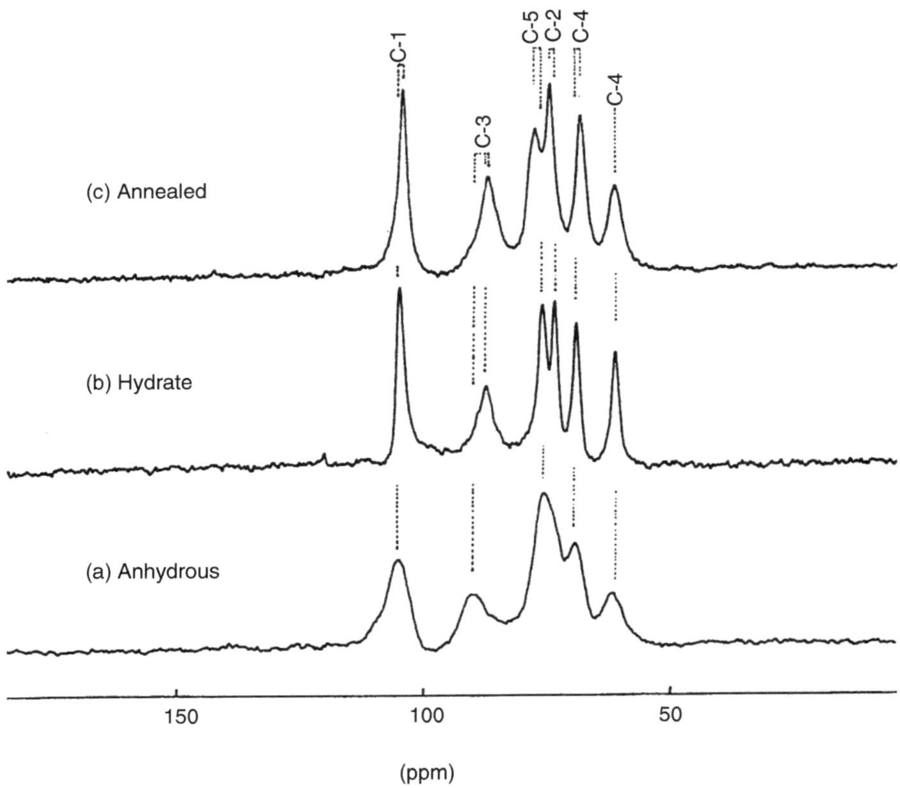

Figure 10.1. ^{13}C CP-MAS NMR spectra of curdlan in anhydrous (a), hydrate (b), and annealed at 150 °C (c).[4]

(87.3 ppm) or the triple helix (86.5 ppm). At first glance, it appears that the ^{13}C chemical shifts of the single helix are very similar to those of the triple helix. Distinction of these two forms, however, is made possible when the peak separation between the C-5 and C-2 carbons is compared (2.0 and 3.2 ppm for the former and the latter, respectively) rather than the C-3 peak position alone. The reason why the C-3 ^{13}C chemical shifts are not always significantly different between the single- and triple-helix form could be ascribed to the fact that the torsion angles between the two types of helices are very similar, as manifested from the lengths of the *c*-axis (fiber axis) being 18.78 Å (triple helix)[18] and 22.8 Å (single helix).[19]

Distinction of the single and multiple helical forms is not easy by the fiber diffraction method, as mentioned already. This problem, however, is easily solved by the ^{13}C NMR approach, if these polymorphic structures can be identified in view of sample history as well as other experimental techniques, and mutual conformational conversions among them can be manipulated by a series of physical treatments under a controlled manner, as illustrated in Figure 10.2.[4] The existence of the single-chain form is straightforward in a sample prepared by either dehydration of the single helix or lyophilization of samples taking the triple helix from DMSO or alkaline solution. This is because even a multiple-stranded (triple) helix can be completely dispersed as a result of the accompanied conformational transition from the triple helix to a random-coil form present in DMSO or alkaline solution. Surprisingly, the triple-helix form is available from lyophilized preparations of fully hydrated, low molecular weight glucans such as laminaran or high molecular weight branched glucans such as schizophyllan from aqueous solution, as judged from the ^{13}C NMR peak positions with reference to those of the annealed curdlan as illustrated in Figure 10.2. This is because these polysaccharides can be readily hydrated under neutral aqueous environment, although the former is soluble in aqueous solution as random coil but the latter is swollen in aqueous solution. The single-helix form of high molecular weight glucan, on the other hand, is readily available from the hydrated single chain form, as judged from their peak positions as compared with those of gel samples, as will be described later, and also due to reversible mutual conversion from the single chain form. It appears that formation of the most energetically stable triple helix may be hampered in the case of more hydrophobic, linear high molecular weight glucans (curdlan) because of an insufficient extent of hydration, resulting in formation of the single helix form. Conversion to the triple helix, however, is made possible only when these single-helical chains can be dispersed at a temperature above 150 °C, followed by slow cooling. Similar diagram was also available from studies on a variety of branched (1 → 3)-β-D-glucans.

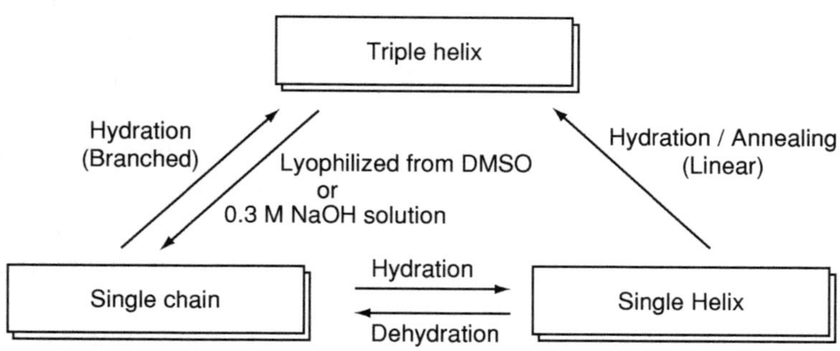

Figure 10.2. Conversion diagram for (1 → 3)-β-D-glucans by a variety of physical treatments.[4]

This approach utilizing cycles of mutual conversion among several conformations proved to be very useful for distinction of the single chain from multiple-stranded chains, and was later extended to various types of polysaccharides such as (1 → 3)-β-D-xylan, (1 → 4)-α-D-glucan, etc. In fact, the primary structure of the cell wall of green algae differs from that of (1 → 3)-β-D-glucan by the absence of the hydroxymethyl group at the C-5 carbon. Indeed, the presence of the triple-helix conformation was originally proposed for (1 → 3)-β-D-xylan,[20,21] and later extended to (1 → 3)-β-D-glucan, because the hydroxymethyl group in both the triple- and single-helical (1 → 3)-β-D-glucan protrudes from the outside of the helices, and does not play an important role in conformational stability.[18,19] For this reason, it is expected that the similar three types of forms in the glucans are also present for (1 → 3)-β-D-xylan, depending on the type of physical treatments, including the manner of isolation. Saitô *et al.* showed a very similar conversion diagram for (1 → 3)-β-D-xylan among the three similar conformations, single chain, single helix, or triple helix.[14] It turns out, however, that the triple-helix form of (1 → 3)-β-D-xylan is much stable than that of (1 → 3)-β-D-glucan, because the former turns out to be more hydrophobic due to the absence of a hydroxymethyl group at the C-5 position. Surprisingly, the triple helix structure of (1 → 3)-β-D-xylan is still retained even if it dissolves in DMSO solution. In such case, it was found that dissolution in aqueous zinc chloride is essential to disperse the triple helix to single chain structure.

10.1.2. (1 → 4)-α-D-Glucan: Amylose and Starch

The two main components of starch, amylopectin and amylose, are higher molecular weight (1 → 4)-α-D-glucans, either highly branched through α-D-(1 → 6)-linkages and linear, or lightly branched, respectively. It has been demonstrated that amylose and starch with α-D-(1 → 4)-linkages exhibit the polymorphs, V, A, B, and C.[22] X-ray diffraction gives two different patterns, A and B, from cereal starch and the starch found in tubers, respectively. As demonstrated in Figure 10.3, these two forms are readily distinguishable from the ^{13}C CP-MAS NMR spectra: the two C-1 peaks of starch A is a triplet, while that of starch B is a doublet.[23–27] The presence of these multiplet patterns was interpreted in terms of the number of chains in the asymmetric unit:[23,24,26] the assigned $P2_1$ space group for starch A has the 2_1 axis perpendicular to the six-residue-per-turn strands of the double helix, and has a twofold axis in the helix direction. Therefore, maltotriose must be taken as the asymmetric unit. On the other hand, the assigned $P3_12_1$ space

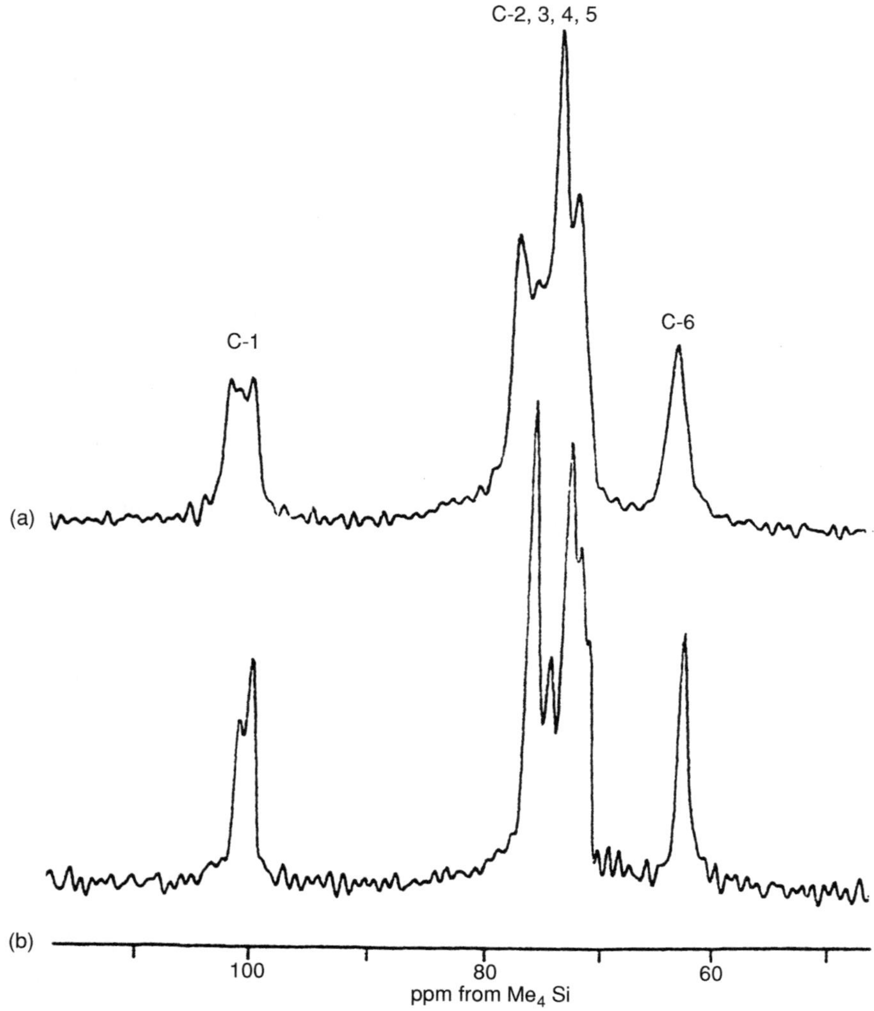

Figure 10.3. ^{13}C CP-MAS NMR spectra of starches hydrated at 100% RH: (a) Nageli amylodextrin (an "A" starch); (b) Lintner starch (a "B" starch) (from Marchessault *et al.*[23]).

group for starch B has the 3_1 helix down the strand. On drying of the starches, the ^{13}C NMR signals are substantially broadened, together with the appearance of signals from the noncrystalline region.[23–27] In particular, Gidley and Bociek[28] demonstrated that the C-1 peak of such an amorphous (1 → 4)-α-D-glucan shows a remarkable large chemical shift range (94–106 ppm), reflecting a wide diversity of local conformations. They showed that such peak positions are well correlated with the sum of contributions from allowed conformations, related to $|\phi| + |\psi|$ angles. These results clearly indicate that water molecules play an important role in maintaining the secondary structure of the starches A and B and amyloses.

The V form exists as complexes with small organic molecules and adopts a left-handed single-helix form. As a result, its C-1 and C-4 signals give rise to single lines, and are displaced downfield from those of the B form (Figure 10.4(c) and (d)).[29] In addition, it was shown that hydration of amorphous amylose of low molecular weight (DP 17) results in complete conversion to the B-type form.[29] This finding is consistent with the previous observation by Senti and Witnaur,[30] indicating that the B-form amylose can be obtained from the V form by hydration. Further, it was shown that the V-form amylose of high molecular weight (DP 1000) complexed with DMSO was converted to the B form by humidification by 96% RH for 12 h, as shown in Figure 10.4(a) and (b). It turned out, however, that this sort of conversion was insufficient (50%) for amylose of intermediate molecular weight (DP 100). The conversion diagram of conformations for amylose among amorphous, V and B forms based on a variety of physical treatments can be summarized in Figure 10.5.[11] Cheetham and Tao[31] examined the effect of hydration up to $\approx 30\%$ moisture for natural maize starches by ^{13}C CP-MAS NMR. The enhanced resolution at higher moisture levels revealed signals that were assigned to the amylose–lipid complex, i.e., V-type amylose. Prolonged treatment of the granule with iodine vapor significantly increased the amount of V-type amylose in the high-amylose samples, but caused a decrease in their degree of crystallinity. Spectral decomposition of the C-1 resonances of the ^{13}C CP-MAS NMR spectra of various starchy substrates (native potato starch, amylopectin, and amylose) following different techniques of treatments (casting, freeze-drying, and solvent exchange) showed the existence of five main types of $\alpha(1 \rightarrow 4)$-linkages.[32]

The B form was initially considered as a single-helical conformation,[5] since the conversion of V to B amylose takes place on humidification.[30] Later, the structure was refined as a right-handed double helix in an anti-parallel fashion.[6] The handedness of the double-helix, however, was further revised as a left-handed one.[33] Nevertheless, it is hardly likely that simple humidification of amylose in a desiccator at ambient temperature causes such an unfolding/folding process leading from the single-stranded helix (V form) to double-stranded helix (B form). In this connection, it was shown that complete dissolution in aqueous solution is an essential requirement for the conversion of the single helix to the triple helix, as found for $(1 \rightarrow 3)$-β-D-glucan.[4] This means that unfolding of the polymer chain followed by refolding in aqueous media is a necessary condition for such conversion.

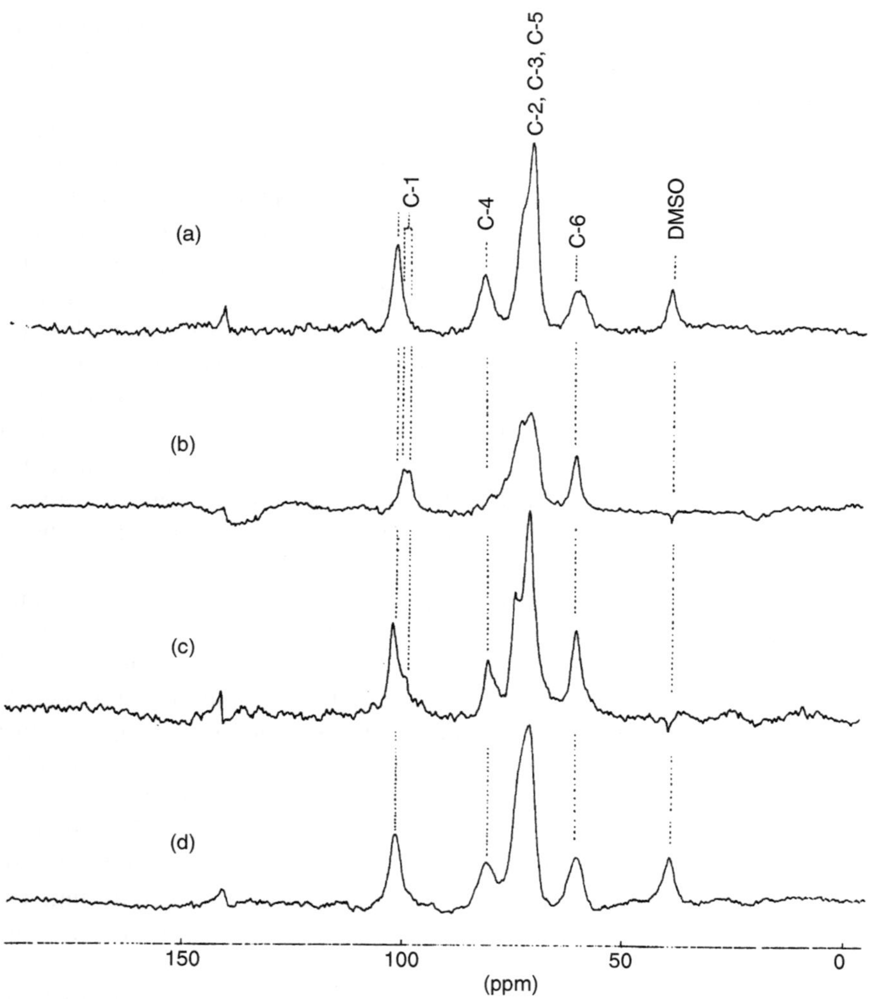

Figure 10.4. ^{13}C CP-MAS NMR spectra of amylose film (DP 1000): (a) anhydrous, (b) hydrated, (c) hydrated iodine complex, and (d) anhydrous iodine complex.[29]

10.1.3. Cellulose

Cellulose is an (1 → 4)-β-D-glucan and is the primary constituent of plant cell walls. Its most common polymorphs, cellulose I and II, are usually identified with the native and regenerated form, respectively. Atalla *et al.*[34] showed that these two polymorphs can easily be distinguished by examination of their ^{13}C NMR spectra. The C-1 peaks for both forms and the C-4 resonance of cellulose II show very definite splittings into two peaks with

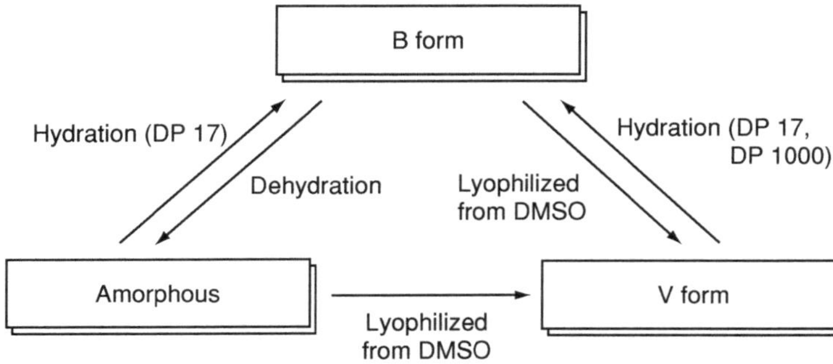

Figure 10.5. Conversion diagram of amylose by various physical treatments.[11]

approximately equal intensities. In addition, both the C-4 and C-6 peaks contain a broad high-field shoulder.[35,36] In contrast to cellulose II, it has been shown that ^{13}C NMR of cellulose I from algae such as *Valonia ventricosa* or from bacteria such as *A. xylinum* exhibits triplet signals for both C-1 and C-4 peaks, as illustrated in Figure 10.6,[39] although that from kraft pulp, lamie fibers, cotton linters, hydrocellulose prepared from cotton linters by acid hydrolysis, or a low DP-regenerated cellulose I does not.[36–40] In an earlier study, Earl and VanderHart suggested a unit cell containing four equivalent glucose units with two different types of glycosidic linkages.[36] Later, Atalla and VanderHart[37,39–41] proposed that native celluloses are composed of more than one crystalline form, I_α and I_β. They came to this conclusion because the multiplet intensities obtained by taking appropriate linear combinations of the spectra of the regenerated cellulose I and of the cellulose from *A. xylinum* are not constant, and not in the ratios of small whole numbers as would be expected if they arose from different sites within a single unit cell.[37,39] The alternative explanation is based on a comparison of NMR and X-ray diffraction data. Cael *et al.*[42] proposed a different interpretation for the multiplet patterns of algal cellulose. The eight-chain unit cell model of cellulose I in *Valonia* derived from electron- and X-ray diffraction data predicts the existence of three distinct molecular chains with different chemical shifts, with a predicted intensity ratio of 4:2:2 or 2: 1:1(see Figure 10.6(g)). This interpretation, however, turned out to be incorrect, in view of the later electron diffraction study leading to conclusions that the two allomorph forms, I_α and I_β, represent two crystalline phases with different crystal habits:[43] the I_α form represents a triclinic phase with one-chain-per-unit cell, while the I_β form represents a monoclinic phase with two-chains-per-unit cell. In the case of cellulose I, the ^{13}C chemical shift ranges of C-1, C-4, and C-6 have been well assigned.[44] The exact origin of each multiplet component, however, is still not well

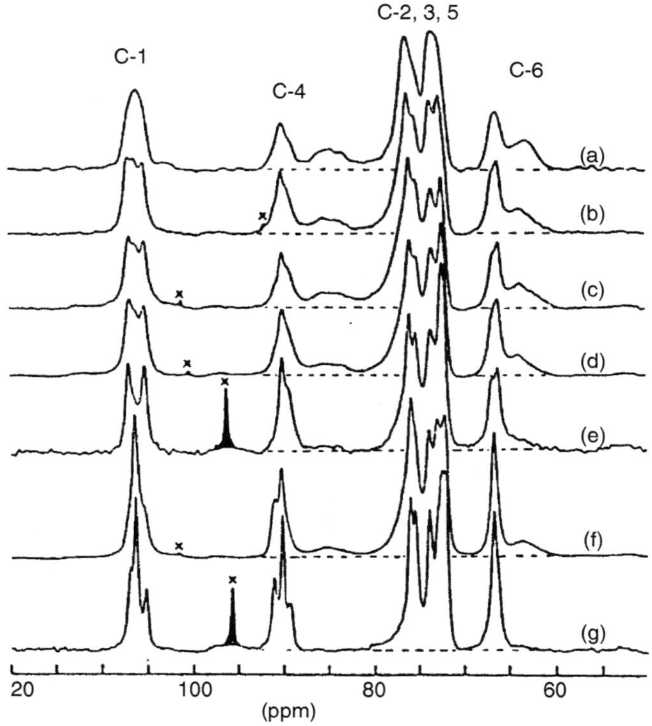

Figure 10.6. ^{13}C CP-MAS NMR spectra of various cellulose I materials (a) Norway spruce kraft pulp; (b) ramie; (c) cotton linters; (d) hydrocellulose from cotton linters; (e) a low-DP generated cellulose I; (f) *A. xylinum* cellulose; and (g) *V. ventricosa* cellulose (from Vander Hart and Attala[39]).

understood. For instance, I_α has a singlet in C-1 and C-6 and a doublet in C-4, while I_β has only doublets. For elucidating the structures of both allomorphs I_α and I_β, it is essential to assign the ^{13}C NMR spectra of each allomorph completely. To this end, Kono *et al.*[45] recorded ^{13}C NMR spectra of ^{13}C-labeled cellulose biosynthesized by *A. xylinum* (ATCC10245) in a culture containing D-[2-^{13}C]glucose or D-[1,3-^{13}C]glycerol as a carbon source. The introduced ^{13}C-labeled carbons from the latter were observed at C-1, C-3, C-4, and C-6, whereas the transitions of the ^{13}C labeling to C-1, C-3, and C-5 from C-2 of the former were observed. The almost pure I_α and I_β cellulose were obtained from cell walls of *Glaucocystis*[46] and the outer skin of Tunicate,[47] respectively. The completely pure forms from any species of plant or animal, however, have not been obtained. Therefore, Kono *et al.*[45] analyzed subspectra of I_α and I_β phases from the original spectrum of *Cladephora* cellulose with simple mathematical treatment such as a linear combination of I_α-rich and I_β-rich cellulose:

$$\text{subspectra}\,(I_\alpha\ \text{phase}) = \text{spectrum}\,(I_\alpha\text{-rich}) - a \times \text{spectrum}\,(I_\beta\text{-rich})$$

$$\text{subspectra}\,(I_\beta\ \text{phase}) = \text{spectrum}\,(I_\beta\text{-rich}) - b \times \text{spectrum}\,(I_\alpha\text{-rich}),$$

where constants *a* and *b* indicate the contents of the I_α phase in I_β-rich cellulose and that of the I_β phase in I_α-rich cellulose, respectively. With the quantitative analysis of the ^{13}C transition ratio and comparing the ^{13}C NMR spectra of *Cladophora* cellulose with those of the ^{13}C-labeled cellulose, the assignment of the cluster of resonances that belong to C-2, C-3, and C-5 of cellulose were performed. All carbons of cellulose I_α and I_β, except for C-1 and C-6 of cellulose I_α, and C-2 of cellulose I_β, were shown in doublet of equal intensity in the ^{13}C NMR spectra of the native cellulose, which suggests that two inequivalent glucopyranose residues were contained in the unit cells of cellulose I_α and I_β allomorphs. These peak assignments for cellulose I_α and I_β were subsequently confirmed by means of ^{13}C homonuclear through-bond correlations, available from 2D refocused ^{13}C INADEQUATE measurements of I_α-rich *Cladophora* and I_β-rich Tunicate celluloses from natural abundance.[48] As illustrated in Figure 10.7, two sets of six carbons of Tunicate cellulose were connected from C-1 though C-6 by the dotted line and six connected by the solid line, each corresponds to the ^{13}C nuclei of magnetically nonequivalent anhydrous glucose residues. In a similar manner, two sets of six carbons are connected for *Cladophora* cellulose (spectrum not shown). The assigned signals thus obtained from cellulose from natural abundance are consistent with the data taken from ^{13}C-labeled preparation from *A. xylilinum.* It is surprising that difference in the ^{13}C chemical shifts between the two types of glucose units is the largest in C-5 carbons (up to 2.5 ppm for cellulose I_α), although the rest is less than 1 ppm. Together with the data of electron diffraction by Sugiyama *et al.*[43] it appears that cellulose I_α consists of a single molecular chain in which two glucose residues are alternatively connected. Cellulose I_β, on the other hand, might be formed by two molecular chains in which uniform center chain and corner chain are present. In other words, cellulose is not simply a homopolymer of glucose but a heteropolymer consisting of four different glucoses. The presence of two glucose units for 11% ^{13}C-labeled cellulose[49] (I_β-rich) is consistent with the data of Kono *et al.*[48] although the detailed assignment of peaks for C-2, C-3, and C-5 is not always consistent with that of Kono *et al.*[45,48] because of severely overlapping peaks in this case. Further, Kono *et al.*[50] performed complete assignment of the ^{13}C resonance for the ring carbons of cellulose acetate polymorphs, CTA I and II.

This kind of detailed assignment was not feasible by the spin diffusion experiment of uniformly ^{13}C-enriched wood sample, although a question concerning the structure of the hemicellulose units is resolved by locating the *O*-acetyl group at the second position of the xylan chain.[51]

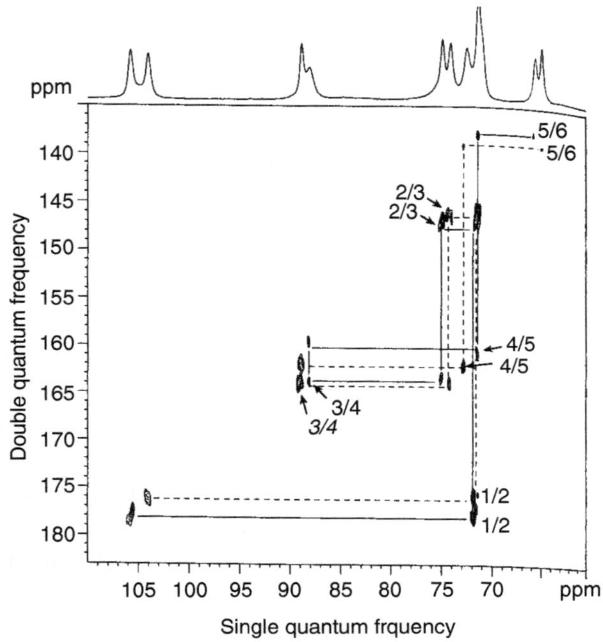

Figure 10.7. 2D refocused ^{13}C INADEQUATE spectrum of Tunicate cellulose. The through-bond connectivities of two magnetically nonequivalent anhydrous glucose residues from cellulose I_β are indicated by the solid and dotted lines (from Kono et al. [48]).

10.1.4. Chitin and Chitosan

Chitin is a (1 → 4)-β-D-2-acetamido-2-deoxyglucan, which is found in insect cuticles, shells of crustaceans, and the cell walls of bacteria and fungi. Chitosan is its deacetylated form as designated by (1 → 4)-β-D-2-amino-2-deoxyglucan. As expected from the similarity of their backbone structures, the fiber repeat of both polymers is identical with that of cellulose. Nevertheless, the ^{13}C NMR spectrum of chitin is much simpler than that of cellulose: the individual carbons give rise to single lines (Figure 10.8).[52,53] Again, the ^{13}C CP-MAS NMR spectrum of the regenerated form of chitin, *N*-acetylchitosan, is the same as that of naturally occurring chitin, implying conformational similarity between the two compounds, although the ^{13}C NMR peaks of the former are much broader than those of the latter.[54]

The ^{13}C NMR spectral patterns of chitosans, on the other hand, are markedly different for the three preparations (Figure 10.8). Chitosans from crab shell, shrimp shell, and the annealed form at higher temperature (at 220 °C) produce different spectra, suggesting the presence of three polymorphs.[49,50] The C-1 and/or C-4 peaks of the polymorphs ''tendon chitosan'' (crab shell) and ''L-2'' (shrimp shell) were split into doublets, whereas those of the ''annealed'' polymorph gave single lines. It is interesting to note that the nonequivalence of two chains in the unit cells of ''tendon chitosan'' and ''L-2'' was made to disappear by the removal of water molecules in the annealing process (''annealed'') or in the formation of metal complexes. Further, it was found that the ^{13}C NMR technique could be applied to analyze conformational changes taking place due to the formation of salts and metal complexes: chitosan forms two types of salts with acids to yield type I (anhydrous) and type II (hydrated) salts. Anions classified as hard Lewis bases, which are always associated with water molecules caused drastic conformational change of chitosan backbone as a relaxed two-fold helix, while anions classified as soft or borderline Lewis bases, which are not hydrated, did not induce a conformational change from an extended two-fold helix.[55]

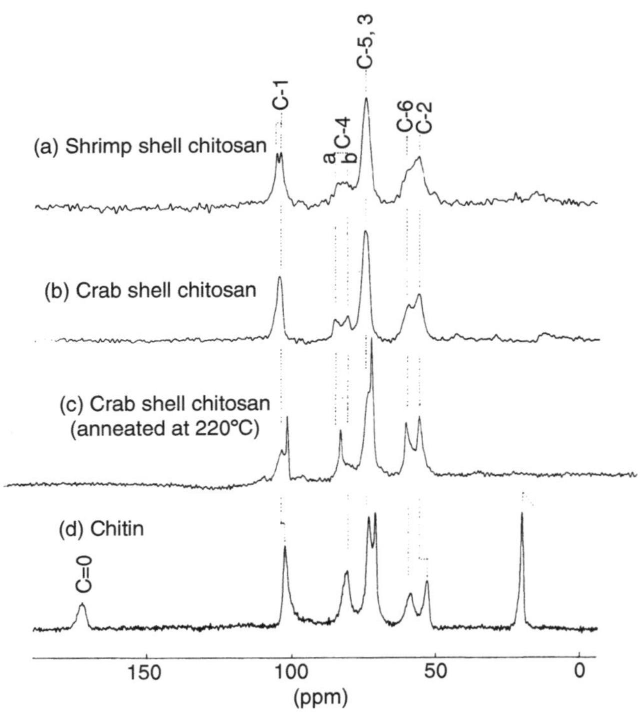

Figure 10.8. ^{13}C CP-MAS NMR spectra of three preparations of chitosan (a–c) and of chitin (d) in the solid state.[53]

10.2. Network Structure, Dynamics, and Gelation Mechanism

The network structure of gels is generally highly heterogeneous from the structural and dynamic point of view. Backbone dynamics of gel network can be very conveniently characterized by means of simple comparative high-resolution ^{13}C NMR measurements by CP-MAS and DD-MAS techniques, which are suitable for recording spectra mainly from the solid-like and liquid-like domains, respectively, depending upon the correlation times of backbone motions, as schematically illustrated in Figure 10.9.[56,57] In principle, the relative proportions of the former and the latter could be simply evaluated by comparison of their ^{13}C NMR peak intensities taking into account the correlation times of the respective correlation times. This kind of two-step model was successfully applied to studies of fibrillation kinetics of human calcitonin, a thyroid peptide hormone with tendency to form amyloid fibrils in concentrated solution.[58,59] In contrast, NMR observation of gel samples is not so simple as expected, because peak-narrowing procedure to achieve high-resolution signals from the solid-like domain fails when any frequency of backbone motions from gel samples interfered with frequencies of either magic angle spinning or proton decoupling, as discussed in Section 3.4. (see gray region in Figure 10.9).

Characterization of high-frequency motions present in swollen polymer chains with correlation times shorter than 10^{-8} s in the liquid-like domain is feasible by measurements of relaxation parameters such as the spin–lattice relaxation times (T_1), spin–spin relaxation times (T_2), and nuclear Overhauser effect (NOE). Surprisingly, it turned out that greater changes in the line widths ($1/\pi T_2$) between the gel state (about 150 Hz) and sol state (14 Hz), for instance, were noted from the liquid-like domain of the gels from chemically cross-linked synthetic polymers[60] and curdlan,[61,62] whereas no significant changes were observed as viewed from the T_1 and NOE values. This is because the T_2 values tend to be affected by the slow motion of long correlation times, while the T_1 and NOE values are mainly determined by fast motions. This means that fast backbone motions of swollen polymer chains visible from ^{13}C NMR signals by solution or DD-MAS spectra are indifferent from the presence or absence of such cross-links. For this reason, segmental motions of backbones of swollen polymer chains were well described by isotropic motions with correlation times of log χ^2 distribution instead of treatment of *single* correlation times.[60–62] In such case, it is necessary to utilize NMR spectrometer capable for signal detection with high-power proton decoupling and magic angle spinning designed for solid state ^{13}C NMR to remove residual dipolar interactions.

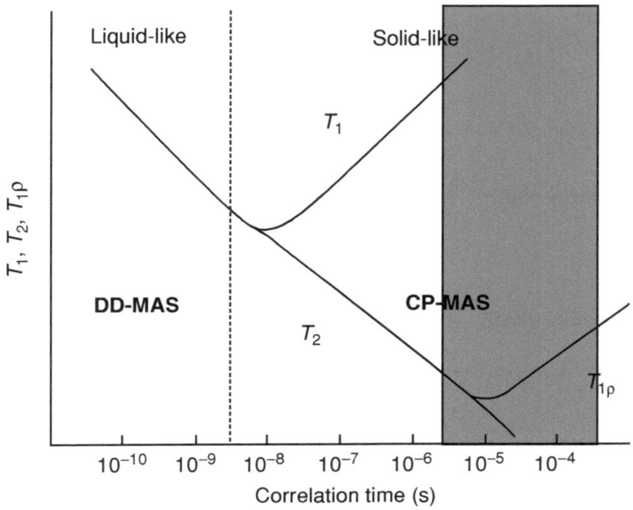

Figure 10.9. Schematic representation of solid-like and liquid-like domains in gel network as detected by CP-MAS and DD-MAS techniques, respectively, and relaxation parameters such as ^{13}C spin–lattice relaxation time (T_1), spin–spin relaxation time (T_2), and ^{1}H spin–lattice relaxation times in the rotational frame ($T_{1\rho}$) to probe their local fluctuation motions. ^{13}C NMR signals from the domain indicated by the gray could be lost due to failure of attempted peak narrowing due to interference of frequency with proton decoupling or magic angle spinning.[57]

As illustrated in Figure 10.10, the ^{13}C NMR spectra of elastic curdlan [(1 → 3)-β-D-glucan] gel recorded by a variety of NMR methods, including broadband decoupling by a solution NMR (the liquid-like domain; B), DD-MAS (intermediate domain; C), and CP-MAS (the solid-like domain; D).[16] The present result arising from the identical spectral features among such domains indicates that the single-helix form is dominant for all of the motionally different domains of curdlan gel. This observation is consistent with a view that the elastic property of curdlan gel is due to the dominant contribution of flexible single helices rather than rigid triple helices present as cross-linked regions. As discussed above, ^{13}C T_1 values of curdlan between sol and gel were unchanged. The amount of the triple-helical chains is thus nominal, if any, as indicated by the arrow of Figure 10.10(d), as far as heating temperature is kept below 80 °C (low-set gel). However, the increased gel strength by heating at a temperature between 80 °C and 120 °C (high-set gel) is well explained in terms of the additional formation of cross-links due to hydrophobic association of the single-helical chains as well as increased proportion of the triple-helical chains.[17] In contrast, the triple helices are dominant in the ^{13}C NMR spectra of branched

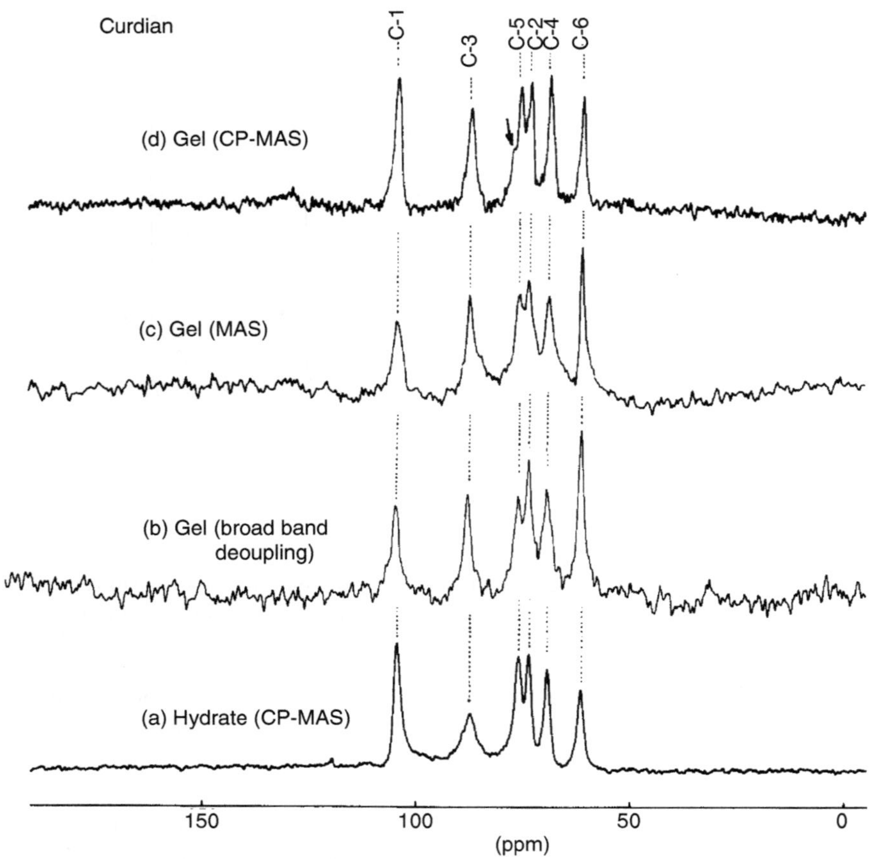

Figure 10.10. ^{13}C NMR spectra of curdlan hydrate (a) and gel (b–d) recorded by different types of experiments. (a) and (d) by CP-MAS; (b) by broad band decoupling; (c) by MAS experiment.[16]

(1 → 3)-β-D-glucans with reference to their characteristic ^{13}C NMR spectral pattern (Figure 10.1(c)).[14,57,63] This means that gelation of the branched glucans proceeds from partial association of the triple-helical chains. This network structure is far from formation of an elastic gel, and seems to be consistent with formation of brittle gel structure.

It was demonstrated that amylose gel contains two kinds of ^{13}C NMR signals: the B-type signals from motionally restricted regions as recorded by CP-MAS NMR and the signals identical to those found in aqueous solution.[64] The latter signals could be ascribed to flexible molecular chains adopting random-coil conformation (liquid-like domain). It turned out that ^{13}C T_1 values of starch gel (33%) as summarized in Table 10.11[13] are different by

Table 10.1. ^{13}C spin–lattice relaxation times of starch gel (33%) by DD- and CP-MAS methods (s)

	C-1	C-4	C-3	C-2	C-5	C-6
Liquid-like domain	0.36	0.29	0.30	0.32	0.29	0.16
Solid-like domain	9.2	11.8	11.9	11.9	11.9	2.1

one order of magnitude between the solid-like and liquid-like domains. This is because the ^{13}C T_1 values of the liquid-like domains are in the vicinity of the T_1 minimum, $\tau_c \sim 10^{-8}$s ($\omega_0\tau_c = 1$), whereas those of the solid-like domain are in the lower temperature side of the T_1 minimum, $\tau_c > 10^{-8}$ s. On the other hand, the former peaks are ascribed to the solid-like domain of cross-links, either double-helical junction zones[64] or aggregated species of single-helical chains as discussed already.[16] So far, two different views have been proposed for gelation mechanism of amylose gels. Miles *et al.*[65] have suggested that amylose gels are formed upon cooling molecularly entangled solutions as a result of phase separation of the polymer-rich phase, whereas Wu and Sarko[6] proposed that gelation occurs through cross-linking by double-helical junction zones. It is therefore important to explore a new means to distinguish the single-helical chains from the multiple-helical chains as secondary structures of amylose. Saitô *et al.*[13] recorded intense ^{13}C NMR signals of freshly prepared starch gel and its retrograded samples after 3 days in a refrigerator, taking random-coil conformation by DD-MAS NMR with and without retrogradation. Further, ^{13}C NMR signals of starch gels recorded by the CP-MAS NMR are ascribed to the presence of B-form chains whose intensities were significantly increased by retrogradation. This finding is consistent with the previous finding that retrogradation results in crystallization as detected by X-ray diffraction.[65] Therefore, it is concluded that retrogradation of starch gel is accompanied by a conformational change from random coil to B-form (single helix).

The well-documented network model of agarose gel[66] arises from the junction zones consisting of aggregated double-helical chains. However, it seems to be very difficult to explain why the intense ^{13}C NMR signals are visible from the liquid-like domain by a solution NMR spectrometer[67] or DD-MAS experiment (top trace in Figure 10.11, a broad peak at 110 ppm being from the probe assembly), in addition to the intense ^{13}C NMR signals available from the solid-like domain by CP-MAS NMR (bottom trace in Figure 10.11). Therefore, it appears that the well-documented model for agarose gel is too much exaggerated and only a small proportion of such double-helical junction zone, if any, is sufficient for gelation, as pointed out already in the cases of the chemically cross-linked synthetic polymers,[60]

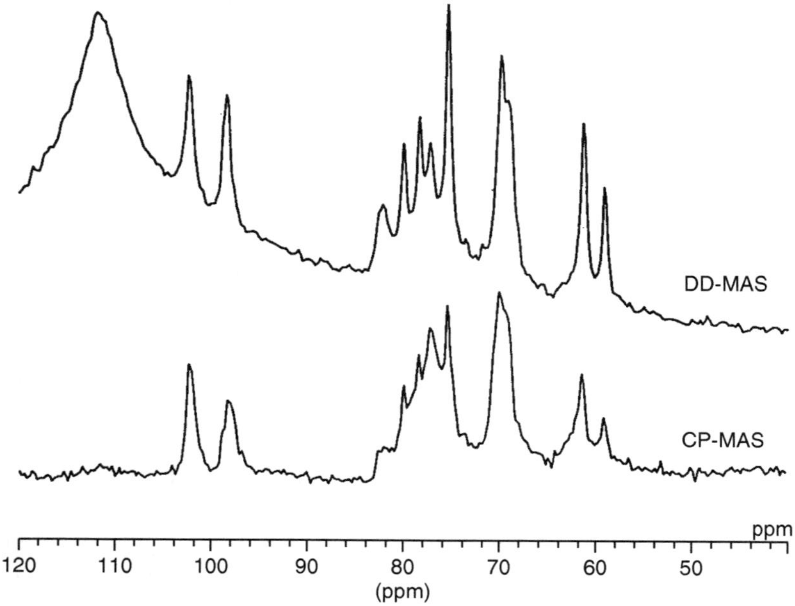

Figure 10.11. ^{13}C NMR spectra of agarose gel. Liquid-like domain (top) and solid-like domain (bottom).[13]

curdlan[61,62] and amylose.[13] In addition, several researchers[8,68–70] questioned the validity of the double-helical junction zones for agarose gel, and proposed an alternative model of gel network containing extended single helices. This is mainly because distinguishing between multiple-stranded helices and nested-single helices are very difficult by fiber X-ray diffraction.[8] In accordance with this, Saitô *et al.*[70] recorded ^{13}C NMR spectra of dried agarose film and its hydrated preparation cast from *N,N*-dimethylacetamide solution at 80 °C under anhydrous conditions with an expectation to increase the proportion of single chain. It was found that the ^{13}C NMR spectrum thus obtained is identical to that obtained from gel sample. Therefore, it is more likely that the network structure of agarose gel consists mainly of the single chain.

κ- and ι-carrageenans are ionic alternating copolymers of 3-linked β-D-galactose-4-sulfate and 4-linked 3, 6-anhydro-α-galactose (in kappa) or its 2 sulfate (in iota) residues, and known to form thermally reversible gels at sufficient concentration in the presence of a variety of cations.[66,71] The sol–gel transition of ι-carrageenan was proposed as a random-coil–double-helix transition because its well-resolved ^{13}C NMR signals at 80 °C by solution NMR disappeared completely at 15 °C, together with adoption of conformationally rigid double-helical form.[67] Domain model was also proposed for gelation of carrageenan, arising from intermolecular double-helix formation, and further association by cation-mediated helix–helix aggregation. Instead of the double helices or intermolecular association of any kind mentioned above, Smidsrød *et al.*[72] proposed an alternative view for gel formation of carrageenan as salt-promoted ''freezing-out'' of linkage conformations. They showed that, upon cooling the solution of oligomeric ι-carrageenan in the presence of lithium iodide from 90 °C to 25 °C, three broad signals originating from C-1 of the two monomeric units of C-3 of the D-galactopyranosyl residues appear at lower fields besides the narrowed peaks from a random-coil chain, as a result of formation of an ordered conformation in which the rotameric forms of the glycosidic linkages are frozen. Therefore, gel formation may be seen as a two-step process, the first involving an intramolecular conformational change and the second involving a decrease in solubility that is ion-dependent.

For a brittle gel sample (5% w/v) of κ-carrageenan as received from Sigma, there appear no ^{13}C NMR signals from flexible portions taking random-coil conformation as recorded by DD-MAS NMR technique, in contrast to the case of agarose gel, although intense signals from the solid-like domain were readily available from CP-MAS method. It turns out that ^{13}C chemical shifts of the gel samples are very close to those of starting powder within the experimental error as a result of taking similar conformations. In contrast, ^{13}C NMR signals from the liquid-like domain were available by DD-MAS NMR for resilient gel of κ-carrageenan in the absence of cations except Na^+ ion, prepared by treatment with Dowex 50W-XB ion exchange resin, followed by addition of NaOH at pH 7.0, as illustrated in Figure 10.12.[57] Inevitably, ^{13}C NMR signal from the solid-like domain recorded by CP-MAS NMR turned out to be less intense (middle trace, in Figure 10.12). In particular, the ^{13}C NMR peaks at the two lowermost regions of the liquid-like domain at 104.5 and 96.6 ppm [C-1 peaks of (1 → 3)-linked β-D-galactosyl and (1 → 4)-linked 3,6-anhydro-α-L-galactosyl residues, respectively] as recorded by DD-MAS NMR are significantly displaced downfield, as compared with those recorded by CP-MAS NMR and also corresponding peaks in the solid state. Further, ^{13}C NMR signals of soft gel from ι-carrageenan (as received) were unavailable from the solid-like domain by CP-MAS NMR method, although the intense ^{13}C NMR were recorded from the liquid-like domain by DD-MAS NMR. This may be caused either by a smaller proportion of the cross-linked region of associated single helices or double-helical junction zones, if any, or suppressed ^{13}C NMR signals by interference of motional frequency with proton decoupling or magic angle spinning. In any case, a possibility that substantial amount of signals could be lost in the swollen gel samples when

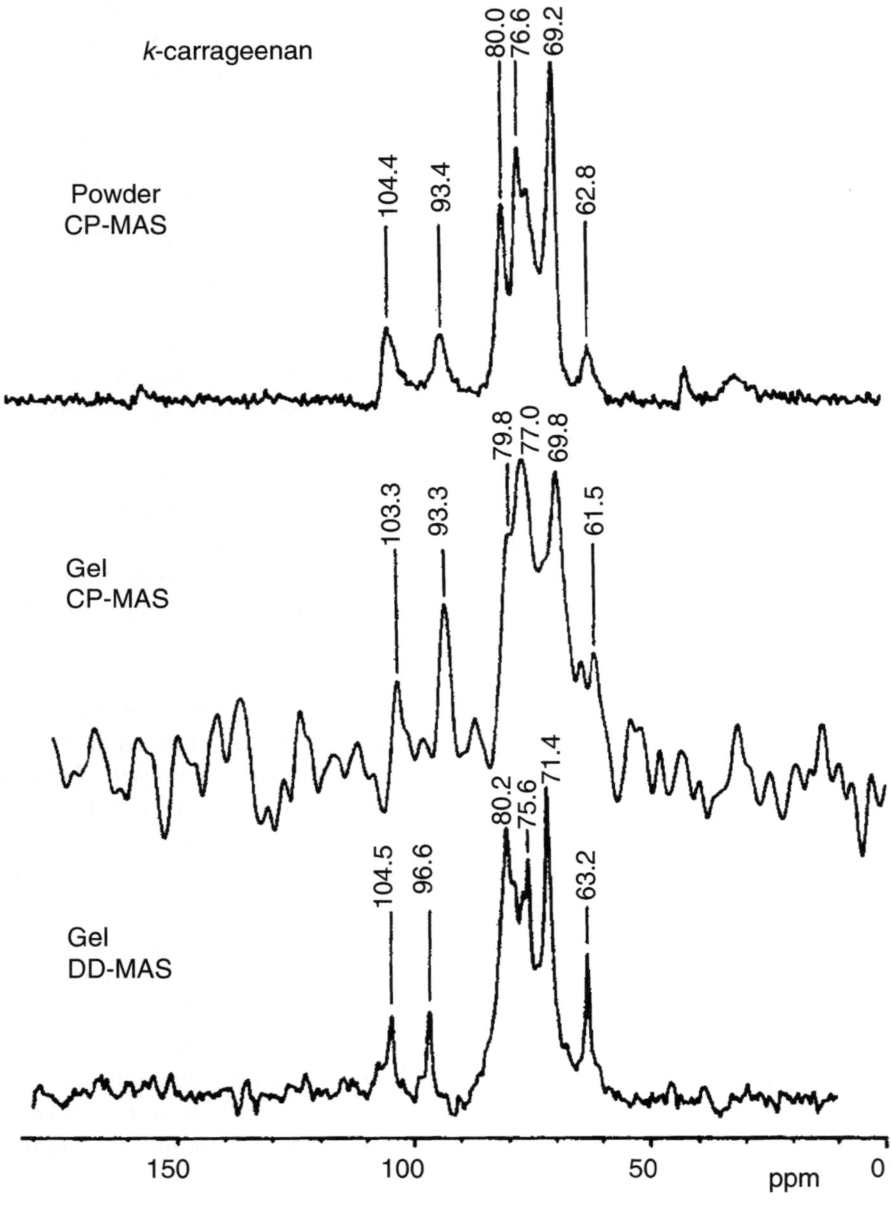

Figure 10.12. ^{13}C CP-MAS NMR spectra of κ-carrageenan in the solid (top),gel (middle), and DD-MAS NMR spectrum of gel (bottom).[57]

polysaccharide backbone undergoing fluctuation motions with the timescale described in Figure 10.9 should be always taken into account.

References

[1] H. Saitô, G. Izumi, T. Mamizuka, S. Suzuki, and R. Tabeta, 1982, *J. Chem. Soc., Chem. Commun.*, 1386–1388.
[2] J. A. Ripmeester, *J. Inclusion Phenom.*, 1986, 4, 129–134.
[3] R. P. Veregin, C. A. Fyfe, R. H. Marchessault, and M. G. Taylor, 1987, *Carbohydr. Res.* 160, 41–56.
[4] H. Saitô, M. Yokoi, and Y. Yoshioka, 1989, *Macromolecules*, 22, 3892–3898.
[5] J. Blackwell, A. Sarko, and R. H. Marchessault, 1969, *J. Mol. Biol.*, 42, 379–383.
[6] H. C. Wu and A. Sarko, 1978, *Carbohydr. Res.*, 61, 7–25.
[7] A. Imberty and S. Perez, 1988, *Biopolymers*, 27, 1205–1221.
[8] S. A. Foord and E. D. T. Atkins, 1989, *Biopolymers*, 28, 1345–1365.
[9] H. Saitô, 1981, *ACS Symp. Ser.*, 150, 125–147.
[10] H. Saitô, 1992, *ACS Symp. Ser.*, 489, 296–310.
[11] H. Saitô, 1995, *Annu. Rep. NMR Spectrosc.*, 31, 157–170.
[12] H. Saitô, 1996, in *Encyclopedia of Nuclear Magnetic Resonance*, D. M. Grant and R. K. Harris, Eds., John Wiley, pp. 3740–3745.
[13] H. Saitô, H. Shimizu, T. Sakagami, S. Tuzi, and A. Naito, 1995, in *Magnetic Resonance in Food Science*, P. S. Belton, I. Delgadillo, A. M. Gil, and G. A. Webb, Eds., Royal Society of Chemistry, London, pp. 257–271.
[14] H. Saitô, J. Yamada, Y. Yoshioka, Y. Shibata, and T. Erata, 1991, *Biopolymers*, 31, 933–940.
[15] Y. Yoshioka, N. Uehara, and H. Saitô, 1992, *Chem. Pharm. Bull.*, 40, 1221–1226.
[16] H. Saitô, Y. Yoshioka, M. Yokoi, and J. Yamada, 1990, *Biopolymers*, 29, 1689–1698.
[17] H. Saitô, R. Tabeta, M. Yokoi, and T. Erata, 1987, *Bull. Chem. Soc. Jpn.*, 60, 4259–4266.
[18] C. T. Chuah, A. Sarko, Y. Deslandes, and R. H. Marchessault, 1983, *Macromolecules*, 16, 1375–1382.
[19] K. Okuyama, A. Otsubo, Y. Fukazawa, M. Ozawa, T. Harada, and N. Kasai, 1991, *J. Carbohydr. Chem.*, 10, 645.
[20] E. D. T. Atkins, K. D. Parker, and R. D. Preston, 1969, *Proc. R. Soc. London, Ser. B*, 173, 209–221.
[21] E. D. T. Atkins and K. D. Parker, 1969, *J. Polym. Sci.*, C28, 69–81.
[22] A. Sarko and P. Zugenmaier, 1980, *ACS Symp. Ser.*, 141, 459–482.
[23] R. H. Marchessault, M. G. Taylor, C. A. Fyfe, and R. P. Veregin, 1983, *Carbohydr. Res.*, 144, C1–C5.
[24] M. J. Gidley and S. M. Bociek, 1985, *J. Am. Chem. Soc.*, 107, 7040–7044.
[25] F. Horii, A. Hirai, and R. Kitamaru, 1986, *Macromolecules*, 19, 930–932.
[26] R. P. Veregin, C. A. Fyfe, R. H. Marchessault, and M. G. Taylor, 1986, *Macromolecules*, 19, 1030–1034.
[27] F. Horii, H. Yamamoto, A. Hirai, and R. Kitamaru, 1987, *Carbohydr. Res.*, 160, 29–40.

[28] M. J. Gidley and S. M. Bociek, 1988, *J. Am. Chem. Soc.*, 110, 3820–3829.
[29] H. Saitô, J. Yamada, T. Yukumoto, H. Yajima, and R. Endo, 1991, *Bull. Chem. Soc. Jpn.*, 64, 3528–3537.
[30] F. R. Senti and L. P. Witnauer, 1948, *J. Am. Chem. Soc.*, 70, 1438–1444.
[31] N. W. H. Cheetham and L. Tao, 1998, *Carbohydr. Polym.*, 36, 285–292.
[32] M. Paris, H. Bizot, J. Emery, J. Y. Buzaré, and A. Buléon, 2001, *Int. J. Biol. Macromol.*, 29, 137–143.
[33] A. Imberty and S. Perez, 1988, *Biopolymers*, 27, 1205–1221.
[34] R. H. Atalla, J. C. Gast, D. W. Sindorf, V. J. Bartuska, and G. E. Maciel, 1980, *J. Am. Chem. Soc.*, 102, 324–3251.
[35] W. L. Earl and D. L. VanderHart, 1980, *J. Am. Chem. Soc.*, 102, 3251–3252.
[36] W. L. Earl and D. L. VanderHart, 1981, *Macromolecules*, 14, 570–574.
[37] R. H. Attala and D. L. VanderHart, 1984, *Science*, 223, 283–285.
[38] J. R. Havens and D. L. VanderHart, 1985, *Macromolecules*, 18, 1663–1676.
[39] D. L. VanderHart and R. H. Attala, 1984, *Macromolecules*, 17, 1465–1472.
[40] D. L. VanderHart and R. H. Attala, 1987, *ACS Symp. Ser.*, 340, 88–116.
[41] R. H. Attala and D. L. VanderHart, 1999, *Solid State Nucl. Magn. Reson.*, 15, 1–19.
[42] J. J. Cael, D. L. W. Kwoh, S. S. Bhattacharjee, and S. L. Patt, 1985, *Macromolecules*, 18, 819–821.
[43] J. Sugiyama, T. Okano, H. Yamamoto, and F. Horii, 1990, *Macromolecules*, 23, 3196–3198.
[44] H. Yamamoto, F. Horii, and H. Odani, 1989, *Macromolecules*, 22, 4130–4132.
[45] H. Kono, S. Yunoki, T. Shikano, M. Fujiwara, T. Erata, and M. Takai, 2002, *J. Am. Chem. Soc.*, 124, 7506–7511.
[46] T. Imai, J. Sugiyama T. Itoh, and F. Horii, 1999, *J. Struct. Biol.*, 127, 248–257.
[47] T. Larson, U. Westermark, and T. Iverson, 1995, *Carbohydr. Res.*, 278, 339–343.
[48] H. Kono, T. Erata, and M. Takai, 2003, *Macromolecules*, 36, 5131–5138.
[49] D. Sakellariou, S. P. Brown, A. Lesage, S. Hediger, M. Bardt, C. A. Meriles, A. Pines, and L. Emsley, 2003, *J. Am. Chem. Soc.*, 125, 4376–4380.
[50] H. Kono, T. Erata, and M. Takai, 2002, *J. Am. Chem. Soc.*, 124, 7512–7518.
[51] M. Bardet, L. Emsley, and M. Vincendon, 1997, *Solid State Nucl. Magn. Reson.*, 8, 25–32.
[52] H. Saitô, R. Tabeta, and K. Ogawa, 1987, *Macromolecules*, 20, 2424–2430.
[53] H. Saitô, R. Tabeta, and K. Ogawa, 1987, in *Industrial Polysaccharides: Genetic Engineering, Structure/Property Relations and Applications*, M. Yalpani, Ed., Elsevier, Amsterdam, pp. 267–280.
[54] H. Saitô, R. Tabeta, and S. Hirano, 1981, *Chem. Lett.*, 1479–1482.
[55] K. Ogawa, T. Yui, and K. Okuyama, 2004, *Int. J. Biol. Macromol.*, 34, 1–8.
[56] H. Saitô, 1981, *Solution Properties of Polysaccharides*, *ACS Symp. Ser.*, 150, 125–147.
[57] H. Saitô, 2004, in *Polysaccharides*: *Structural Diversity and Functional Versatility*, Second Edition, S. Dumitriu, Ed., Marcel Dekker, New York, pp. 253–266.
[58] M. Kamihira, Y. Oshiro, S. Tuzi, A. Y. Nosaka, and H. Saitô, 2003, *J. Biol. Chem.*, 278, 2859–2865.
[59] M. Kamihira, A. Naito, S. Tuzi, A. Y. Nosaka, and H. Saitô, 2000, *Protein Sci.*, 9, 867–877.
[60] K. Yokota, A. Abe, S. Hosaka, I. Sakai, and H. Saitô, 1978, *Macromolecules*, 11, 95–100.

[61] H. Saitô, T. Ohki, and T. Sasaki, 1977, *Biochemistry*, 16, 908–914.
[62] H. Saitô, E. Miyata, and T. Sasaki, 1978, *Macromolecules*, 11, 1244–1251.
[63] H. Saitô, 1992, *Viscoelasticity of Biomaterials*, *ACS Symp. Ser.*, 489, 296–310.
[64] M. J. Gidley, 1989, *Macromolecules*, 22, 351–358.
[65] M. J. Miles, V. J. Morris, and S. G. Ring, 1985, *Carbohydr. Res.*, 135, 257–269.
[66] S. Arnott, W. E. Scott, D. A. Rees, and C. G. A. McNab, 1974, *J. Mol. Biol.*, 90, 253–267.
[67] F. M. Nicolaisen, I. Meyland, and K. Schaumburg, 1980, *Acta Chem. Scand.*, B34, 579–583.
[68] M. R. Letherby and D. A. Young, 1981, *J. Chem. Soc. Faraday I*, 77, 1953–1966.
[69] I. T. Norton, D. M. Goodall, K. R. J. Austen, E. R. Morris, and D. A. Rees, 1986, *Biopolymers*, 25, 1009–1029.
[70] H. Saitô, M. Yokoi, and J. Yamada, 1990, *Carbohydr. Res.*, 199, 1–10.
[71] E. R. Morris, D. A. Rees, and G. Robinson, 1980, *J. Mol. Biol.*, 138, 349–362.
[72] O. Smidsrød, I. -I. Andresen, H. Grasdalen, B. Larsen, and T. Painter, 1980, *Carbohydr. Res.*, 80, C11–C16.

Chapter 11

POLYPEPTIDES AS NEW MATERIALS

11.1. Liquid-Crystalline Polypeptides

It is well known that poly(γ-benzyl L-glutamate) (PBLG) forms the lyotropic liquid-crystalline state in good solvents such as dichloromethane, 1,4-dioxane, chloroform, and so on,[1–5] and poly(L-glutamates) (PGs) with long alkyl side chains form thermotropic liquid crystals (LCs).[6] If these polypeptides are placed in a magnet, they orient to the direction of the magnetic field and then form the nematic LCs.[7–12] Such findings lead to a large development in one of the very important fields in polymer science. In this section, we are concerned with the structural and dynamic characterization of the above-mentioned polypeptides in the liquid-crystalline state by solid state NMR.[13–23]

11.1.1. Liquid-Crystalline PLGs with Long *n*-Alkyl or Oleyl Side Chains

A series of α-helical PLGs with *n*-alkyl side chains of various lengths (chain lengths $n = 10$–18) form a crystalline phase composed of paraffin-like crystallites together with the α-helical main chain packing into a characteristic structure.[6,13–19] The polymers form thermotropic LCs by melting of the side-chain crystallites at temperatures above ca. 50 °C. The long *n*-alkyl side chains play a role of solvent to form LCs. In order to obtain the detailed information about the structure and dynamics of these LCs, it is essential to study the structure and motion of the main chains and side chains at various temperatures.

11.1.1.1. POLG in the Thermotropic Liquid-Crystalline State

^{13}C CP-MAS NMR spectra of poly(γ-octadecyl L-glutamate) (POLG; or PG-18 where the number of carbon atoms in the *n*-alkyl group follows the letters

PLG) with *n*-octadecyl side chains in the liquid-crystalline state are illustrated in Figure 11.1 as a function of temperature. Assignment of peaks for the amide CO, ester CO, C_α and C_β carbons is straightforward with reference to the data of PBLG.[13–18] The assignment of C_γ peak is easily made at high temperature because the overlapping interior CH_2 peak is shifted upfield. The main and small peaks for the interior CH_2 carbons are designated by I and A, respectively: The peak I is ascribed to the all-*trans* zigzag conformation in the crystalline state, while the peak A arises from the CH_2 carbons in the noncrystalline state.

At room temperature, POLG takes the right-handed α-helix conformation in view of the ^{13}C chemical shifts of the amide CO, C_α, and C_β, carbons. This means that the inner part of side chain has the same conformation as PBLG. The peaks for the outer part of side chain appear in the vicinity of 30 ppm. The ^{13}C chemical shift data of *n*-paraffins, cyclic paraffins, and polyethylene can be used to characterize the conformation and crystal structure of the side chain CH_2 carbons. The CH_2 resonance in paraffins appears at higher field by 4–6 ppm if any carbon atom three bonds away is in a *gauche* conformation, as compared to that of *trans* conformation. In

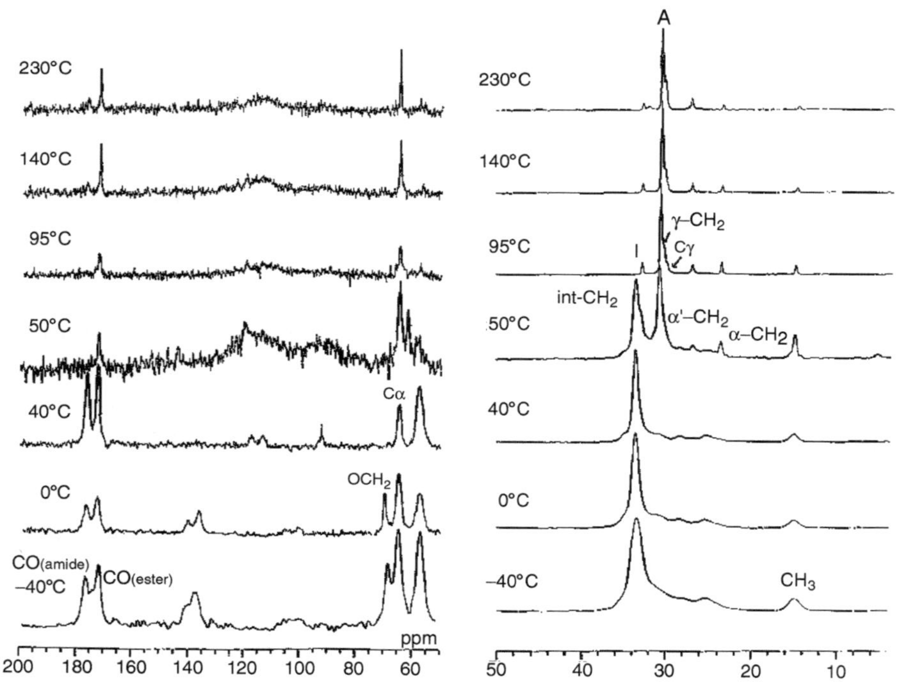

Figure 11.1. 67.5 MHz ^{13}C CP-MAS NMR spectrum of poly(γ-*n*-octadecyl L-glutamate) as a function of temperature (from Yamanobe *et al.*[13] and Katoh *et al.*[18]).

fact, the CH_2 resonance of paraffins in liquid or in solution and of polyethylene in a noncrystalline state appears at higher field by 2–4 ppm than that in the crystalline state. Therefore, the peak I arises from the CH_2 carbons of the all-*trans* zigzag conformation in the crystalline state, and the peak A arises from the CH_2 carbons undergoing fast exchange between the *trans* and *gauche* conformation at room temperature (27 °C) in view of their chemical shifts.[24] The peak I of POLG disappears at temperatures above 50 °C and the intensity of peak A increases noticeably (Figure 11.1), arising from the melting of side-chain crystallites. At the same time, the peak A is gradually displaced upfield from 30.6 to 30.2 ppm due to the increased *gauche* population according to the γ-effect.[25]

At temperatures about 50 °C, the main-chain peaks are suppressed due to failure of proton decoupling due to interference of fluctuation frequency with frequency of proton decoupling,[26] as discussed in Section 3.3. This means that the main chain of POLG undergoes fluctuation frequency in the order of 60 kHz.

11.1.1.2. POLLG (PG with Oleyl Side Chains) in the Liquid-Crystalline State[15]

The melting temperature of unsaturated side chains of poly(γ-oleyl L-glutamate) (POLLG) is much lower than that of saturated side chains of POLG (PG-18). Obviously, double bond placed in the central part of the oleyl group interrupts the crystallization of the side chains. Thus, the liquid-crystalline nature of α-helical rods may be maintained all the way to −40 °C. The ^{13}C CP-MAS NMR spectrum of POLLG in the liquid-crystalline state is very similar to that of POLG (spectra not shown), except for the interior CH_2 peak at 30.7 ppm. The side chains are melting at room temperature, while the main chain is in an α-helix. The amide CO peak disappears at −20 °C. Therefore, the main chain in the liquid-crystalline state undergoes reorientation at a frequency of 60 kHz at −20 °C.

11.1.1.3. Dynamic Feature of POLG and POLLG

In order to understand the dynamic feature of their thermotropic behavior, proton NMR relaxation times such as ^{1}H T_1 and $T_{1\rho}$ were recorded at frequency of 90 MHz and the locking field B_1 at 1 mT over a wide range of temperatures, as shown in Figures 11.2 and 11.3, respectively.[16] According to the BPP theory,[27] T_1 passes through a minimum and increases again when the correlation time τ_c for molecular motion increases further. Elevated

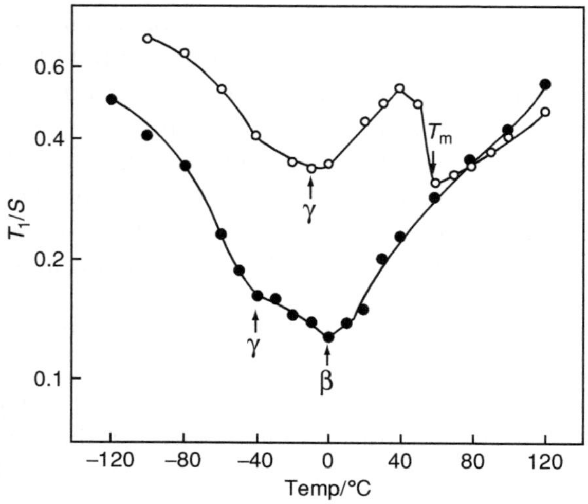

Figure 11.2. Temperature dependence of T_1 for poly(γ-*n*-octadecyl L-glutamate) (○) and poly(γ-oleyl L-glutamate)(●): β, β-relaxation; γ, γ-relaxation; T_m, melting point of side-chain crystallite (from Mohanty *et al.*[16]).

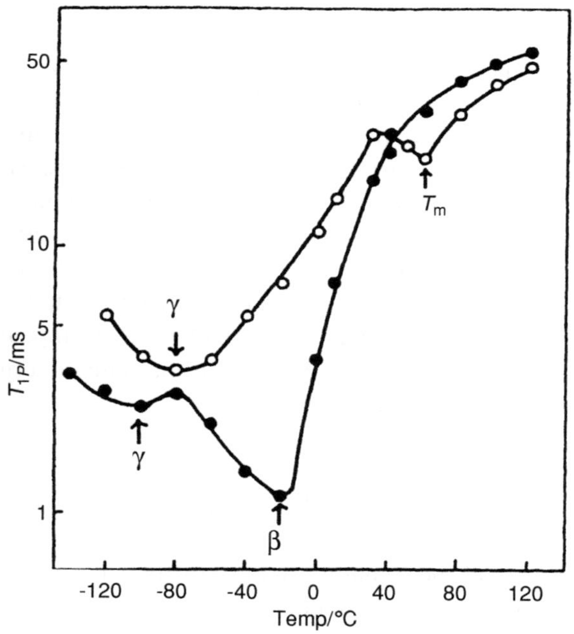

Figure 11.3. Temperature dependence of $T_{1\rho}$ for poly(γ-*n*-octadecyl L-glutamate) (○) and poly(γ-oleyl L-glutamate)(●): β, β-relaxation; γ, γ-relaxation; T_m, melting point of side-chain crystallite (from Mohanty *et al.*[16]).

temperature leads to an increased molecular motion of polymers in the solid state (the decrease of τ_c), and so T_1 and $T_{1\rho}$ decrease. A minimum in T_1 is reached at $\omega_0\tau_c = 2\pi\nu_c\tau_c = 1$, where ω_0 and ν_c are the resonance frequencies in radians per second and in Hz, respectively, and T_1 again increases. From the T_1 minimum, we can obtain the correlation time τ_c for molecular motion at MHz frequencies. On the other hand, $T_{1\rho}$ shows T_1-like behavior against temperature, but a minimum in $T_{1\rho}$ is reached at $\omega_1\tau_c = 1$, where $\omega_1/2\pi = \gamma B_1/2\pi = 42.6$ kHz.[28] From the $T_{1\rho}$ minimum, we can obtain the correlation time τ_c for molecular motion at kHz frequencies. Thus, $T_{1\rho}$ behaves similarly to T_2 and is more sensitive to lower frequency motions.

As seen from Figure 11.2, T_1 of PLG-18 decreases from 700 to 350 ms as the temperature is raised from −100 to −10 °C. This means that the molecular motion is in the slow molecular-motion region; i.e., $\omega\tau_c \gg 1$. At temperatures above −10 °C, T_1 increases from 350 to 550 ms as the temperature is raised from −100 to 45 °C. This means that molecular motion is nearly in the extreme narrowing region ($\omega\tau_c \ll 1$). The relaxation arises from the side-chain motion, which corresponds to the γ-relaxation observed in viscoelastic measurements. This can be justified from the Arrhenius plot as shown in Figure 11.4. However, T_1 decreases from 550 to 327 ms as the temperature is raised from 45 to 60 °C and again increases from 327 to 403 ms as the temperature is further raised from 60 to 120 °C. The minimum at the lower temperature depends on the observed frequency, but does not

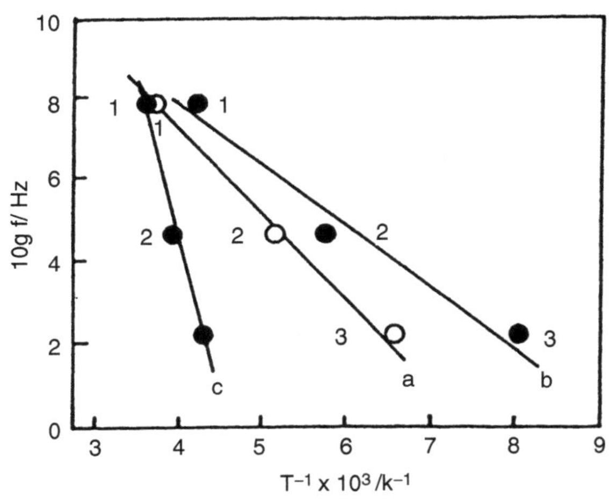

Figure 11.4. Arrhenius plots of log *f* in poly(γ-*n*-octadecyl L-glutamate) (a) and poly(γ-oleyl L-glutamate) (b and c) against the inverse of the absolute temperature. *f* is the frequency in Hz. (a and b) From γ-relaxation. (c) From β-relaxation. 1 and 2 were obtained from T_1 and $T_{1\rho}$, and 3 was obtained from the viscoelastic data (from Mohanty *et al.*[16]).

depend on frequency at higher temperature (see $T_{1\rho}$). As shown in Figure 11.3, the first minimum comes from relaxation and the second one comes from the first-order melting transition.

Two distinct minima are observed in $T_{1\rho}$. $T_{1\rho}$ decreases from 6.0 to 3.0 ms as the temperature is increased from −120 to – 80 °C and increases from 3.0 and 29 ms through the first minimum as the temperature is further increased. Again, the $T_{1\rho}$ value decreases from 29 to 21 ms as the temperature is increased from 30 to 60 °C and increases 21 to 41 ms through the second minimum as the temperature is further increased from 60 to 120 °C. Below −80 °C, the γ-relaxation comes from the molecular motion corresponding to the rotation of methyl groups in the side chains at a frequency below ca. 40 kHz seen from the $T_{1\rho}$ minimum. The activation energy, ΔE, for the γ-relaxation can be determined by using $\tau_c = \tau_0 \exp(-\Delta E/kT)$, where τ_0 is the prefactor, k is the Boltzmann constant, and T is the absolute temperature. An accurate correlation time τ_c can be estimated using $\tau_c = 1/2\pi\nu_c$. Therefore, the activation energy ΔE was determined from the plots of log f (the frequency in Hz) against $1/T$ as shown in Figure 11.4, where the values of f for the T_1 and $T_{1\rho}$ minima are 90 MHz and 42.6 kHz, respectively. The activation energy of PLG-18 is 10 kCal/mol. This is a reasonable value for the γ-relaxation and agrees with that (11 kCal/mol) obtained from the mechanical relaxation by viscoelastic measurements.[6]

The $T_{1\rho}$ values of POLLG are plotted against temperature in the temperature range from −120 to 120 °C as shown in Figure 11.2. T_1 decreases from 504 to 138 ms as the temperature is increased from −120 to 0 °C. The T_1 value increases from 138 to 561 ms as the temperature is further increased from 0 to 120 °C. Only one minimum is observed in the temperature range from −120 to 120 °C. However, if we look at the relaxation curve carefully, there may be two overlapping relaxations in the vicinity of −40 and 0 °C. The separation of the two minima is unclear. In order to clarify this problem, the $T_{1\rho}$ values are plotted against temperature as shown in Figure 11.3. $T_{1\rho}$ decreases from 3.5 to 2.5 ms as the temperature is increased from −120 to− 100 °C and increases from 2.5 to 2.9 ms through the first minimum as the temperature is further increased from −100 to −80 °C. The $T_{1\rho}$ values decreases from 2.9 to 1.2 ms as the temperature is increased from −80 to −20 °C and increases from 1.2 to 49 ms through the second minimum as the temperature is increased −20 to 120 °C. Two distinct minima clearly appear. The first minimum at about −100 °C is the γ-relaxation and the second one at about 30 °C is the β-relaxation. This explanation is similar to the case of PG-18.

When the experimental data on the two minima in the relaxation curve are used, the activation energies for both relaxations were determined by plotting log f versus the inverse of the absolute temperature (Figure 11.4(b)). The activation energy for the γ-relaxation so determined was about 7 kCal/mol, which is smaller than that of POLG. This means that the oleyl side chains are more flexible than the *n*-octadecyl side chains. This is consistent with the result that POLLG is in a liquid-crystalline state over a wide range of temperatures. On the other hand, β-relaxation arises from the overall motion of the side chains. The activation energy found for the β-relaxation is about 37 kCal/mol (Figure 11.4(c)). This value is larger than that for the γ-relaxation but is reasonable compared with the β-relaxation of other polymers. From the above results for PG-18 and POLLG, the existence of an unsaturated double bond in the oleyl groups of the latter polypeptide leads to substantially different molecular motion compared with the former polypeptide with *n*-alkyl groups having the same number of carbons and produces notable differences in their physical properties.

In addition, the diffusional behavior of PG with long *n*-alkyl side chains in the thermotropic liquid-crystalline state and lyotropic liquid-crystalline state has been elucidated by the determination of the diffusion coefficients by high-field gradient NMR method[19–23] by aid of diffusion theory of polymer chains.[29,30]

11.1.2. PBLG in the Lyotropic Liquid-Crystalline State

As mentioned above, PBLG forms the lyotropic liquid-crystalline state in good solvents such as dichloromethane, dioxane, chloroform.[1–5] and when the PBLG liquid-crystalline solution is placed in a strong magnetic field, the main chain can be oriented in the direction of the magnetic field.[7–12] The molecular motion of solvent and PBLG in the liquid-crystalline solution is considerably influenced by intermolecular interactions. By reacting highly oriented PBLG chains with a cross-linker in a strong magnetic field, cross-linked PBLG gel with the highly oriented α-helical chains is expected to be prepared. In the liquid-crystalline state and the gel state, the orientation of the PBLG chains can be determined by using high-resolution solid state ^{13}C NMR. As the PBLG chain is undergoing very slow rotation or libration, the ^{13}C NMR spectrum of the polymer cannot be obtained by conventional solution ^{13}C NMR measurements because of the difficulty of the removal of dipolar interactions.

11.1.2.1. Orientation of PBLG in the Liquid-Crystalline State

For any specified orientation of PBLG, the coordinate system may be oriented such that the magnetic field is not parallel to one of the principal axes in the chemical shift tensor components (δ_{11}, δ_{22}, and δ_{33}).[19] In this case the chemical shift (δ_{ZZ}) depends on the angles that specify the orientation of the magnetic field direction relative to the principal axes, as well as the absolute magnitudes of the principal values. Thus, the chemical shift (δ_{ZZ}) is expressed as discussed in Section 2.1 (Eq. (2.6)). The hydrogen bond direction is along the α-helical chain.[19,31] The principal value of δ_{22} is almost directed along the >C=O bond and it is almost along the α-helical chain axis. Therefore, for an oriented α-helical polypeptide chain, we have

$$\delta_{ZZ} = \delta_{22} \tag{11.1}$$

when $\alpha = 90°$ and $\beta = 90°$ in Eq. (2.6). Therefore, the observed chemical shift δ_{obs} becomes δ_{22}, and thus $\delta_{obs} \neq \delta_{ZZ}$ because there exists the angle σ between the α-helical chain axis and the magnetic field direction, which may take any specified value. In the case of some biomolecules, it is reported that the average angle σ is about 16.85°.[31,32] Therefore, the order parameter S of the α-helical polypeptide chains with respect to the applied magnetic field can be expressed by

$$S = (3\cos^2\sigma - 1)/2 \tag{11.2}$$

Under the MAS experiments, the observed chemical shift is obtained as an average of the three chemical shift tensor as shown in Eq. (2.3).

11.1.2.2. ^{13}C Chemical Shift of the Amide Carbonyl and the Order Parameter S of PBLG in the Liquid-Crystalline State

The static ^{13}C CP NMR spectrum of 25 wt% PBLG/dioxane liquid-crystalline solution is shown in Figure 11.5(a). The assignment of the ^{13}C peaks can be made straightforwardly by reference data on solid PBLG.[33] The amide and ester carbonyl carbons of PBLG appear at 193.2 and 171.8 ppm, respectively. The C_α, C_{phenyl}, C_1(phenyl), C_{OCH2}, C_β, and C_γ carbons appear at 54.3, 145.6, 133.2, 68.8, 34.9, and 27.8 ppm, respectively. The ^{13}C CP-MAS NMR spectrum of PBLG in the solid state is shown in Figure 11.5(c). The ^{13}C chemical shift positions of the individual carbons in solid PBLG are very different from those in the liquid-crystalline state. Also, the ^{13}C CP-MAS NMR spectrum of PBLG in the liquid-crystalline state is shown in Figure 11.5(b). The ^{13}C chemical shift positions of the individual carbons in the liquid-crystalline state are very close to those in the solid state. These chemical shift values are summarized in Table 11.1 together with the ^{13}C chemical shift values in the liquid-crystalline state as obtained by the static CP method.

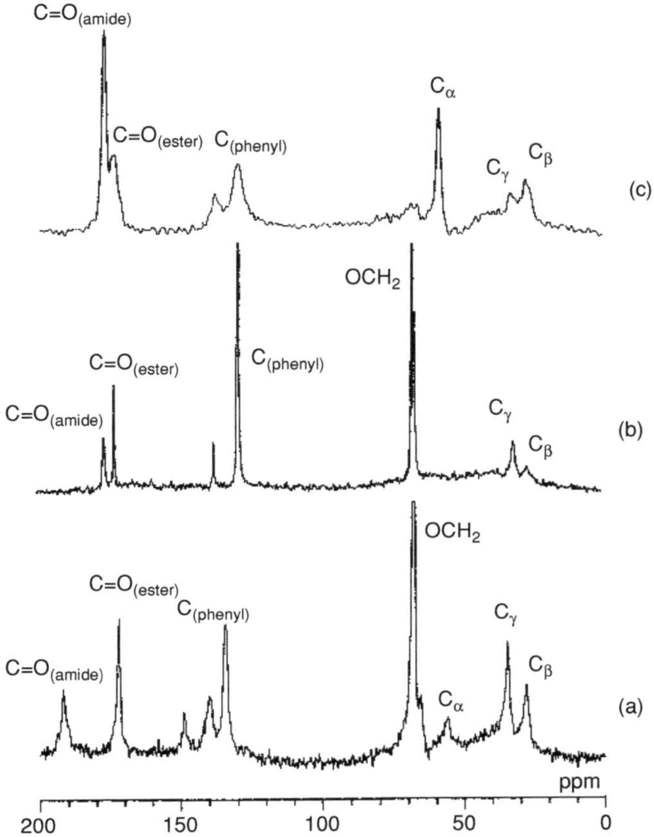

Figure 11.5. Static ^{13}C CP NMR spectrum (a) and ^{13}C CP-MAS NMR spectrum (b) of a 25 wt% PBLG/dioxane liquid-crystalline solution at 67.8 MHz and 300 K, and ^{13}C CP-MAS NMR spectrum of PBLG in the solid state at 67.8 MHz (c) at 300 K (from Zhao *et al.*[19]).

Table 11.1. ^{13}C NMR chemical shifts of PBLG in the solid state and in liquid-crystalline state in 1,4-dioxane.

	^{13}C NMR chemical shifts (ppm)							
Sample	C=O(amide)	C=O(ester)	C_α	C_β	C_γ	OCH_2	Phenyl C_1	Phenyl
Solid state								
CP-MAS	176.2	172.4	57.5	26.2	[a]	66.7	[a]	128.9
TOSS	176.3	173.1	57.3	26.5	31.1	[a]	136.7	129.5
Liquid-crystalline state								
CP-MAS	176.3	172.5	[a]	[a]	31.3	67.4	137.3	128.7
CP (static)	193.2	171.7	55.6	27.4	34.3	68.1	[a]	139.3

[a]The chemical shift cannot be read because of overlapping.

The isotropic ^{13}C chemical shift values for the amide carbonyl carbons of PBLG in the solid state (α-helix) and in the liquid-crystalline state are 176.4 and 176.2 ppm, respectively. These chemical shift values are almost the same, and this means that the main chain of PBLG in the liquid-crystalline state takes the α-helical conformation. The ^{13}C signal of the amide CO carbon in the liquid-crystalline state, as determined by static CP method, appears at lower field by about 17 ppm than that in the solid state determined by the CP-MAS method. What is the cause of such shift? The α-helical chain is aligned in the magnetic field direction and then the hydrogen bonded carbonyl axis is oriented to the magnetic field direction. The direction of the principal value δ_{22} of the carbonyl carbon is oriented to the carbonyl bond axis. Therefore, the ^{13}C chemical shift of the carbonyl carbon in the liquid-crystalline state is different from the isotropic ^{13}C chemical shift of solid PBLG. This becomes a very important factor in determining the carbon parameter of PBLG in the liquid-crystalline state as described below. In order to elucidate the orientation of PBLG in the liquid-crystalline state through the observation of the ^{13}C chemical shift of the amide carbonyl carbon, the ^{13}C chemical shift tensor components are needed. For this, the slow MAS experiments were carried out to determine the chemical shift tensor components with high accuracy.

The ^{13}C chemical shift tensor components of the amide carbonyl carbon of PBLG in the liquid-crystalline state (25 wt% PBLG/dioxane liquid-crystalline solution) are determined by ^{13}C CP-MAS NMR spectrum at the slow-MAS rate of 1.9 kHz, which has the spinning sidebands in the carbonyl and phenyl carbon region. The intensities of the spinning sidebands of the amide carbonyl carbon are determined by the computer-fitting. The three principal values of the ^{13}C chemical shift tensor were obtained to be $\delta_{11} = 238.7 \pm 1.8$ ppm, $\delta_{22} = 198.5 \pm 1.0$ ppm, and $\delta_{33} = 92.3 \pm 1.8$ ppm, on the basis of the intensities of spinning sidebands (± 1, ± 2, and ± 3) of the amide CO carbon.[34] The chemical shift value δ_{iso} of the amide CO carbon in the liquid-crystalline state is very close to the δ_{22} value. This means that the principal axis of the amide CO carbon tensor component, δ_{22}, is almost oriented to the magnetic field direction.

As mentioned above, the ^{13}C chemical shift, δ_{obs}, of the amide carbonyl carbon of PBLG in the liquid-crystalline state obtained by the static CP method is 193.2 ppm, which is close to the δ_{22} value (198.5 ppm) of solid PBLG. The former appears at higher field by 5.3 ppm than does the latter. If a PBLG chain is completely oriented in the magnetic field direction, the direction of δ_{22} agrees with the magnetic field direction, and the order parameter S becomes 1. The observed chemical shift difference (5.3 ppm) between δ_{22} and δ_{obs} shows that a PBLG chain axis is deviated from the magnetic field direction, and thus the direction of the δ_{22} is deviated from the magnetic field direction. This shows that the order parameter S becomes smaller than 1 and the PBLG chain is fluctuating around the magnetic field direction.

Suppose that this fluctuation obeys the Gaussian distribution $f(\varphi,\theta)$, the (x, y, z) coordinates in the molecular system are transformed to the laboratory coordinates. Then, from this situation and Eq. (11.1), the chemical shift δ is expressed by[19]

$$\begin{aligned}\delta &= \iint \delta(\varphi,\theta)f(\varphi,\theta)d\varphi d\theta \\ &= \int_{-\pi/2}^{+\pi/2}\int_{-\pi}^{+\pi} \{\delta_{11}\cos^2(\varphi+\varphi_0)\sin^2(\theta+\theta_0)+ \\ &\quad \delta_{22}\sin^2(\varphi+\varphi_0)+\delta_{33}\cos^2(\theta+\theta_0)\} \\ &\quad (1/2\pi\sigma^2)\exp[-(\varphi^2+\theta^2)/2\sigma^2]d\varphi d\theta\end{aligned} \quad (11.3)$$

where φ_0 and θ_0 are the average values of the angles φ and θ, respectively, and σ is the standard deviation of Gaussian distribution. By substituting the δ_{obs} into Eq. (11.3), we can obtain the relationship between the angles (φ_0 and θ_0) and the standard deviation σ of the Gaussian distribution (not shown).

By substituting this value (193.2 ppm) into Eq. (11.3), the order parameter S can be obtained to be 0.875 ± 0.025. For the oriented liquid-crystalline PBLG solution system, the relationship between the order parameter S and the value of δ_{obs} can be expressed (as shown in Figure 11.6) by

$$S = 0.024 \times \delta_{obs} - 3.758 \quad (11.4)$$

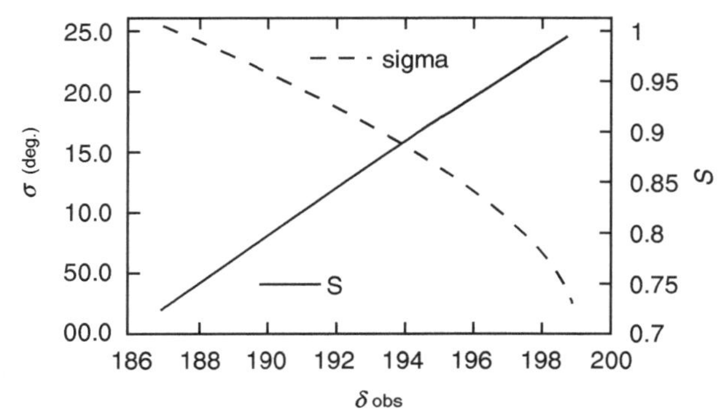

Figure 11.6. The relationship between δ_{obs} (the observed ^{13}C chemical shift of the main-chain carbonyl carbon in PBLG) and the standard deviation σ in the Gaussian distribution, and between δ_{obs} and the order parameter S (from Zhao *et al.*[19]).

From this relation, if we obtain the δ_{obs} value, the order parameter S can be determined. In the completely oriented PBLG system ($S = 1$), $\delta_{obs} = \delta_{22} =$ 198.5 ppm. However, in a real system, as the main chain with the α-helical conformation is fluctuating around the magnetic field, δ_{obs} appears at upper-field than δ_{22}.

11.1.2.3. The Degree of Orientation of PBLG Chains During the Cross-linking Reaction by Using the ^{13}C Chemical Shift Tensor Components

As described above, the cross-linking reaction of highly oriented PBLG chains with 1,6-diaminohexane as cross-linker as a function of the reaction time during the PBLG gel formation can be studied through the observation of the amide carbonyl carbon of PBLG chains by static ^{13}C CP NMR.[19] As the reaction time is increased at 25 °C, the spectrum becomes broad with an increase in time and after the reaction time of 14 h, the signals disappear completely. These phenomena can be explained by the following discussions. The heat generated by the ester–amide exchange reaction leads to an increase in the molecular motion of PBLG. Thus, the CP efficiency for the signal intensity is reduced and the signals disappear gradually. From the ^{13}C chemical shift value of the main-chain carbonyl carbon of PBLG in the oriented gel, the degree of orientation for PBLG is greatly reduced. This shows that the orientation of the PBLG chains becomes isotropic.

On the other hand, after the sample was placed outside the magnetic field at 15 °C, the observed spectrum becomes broad slowly. This shows that at lower temperatures the oriented liquid-crystalline structure is not so strongly disturbed by heat generated from the cross-linking reaction. When the cross-linker with a shorter *n*-alkyl chain is used, the oriented liquid-crystalline structure is more strongly formed. The observed ^{13}C chemical shift value, δ_{obs}, of the main-chain carbonyl carbon of PBLG in the oriented gel is obtained as a function of the cross-linking reaction time. As seen from this figure, the order parameter S is increased with an increase in the cross-linking reaction time at the initial time of the gel formation. However, the order parameter S of the PBLG network chains in the gel decreases with an increase in the cross-linking reaction time after the reaction system was taken from the magnetic field. The order parameter S decreases from 0.86 to 0.81 as shown in Figure 11.7. As seen from these experiments, the reasonable reaction conditions to obtain the oriented polypeptide gel are as follows: a cross-linker with a short *n*-alkyl chain, such as 1,2-diaminoethane, should be used and a lower reaction temperature such as 15 °C should be

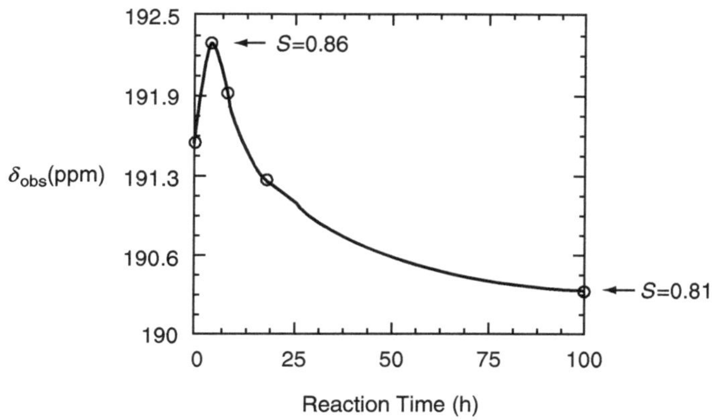

Figure 11.7. The plots of the observed chemical shifts, δ_{obs} (ppm), of the main-chain carbonyl carbon in a PBLG gel system as a function of the cross-linking reaction time (from Zhao *et al.*[19]).

chosen. The ^{13}C chemical shift tensor components can be used to monitor the degree of orientation for PBLG during the cross-linking reaction.[35]

11.2. Blend System

Synthetic homopolypeptides take some specified conformations such as the α-helix, β-sheet, etc. These individual conformations are transformed into other conformations under certain conditions such as temperature and quenching.[36–42] For example, the main chain of poly(β-benzyl L-aspartate) takes a right-handed α (α_R)-helix form within the temperature range from room temperature to 117 °C and is transformed to the left-handed α (α_L)-helix form, the ω-helix form, the β-sheet form at temperatures above 117°C. On the other hand, copolymers of L-Ala and (Gly, [(Ala, Gly)$_n$], take the right-handed α-helix, β-sheet, and 3_1-helix forms in the solid state are obtained by changing the mixture ratio or by the solvent treatment.[43–46]

These transformations arise from the energetic stability caused by intra-molecular or intermolecular hydrogen bond interactions. Thus, by the balance of intramolecular and intermolecular hydrogen bond interactions in polypeptide blends, it is expected that the strength of intermolecular interaction in the blends is different from those in homopolypeptides and then new conformations can be formed by intermolecular hydrogen bond interactions that do not exist originally in homopolypeptides. There are many studies on intermolecular hydrogen bond interactions in homopolypeptides and copolypeptides in the solid state, but to the best of our knowledge, there is little study on intermolecular HB interactions in polypeptide blends except for our previous studies.

Solid state NMR chemical shift is a very useful means to analyze their conformations or conformational changes as described above. Also, the relaxation times can be used to provide information on the dynamics. Especially, spin–lattice relaxation time in the rotating frame of ^{1}H (^{1}H $T_{1\rho}$) is very sensitive to the domain size of individual polymers in polymer blends through the spin-diffusion process and thus can be used to study the miscibility of polymer blends. Since the efficiency of spin diffusion is governed by dipole–dipole interactions, knowledge of the rate of spin diffusion among proton spins of individual polymers in polymer blends would provide useful information about domain sizes in the region of 1.7–5.5 nm.[47–69] The ^{13}C–^{1}H HETCOR spectrum often has multiple proton cross peaks for each carbon, and these cross peaks can be extremely helpful for assignment of the spectrum.[70] Thus, this method can be also used to characterize the structure of polymers in the solid state.[71–86]

From such a background, some kinds of polypeptide blend samples have been studied by solid state NMR.[87–91] It is well known that two kinds of homopolypeptides do not blend by conventional method that polypeptides are dissolved in strong-acidic solvents such as trifluoro acetic acid, dichloroacetic acid and precipitated by water or methanol.

11.2.1. Blend Preparation

On the contrary, let us introduce the blend method for preparing polypeptide blend consisting of two kinds of homopolypeptides developed by Murata *et al.*[88] The Muratas' method leads to a breakthrough for preparing polypeptide blends. Helical polypeptide and β-sheet polypeptide are dissolved in trifluoroacetic acid (TFA) with a 2.0 wt/wt% amount of H_2SO_4. Next, a mixture of helical polypeptide and β-sheet polypeptide with various ratios of 80/20, 50/50, and 20/80 (wt/wt%) are dissolved in TFA with a 2.0 wt/wt% amount of H_2SO_4. The solution is added to alkaline water at room temperature and then the precipitated mixture sample is washed by water and dried under vacuum at temperatures from 308 to 318 K, respectively.

11.2.2. Conformational Characterization

As NMR methodology for elucidating conformational stability in the polypeptide blends, the conformation-dependent ^{13}C NMR chemical shift for polypeptides in the solid state has been reported.[87–91] It has been elucidated that the ^{13}C NMR chemical shifts of a number of polypeptides in the solid state, as determined by the ^{13}C CP-MAS method, are significantly displaced, depending on their particular conformations such as α-helix, 3_1-helix, or β-sheet form.

Figure 11.8 shows the ^{13}C CP-MAS NMR spectra of PLA, poly(L-valine) (PLV), and PLA/PLV blend samples with mixture ratio of 80/20, 50/50, and 20/80 (wt/wt%). Here, samples of PLA (α-helix) and PLV (β-sheet) were treated with the same condition as a mixture of PLA/PLV blend samples. The three intense peaks of PLA at 176.7, 53.2, and 16.0 ppm assigned to the C=O, C_α, and C_β carbons, respectively, of the right-handed α-helix form (Figure 11.8(a)). On the other hand, the four intense peaks of PLV at 172.3, 58.6, 32.9, and 19.0 ppm can be assigned to the C=O, C_α, C_β, and C_γ carbon, respectively, of the β-sheet form (Figure 11.8(e)). The observed ^{13}C CP-MAS NMR spectra of the PLA/PLV blend samples with various ratios of 80/20, 50/50, and 20/80 (wt/wt%) obtained by the Muratas' method are shown in Figure 11.8 (b)–(d), respectively. In these ^{13}C CP-MAS spectra, a new asterisked peak for the C_α carbon of PLA appears clearly at 49.1–49.6 ppm. This peak can be assigned to the C_α carbon of PLA with the β-sheet form by using the reference data as shown in Table 6.1. In order to clarify the appearance of this new peak, the carbonyl carbon region and the C_α, C_β and C_γ carbon region in the spectrum of PLA/PLV blend, with a mixture ratio of 50/50, were expanded as shown in Figure 11.8. By computer-fitting, the observed spectrum was decomposed as a sum of Lorentzian lines, and then the components of the α-helix and β-sheet forms for PLA and PLV were determined. If we look at the spectra carefully, another new peak of the C_β carbon of PLA appears at about 21.1 ppm, in addition to an intense peak assigned to the α-helix form (16.0 ppm), and can be assigned to the β-sheet form (21.1 ppm) by using reference data. These results show that the α-helix form of PLA in the PLA/PLV blend samples are partially transformed to the β-sheet form.

It is very significant to state that only PLA was treated by the TFA-alkaline water, and PLA has not changed its conformation. Nevertheless, when PLA/PLV blend samples were prepared by the same treatment, the β-sheet form in PLA was formed. The origin of the formation of the β-sheet form in PLA comes from the existence of PLV. Therefore, the β-sheet form of PLA in the PLA/PLV blends is incorporated into the PLV with the β-sheet form, by forming intermolecular interactions with PLV, and another component of PLA remains in the α-helix form. The β-sheet form of the PLA chains having intermolecular interactions with PLV chains is much more stable than the α-helix form of the PLA chains themselves. Thus, the generation of the same conformations of PLA in PLA/PLV blends may

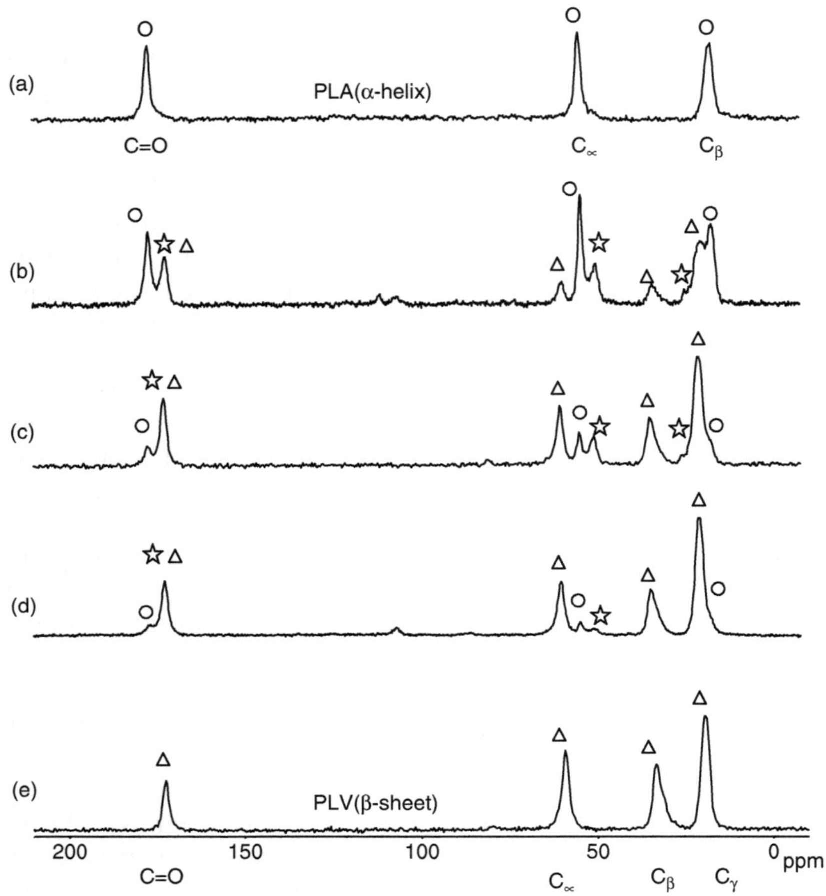

Figure 11.8. ^{13}C CP-MAS NMR spectra of PLA (○), PLV (Δ), and PLA/PLV blend samples which were prepared by adding their TFA solutions with a 2.0 wt/wt% amount of H_2SO_4 to alkaline water. Homopolypeptides of PLA (α-helix) and PLV (β-sheet) are prepared using same condition as PLA/PLV (80/20, 50/50, 20/80) blend samples. The symbols of stars show the new signals that were produced by this blend condition. (a) PLA, (b) PLA/PLV (80/20), (c) PLA/PLV (50/50), (d) PLA/PLV (20/80), and (e) PLV (from Murata *et al.*[88]).

be closely associated with changes in the strength of intermolecular interaction between polypeptides, which comes from rapid environmental changes that occur by adding the TFA solution of PLA/PLV blends with a 2.0 wt/wt% of H_2SO_4 to alkaline water. These results are in the similar situation as reported previously[92] that the conformation of the minor L-Ala residue component in the major β-benzyl L-Asp residue component in the copolypeptides depends on the conformation of the major β-benzyl L-Asp residue component, and when the L-Ala residues were hydrogen-bonded with the β-benzyl L-Asp residues with the β-sheet form, the L-Ala residues are incorporated into the β-sheet form. Similar experimental results on PLA/poly(L-isoleucine) (PLIL) blends, PG/PLV blends, and PDA/PLV blends have been obtained.

11.2.3. Domain Size Analysis

In all the polypeptide blend samples, the proton spin diffusion between each homopolypeptide occurs on the ^{1}H $T_{1\rho}$ timescale. The maximum effective diffusion distance was obtained from these ^{1}H $T_{1\rho}$ values. The maximum effective diffusion distance L of the proton spin diffusion is expressed by the following equation:

$$L = (6D_{\text{spin}}t)^{1/2} \quad (11.5)$$

where ^{1}H $T_{1\rho}$ is used for t. Although the value of D_{spin} may somewhat depend on the different proton densities in the blend systems, the average D_{spin} has been used in the analysis of polymer blend systems. In general, in the $T_{1\rho}$ experiments on polymer blends the 10^{-12} cm^2/s as the averaged value has been used for determining qualitatively or semi-quantitatively the domain size of the blend systems.[59,92–95] By substituting the ^{1}H $T_{1\rho}$ values of 16, 18, 15, and 16 ms averaged over all of the protons for PLA/PLV (50/50), PLA/PLIL (50/50), PG/PLV (50/50), and the PDA/PLV (50/50) blend samples, respectively, into Eq. (11.5), we can approximately estimate L to be about 3 nm as the domain size. This shows that the domain size of the individual polypeptides in the blend is not so large, and they are miscible at the molecular level. This equation should be used for qualitative or semi-quantitative discussion of the miscibility of polymer blends as suggested by many reports.

It is noted that PLA with the β-sheet form in the PLA/PLV blends is incorporated into PLV in the β-sheet form and then takes the β-sheet form by making up intermolecular interactions with PLV. However, the other component of PLA takes the α-helix form. In addition, proton spin diffusion between the PLAs in the α-helix and β-sheet forms and PLV in the β-sheet form occurs on the ^{1}H $T_{1\rho}$ timescale and so PLA and PLV are miscible at the molecular level with a domain size of about 3 nm. The other three kinds of PLA/PLIL, PG/PLV, and PDA/PLV blend samples are also miscible at the molecular level with a domain size of about 3 nm.

11.2.4. ^{13}C–^{1}H HETCOR Spectral Analysis and Structural Characterization

HETCOR spectrum often has multiple proton cross peaks for each carbon, and these cross peaks can be extremely helpful for spectral assignment and is used to characterize the structure of polymers in the solid state.[70–86]

2D FSLG ^{13}C–^{1}H HETCOR spectrum with a short contact time (0.2 ms) of PG/PLV (50/50) blend is shown in Figure 11.9.[90] The horizontal axis (F_2) corresponds to the ^{13}C (chemical shift range: 3 to +182 ppm), and the vertical axis (F_1) corresponds to the ^{1}H (chemical shift range: −1 to +12 ppm). 1D^{13}C CP-MAS spectrum for the horizontal (^{13}C) axis is shown at the top of this figure, where the intense correlations arise from the dipolar coupling between the carbons and their directly bonded protons. The corresponding signals for the C_α and H_α (C_α–H_α) dipolar coupling in PG

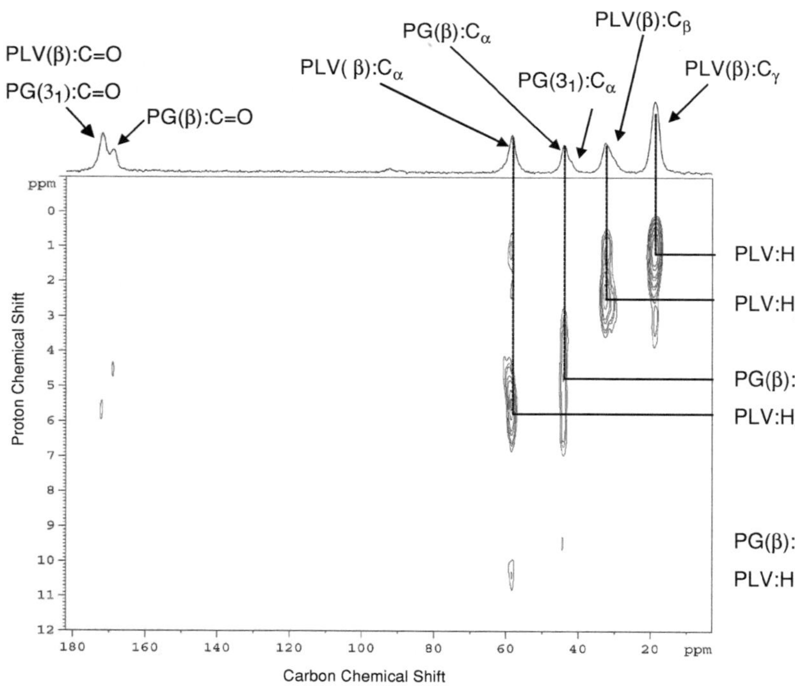

Figure 11.9. The 2D FSLG ^{13}C–^{1}H HETCOR spectra of PG/PLV (50/50) blend sample with contact time of 0.2 ms (from Murata *et al.*[90]).

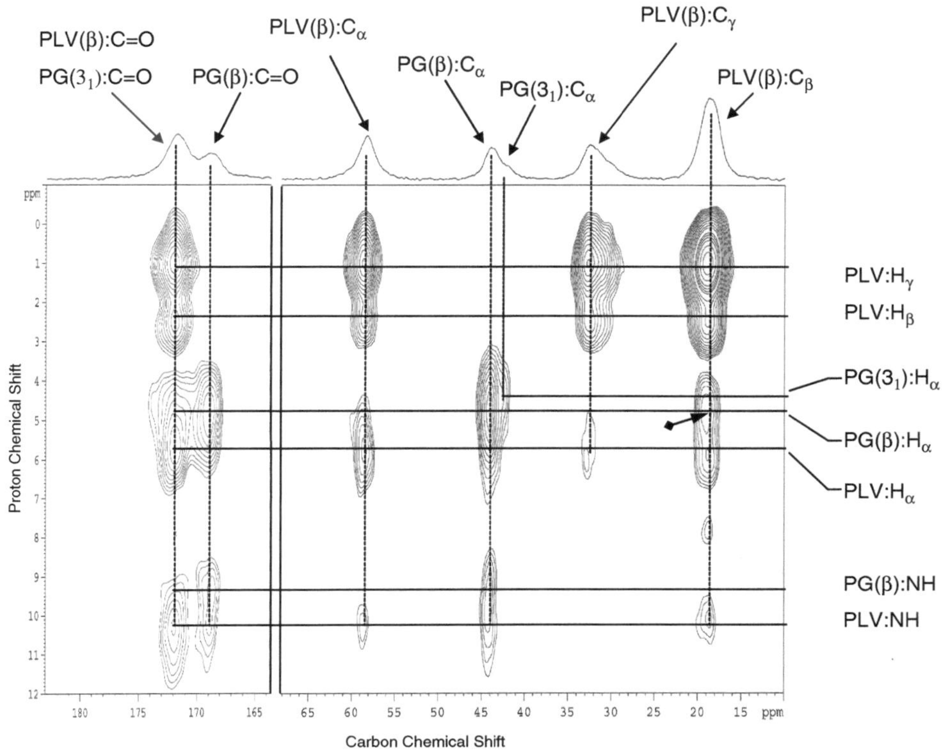

Figure 11.10. 2D FSLG ^{13}C–^{1}H HETCOR spectra for the carbonyl carbon region and for the C_α, C_β, and C_γ carbons region of PG/PLV(50/50) blend sample with contact time of 1.5 ms (from Murata *et al.*[90]).

(β-sheet) and for the C_α–H_α, C_β–H_β, and C_γ–H_γ dipolar couplings in PLV appear. Further, the other weak correlation peaks are observed for the C_α/H_γ, C_β/H_γ, C_α/NH, and C=O/H_α dipolar couplings in PLV, and for the C_α /NH and C=O/H_α dipolar couplings in PG (β-sheet).

In the ^{1}H spectrum, peaks at 9.3 and 4.7 ppm can be assigned to the NH and H_α protons of PG in the β-sheet form, respectively, and further peaks at 10.3, 5.6, 2.3, and 1.1 ppm to the NH, H_α, H_β, and H_γ protons of PLV in the β-sheet form, respectively. The ^{1}H peaks for PG in the 3_1-helix form by the FSLG HETCOR method with contact time of 0.2 ms because the 3_1-helix form in PG is the minor component. The FSLG ^{13}C–^{1}H HETCOR spectrum with contact time of 0.5 ms (not shown) shows the correlation signal between the C_α carbon and the H_α proton of PG in the 3_1-helix form, and the ^{1}H chemical shift of the H_α proton is 4.3 ppm. Each of the ^{1}H signal has large chemical shift difference enough to analyze corresponding correlation signals. Further, correlation signals for the C=O/NH, C=O/H_β C=O/H_γ, and C_γ–H_α dipolar couplings in PLV, and for the C=O–NH dipolar coupling in PG with the β-sheet form are observed. In addition, it is very significant to show that the intermolecular correlation signals for the PG (β-sheet)C_α/PLV(β-sheet)H_αPG(β-sheet)C_α/PLV(β-sheet)NH, PG(β-sheet) C=O/PLV(β-sheet)H_α and PG(β-sheet)C=O/PLV(β-sheet)NH dipolar couplings are observed.

On the other hand, 2D FSLG ^{13}C–^{1}H HETCOR spectrum with long contact time of 1.5 ms (Figure 11.10) shows a significant and new signal. The new correlation signal appeared between the C_γ carbon of PLV (β-sheet) at 19 ppm and the H_α proton of PG (β-sheet) at 4.7 ppm as shown by the arrow. In order to get more detailed information, ^{1}H slice objection of the H_α (4.7 ppm) and NH (9.3 ppm) protons of PG (β-sheet), H_γ (1.1 ppm), and H_β (2.3 ppm) protons of PLV (β-sheet), and that of the H_α (5.7 ppm) and NH (10.3 ppm) protons of PLV (β-sheet) can be represented (not shown). It can show intermolecular correlations. For example, the slice spectra of the H_α proton for PG (β-sheet) show the intermolecular correlations with the C=O, C_α and side-chain carbons of PLV (β-sheet), and also the sliced spectra of the H_α proton for PLV (β-sheet) show the intermolecular correlations with the C=O and C_α carbons of PG (β-sheet). Similarly, the sliced spectra of the NH proton for PG (β-sheet) show the intermolecular correlations with the C=O, C_α and side-chain carbons of PLV (β-sheet), and those of the NH proton of PLV (β-sheet) show the intermolecular correlations with the C=O and C_α carbons of PG (β-sheet).

As seen from these results, the helix and β-sheet forms of PLA, PG, or PDA in the blend samples are incorporated into the PLIL or PLV in the β-sheet form and then takes the β-sheet form by forming hydrogen bonds with the β-sheet form the PLIL or PLV. This means that polypeptide chains by changing from the helix form to the β-sheet form hydrogen-bonded with β-sheet type polypeptide chains are energetically more stable than the helix form of the polypeptide chains hydrogen-bonded by themselves.

References

[1] C. Robinson, 1956, *Trans. Faraday Soc.*, 52, 571–592.
[2] C. Robinson and J. C Ward, 1957, *Nature*, 180, 1183–1184.
[3] C. Robinson and J. C Ward, 1958, *Discuss. Faraday Soc.*, 25, 29–42.
[4] C. Robinson, 1961, *Tetrahedron*, 13, 219–234.
[5] A. Eliot and E. T. Ambrose, 1958, *Discuss. Faraday Soc.*, 29, 246–251.

[6] J. Watanabe, H. Ono, I. Uematsu, and A. Abe, 1985, *Macromolecules*, 18, 2141–2148.
[7] S. Sobajima, 1967, *J. Phys. Soc. Jpn.*, 23, 1070–1078.
[8] M. Panar and W. D. Phillips, 1968, *J. Am. Chem. Soc.*, 90, 3880–3882.
[9] R. D. Orwell and R. L. Vold, 1971, *J. Am. Chem. Soc.*, 93, 5335–5338.
[10] E. T. Samulski and A. V. Tobolsky, 1969, *Mol. Cryst. Liq. Cryst.*, 7, 433–742.
[11] B. M. Fung, M. J. Gerace, and L. S. Gerace, 1970, *J. Phys. Chem.*, 74, 83–87.
[12] I. Ando, T. Hirai, Y. Fujii, and A. Nishioka, 1983, *Makromol. Chem.*, 184, 2581–2592.
[13] T. Yamanobe, M. Tsukahara, T. Komoto, J. Watanabe, I. Ando, I. Uematsu, K. Deguchi, T. Fujito, and M. Imanari, 1988, *Macromolecules*, 21, 48–50.
[14] M. Tsukahara, T. Yamanobe, T. Komoto, J. Watanabe, and I. Ando, 1987, *J. Mol. Struct.*, 159, 345–353.
[15] B. Mohanty, T. Komoto, J. Watanabe, I. Ando, and T. Shiibashi, 1989, *Macromolecules*, 22, 4451–4455.
[16] B. Mohanty, J. Watanabe, I. Ando, and K. Sato, 1990, *Macromolecules*, 23, 4908–4911.
[17] T. Yamanobe, H. Tsukamoto, Y. Uematsu, I. Ando, and I. Uematsu, 1993, *J. Mol. Struct.*, 295, 25–37.
[18] E. Katoh, H. Kurosu, and I. Ando, 1994, *J. Mol. Struct.*, 318, 123–131.
[19] C. Zhao, H. Zhang, T. Yamanobe, S. Kuroki, and I. Ando, 1999, *Macromolecules*, 32, 3389–3395.
[20] C. Zhao, S. Kuroki, and I. Ando, 2000, *Macromolecules*, 33, 4486–4489.
[21] Y. Yin, C. Zhao, S. Kuroki, and I. Ando, 2000, *J. Chem. Phys.*, 113, 7635–7639.
[22] Y. Yin, C. Zhao, S. Kuroki, and I. Ando, 2002, *Macromolecules*, 35, 2335–2338.
[23] Y. Yin, C. Zhao, A. Sasaki, H. Kimura, S. Kuroki, and I. Ando, 2002, *Macromolecules*, 35, 5910–5961.
[24] W. L. Earl and D. L. VanderHart, 1979, *Macromolecules*, 12, 762–767.
[25] A. E. Tonelli and F. C. Schilling, 1981, *Acc. Chem. Res.*, 14, 233–238.
[26] W. P. Rothwell and J. S. Waugh., 1981, *J. Chem. Phys.*, 74, 2721–2732.
[27] N. Bleombergen, E. M. Purcell, and R. V. Pound, 1948, *Phys. Rev.*, 73, 679–712.
[28] I. Ando and T. Asakura, Eds., 1998, *Solid State NMR of Polymers*, Elsevier Science, Amsterdam.
[29] J. G. Kirkwood, 1954, *J. Polym. Sci.*, 12, 1–14.
[30] M. Doi and S. F. Edwards, 1986, *The Theory of Polymer Dynamics*, Chapter 8, Clarendon Press, Oxford.
[31] N. Asakawa, S. Kuroki, H. Kurosu, I. Ando, A. Shoji, and T. Ozaki, 1992, *J. Am. Chem. Soc.*, 114, 3261–3265.
[32] C. J. Hartzell, M. Whitfield, T. G. Oas, and G. P. Drobny, 1987, *J. Am. Chem. Soc.*, 109, 5966–5969.
[33] A. Shoji, T. Ozaki, H. Saitô, R. Tabeta, and I. Ando, 1984, *Macromolecules*, 17, 1472–1479.
[34] J. Herzfeld and A. E. Berger, 1980, *J. Chem. Phys.*, 73, 6021–6030.
[35] Y. Yamane, M. Kanekiyo, S. Koizumi, C. Zhao, S. Kuroki, and I. Ando, 2004, *J. Appl. Polymer Sci.*, 92, 1053–1060.
[36] E. M. Bradbury, A. R. Downie, A. Elliott, and W. E. Hanby, 1960, *Proc. R. Soc. London, Ser. A*, 259, 110–128.
[37] R. H. Karlson, K. S. Norland, G. D. Fasman, and E. R. Bluot, 1960, *J. Am. Chem. Soc.*, 82, 2268–2275.

[38] B. R. Malcom, 1970, *Biopolymers*, 9, 911–922.
[39] H. Kyotani and H. Kanetsuna, 1972, *J. Polym. Sci., Part B:Polym. Phys.*, 10, 1931–1939.
[40] H. Saitô, R. Tabeta, I. Ando, T. Ozaki, and A. Shoji, 1983, *Chem. Lett.*, 1437–1440.
[41] M. Okabe, T. Yamanobe, T. Komoto, J. Watanabe, I. Ando, and I. Uematsu, 1989, *J. Mol. Struct.*, 213, 213–220.
[42] T. Akieda, H. Mimura, S. Kuroki, H. Kurosu, and I. Ando, 1992, *Macromolecules*, 25, 5794–5797.
[43] J. C. Andries, J. M. Anderson, and A. G. Walton, 1971, *Biopolymers*, 10, 1049–1057.
[44] H. Saitô, T. Tabeta, A. Shoji, T. Ozaki, I. Ando, and T. Miyata, 1984, *Biopolymers*, 23, 2279–2297.
[45] H. Saitô, T. Tabeta, T. Asakura, Y. Iwanaga, A. Shoji, T. Ozaki, and I. Ando, 1984, *Macromolecules*, 17, 1405–1412.
[46] S. Ando, T. Yamanobe, I. Ando, A. Shoji, T. Ozaki, T. Tabeta, and H. Saitô, 1985, *J. Am. Chem. Soc.*, 107, 7648–7652.
[47] E. O. Stejskal, J. Schaefer, M. D. Sefcik, and R. A. Mckay, 1981, *Macromolecules*, 14, 275–279.
[48] L. C. Dickinson, H. Yang, C. W. Chu, R. S. Stein, and J. C. W. Chien, 1987, *Macromolecules*, 20, 1757–1760.
[49] C. W. Chu, L. C. Dickinson, and J. C. W. Chien, 1988, *Polym. Bull.*, 19, 265–268.
[50] J. F. Parmer, L. C. Dickinson, J. C. W. Chien, and R. S. Porter, 1989, *Macromolecules*, 22, 1078–1083.
[51] X. Zhang, A. Natansohn, and A. Eisenberg, 1990, *Macromolecules*, 23, 412–416.
[52] C. Marco, J. G. Fatou, M. A. Gomez, H. Tanaka, and A. E. Tonelli, 1990, *Macromolecules*, 23, 2183–2188.
[53] C. W. Chu, L. C. Dickinson, and J. C. W. Chien, 1990, *J. Appl. Polym. Sci.*, 41, 2311–2325.
[54] B. Mohanty, J. Watanabe, I. Ando, and K. Sato, 1990, *Macromolecules*, 23, 4908–4911.
[55] L. Jong, E. M. Pearce, T. K. Kwei, and L. C. Dickinson, 1990, *Macromolecules*, 23, 5071–5074.
[56] J.-F. Masson and R. St. J. Manley, 1992, *Macromolecules*, 25, 589–592.
[57] C. Brosseau, A. Guillermo, and J. P. Cohen-Addad, 1992, *Polymer*, 33, 2076–2083.
[58] C. Brosseau, A. Guillermo, and J. P. Cohen-Addad, 1992, *Macromolecules*, 25, 4535–4540.
[59] J. Clauss, K. Schmidt-Rohr, and H. W. Spiess, 1993, *Acta. Polym.*, 44, 1–17.
[60] S. Schantz and N. Ljungqvist, 1993, *Macromolecules*, 26, 6517–6524.
[61] T. K. Kwei, Y. K. Dai, X. Lu, and R. A. Weiss, 1993, *Macromolecules*, 26, 6583–6588.
[62] A. Asano, K. Takegoshi, and K. Hikichi, 1994, *Polymer*, 35, 5630–5636.
[63] D. E. Demco, A. Johansson, and J. Tegenfeldt, 1995, *Solid State Nucl. Magn. Reson.*, 4, 13–38.
[64] T. Miyoshi, K. Takegoshi, and K. Hikichi, 1996, *Polymer*, 37, 11–18.
[65] D. L. VanderHart and G. B. McFadden, 1996, *Solid State Nucl. Magn. Reson.*, 7, 45–66.
[66] S.-Y. Kwak and N. Nakajima, 1996, *Macromolecules*, 29, 3521–3524.

[67] S.-Y. Kwak, J.-J. Kim, and U. Y. Kim, 1996, *Macromolecules*, 29, 3560–3564.
[68] M. Guo, 1997, *Macromolecules*, 30, 1234–1235.
[69] A. Asano and K. Takegoshi, 1998, *Solid State NMR of Polymers*, Chapter 10, I. Ando and T. Asakura, Eds., Elsevier Science, Amsterdam.
[70] B.-J. Van Rossum, H. Forester, and H. J. M. de Groot, *J. Magn. Reson.*, 124, 516–519.
[71] P. Caravatti, G. Bodenhausen, and R. R. Ernst, 1982, *Chem. Phys. Lett.*, 89, 363–367.
[72] P. Caravatti, L. Braunschweiler, and R. R. Ernst, 1983, *Chem. Phys. Lett.*, 100, 305–310.
[73] D. P. Burum and A. Bielecki, 1991, *J. Magn. Reson.*, 94, 645.
[74] A. Bielecki, D. P. Burum, D. M. Rice, and F. E. Karasz, 1991, *Macromolecules*, 24, 4820–4822.
[75] C. E. Bronnimann, C. F. Ridenour, D. R. Kinney, and G. E. Maciel, 1992, *J. Magn. Reson.*, 97, 522–534.
[76] S. Kaplan, 1993, *Macromolecules*, 26, 1060–1064.
[77] C. H. Wu, A. Ramamoorthy, and S. J. Opella, 1994, *J. Magn. Reson.*, 109(A), 270–272.
[78] K. Takegoshi and K. Hikichi, 1994, *Polym. J.*, 26, 1377–1380.
[79] J. L. White and P. A. Mirau, 1994, *Macromolecules*, 27, 1648–1650.
[80] S. Li, D. M. Rice, and F. E. Karasz, 1994, *Macromolecules*, 27, 2211–2218.
[81] S. Li, D. M. Rice, and F. E. Karasz, 1994, *Macromolecules*, 27, 6527–6531.
[82] P. A. Mirau and J. L. White, 1994, *Magn. Reson. Chem.*, 32, S23–S29.
[83] Z. Gu, C. F. Ridenour, C. F. Bronnimann, T. Iwashita, and A. McDermott, 1996, *J. Am. Chem. Soc.*, 118, 822–829.
[84] J. J. Balbach, Y. Ishii, O. N. Antzutkin, R. D. Leapman, N. W. Rizzo, F. Dyda, J. Reed, and R. Tycko, 2000, *Biochemistry*, 39, 13748–13759.
[85] B.-J. van Rossum, C. P. de Groot, V. Ladizhansky, S. Vega, and H. J. M. de Groot, 2000, *J. Am. Chem. Soc.*, 122, 3465–3472.
[86] B.-J. van Rossum, E. A. M. Schulten, J. Raap, H. Oschkinat, and H. J. M. de Groot, 2002, *J. Magn. Reson.*, 155, 1–14.
[87] J. Nakano, S. Kuroki, I. Ando, T. Kameda, H. Kurosu, T. Ozaki, and A. Shoji, 2000, *Biopolymers*, 54, 81–88.
[88] K. Murata, S. Kuroki, H. Kimura, and I. Ando, 2002, *Biopolymers*, 64, 26–33.
[89] K. Murata, S. Kuroki, and I. Ando, 2002, *Polymer*, 43, 6871–6878.
[90] K. Murata, H. Kono, E. Katoh, S. Kuroki, and I. Ando, 2003, *Polymer*, 44, 4021–4207.
[91] K. Murata, S. Kuroki, E. Katoh, and I. Ando, 2003, *Annu. Rep. NMR Spectrosc.*, 51, 1–57.
[92] S. Tuzi, T. Komoto, I. Ando, H. Saitô, A. Shoji, and T. Ozaki, 1987, *Biopolymers*, 26, 1983–1992.
[93] J. R. Havens and D. L. VanderHert, 1985, *Macromolecules*, 18, 1663–1676.
[94] R. A. Assink, 1978, *Macromolecules*, 11, 1233–1237.
[95] D. C. Douglass and G. P. Jones, 1966, *J. Chem. Phys.*, 45, 956–963.

Chapter 12
GLOBULAR PROTEINS

Three-dimensional (3D) structural determination of globular proteins, based on NMR constraints available from microcrystalline samples instead of single crystals essential for X-ray diffraction, is undoubtedly one of the most challenging goals for current development of solid state NMR methodology as applied to structural biology. In the solid sample, uniform and the highest level of ^{13}C and ^{15}N enrichment is a prerequisite for an unbroken sequence of backbone assignments and sensitivity improvement, respectively. In contrast to solution NMR, it is anticipated that resulting line broadenings caused by dipolar and *J*-couplings may be a serious obstacle for further structural analysis, if their spectra were recorded by a conventional CP-MAS technique with low spinning rate. This problem can be overcome by fast MAS ($\nu_{MAS} \gg 8$ kHz), high-power proton decoupling ($\omega_I/2\pi \gg 70$ kHz) with an efficient decoupling scheme such as TPPM[2] or XiX[3] decoupling, and homonuclear ^{13}C–^{13}C *J*-decoupling, which cannot be eliminated by fast MAS alone, at higher magnetic field to prevent the appearance of strong coupling, together with the low-power ^{15}N broadband decoupling to remove weak scalar ^{15}N–^{13}C couplings.[1]

12.1. (Almost) Complete Assignment of ^{13}C NMR Spectra of Globular Proteins

Indeed, the advantage of relatively fast MAS (30 kHz using a 2.5-mm o.d. rotor) is evident as manifested from improved spectral resolution compared to MAS at 12 kHz[4] for the model protein Crh[5] (catabolite repression histidine-containing phosphocarrier protein; 2 × 10.4 kDa or dimeric form with 93 residues) (Figure 12.1). The *S/N* ratio can be significantly improved by taking advantage of optimized pulse sequence available with rapid MAS. In fact,

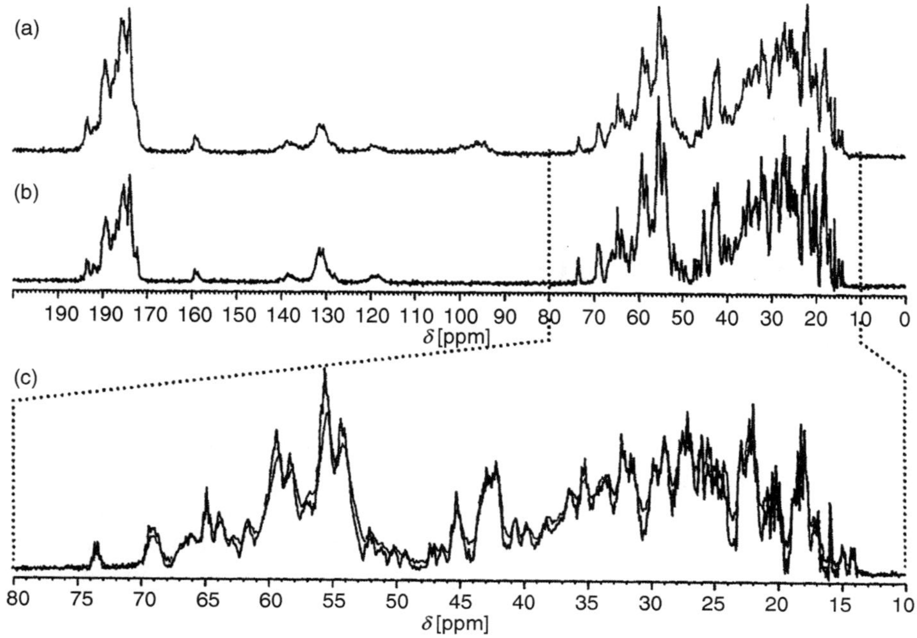

Figure 12.1. 1D ^{13}C NMR spectra of Crh (a) at 12 kHz MAS frequency and 80 kHz decoupled field strength using TPPM decoupling, (b) at 30 kHz MAS frequency at 150 kHz XiX decoupling. (c) Comparison of the aliphatic region of two ^{13}C spectra: 12 kHz MAS and 30 kHz MAS spectrum (from Ernst *et al.*,[4]).

going from a 4- to 2.5-mm o.d. rotor increases the maximum spinning frequency from about 18 to 30 kHz and the maximum rf field amplitude on the proton channel increases from about 100 to 150 kHz. Further, they showed that the *S/N* ratio of the cross-peaks in a homonuclear 2D shift-correlation experiment using DREAM[6,7] polarization transfer sequence with a 2.5-mm o.d. rotor is almost two times higher than the cross-peaks in a PDSD[8,9] experiment using a full 4-mm o.d. rotor, owing to primarily based on the higher polarization transfer efficiency of the DREAM sequence. The experimental aspect of the PDSD and DREAM experiments was already discussed as shown in Figures 5.7 and 5.8, respectively. At the maximum rotor speed for a given rotor diameter, heating by 20–30 K is observed and must be compensated by cooling, however.

Such achieved spectral resolution permits partial or almost complete assignment of individual NMR peaks of relatively small globular proteins, including SH_3 domain of α-spectrin (7.2 kDa or 62 residues),[10–12] basic pancreatic trypsin inhibitor (BPTI; 6.5 kDa or 58 -residues),[13] ubiquitin (8.6 kDa or 76 residues),[1,14,15] Crh,[4,5] reassembled thioredoxin,[16] etc. to respective amino acid residues based on the shift correlations using through space (PDSD and DREAM) and through bond (TOBSY)[17] connectivities, and ^{15}N–^{13}C and ^{15}N–^{13}C–^{13}C correlations including the intra-residue (NCA or CAN) and inter-residue (NCO, NCOCX, N(CO)CA) and, NCACX and N(CA)CB (Figure 5.10), as described already in Section 5.2.3 together with the illustrative examples for such correlations. Even in such microcrystalline globular proteins, however, all the ^{13}C or ^{15}N NMR signals are not always fully visible. For instance, signals are absent in the α-spectrin SH3 domain from the first six N-terminal and the last C-terminal residues, from V 46, N 47, from tyrosine side chains of Y 13 and Y 15 and from the side chain of K 39.[10] In solution, the N-terminus is flexible, leading to stronger signals than observed from the globular part of the protein. One reason for the absence of such signals in the solid may be flexibility of the N-terminus, which could be on a timescale that interferes with the proton decoupling. In fact, ^{13}C NMR signals of densely ^{13}C-labeled bR from PM labeled with [1,2,3-^{13}C]Ala were substantially broadened by accelerated transverse relaxation rate[18] caused by the increased number of relaxation pathways through a number of ^{13}C–^{13}C homonuclear dipolar interactions and scalar *J*-couplings in the presence of intermediate or slow fluctuation motions with correlation times in the order of 10^4–10^5 Hz.[19,20] Another explanation for the missing signals for α-spectrin SH3 domain relates to the occurrence of heterogeneous broadening, which may result from a multitude of conformers that are ''frozen out'' upon precipitation during the preparation of the solid sample.[10] This leads to increased chemical shift dispersion for the signals of the involved residues and cannot be removed by decoupling techniques.

12.2. 3D Structure: α-Spectrin SH3 Domain

In contrast to solution NMR, it is not easy to achieve determination of 3D structure for globular proteins based on NMR constraints available from microcrystalline preparations, even if their assignment of peaks were completed as demonstrated earlier. This is because a suitable methodology for the collection of a large number of distance restraints in the range of 2–7 Å from a small number of samples and with minimal experimental effort is still lacking[11] necessary for structure calculation based on MD simulation, although very accurate distance or torsion angle constraints are available in the solid state based on REDOR- or RR-based measurements or a variety of recently developed pulse techniques, as discussed in Sections 6.2 and 6.3. In fact, it is too time-consuming, however, and requires a huge number of samples if one attempt to determine such accurate interatomic distances for globular proteins. Instead, it is inevitable to rely on less accurate means to determine distances based on, for instance, PDSD measurements for defining 3D structure. The measurement of such restraints is hindered in fully ^{13}C-enriched samples by dipolar truncation effects.[21,22] This is especially true at longer distances where weaker couplings are ‘‘truncated’’ by the presence of much stronger couplings.

To reduce the dipolar truncation effects, a set of differently labeled samples of α-spectrin SH3 domain was prepared[11] using [1,3-^{13}C] and [2-^{13}C]glycerol as carbon sources, discussed in Section 5.1, which enabled the semiquantitative interpretation of cross-peak intensities. In fact, the alternating labeling pattern in [1,3-^{13}C]glycerol (1,3-SH3)- and [2-^{13}C]glycerol (2-SH3)-labeled SH3, instead of a uniformly ^{13}C-labeled preparation (U-SH3), allows the observation of long-range interactions, while relayed polarization transfer available from fully labeled preparation is blocked. As expected, it was shown that 2D ^{13}C–^{13}C PDSD spectra of 2-SH3 and 1,3-SH3 spectra are simplified owing to the less extensive labeling compared with those of U-SH3, even if they were recorded with a long mixing time of 500 ms to establish long-range correlations. To extract distance restraints from the cross-peak intensities, a set of spectra with mixing times of 50, 100, 200, and 500 ms was recorded in order to categorize the carbon–carbon distance restraints, empirically in four restraint classes as a similar manner to the treatment of NOE data in solution NMR. The reference distance for the first class was definitely by cross-peaks between sequential C_α atoms ($\approx$ 3.8 Å) that appeared within 50 ms, whereas interaction between sequential C_α and C_β ($\approx$ 4.6 Å) appeared first at 100 ms (the second class). The third class contained sequential C_β–C_β interactions ($\approx$ 5.8 Å). All other interactions to the fourth class, with distances in the range of 2.5–7.5 Å. The lower bound was always kept at 2.5 Å, to account for an apparent lower signal intensity due to incomplete suppression of dipolar truncation effects and/or fractional labeling. On the basis of experimental protocol from this procedure, together with six nitrogen–nitrogen restraints and the data of dilution experiment containing 80% unlabeled material to reduce the effect of intermolecular contact from surface, a list of 292 inter-residue distance restraints for residues 7–61 was generated and subjected to a structure calculation protocol with the program CNS version 1.0.[23] The 15 lowest energy structures were selected out of 200, which represent the fold of the SH3 domain. The C_α coordinates of the regular structure elements show a RMS deviation of 1.6 Å to the average structure and of 2.6 Å to the X-ray structure.

Further, 3D ^{15}N–^{13}C–^{13}C dipolar correlation experiments were applied to [[2-^{13}C]glycerol,^{15}N]- and [[1,3-^{13}C]glycerol, ^{15}N]-labeled α-spectrin SH3 domain to resolve overlap of signals by adding a ^{15}N dimension and allow the identification of backbone carbon–carbon restraints of the C_α–CO,

CO–C_α, C_α–C_α, CO–CO type and restraints involving side-chain carbons, which were not accessible in the 2D experiments.[24] Additional restraints for confining the structure were obtained from ϕ and ψ backbone torsion angles of 29-residues, derived from C_α, C_β, CO, NH, and H_α chemical shifts using the program TALOS[25] based on the concept of the conformation-dependent displacements of peaks, as discussed in Section 6.4. They have performed two structure calculations. For a first calculation, only distance restraints were used. By selecting the 10 lowest-energy structure after structural calculation using a list of 889 inter-residue restraints, the C_α coordinates of the regular structure element exhibited an RMSD of 1.1 Å with respect to the average structure and of 1.6 Å to the X-ray structure, as illustrated in Figure 12.2[24]. By adding the 58 torsion-angle restraints from TALOS in a second structure calculation, however, the RMSD of the C_α coordinates of the regular structure elements with respect to the average structure is reduced to 0.7 Å. This structure is also closer to the X-ray structure, with a deviation of 1.2 Å.

The above-mentioned protocol based on PDSD measurement, however, is still premature as a novel means to determine 3D structure of globular protein in the solid state. This is because there remains a question whether the intensities of cross-peaks available from PDSD experiments at prolonged mixing time in the solid can be straightforwardly related to individual interatomic distances as encountered for those of NOESY measurements

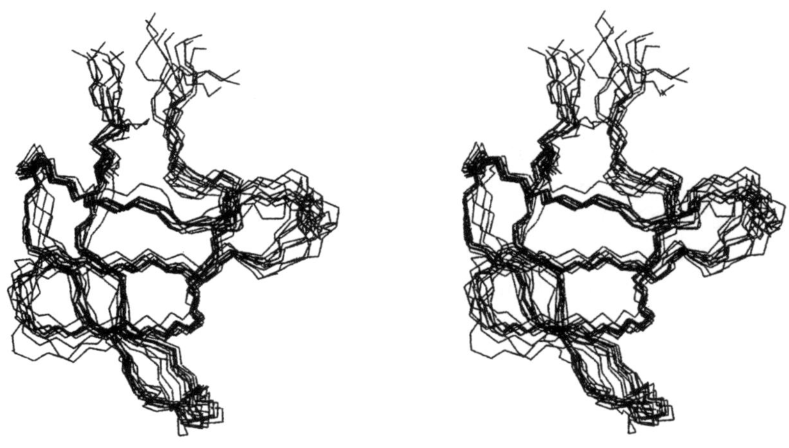

Figure 12.2. Solid-state structure of the α-spectrin SH3 domain. Stereo view of the 10 lowest-energy structures, representing the fold of α-spectrin SH3 domain. For comparison, the X-ray structure is included displayed and overlaid with the family of 10 solid-state structures by fitting the backbone C_α atoms to the average solid state structure (from Castellani *et al.*,[24]).

or not. In fact, theoretical background of PDSD measurements is rather complicated and varied substantially depending upon respective experimental condition.[26–28] In general, transition probability $P_{\mathrm{if}}(t)$ between two nuclei A and B due to spin diffusion for static sample is given by:

$$P_{\mathrm{if}}(t) = 1/2\pi g_0^{\mathrm{AB}}(\omega_{\mathrm{A}} - \omega_{\mathrm{B}})\omega_{\mathrm{D}}^2 t \tag{12.1}$$

where g_0^{AB}, $(\omega_{\mathrm{A}} - \omega_{\mathrm{B}})$, and ω_{D} are zero-quantum line shape, chemical shift difference, and dipolar coupling between A and B nuclei, respectively. In particular, the last term is

$$\omega_D \propto (3\cos^2\theta - 1)/r^3 \tag{12.2}$$

where θ and r are the angle between the applied magnetic field and vector connecting between A and B nuclei and its distance, respectively. This formula is for static sample. It is certain, therefore, that the intensities of the cross-peaks in the PDSD experiment under MAS are not necessarily a simple function of the distance r between A and B alone. In fact, the structure of α-spectrin SH3 domain is a sole 3D structure as a globular protein revealed by constraints from solid state NMR which is registered in the protein data bank (PDB), although 14 membrane proteins and 3592 proteins, peptides, and viruses have been determined by constraints from solid state and solution NMR, respectively, currently available from the PDB bank. Undoubtedly, more theoretical and experimental evaluations may be required prior to further application of this approach based on PDSD measurements.

The simultaneous measurement was demonstrated for several backbone torsion angles ψ in the uniformly ^{13}C, ^{15}N-labeled α-spectrin SH3 domain using two different 3D ^{15}N–^{13}C–^{13}C–^{15}N dipolar chemical shift MAS NMR experiments.[29] While INADEQUATE NCCN experiment provides better *S/N* and is better compensated with respect to the homonuclear *J*-coupling effects, the NCOCA NCCN scheme has the advantage of having better spectral resolution, which result in the larger number of ψ torsion angle constraints that can be determined. For the case of α-spectrin SH3 domain, they determined 13 ψ angle constraints with INADEQUATE experiment and 22 ψ were measured in the NCOCA NCCN experiment.

12.3 Ligand-Binding to Globular Protein

The application of NMR as a tool for the characterization of protein–ligand complexes is a powerful methodology for drug discovery process. Anti-apoptotic protein Bcl-xL (MW ≈ 20 kDa) acts as an apoptosis (programmed cell death) antagonist, while other proteins such as Bak or Bax, promote cell death, including apoptosis following exposure to chemotherapeutic drugs. ^{13}C–^{13}C 2D solid-state NMR of uniformly and selectively ^{13}C-labeled Bcl-xL were recorded[30] using DARR[31,32] for a homonuclear magnetization transfer. The suppression of recoupling of weaker dipolar coupling by stronger ones (the dipolar truncation) by DARR becomes less prominent due to the orientation selectivity.[29] To obtain a detailed site-specific ligand-binding site, it is important to obtain sufficient resolution to resolve a large number of protein resonance. Some amino acid types, such as Ile, are fully resolved in the 2D spectrum, while other regions (e.g., Ala C_α-C_β) are still very congested. In particular, they showed that $C_{\delta 1}$ methyl groups of Ile 118 and Ile 144 as assigned on the basis of assigned solution NMR data were significantly shifted upon addition of the peptide Bak. The agreement with prior solution state NMR results indicates that the binding pocket in solid and liquid samples is similar for this protein.

Ribulose-1,5-bisphosphate carboxylase/oxygenase (Rubisco) is a hexadecamer of approximately 550 kDa in most organisms. REDOR and TEDOR measurements were performed to obtain the average internuclear distance between the 99% $^{13}CO_2$-labeled activator carbamino carbon to the phosphate phosphorus nuclei of active-site-bound 2-carboxy-D-arabinitol 1,5-bisphosphate (CABP), in freeze-quenched, lyophilized samples of confrey Rubisco.[33] The distance 7.5 $\pm$ 0.5 Å determined by NMR is in agreement with that of 7.7 Å inferred from the crystal structure coordinates for spinach Rubisco–CABP–CO_2–Mg^{2+} quaternary complex.

The 46-kDa enzyme 5-enolpyruvylshikimate-3-phosphate (EPSP) synthase catalyzes the condensation of shikimate-3-phosphate (S3P) and phosphoenolpyruvate (PEP) to form EPSP. The reaction is inhibited by the commercial herbicide $HO_3PCH_2NHCH_2COOH$ (Glp), which in the presence of S3P, binds to EPSP synthase to form a stable ternary complex: EPSP–S3P–Glp. Distances of ^{15}N labels in lysine, arginine, and histidine residues of EPSP synthase to a ^{13}C label in Glp, and to the ^{31}P in S3P and Glp were determined by REDOR.[34] Three lysine and four arginine residues are claimed to be in the proximity of the phosphate group of S3P and the carboxyl and phosphonate groups of Glp. DRAMA was used to determine the single ^{31}P–^{31}P distance, while REDOR was used to determine one ^{31}P–^{15}N distance and five ^{31}P–^{13}C distances.[35] MD simulations of an S3P–Glp complex based on such restraints suggest that Glp is unlikely to bind in the same fashion as PEP. Further, a trifluoromethyl-substituted shikimate-based bisubstrate inhibitor (SBBI) complexed to [$^{15}N_2$]Arg-EPSP, which is closer mimic of S3P–PEP tetrahedral intermediate than S3P–Glp was further utilized to characterize the conformation of a bound trifluoromethylketal, shikimate-based bisubstrate based on a combination of

$^{15}N\{^{19}F\}$, $^{31}P\{^{15}N\}$, and $^{31}P\{^{19}F\}$ REDOR measurements.[36] There is general agreement between the REDOR model and the crystal structure with respect to the global folding of the two domains of EPSP synthase and the relative positioning of S3P and Glp in the binding pocket.[37] However, some of the REDOR data are in disagreement with predictions based on the crystallography. These discrepancies may be attributed to the use of lyophilized samples for REDOR experiments instead of crystalline preparation, and also the effect of multiple-spin systems present for data analysis, as discussed in Section 6.2.7. REDOR NMR was used to investigate the conformation of a ^{13}C-, ^{15}N-, ^{19}F-labeled inhibitor bound to human factor Xa, a 45-kDa enzyme belonging to the serine protease class.[38,39]

References

[1] S. K. Straus, T. Bremi, and R. R. Ernst, 1996, *Chem. Phys. Lett.*, 262, 709–715.

[2] A. E. Bennett, C. M. Riensra, M. Auger, K. V. Lakshmi, and R. G. Griffin, 1995, *J. Chem. Phys.*, 103, 6951–6958.

[3] A. Detken, E. H. Hardy, M. Ernst, and B. H. Meier, 2002, *Chem. Phys. Lett.*, 356, 298–304.

[4] M. Ernst, A. Detken, A. Böckmann, and B. H. Meier, 2003, *J. Am. Chem. Soc.*, 125, 15807–15810.

[5] A. Böckmann, A. Lange, A. Galinier, S. Luca, N. Giraud, M. Juy, H. Heise, R. Montserret, F. Penin, and M. Baldus, 2003, *J. Biomol. NMR*, 27, 323–339.

[6] R. Verel, M. Baldus, M. Ernst, and B. H. Meier, 1998, *Chem. Phys. Lett.*, 287, 421–428.

[7] R. Verel, M. Ernst, and B. H. Meier, 2001, *J. Magn. Reson.*, 150, 81–99.

[8] N. M. Szeverenyi, M. J. Sullivan, and G. E. Maciel, 1982, *J. Magn. Reson.*, 47, 462–475.

[9] M. Ernst and B. H. Meier, 1998, in *Solid State NMR of Polymers*, I. Ando and T. Asakura Eds., Elsevier, The Netherlands, pp. 83–121.

[10] J. Pauli, M. Baldus, B. van Rossum, H. de Groot, and H. Oschkinat, 2001, *Chembiochem*, 2, 272–281.

[11] F. Castellani, B. van Rossum, A. Diehl, M. Schubert, K. Rehbein, and H. Oschkinat, 2002, *Nature*, 420, 98–102.

[12] F. Castellani, B. J. van Rossum, A. Diehl, K. Rehbein, and H. Oschkinat, 2003, *Biochemistry*, 42, 11476–11483.

[13] A. McDermott, T. Polenova, A. Bockmann, K. W. Zilm, E. K. Paulson, R. W. Martin, and G. T. Montelione, 2000, *J. Biomol. NMR*, 16, 209–219.

[14] T. I. Igumenova, A. J. Wand, and A. E. McDermott, 2004, *J. Am. Chem. Soc.*, 126, 5323–5331.

[15] T. I. Igumenova, A. E. McDermott, K. W. Zilm, R. W. Martin, E. K. Paulson, and A. J. Wand, 2004, *J. Am. Chem. Soc.*, 126, 6720–6727.

[16] D. Marulanda, M. L. Tasayco, A. McDermott, M. Cataldi, V. Arriaran, and T. Polenova, 2004, *J. Am. Chem. Soc.*, 126, 16608–16620.

[17] M. Baldus and B. H. Meier, 1996, *J. Magn. Reson.* A121, 65–69.
[18] S. Yamaguchi, S. Tuzi, K. Yonebayashi, A. Naito, R. Needleman, J. K. Lanyi, and H. Saitô, 2001, *J. Biochem. (Tokyo)*, 129, 373–382.
[19] W. P. Rothwell and J. S. Waugh, 1981, *J. Chem. Phys.*, 75, 2721–2732.
[20] A. Naito, A. Fukutani, M. Uitdehaag, S. Tuzi, and H. Saitô, 1998, *J. Mol. Struct.*, 441, 231–241.
[21] P. Hodgkinson and L. Emsley, 1999, *J. Magn. Reson.*, 139, 46–59.
[22] S. Kiihne, M. A. Mehta, J. A. Stringer, D. M. Gregory, J. C. Shiels, and G. P. Drobny, 1998, *J. Phys. Chem.*, A102, 2274–2282.
[23] A. T. Brünger, P. D. Adams, G. M. Clore, W. L. Delano, P. Gros, R. W. Grosse-Kunstleve, J.-S. Jiang, J. Kuszewski, M. Nilges, N. S. Pannu, R. J. Read, L. M. Rice, T. Simonson, and G. L. Wallen, 1998, *Acta Crystallogr.*, D54, 905–921.
[24] F. Castellani, B.-J. van Rossum, A. Diehl, K. Rehbein, and H. Oschkinat, 2003, *Biochemistry*, 42, 11476–11483.
[25] G. Cornilescu, F. Delaglio, and A. Bax, 1999, *J. Biomol. NMR*, 13, 289–302.
[26] A. Abragam, *Principle of Nuclear Magnetism*, Clarendon Press, Oxford, 1961.
[27] K. Schmidt-Rohr and H. W. Spiess, *Multidimensional Solid-state NMR and Polymers*, Academic Press, New York, 1994.
[28] M. Ernst and B. H. Meier, 1998, Spin Diffusion in Solids, in *Solid State NMR of Polymers*, I. Ando and T. Asakura, Eds., Elsevier, The Netherlands, pp. 85–121.
[29] V. Ladizhansky, C. P. Jaroniec, A. Diehl, H. Oschkinat, and R. G. Griffin, 2003, *J. Am. Chem. Soc.*, 125, 6827–6833.
[30] S. G. Zech, E. Olejniczak, P. Hajduk, J. Mack, and A. McDermott, 2004, *J. Am. Chem. Soc.*, 126, 13948–13953.
[31] K. Takegoshi, S. Nakamura, and T. Terao, 2001, *Chem. Phys. Lett.*, 344, 631–637.
[32] K. Takegoshi, S. Nakamura, and T. Terao, 2003, *J. Chem. Phys.*, 118, 2325–2341.
[33] D. D. Mueller, A. Schmidt, K. L. Pappan, R. A. McKay, and J. Schaefer, 1995, *Biochemistry*, 34, 5597–5603.
[34] L. M. McDowell, A. Schmidt, E. R. Cohen, D. R. Studelska, and J. Schaefer, *J. Mol. Biol.*, 1996, 256, 160–171.
[35] L. M. McDowell, M. Lee, R. A. McKay, K. S. Anderson, and J. Schaefer, 1996, *Biochemistry*, 35, 3328–3334.
[36] L. M. McDowell, D. R. Studelska, B. Poliks, R. D. O'Connor, and J. Schaefer, 2004, *Biochemistry*, 43, 6606–6611.
[37] L. M. McDowell, B. Poliks, D. R. Studelska, R. D. O'Connor, D. D. Beusen, and J. Schaefer, 2004, *J. Biomol. NMR*, 28, 11–29.
[38] L. M. McDowell, M. A. McCarrick, D. R. Studelska, W. J. Guilford, D. Arnaiz, J. L. Dallas, D. R. Light, M. Whitlow, and J. Schaefer, 1999, *J. Med Chem.*, 42, 3910–3918.
[39] L. M. McDowell, M. A. McCarrick, D. R. Studelska, R. D. O'Connor, D. R. Light, W. J. Guilford, D. Arnaiz, M. Adler, J. L. Dallas, B. Poliks, and J. Schaefer, 2003, *J. Med. Chem.*, 46, 359–363.

Chapter 13

MEMBRANE PROTEINS I: DYNAMIC PICTURE

Membrane proteins, which are integral parts of biological membranes, have at least one segment of peptide chain traversing the lipid bilayer, and are known to play an essential role in a variety of biological functions, such as transport of appropriate molecules into or out of the cell, receiving and transducing chemical signals from the environment, catalysis of chemical reactions, proton or ion channel, etc.[1] Many integral membrane proteins in the membrane environment are known to assemble into oligomeric complexes rather than monomers to form tertiary and quaternary structures necessary for biological functions.[2,3] Such oligomeric complexes have been confirmed in view of the 3D pictures of membrane proteins revealed either from 2D or 3D crystals, including light-driven proton pump bR,[4–8] chloride pump halorhodopsin[9,10] and phototaxis receptor sensory rhodopsin II (phoborhodopsin),[11–13] photosynthetic reaction center,[14] light-harvesting complex,[15,16] cytochrome *c* oxidase,[17,18] potassium and mechanosensitive channels,[19,20] bovine rhodopsin,[21] calcium pump of sarcoplasmic reticulum,[22] etc.

Nevertheless, it should be taken into account that membrane proteins are far from rigid body as conceived, at least at *ambient temperature*, in spite of currently available 3D structural model, revealed by cryo-electron microscope or X-ray diffraction studies on ''so called'' crystalline preparations. Instead, they are rather flexible, undergoing various kinds of molecular motions with correlation times in the order of 10^{-2}–10^{-8} s, depending upon portions under consideration, especially at loops and N- or C-terminal residues.[23,24] This picture is readily available from site-directed solid state ^{13}C NMR studies on bR,[23,24] *Natronobacterium pharaonis* phoborhodopsin (*p*pR or sensory rhodopsin II),[25] its transducer (*p*HtrII),[26] and *E. coli* diacylglycerol kinase (DGK)[27] to be described below. This view is also readily recognized by the fact that their hydrophilic loops or N- or C-terminal residues are fully exposed to isotropic aqueous phase, although their hydrophobic transmembrane α-helices are embedded within the liquid crystalline environment of lipid bilayer. This dynamic picture turned out to be more pronounced especially when they are present as monomers in lipid bilayers instead of PM, as demonstrated for reconstituted bR in lipid bilayers.[28,29] The biological activity of bR reconstituted in lipid bilayers turned out to be native-like for proton pump in spite of the monomeric form.[30] Even at cryo-temperature, the 3D pictures available from diffraction studies are known to lack several residues located at surface areas, and may vary substantially among preparations exhibiting polymorphism, because of their flexibility and conformational variability.

Undoubtedly, such a detailed dynamic picture of membrane proteins especially at the flexible surface residues cannot be obtained by diffraction or other spectroscopic means and is available only from site-directed solid state NMR study on fully hydrated membrane proteins at ambient temperature, which is relevant to their biological functions. This picture, however, seems to be obviously unfavorable even if several solid state NMR experiments were carried out on membrane proteins to reveal their 3D structures based on either orientational or distance constraints. To minimize such unfavorable fluctuation motions with correlation times shorter than 10^{-8} s as least as possible, NMR measurements using mechanically oriented system or REDOR experiments are commonly carried out either at lower temperature or hydration condition up to 70% RH, respectively, or both. In fact, very similar spectral changes were noted as manifested from ^{13}C NMR spectra of [3-^{13}C]Ala-bR recorded at low temperature or lower hydration conditions.[24,31] Still, no sufficient structural information related to 3D structure is available from the loops or N- or C-terminal residues under such conditions.

13.1. Bacteriorhodopsin

bR is the sole membrane protein of seven α-helical transmembrane chains (Figure 13.1(b)) present in the PM of *Halobacterium salinurum*. This is active as a light-driven proton pump that translocates protons from the inside to outside of the cell, through photoisomerization of retinal from the all-*trans*, 15-*anti* to the 13-*cis*, 15-*anti* form covalently linked to Lys^{216} (helix G) of single-chain polypeptide of 248 amino acid residues (26 kDa) through a protonated Schiff base (see Figure 2.8).[32–35] In addition to its own right to gain insight into molecular mechanism of proton pump, bR has been also considered as a prototype of a variety of G-protein coupled receptors (GPCRs) in view of similarity in the transmembrane α-helical arrangement in the membrane, and as an exceptionally easy protein for handling, in

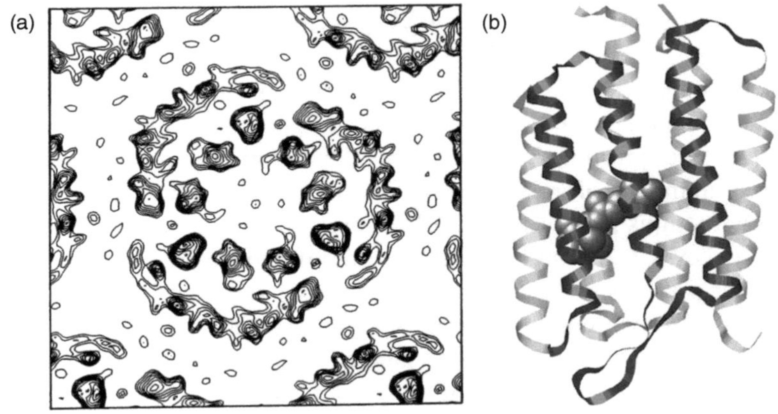

Figure 13.1. A top view of hexagonal packing of bacteriorhodopsin in 2D crystalline lattice (a) and its 3D structure revealed by X-ray diffraction study on 3D crystal (1.5 Å resolution (b)).[24]

relation to large-scale preparation, protein folding, and formation of 2D crystalline lattice, as far as it is expressed from *Halobacteria*.

This molecule in the membrane is known to form hexagonal arrays[36] (Figure 13.1(a)), leading to a 2D crystal lattice through its oligomerization to the trimeric form,[37] rather than monomeric form. Its 3D structure is now available at various degrees of resolution from a cryo-electron microscopy study on a 2D crystal and X-ray diffraction studies on 3D crystals[4–8,38–42] based on a novel crystallization concept using the lipidic cubic phase[43] or vesicle fusion.[44] It is notable that several residues from the surface are missing from the 3D structure when it is compared with a schematic representation of secondary structure based on such studies, as illustrated in Figure 2.8. The surface structure of the 3D structure, however, is substantially modified by crystallographic contacts and may not represent the true conformational state *in vivo*,[45,46] on the basis of a comparative study of these data as well as from atomic force microscopic (AFM) observations. Such a structure from 2D crystals is also modified by freezing and interaction with the cryoprotectant.[46] For this reason, site-directed ^{13}C solid state NMR approach on bR in 2D crystal is obviously very suitable to bridge the gaps between the pictures of rigid body available from diffraction studies at low temperature and those of flexible molecules by solid state NMR at ambient temperature.

13.1.1. Site-Directed ^{13}C NMR Approach

^{13}C NMR signals of densely ^{13}C-labeled proteins such as [1,2,3-$^{13}C_3$]Ala-labeled bR were substantially broadened due to the accelerated transverse relaxation rate promoted by inherent fluctuation motions coupled with increased numbers of relaxation pathways through ^{13}C–^{13}C dipolar and scalar interactions (see Figure 5.2).[47] Therefore, application of multidimensional ^{13}C NMR techniques which are effective for resolution enhancement in globular proteins as described in Section 5.2 and Chapter 12 will not be always an effective remedy to overcome the problem to resolve overlapped signals arising from uniformly ^{13}C-labeled proteins, as far as intact whole membrane proteins are concerned.

The site-directed ^{13}C NMR described here proves to be an alternative means suitable for revealing local conformation and dynamics of ^{13}C-labeled proteins with a single source of ^{13}C-labeled amino acid, because this approach is free from such an undesirable line broadenings by the accelerated transverse relaxation times. In fact, the ^{13}C NMR spectra of [3-^{13}C]Ala-bR recorded by DD-MAS and CP-MAS techniques exhibit well-resolved signals up to 12 peaks under *fully hydrated* condition, as already illustrated in Figure 2.7(a) and (b), respectively. Further, ^{13}C NMR signals are almost fully visible for [1-^{13}C]Val-labeled bR (see Figure 5.6 and Table 13.1), as far as their 2D crystalline preparations are concerned.[48,49] The relative peak intensities of these peaks are indeed consistent with the predicted amounts of amino acid residues under consideration.[49] These ^{13}C NMR signals were initially classified as the portions of the transmembrane α-helices, loops, C-terminal α-helix (as viewed from Ala residue only), and random coil with reference to the conformation-dependent displacements of ^{13}C NMR signals (Tables 6.1 and 6.2).[51–53] They were subsequently assigned to individual

Table 13.1. ^{13}C NMR peak intensities of ^{13}C-labeled bR recorded by CP-MAS NMR[a]

	2D crystal			Monomer		
	[3-^{13}C] Ala-labeled	[1-^{13}C] Ala-labeled	[1-^{13}C] Val-labeled	[3-^{13}C] Ala-labeled	[1-^{13}C] Ala-labeled	[1-^{13}C] Val-labeled
C-terminal α-helix	+[b]	+[b]	c	+[b]	+[b]	c
Loop	+	−	+	−	−	±
Transmembrane α-helix	+	±	+	±	−	±

[a] key: +; fully visible, ± ; partially suppressed, –; completely suppressed.
[b] These peaks are also visible by DD-MAS NMR.
[c] No Val residue is present in the C-terminal α-helix

residues by a site-directed manner with reference to the reduced peak intensities of the site-directed mutants as compared to those of fully visible wild-type (as illustrated in Figures 5.4 and 5.5), if an appropriate mutant is available.[48,49]

For instance, the three intense peaks in the DD-MAS NMR spectrum of [3-^{13}C]Ala-bR marked by gray (Figure 2.7 (a)), which are ascribed to Ala 228 and 233, Ala 240, 244–246, and Ala 235 from the high field to the low field located at the C-terminal residues at the surface, are preferentially suppressed in the CP-MAS NMR (Figure 2.7 (b)) because of their inefficient CP rate arising from their fast fluctuation motions with correlation times shorter than 10^{-8} s. In this way, over 60% of the ^{13}C NMR signals from [3-^{13}C]Ala- and [1-^{13}C]Val-labeled bR (Figure 5.6) have been assigned as already discussed in Section. 5.1.2, and utilized to characterize conformation and dynamics of bR at ambient temperature. It is noteworthy, however, that such excellent spectral resolution could be easily deteriorated to yield only three overlapped peaks, when RH of samples was reduced to lower than 90%.[50]

It is surprising to note, however, that ^{13}C NMR signals of fully hydrated ^{13}C-labeled bR are not always fully visible in spite of 2D crystalline preparations used for NMR measurements: several signals are suppressed, depending upon the positions where ^{13}C-labeled amino acid probes are located,[47,49] types of ^{13}C-labeled amino acids,[49] ground state or photo-intermediate,[54,55] ionic strength,[56–58] pH,[57,58] temperature,[57–59] site-directed mutants,[28,60]etc. that are related to respective local dynamics to be described below. In such cases, ^{13}C NMR peaks from the transmembrane α-helices and loops of [1-^{13}C]Ala-labeled bR are partially and completely suppressed, respectively, as summarized in Table 13.1. As to the type of ^{13}C-labeled amino acid residues mentioned above, ^{13}C NMR signals of surface area from 2D crystalline preparations of [1-^{13}C]Gly-, Leu-, Phe-, and Trp-labeled bR are almost completely suppressed as will be discussed in the next section,[49] although ^{13}C NMR signals from [1-^{13}C]Val and Ile turn out to be fully visible.[49]

13.1.2. Evaluation of ^{13}C NMR Signals from Residues Located at Surface Area

^{13}C NMR signals of residues located at the loop and transmembrane α-helices near the surface could be preferentially broadened by dipolar interaction between ^{13}C nuclei under consideration and Mn^{2+} ion bound to the negatively charged residues of bR and lipids at membrane surface.[49,61] Therefore, the observed peak intensities with (I) and without (I_0) added Mn^{2+} ion can be compared to estimate the relative contribution of ^{13}C NMR signals arising from the residues located within 8.7 Å from the negatively charged surface as $1-I/I_0$.[49] These data for [1-^{13}C]Val- and Ile-labeled bR are in good agreement with the amounts of the respective residues estimated from the expected number of these residues within 8.7 Å from the negatively charged residues at the membrane surfaces, as summarized in Table 13.2. The relative contributions of the ^{13}C NMR signals from the surface areas turned out to be surprisingly lower than the expected values: they are 0.14, 0, 0.11 for [1-^{13}C]Gly-, Ala-, and Leu-bR, respectively, while the expected data based on predicted number of these residues from the surface areas are 0.67, 0.62, and 0.49, respectively. These findings indicate that ^{13}C NMR signals from the surface areas of bR labeled with these amino acids are substantially suppressed, even in the absence of Mn^{2+} ion, owing to interference of incoherent low-frequency fluctuation motions with the coherent frequency of MAS, as demonstrated for [1-^{13}C]Ala-bR. This is true for the ^{13}C NMR peak intensities of [1-^{13}C]Phe- and Trp-bR: they are lower than the predicted relative amounts of these residues at the membrane surfaces.

The presence of such low-frequency, *residue-specific dynamics* leading to completely or partially suppressed peaks in the absence of Mn^{2+} ion is well related to the possibility of conformational fluctuations caused by the time-dependent deviation from the torsion angles corresponding to the minimum energy of a particular conformation. Naturally, it is conceivable that such conformational space allowed for fluctuation motion may be limited to a very narrow area for Val or Ile residues with bulky side chains at C_α, together with limited χ_1 rotation around the C_α–C_β bond as shown by C_α–$C_\beta H(X)(Y)$ where X and Y are substituents on C_β. In contrast, this minimum may be rather shallow for Gly residues in view of the widely allowed

Table 13.2. Comparison of relative proportion of the ^{13}C NMR signals of [1-^{13}C]amino acid-labeled bR from the surface areas as estimated from the ^{13}C NMR intensity ratio with (I) and without (I_0) Mn^{2+} ion[49]

	Estimated from ^{13}C signals ($1-I/I_0$)	Predicted amounts of residues near the surface (8.7 Å from the membrane surface)	Suppressed ^{13}C NMR peaks from the surface areas caused by slow motions
Gly	0.14	0.67	Suppressed
Ala	≈ 0	0.62	Almost completely suppressed
Leu	0.11	0.49	Suppressed
Phe	0.24	0.55	Suppressed
Trp	0.24	0.38	Suppressed
Val	0.41	0.38	None
Ile	0.50	0.56	None

conformational space. Therefore, it is plausible that the above-mentioned low-frequency, residue-specific backbone dynamics are present for Ala, Leu, Phe, and Trp residues, because backbone dynamics in these systems could be coupled with a possible rotational motion of the χ_1 angle around the C_α–C_β bond, as schematically represented by the C_α–$C_\beta H_2$–Z system, where Z is H, isopropyl, phenyl, or indole.

Mn^{2+}-induced suppression of peaks is expressed as a function of distances between the electron of Mn^{2+} ion and nuclear spin r and correlation time of rotational reorientation of the spin pair (τ_r), according to Solomon–Bloembergen equation:[62,63]

$$1/T_{2c} = [S(S+1)\gamma_c^2 g^2 \beta^2/(15r^6)][(4\tau_{c1} + 3\tau_{c1})/(1+\omega_c^2\tau_{c1}^2) + 13\tau_{c2}/(1+\omega_e^2\tau_{c2}^2)] \quad (13.1)$$

$$1/\tau_{c1} = 1/T_{1e} + 1/\tau_r + 1/\tau_m, \; 1/\tau_{c2} = 1/T_{2e} + 1/\tau_r + 1/\tau_m,$$

where T_{2c} is the spin–spin relaxation time of a ^{13}C nucleus, r is the distance between the ^{13}C nucleus and Mn^{2+} ion, S is the total electron spin, ω_c and ω_e are the nuclear and electronic Larmor precession frequencies, γ_c is the gyromagnetic ratio of ^{13}C, g is the g factor of Mn^{2+}, β is the Bohr magneton, T_{1e} and T_{2e} are the spin–lattice and spin–spin relaxation times of an electron, respectively, τ_r is the rotational correlation time of the Mn^{2+}–bR complex, and τ_m is the lifetime of the Mn^{2+} complex. The g factor and T_{1e} of Mn^{2+} are assumed to be identical to those of an aqua ion. τ_m is assumed to be longer than T_{1e} (3×10^{-9}s). When τ_r is longer than T_{1e}, the line width corresponding to the calculated T_{2c} value becomes greater than 100 Hz in the area within 8.7 Å from Mn^{2+} ion at negatively charged amino acid residues. It turned out that the prediction based on this equation is consistent with the recent experimental data utilizing 40 μM Mn^{2+} ion to select the three ^{13}C NMR peaks from [1-^{13}C]-labeled Pro residues located in the inner part of the transmembrane α-helices, Pro 50, 91, and 186.[64]

13.1.3. Dynamic Pictures: 2D Crystals

The present findings indicate that fully hydrated membrane proteins are flexible at ambient temperature, undergoing fluctuation motions with a variety of correlation times, depending upon the sites of interest and the manner of sample preparations either from 2D crystalline or monomeric forms. Indeed, ^{13}C NMR signals from the N- or C-terminal ends of [3-^{13}C]Ala-labeled bR cannot be recorded by CP-MAS NMR technique, because the dipolar interactions for these portions essential for the NMR observation are averaged out due to isotropic fluctuation motions with correlation time in the order of 10^{-8} s. In addition, specific portions undergoing local fluctuation motions in the order of 10^{-4}–10^{-5} s can be located either in the loops or the transmembrane α-helices near the membrane surface, even though specifically suppressed peaks are present due to the presence of fluctuation frequency interfering with frequency of either proton decoupling or MAS,[65,66] as summarized in Table 13.1 for a 2D crystal.

It is expected that such fluctuation motions could be slowered when spectra were recorded either at lower temperature or lower state of hydration. Unfortunately, spectral resolution was deteriorated when ^{13}C NMR spectra were recorded at such conditions.[50,53,59] Instead, the ^{13}C NMR signals from the C-terminal groups, including the C-terminal α-helix are completely suppressed at temperatures below −20°C owing to slowed fluctuation motion with a correlation time in the order of 10^{-5} s that interferes with the proton decoupling frequency (Figure 13.2). Furthermore, the rest of the peaks from the loops and transmembrane α-helices are substantially broadened to the extent to give featureless ^{13}C NMR peaks at temperatures below −20°C. This finding can be interpreted as indicating that the well-resolved ^{13}C NMR peaks of bR in a 2D crystal at ambient temperature are achieved as a result of fast chemical exchange with a rate constant of 10^2/s, among peptide chains taking slightly different conformations with chemical shift differences in the order of 10–10^2 Hz.[50] Therefore, broadened ^{13}C NMR spectra recorded at a temperature below −20°C (Figure 13.2) arose from superimposed peaks from their different local conformations, based on the conformation-dependent displacements of peaks.

It is noted that the ^{13}C NMR signal of Ala 184 from [3-^{13}C]Ala-bR is resonated at the peak position of anomalously lower field, 17.27 ppm, within the area of the loop region (see Figure 2.7), in spite of its local conformation as a transmembrane α-helix which should be resonated at a position higher than 16.88 ppm. Contributions of the two superimposed peaks, Ala 103 and 160, were successfully removed by accelerated spin–spin relaxation rate in the presence of surface-bound Mn^{2+} ions.[61] Thus, assignment of Ala 184 as the remaining peak was confirmed by the fact that this peak at 17.2 ppm is absent in A184G mutant. Further, this assignment was also confirmed by the finding that the Ala 184 peak at 17.2 ppm is displaced upfield when such a kinked structure was removed by replacement of Pro 186 with Ala. This remarkable upfield displacement of the peak is ascribed to changes of the local torsion angles of peptide unit in Ala 184, located at the kink of the helix F induced by Pro 186, because the replacement of Pro 186 by Ala is expected to restore the kink to the normal α-helix and bring back the torsion angles of Ala 184 to those of the typical α-helix. Such anomaly in the peak position caused by the conformational alteration

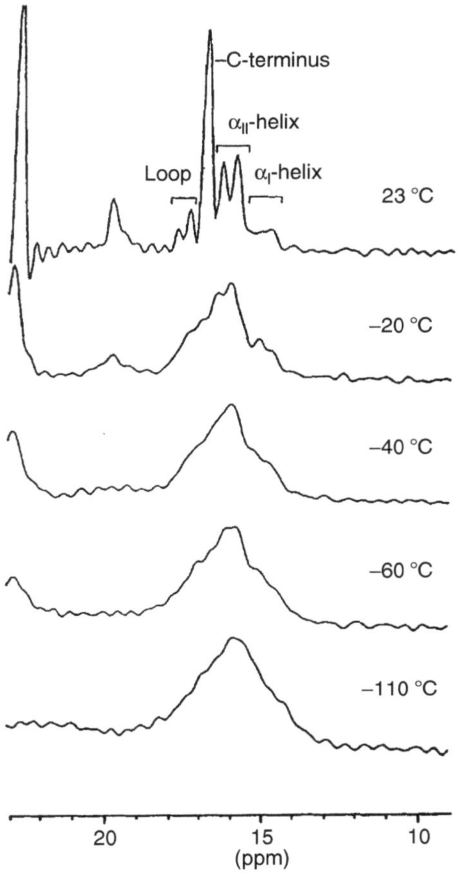

Figure 13.2. Temperature-dependent change in the ^{13}C DD-MAS NMR spectra of [3-^{13}C]Ala-labeled bR.[24]

of the transmembrane helices due to the presence of a kinked structure was also confirmed by examination of [3-^{13}C]Ala^{184}-, [1-^{13}C]Val^{187}-labeled wild-type and P186 mutant of transmembrane fragment F (164–194) of bR incorporated into lipid bilayer.[67]

13.1.4. Dynamic Pictures: Distorted 2D Crystals or Monomer

Backbone dynamics could be substantially modified when 2D lattice assembly is distorted or disrupted as in bO prepared from either

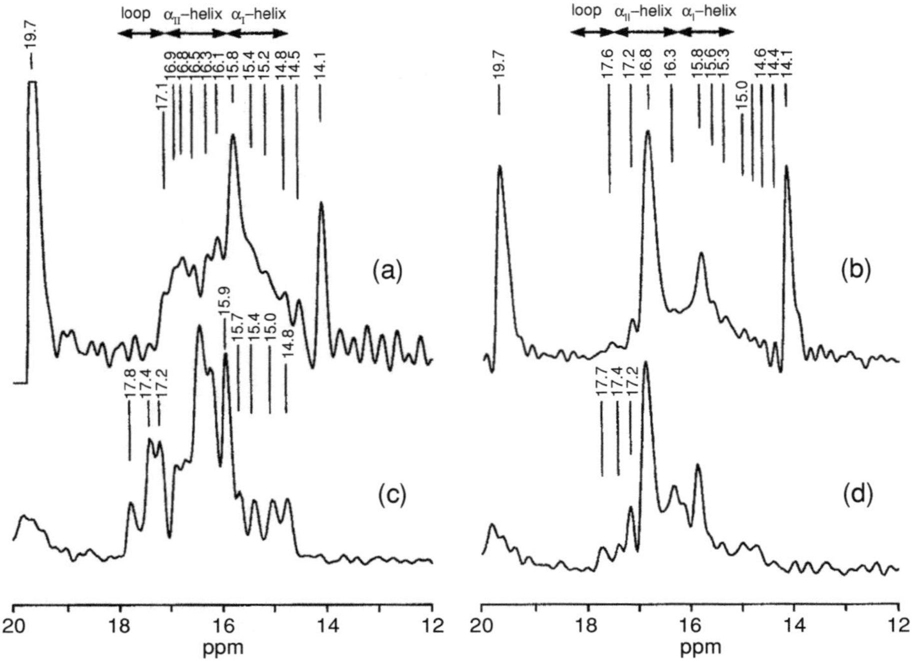

Figure 13.3. ^{13}C CP-MAS (left) and DD-MAS NMR spectra of [3-^{13}C]Ala-labeled bR reconstituted in egg PC bilayer (a and b; 1:50 mol ratio) and from PM (c and d). The intense peaks at 19.7 and 14.1 ppm are ascribed to lipid methyl groups from *Halobacteria* and egg PC, respectively.[29]

hydroxylamine-treated bR or retinal-deficient E1001 strain in which helix–helix interactions are substantially modified due to lack of retinal,[59] and W80L and W12L mutants in which the side chain of one of two Trp residues oriented outward from the transmembrane α-helices at the interface for lipid–protein interactions is absent in these mutants.[28] In addition, ^{13}C NMR signals of loops and some transmembrane α-helices near the surface turned out to be completely suppressed for reconstituted [3-^{13}C]Ala-bR in lipid bilayer (Figure 13.3)[29] as a result of taking *monomeric* form rather than naturally occurring 2D crystal (Table 13.1 and Figure 13.4).[28,29,59,67a] Such suppressed ^{13}C NMR peaks from [3-^{13}C]Ala-labeled bR is obviously caused by the presence of low-frequency fluctuation motions (10^5 Hz) interfering with the frequency of proton decoupling in distorted trimeric form or monomer. Further, one should also anticipate that ^{13}C NMR signals are almost completely suppressed for fully hydrated *monomeric* [1-^{13}C]Gly-, Ala-, Leu-, Phe-, and Trp-labeled bR in lipid bilayers,[49] because even the transmembrane α-helices acquire accelerated fluctuation motions with the correlation times

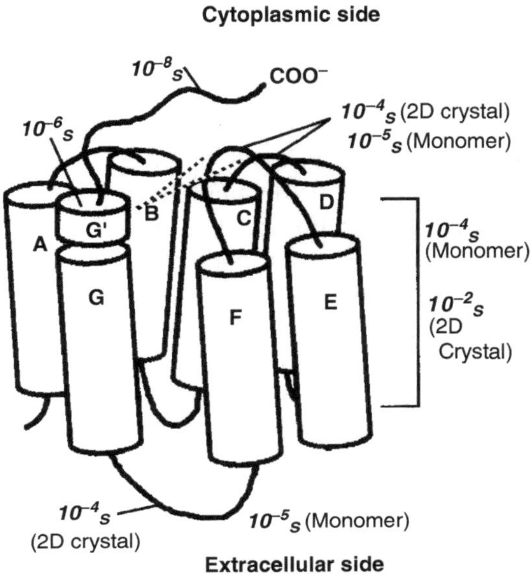

Figure 13.4. Schematic representation of the location of the C-terminal α-helix (helix G′ protruding from the membrane surface) and interaction with the C–D and E–F loops (dotted lines) (cytoplasmic surface complex) and the manner of fluctuation with different correlation times. Note that correlation times for the fluctuation motions of the transmembrane α-helices differ substantially between preparations whether this molecule is present in 2D crystal or monomer.[47, 67a]

being in the order of 10^{-4} s, in the absence of such specific protein–protein interaction essential for the formation of the trimeric structure leading to the 2D crystal and regulation of such backbone dynamics.[65]

The first proton transfer of bR occurs from the protonated Schiff base (SB) to the anionic Asp 85 in the L-to-M reaction.[32–36] Protonation of Asp 85, then, induces the release of a proton from the proton release groups involving Glu 194 and Glu 204. An M-like state was demonstrated to arise for D85N mutant of bR without illumination,[68] because the lowered pK_a of SB allows its deprotonation at alkaline condition (pH $\approx$ 10) at ambient temperature.[69] Kawase *et al.*[54] showed that the ^{13}C NMR peak intensities of three to four [3-^{13}C]Ala-labeled residues from the transmembrane α-helices, including Ala 39, 51, and 53 (helix B), and 215 (helix G), were suppressed in D85N and D85N/D96N in both CP-MAS and DD-MAS spectra, irrespective of the pH. This is due to acquisition of intermediate time range motions at such portions, with correlation time in the order of 10^{-4}–10^{-5} s. Greater changes were achieved, however, at pH 10, which indicate large-amplitude motions of transmembrane helices upon deprotonation of SB and the formation of the M-like state. The spectra detected more rapid motions in the extracellular and/or cytoplasmic loops, with correlation times decreasing from 10^{-4} to 10^{-5} s. This spectral change is very similar to that found for bO, consistent with the previous observation by IR measurement,[70] because constraints from the protonated SB are relaxed in D85N at higher pH and removed permanently in bO. In particular, D85N acquired local fluctuation motion with a frequency of 10^4 Hz in the transmembrane B α-helix, concomitant with deprotonation of SB in the M-like state at pH 10, as manifested from the suppressed ^{13}C NMR signal of [1-^{13}C]-labeled Val 49 residue.[55] Local dynamics at Pro 50 with Val 49 as the neighbor turned out to be unchanged, irrespective of the charged state of SB as viewed from the ^{13}C NMR of [1-^{13}C]-labeled Pro 50. This means that the transmembrane B α-helix is able to acquire the fluctuation motion with a frequency of 10^4 Hz beyond the kink at Pro 50 in the cytoplasmic side, as schematically illustrated in Figure 13.5. Concomitantly, fluctuation motion at the C helix with frequency in the order of 10^4 Hz was found to be prominent, due to deprotonation of SB at pH 10, as viewed from the suppressed ^{13}C NMR signal of Pro 91. Accordingly, a novel mechanism for proton uptake and transport was proposed[55] on the basis of a dynamic aspect that a transient environmental change from a hydrophobic to hydrophilic nature of Asp 96 and SB is responsible for the reduced pK_a which makes proton uptake efficient as a result of fluctuation motion at the cytoplasmic side of the transmembrane B and C α-helices.

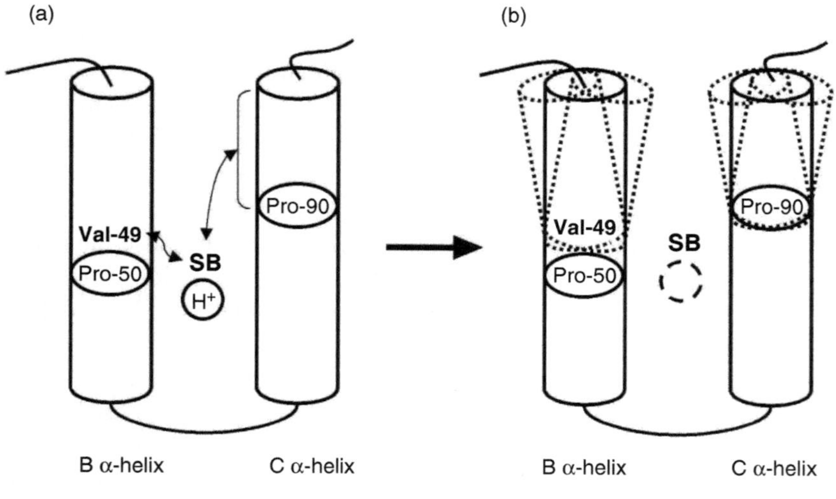

Figure 13.5. Schematic representation of the dynamic behavior of the B and C α-helices of D85N mutant accompained by protonation of the Schiff base, as viewed from the^{13} CNMR spectral behavior of Val 49 and Pro 91, described in the text. (a); ground state (pH 7); (b); M-like state at pH 10.[55]

13.1.5. Surface Structure

Surface structures are undoubtedly very important for a variety of membrane proteins in view of their role in biological responses initiated either at the cytoplasmic or extracellular surfaces. For instance, proton uptake by bR is initiated at the cytoplasmic surface, and protons are released at the extracellular surface driven by photocycle of retinal. Such surface structures of bR revealed by diffraction studies, however, are still obscured or inconsistent among a variety of the 3D structures so far available from cryo-electron microscope and X-ray diffraction studies.[4–8,37–42] This is because it can be easily altered by a variety of intrinsic or environmental factors such as temperature, pH, ionic strength, site-directed mutagenesis, crystallographic contact, etc.[23,24,45–49] In this connection, it is emphasized that the current site- directed ^{13}C NMR approach is an unrivaled means to clarify these problems to be described below. As an alternative means, fluorescence,[71,72] spin-labeling,[73] and heavy-atom labeling,[74,75] and AFM techniques[45,46] have been utilized to probe such structures at ambient temperature, although these techniques are not always free from plausible perturbations due to steric hindrance by the introduced probes, except in the case of AFM.

^{13}C NMR studies on [3-^{13}C]Ala-bR showed that the C-terminal residues, 226–235, participate in the formation of the C-terminal α-helix protruding from the membrane surface (helix G′ in Figure 13.4), as revealed by the peak position of 15.91 ppm at ambient temperature with reference to the conformation-dependent ^{13}C chemical shifts of Ala C_β peak,[57,59,76] as described in Section 6.4. The existence of the α-helix form in this region was also confirmed in view of the corresponding conformation-dependent displacements of the C_α and C=O peaks from [2-^{13}C]- and [1-^{13}C]Ala-labeled bR, respectively.[47] This α-helix was either only partially visible or completely invisible by X-ray diffraction studies, depending upon the type of crystalline preparations used[4–7,44]due to the presence of motions in this region with correlation times in the order of 10^{-6} s at ambient temperature (Figure 13.4), as judged from the carbon spin–lattice relaxation times, T_1^C, and spin–spin relaxation times, T_2^C, under CP-MAS condition.[47] The existence of this sort of the cytoplasmic α-helices has been more clearly identified for *p*pR[11–13,25]and bovine rhodopsin[21] by X-ray diffraction and ^{13}C NMR spectra.

It is interesting to note that the above-mentioned two types of surface structures, the C-terminal α-helix and cytoplasmic loops, are not present independently but are held together to form *cytoplasmic surface complex* as schematically illustrated in Figure 13.4 which is stabilized by salt bridges and/or cation-mediated linkages of a variety of side chains, although they are still undergoing fluctuation motions with frequencies in the order of 10^4–10^6 Hz. This view was substantiated by significant ^{13}C NMR spectral changes of [3-^{13}C]Ala-labeled bR and its mutants, as a function of ionic strength, temperature, pH, truncation of the C-terminal α-helix, and site-directed mutation at cytoplasmic loops.[57] For instance, increased ionic strength from 10 to 100 mM NaCl causes simultaneous changes of the downfield displacement of Ala 103 signal of the C–D loop and the reduced peak intensity of the C-terminal α-helix. This finding together with other spectral changes caused by temperature and pH variation leads to a possibility of the cytoplasmic surface complex as illustrated in Figure 13.4 (dotted). This view is consistent with the following experimental finding on blue membranes in which surface-bound cations of bR are completely removed by treatment with cation exchange resin (deionized blue) or neutralization of surface charge at pH 1.2 (acid blue)[56]: ^{13}C NMR signals from such loops are completely suppressed by accelerated fluctuation motions with correlation time in the order of 10^{-5} s and the ^{13}C chemical shift of the C-terminal α-helix was displaced upfield. In addition, partial neutralization of Glu and Asp residues located at the extracellular residues such as E194Q/E204Q (2 Glu), E9Q/E194Q/E204Q (3 Glu), and E9Q/E74Q/E194Q/E204Q (4 Glu) caused disorganized trimeric form[60] to result in global fluctuation motions at these loop regions.

Several pieces of evidence have been presented as to how specific surface electrical charges at the cytoplasmic surfaces are involved in efficient proton uptake during the photocycle in relation to the biological significance of the above-mentioned surface complex. In fact, the surface-exposed amino acids Asp 35 (A–B loop), Asp 102 and Asp 104 (C–D loop), and Glu 161 (E–F loop) seem to efficiently collect protons from the aqueous bulk phase and funnel them to the entrance of the cytoplasmic proton pathway.[77] In addition, a dominant "proton binding cluster" was shown which consists of Asp 104, Glu 160, and Glu 234 (C-terminal α-helix), together with Asp 36 as a mediator to deliver the proton to a channel.[78,79]

13.1.6. Long Distance Interaction Among Residues for Information Transfer

Proton transfer in bR is activated by photoisomerization of all *trans* retinal to the 13-*cis* form, followed by proton transfer from the retinal SB to Asp 85, release of a proton from residues or water molecule(s) at the extracellular surface, and uptake from cytoplasmic surface through reprotonation of the SB by Asp 96, resulting in proton transfer from the cytoplasmic to extracellular side.[80] In fact, it appears that this process proceeds with induced conformation and/or dynamics changes at both the extracellular and cytoplasmic side owing to protonation of Asp 85.[81,82] This means that the information of the protonation at Asp 85 should be transmitted to both extracellular and cytoplasmic regions through specific interactions. Tanio *et al.*[83,84] recorded ^{13}C NMR spectra of a variety of site-directed mutants in

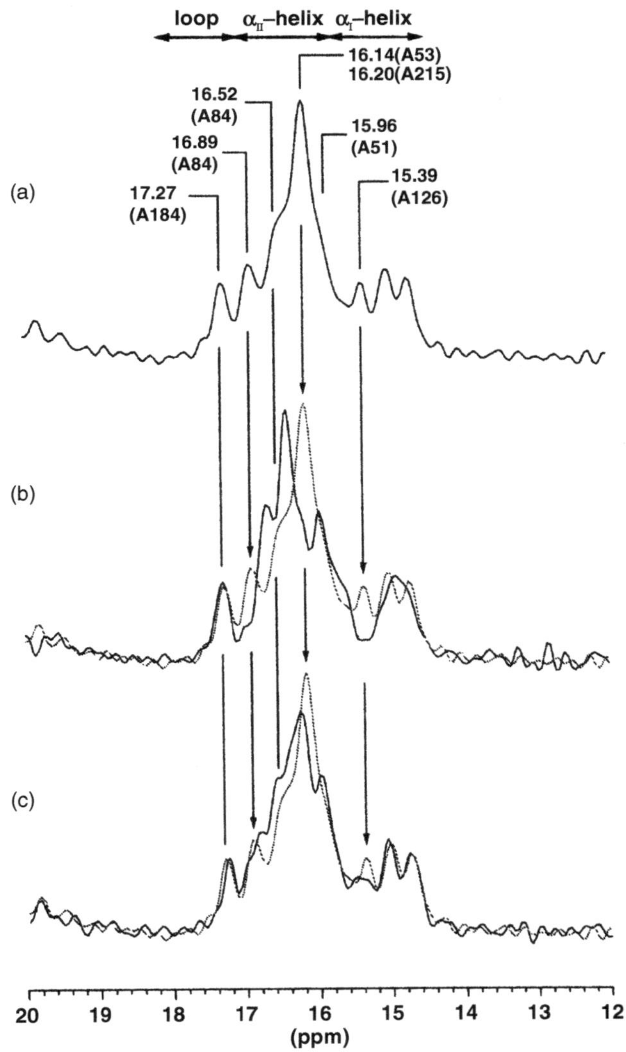

Figure 13.6. ^{13}C CP-MAS NMR spectra of [3-^{13}C]Ala-labeled wild-type (a), E204Q (b), and E204D (c) mutants. All spectra were recorded in the presence of 40 μM Mn^{2+} ion. Dotted spectra in the traces (b) and (c) are from 40 μM Mn^{2+}-treated wild-type.[85]

order to clarify how such interactions, if any, could be modified by changes of electric charge or polarity in mutants. This is based on the expectation that such interaction should also exist in bR even in the unphotolyzed state, among backbone, side chains, bound water molecules, etc. They found that there is indeed a long-distance interaction between Asp 96 and extracellular surface through Thr 46, Val 49, Asp 85, Arg 82, Glu 204, and Glu 194 in the unphotolyzed state. For instance, conformational changes were induced at the extracellular region through a reorientation of Arg 82 when Asp 85 was uncharged, as manifested from the recovery of the missing Ala 126 signal of D85N in the double mutant, D85N/R82Q.[84] The underlying spectral change might be interpreted in terms of the presence of perturbed Ala 126 mediated by Tyr 83, which is located between Arg 82 and Ala 126.[41] It is possible that disruption of the above-mentioned interactions after protonation of Asp 85 in the photocycle could cause the same kind of conformational change as detected from the [3-^{13}C]Ala-labeled peaks of Ala 196 and Ala 126,[84] and also the [1-^{13}C]Val-labeled peaks of Val 49 and 199.[83] Further, it was proposed that charged state of surface residues, especially at the side chain of extracellular Glu residues, Glu 194 and 204, could be transmitted to the inner part of the helices such as Ala 53, 84, and 215 to alter the local conformations of transmembrane α-helices near SB through side-chain interactions, as viewed from the respective displacement of [3-^{13}C]Ala-labeled peaks,[85] as shown in Figure 13.6. It was also analyzed how information of the protonation at Asp 85 from helix C is initially transmitted to helices B (Val 49) and G (Val 213) through modified helix–helix interaction through the side chains of Arg 82.[84,85]

This kind of long-distance information transfer could also be relevant in general to signal transduction systems such as G-protein coupled or other receptor molecules.

13.2. Phoborhodopsin and Its Cognate Transducer

Halobacteria contain a family of four retinal proteins, bR, halorhodopsin (hR), sensory rhodopsin I (sR I), and phoborhodopsin II (pR or sensory rhodopsin II, sR II), which carry out two distinct functions through a common photochemical reaction. In particular, bR and hR are light-driven ion pumps transporting proton and chloride, respectively,[86] while the two sensory rhodopsins are photoreceptors active for positive and negative phototaxis, respectively.[87] Here, *p*pR as a pigment protein

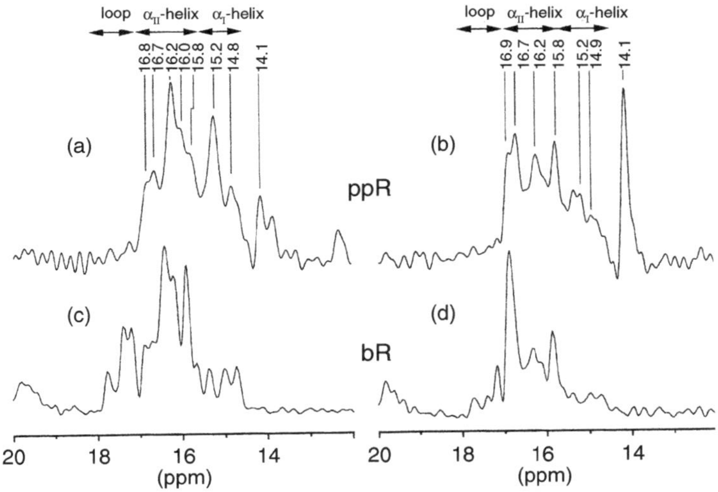

Figure 13.7. ^{13}C CP-MAS (left) and DD-MAS (right) NMR spectra of [3-^{13}C]Ala-labeled *p*PR (a), (b) in egg PC bilayer as compared with those of [3-^{13}C]Ala-labeled bR (c), (d) from PM.[25]

from *Natronobacterium pharonis* corresponds to pR of *H. salinurum* with 50% homology of amino acid sequence.[88] These two photoreceptors with proton transport activity are activated to yield signaling for phototaxis, through their plausible conformational changes, after receiving incoming light by tightly complexing to their respective cognate transducers consisting of two transmembrane helices (TM1 and TM2), HtrI and HtrII, respectively.[89]

It is anticipated that *p*pR reconstituted in egg PC bilayer at ambient temperature is present as a monomer in view of the comparative site-directed ^{13}C NMR data with those of reconstituted bR and its mutants in lipid bilayers[28,29] as described in Section 13.1.4, in spite of the revealed oligomeric structures by cryo-electron microscopic or X-ray diffraction studies on 2D or 3D crystals at low temperature.[11–13] In particular, Arakawa *et al.*[25] showed that ^{13}C NMR signals of the loop regions were absent for [3-^{13}C]Ala-labeled *p*pR (Figure 13.7), in a similar manner as observed for reconstituted bR in lipid bilayers (Figure 13.3). Instead, seven ^{13}C NMR signals were resolved for the transmembrane α-helices, consisting of the normal α_I-helices and α_{II}-helices with low-frequency fluctuation,[90] besides the peak at 14.1 ppm arising from egg PC, with reference to the peak positions of bR. In contrast to bR, the cytoplasmic α-helix of *p*pR is too flexible for uncomplexed state with fluctuation frequencies $> 10^8$ Hz to be observed by the CP-MAS NMR. It is interesting to note that distinct dynamics change for *p*pR is accompanied by complex formation with the truncated cognate transducer *p*HtrII (1–159), as revealed by the increased peak intensity at 15.9 ppm ascribable to the C-terminal α-helix by ^{13}C CP-MAS NMR, together with the improved spectral resolution for signals from whole area. This observation is consistent with a view that the accompanied change in the ''effective molecular mass'' from the system of 7 transmembrane α-helices (*p*pR alone) to 18 transmembrane α-helices of the complex $[2 \times (7 + 2)]$ transmembrane α-helices as 2:2 complex results in a change of fluctuation frequency from 10^4–10^5 to 10^4 Hz.[25,91] In particular, the increased intensity in the CP-MAS NMR spectrum at 15.9 ppm is caused by efficient magnetization owing to more immobilized cytoplasmic α-helix due to the complex formation. Therefore, it appears that the mutual interaction among the extended TM1 and TM2 helices of *p*HtrII beyond the surface and the cytoplasmic α-helix of *p*pR play an important role for stabilization of the complex. ^{13}C NMR signals of [1-^{13}C]Val-labeled *p*pR, on the other hand, were visible from the loop region, regardless of the free and complexed states. The peak intensities of the transmembrane α-helices of [1-^{13}C]Val-labeled *p*pR were appreciably suppressed in the CP-MAS NMR spectrum as compared with those of free *p*pR, leaving no spectral change in the loop region.

The cognate transducer *p*HtrII (1–159) consists of the two transmembrane α-helices with the C-terminal residue protruding from the cytoplasmic membrane surface, which presumably assumes the coiled-coil form responsible for phototaxis.[92] As demonstrated in Figure 13.8, the intense ^{13}C DD-MAS NMR spectrum (dotted trace) of [3-^{13}C]Ala-labeled cognate transducer complexed with *p*pR in egg PC bilayer was superimposed upon the corresponding ^{13}C CP-MAS NMR spectrum (solid trace).[26,92] The intense, low-field α_{II}-helical ^{13}C DD-MAS NMR peaks of *p*HtrII (1–159) resonated at 16.6 and 16.3 ppm are ascribed to [3-^{13}C]Ala residues located at the coiled-coil portion protruding from the membrane surface, in view of the conformation-dependent displacement of peaks,[23,24,51–53] together with their unique backbone dynamics by which they are suppressed in the CP-MAS NMR. It is also interesting to note that the low-field α-helical peaks of *p*HtrII (1–159) are not always fully visible in the absence of *p*pR, because the fluctuation frequency is close to the proton decoupling (10^5 Hz) to result in suppressed peaks.[26,92] The high-field envelope peaks at 15.5 ppm are ascribed to Ala residues in the transmembrane α-helices in view of the conformation-dependent displacement of peaks.[23,24,51–53] In contrast, the ^{13}C NMR signals of [1-^{13}C]Val-labeled *p*HtrII (1–159) were visible in the absence of *p*pR, but almost suppressed in the presence of *p*pR.[26] This means that such spectral change due to complex formation was interpreted in terms

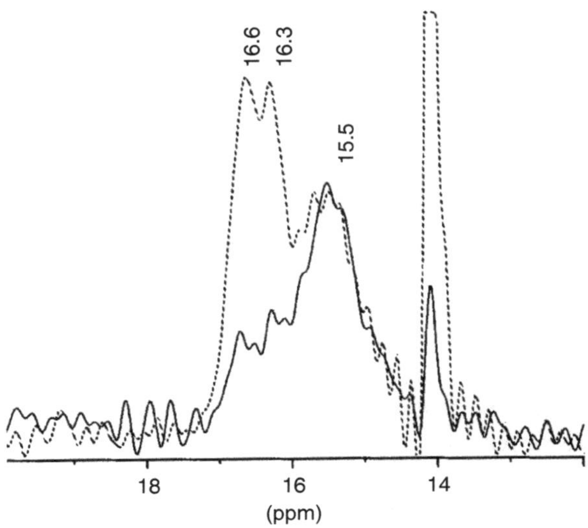

Figure 13.8. ^{13}C DD-MAS (dotted trace) and CP-MAS (solid trace) spectra of [3-^{13}C]Ala-labeled transducer *p*HtrII (1–159) in the presence of *pp*R in egg PC bilayer.[26]

of induced dynamics change caused by the presence of fluctuation motion with frequency in the order of 10^4 Hz which could interfere with frequency of MAS.[26] In such case, the accompanied spectral change should be more significant when the effective molecular mass was changed from the system of two-transmembrane α-helices of the transducer alone to the above-mentioned 18 transmembrane α-helices of *p*HtrII (1–159)–*pp*R complex. The observed dynamics change in the transmembrane α-helices, however, turned out to be from 10^3 Hz of the uncomplexed to 10^4 Hz in the complexed states.[26] It is also noted that the fluctuation frequency of the C-terminal α-helix is increased in spite of the involvement of this moiety in the complex formation with *pp*R. This kind of obvious contradiction can be compromised by a view that the uncomplexed state of *p*HtrII (1–159) is not necessarily present as a monomer but present as aggregated state. Therefore, it is interesting to note that the increased drastic peak intensity at the C-terminal α-helix in the complex is caused by the direct involvement of this portion to the complex, in spite of the presence of this portion outside the membrane surface. For this reason, the formation of the protein complex could be more clearly distinguished by means of the dynamics changes as reflected in the ^{13}C NMR line widths and peak intensities.

13.3. Diacylglycerol Kinase

Diacylglycerol kinase (DGK) from *E. coli* is a small, 121 amino acid, integral membrane protein that catalyzes the conversion of diacylglycerol and MgATP to phosphatic acid and MgADP. It has been shown that DGK is homotrimeric both in micellar system and 1-palmitoyl-2-oleyl-sn-glycero-3-phosphocholine (POPC) vesicles.[93,94] The topology of DGK as revealed by sequence data and β-lactamase and β-galactosidase fusion experiments consists of three transmembrane domains and two cytoplasmic domains.[95,96]

^{13}C NMR spectra of [3-^{13}C]Ala-, [1-^{13}C]Val-labeled *E. coli* DGK in DM or reconstituted in POPC and DPPC bilayers were recorded using CP-MAS and DD-MAS methods.[27] Surprisingly, the ^{13}C NMR spectra of [1-^{13}C]Ala-labeled DGK recorded by both the methods were broadened to yield rather featureless peaks at physiological temperatures, both in DM solution or lipid bilayers at liquid crystalline phase. The ^{13}C NMR spectra of [1-^{13}C]Ala-labeled DGK recorded by the DD-MAS ((a)–(c)) and CP-MAS ((e)–(f)) methods in POPC bilayers are substantially broadened as far as the samples are retained in lipids of liquid crystalline phase at temperatures between 20° C and −5° C, as shown in Figure 13.9. The intense peaks at 16.7–16.8 ppm are more pronounced in the DD-MAS spectra and ascribed to Ala residues involved in the α-helix form of the N-terminal region protruding from the membrane surface. This is because they can be assigned to Ala residues located at the more flexible N-terminal α-helical regions anchored at the membrane surface rather than the transmembrane α-helices aligned with the membrane normal. The spectral resolution of the CP-MAS NMR spectrum, however, was substantially improved to yield four peaks 14.5, 15.5, 16.0, and 16.7 ppm at −10 °C just below the phase-transition temperature of POPC bilayer (−5 °C). This situation is also realized in DPPC bilayer at which gel-to-liquid crystalline phase transition occurs at 41 °C: the well-resolved spectra consisting of four peaks were observed at temperature below 35 °C but the spectrum was broadened at 45 °C of liquid crystalline phase. It is also notable that ^{13}C NMR spectra of [1-^{13}C]Val-labeled DGK were completely suppressed at temperatures corresponding to the liquid crystalline phase of both POPC and DPPC bilayers. This means that the broadened or completely suppressed peaks of the ^{13}C NMR spectra of [3-^{13}C]Ala- and [1-^{13}C]Val-labeled DGK in the lipid bilayers of the liquid crystalline phase are ascribed to interference of motional frequencies of DGK with frequencies of proton decoupling or MAS (10^5 or 10^4 Hz, respectively). While DGK can be tightly packed in gel-phase lipids, DGK is less tightly packed at physiological temperatures, where it becomes more mobile. The fact that the enzyme activity is low under conditions where motions are restricted

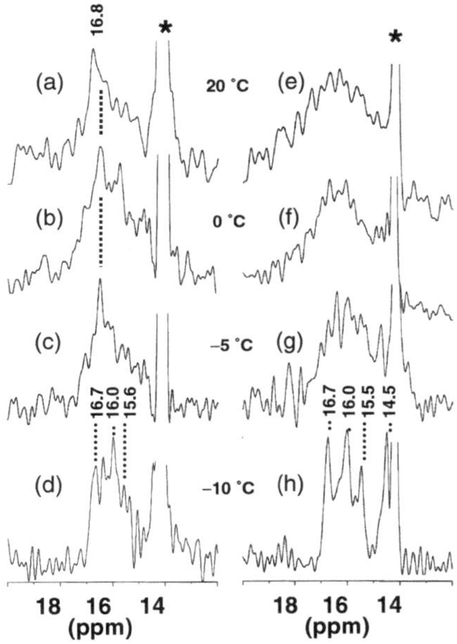

Figure 13.9. ^{13}C DD-MAS (left) and CP-MAS (right) NMR spectra of [3-^{13}C]Ala-labeled DGK in POPC bilayer at various temperatures. The peaks labeled with asterisks are ascribed to the methyl peak from POPC.[27]

and high when conformational fluctuations can occur suggests that acquisition of low-frequency backbone motions, on the microsecond to millisecond timescale, may facilitate the efficient enzymatic activity of DGK. Clearly, the present observations demonstrate that site-directed ^{13}C NMR approach is very useful for probing the conformation and dynamics of membrane proteins in a membrane environment.

More detailed account for the site-directed ^{13}C NMR on membrane proteins are described elsewhere.[67a]

References

[1] C. Branden and J. Tooze, 1999, *Introduction to Protein Structure*, Second Edition, Garland Publishing, New York.

[2] B. J. Bormann and D. M. Engelman, 1992, *Annu. Rev. Biophys. Biomol. Struct.*, 21, 23–242.

[3] M. H. B. Stowell and D. C. Rees, 1995, *Adv. Protein Chem.*, 46, 279–311.

[4] N. Grigorieff, T. A. Ceska, K. H. Dowing, J. M. Baldwin, and R. Henderson, 1996, *J. Mol. Biol.*, 259, 393–421.
[5] E. Pebay-Peyroula, G. Rummel, J. P. Rosenbusch, and E. M. Laudau, 1997, *Science*, 277, 1676–1681.
[6] H. Luecke, H. T. Richter, and J. K. Lanyi, 1998, *Science*, 280, 1934–1937.
[7] L. Essen, R. Siegert, W. D. Lehmann, and D. Oesterhelt, 1998, *Proc. Natl. Acad. Sci. USA*, 95, 11673–11678.
[8] H. Sato, K. Takeda, K. Tani, T. Hino, T. Okada, M. Nakasako, N. Kamiya, and T. Kouyama, 1999, *Acta Crystallogr. D*, 55, 1251–1256.
[9] W. A. Havelka, R. Henderson, J. A. Heymann, and D. Oesterhelt, 1993, *J. Mol. Biol.*, 234, 837–846.
[10] M. Kolbe, H. Besir, L.-O. Essen, and D. Oesterhelt, 2000, *Science*, 288, 1390–1396.
[11] E. R. S. Kunji, E. N. Spudich, R. Grisshammer, R. Henderson, and J. L. Spudich, 2001, *J. Mol. Biol.*, 308, 279–293.
[12] H. Luecke, B. Schobert, J. K. Lanyi, E. N. Spudich, and J. L. Spudich, 293, *Science*, 1499–1503.
[13] A. Royant, P. Nollert, K. Edman, R. Neutze, E. M. Landau, E. Pebay-Peyroula, and J. Navaro, 2001, *Proc. Natl. Acad. Sci. USA*, 98, 10131–10136.
[14] J. Deisenhofer, O. Epp, K. Miki, R. Huber, and H. Michel, 1985, *Nature*, 318, 618–624.
[15] W. K. Kuhlbrandt, D. N. Wang, and Y. Fujiyoshi, 1994, *Nature*, 367, 614 – 621.
[16] G. McDermott, S. M. Prince, A. A. Freer, A. M. Hathornthwaite-Lawless, M. Z. Papiz, R. J. Cogdell, and N. W. Isaacs, 1995, *Nature*, 374, 517–521.
[17] S. Iwata, C. Ostermeier, B. Ludwig, and H. Michel, 1995, *Nature*, 376, 660 – 669.
[18] T. Tsukihara, H. Aoyama, E. Yamashita, T. Tomizaki, H. Yamaguchi, K. Shinzawa-Itoh, R. Nakashima, R. Yaono, and S. Yoshikawa, 1996, *Science*, 272, 1136–1144.
[19] D. A. Doyle, J. M. Cabral, R. A. Pfuetzner, A. Kuo, J. M. Gulbiss, S. L. Cohen, B. T. Chait, and R. MacKinnon, 1998, *Science*, 280, 69–77.
[20] G. Chang, R. H. Spencer, A. T. Lee, M. T. Barclay, and D. C. Rees, 1998, *Science*, 282, 2220–2226.
[21] K. Palczewski, T. Kumasaka, T. Hori, C. A. Behnke, H. Motoshima, B. A. Fox, I. Le Trong, D. C. Teller, T. Okada, R. E. Stenkamp, M. Yamamoto, and M. Miyano, 2000, *Science*, 289, 739–745.
[22] C. Toyoshima, M. Nakasako, H. Nomura, and H. Ogawa, 2000, *Nature*, 405, 647– 655.
[23] H. Saitô, S. Tuzi, S. Yamaguchi, M. Tanio, and A. Naito, 2000, *Biochim. Biophys. Acta*, 1460, 39– 48.
[24] H. Saitô, S. Tuzi, M. Tanio, and A. Naito, 2002, *Annu. Rep. NMR Spectrosc.*, 47, 39–108.
[25] T. Arakawa, K. Shimono, S. Yamaguchi, S. Tuzi, Y. Sudo, N. Kamo, and H. Saitô, 2003, *FEBS Lett.*, 536, 237–240.
[26] S. Yamaguchi, K. Shimono, Y. Sudo, S. Tuzi, A. Naito, N. Kamo, and H. Saitô, 2004, *Biophys. J.*, 86, 3131–3140.
[27] S. Yamaguchi, S. Tuzi, J. U. Bowie, and H. Saitô, 2004, *Biochim. Biophys. Acta*, 1698, 97–105.
[28] H. Saitô, T. Tsuchida, K. Ogawa, T. Arakawa, S. Yamaguchi, and S. Tuzi, 2002, *Biochim. Biophys. Acta*, 1565, 97–106.
[29] H. Saitô, K. Yamamoto, S. Tuzi, and S. Yamaguchi, 2003, *Biochim. Biophys. Acta*, 1616, 127–136.

[30] K. S. Huang, H. Bayley, M. J. Liao, E. London, and H. G. Khorana, 1981, *J. Biol. Chem.*, 256, 3802–3809.

[31] S. Tuzi, A. Naito, and H. Saitô, 1996, *Eur. J. Biochem.*, 239, 294–301.

[32] W. Stoekenius and R. A. Bogomolni, 1982, *Annu. Rev. Biochem.*, 51, 587–616.

[33] R. A. Mathies, S. W. Lin, J. B. Ames, and W. T. Pollard, 1991, *Annu. Rev. Biophys. Biophys. Chem.*, 20, 491–518.

[34] J. K. Lanyi, 1997, *J. Biol. Chem.*, 272, 31209–31212.

[35] J. K. Lanyi, 2000, *Biochim. Biophys. Acta*, 1459, 339–345.

[36] A. E. Blaurock and W. Stoeckenius, 1971, *Nat. New Biol.*, 233, 152–155.

[37] J. M. Baldwin, R. Henderson, E. Beckman, and F. Zemlin, 1988, *J. Mol. Biol.*, 202, 585–591.

[38] R. Henderson, J. M. Baldwin, T. A. Ceska, F. Zemlin, E. Beckmann, and K. H. Dowing, 1990, *J. Mol. Biol.*, 213, 899–929.

[39] Y. Kimura, D. G. Vassylyev, A. Miyazawa, A. Kidera, M. Matsushima, K. Mitsuoka, K. Murata, T. Hirai, and Y. Fujiyoshi, 1997, *Nature*, 389, 206–211.

[40] K. Mitsuoka, T. Hirai, K. Murata, A. Miyazawa, A. Kidera, Y. Kimura, and Y. Fujiyoshi, 1999, *J. Mol. Biol.*, 286, 861–882.

[41] H. Luecke, B. Schobert, H. -T. Richter, J. -P. Cartailler, and J. K. Lanyi, 1999, *J. Mol. Biol.*, 291, 899–911.

[42] H. Belrhali, P. Nollert, A. Royant, C. Menzel, J. P. Rosenbusch, E. M. Landau, and E. Pebay-Peyroula, 1999, *Structure*, 7, 909–917.

[43] E. M. Landau and J. P. Rosenbusch, 1996, *Proc. Natl. Acad. Sci. USA*, 93, 14532–14535.

[44] K. Takeda, H. Sato, T. Hino, M. Kono, K. Fukuda, I. Sakurai, T. Okada, and T. Kouyama, 1998, *J. Mol. Biol.*, 283, 463–474.

[45] J. B. Heymann, D. J. Muller, E. M. Landau, J. P. Rosenbusch, E. Pebay-Peyroula, G. Buldt, and A. Engel, 1999, *J. Struct. Biol.*, 128, 243–249.

[46] D. Muller, J. B. Heymann, F. Oesterhelt, C. Moller, H. Gaub, G. Buldt, and A. Engel, 2000, *Biochim. Biophys. Acta*, 1460, 27–38.

[47] S. Yamaguchi, S. Tuzi, K. Yonebayashi, A. Naito, R. Needleman, J. K. Lanyi, and H. Saitô, 2001, *J. Biochem. (Tokyo)*, 129, 373–382.

[48] S. Tuzi, S. Yamaguchi, A. Naito, R. Needleman, J. K. Lanyi, and H. Saitô, 1996, *Biochemistry*, 35, 7520–7527.

[49] H. Saitô, J. Mikami, S. Yamaguchi, M. Tanio, A. Kira, T. Arakawa, K. Yamamoto, and S. Tuzi, 2004, *Magn. Reson. Chem.*, 42, 218–230.

[50] S. Tuzi, A. Naito, and H. Saitô, 1993, *Eur. J. Biochem.*, 218, 837–844.

[51] H. Saitô, 1986, *Magn. Reson. Chem.*, 24, 835–852.

[52] H. Saitô and I. Ando, 1989, *Annu. Rep. NMR Spectrosc.*, 21, 209–290.

[53] H. Saitô, S. Tuzi, and A. Naito, 1998, *Annu. Rep. NMR Spectrosc.*, 36, 79–121.

[54] Y. Kawase, M. Tanio, A. Kira, S. Yamaguchi, S. Tuzi, A. Naito, M. Kataoka, J. K. Lanyi, R. Needleman, and H. Saitô, 2000, *Biochemistry*, 39, 14472–14480.

[55] A. Kira, M. Tanio, S. Tuzi, and H. Saitô, 2004, *Eur. Biophys. J.*, 33, 580–588.

[56] S. Tuzi, S. Yamaguchi, M. Tanio, H. Konishi, S. Inoue, A. Naito, R. Needleman, J. K. Lanyi, and H. Saitô, 1999, *Biophys. J.*, 76, 1523–1531.

[57] S. Yamaguchi, K. Yonebayashi, H. Konishi, S. Tuzi, A. Naito, J. K. Lanyi, R. Needleman, and H. Saitô, 2001, *Eur. J. Biochem.*, 268, 2218–2228.

[58] K. Yonebayashi, S. Yamaguchi, S. Tuzi, and H. Saitô, 2003, *Eur. Biophys. J.*, 32, 1–11.

[59] S. Yamaguchi, S. Tuzi, M. Tanio, A. Naito, J. K. Lanyi, R. Needleman, and H. Saitô, 2000, *J. Biochem. (Tokyo)*, 127, 861–869.
[60] H. Saitô, S. Yamaguchi, K. Ogawa, S. Tuzi, M. Márquez, C. Sanz, and E. Padrós, 2004, *Biophys. J.*, 86, 1673–1681.
[61] S. Tuzi, J. Hasegawa, R. Kawaminami, A. Naito, and H. Saitô, 2001, *Biophys. J.*, 81, 425–434.
[62] I. Solomon, 1955, *Phys. Rev.*, 99, 559–565.
[63] N. Bloembergen, 1957, *J. Chem. Phys.*, 27, 572–573.
[64] S. Tuzi, A. Naito, and H. Saitô, 2003, *J. Mol. Struct.*, 654, 205–214.
[65] D. Suwelack, W. P. Rothwell, and J. S. Waugh, 1980, *J. Chem. Phys.*, 73, 2559–2569.
[66] W. P. Rothwell and J. S. Waugh, 1981, *J. Chem. Phys.*, 74, 2721–2732.
[67] H. Saitô, S. Yamaguchi, H. Okuda, A. Shiraishi, and S. Tuzi, 2004, *Solid State Nucl. Magn. Reson.*, 25, 5–14.
[67a] H. Saitô, 2006, *Annu. Rep. NMR Spectrosc.*, 57, 101–178.
[68] M. Kataoka, H. Kamikubo, F. Tokunaga, L. S. Brown, Y. Yamazaki, A. Maeda, M. Sheves, R. Needleman, and J. K. Lanyi, 1994, *J. Mol. Biol.*, 243, 621–638.
[69] L. S. Brown, H. Kamikubo, L. Zimanyi, M. Kataoka, F. Tokunaga, P. Verdegem, J. Lugtenburg, and J. K. Lanyi, 1997, *Proc. Natl. Acad. Sci. USA*, 94, 5040–5044.
[70] G. J. Ludlam and K. J. Rothschild, 1997, *FEBS Lett.*, 407, 285–288.
[71] R. Renthal, N. Dawson, J. Tuley, and P. Horowitz, 1983, *Biochemistry*, 22, 5–12.
[72] J. Marque, K. Kinoshita, Jr., R. Govindjee, A. Ikegami, T. G. Ebrey, and J. Otomo, 1986, *Biochemistry*, 25, 5555–5559.
[73] H. J. Steinhoff, R. Mollaaghababa, C. Altenbach, K. Hideg, M. Krebs, H. G. Khorana, and W. L. Hubbell, 1994, *Science*, 266, 105–107.
[74] W. Behrens, U. Alexiev, R. Mollaaghababa, H. G. Khorana, and M. P. Heyn, 1998, *Biochemistry*, 37, 10411–10419.
[75] M. P. Krebs, W. Behrens, R. Mollaaghababa, H. G. Khorana, and M. P. Heyn, 1993, *Biochemistry*, 32, 12830–12834.
[76] S. Tuzi, S. Naito, and H. Saitô, 1994, *Biochemistry*, 33, 15046–15052.
[77] J. Riesle, D. Oesterhelt, N. A. Dencher, and J. Heberle, 1996, *Biochemistry*, 35, 6635–6643.
[78] S. Checover, E. Nachliel, N. A. Dencher, and M. Gutman, 1996, *Biochemistry*, 36, 13919–13928.
[79] S. Checover, Y. Marantz, E. Nachliel, M. Gutman, M. Pfeiffer, J. Tittor, D. Oesterhelt, and N. A. Dencher, 2001, *Biochemistry*, 40, 4281–4292.
[80] A. H. Dioumaev, H.-T. Richter, L. S. Brown, M. Tanio, S. Tuzi, H. Saitô, Y. Kimura, R. Needleman, and J. K. Lanyi, 1998, *Biochemistry*, 37, 2496–2505.
[81] Y. Yamazaki, M. Hatanaka, H. Kandori, J. Sasaki, W. F. Karstens, J. Raaps, J. Lugtenburg, M. Bizounok, J. Herzfeld, R. Needleman, J. K. Lanyi, and A. Maeda, 1995, *Biochemistry*, 34, 7088–7093.
[82] Y. Yamazaki, S. Tuzi, H. Saitô, H. Kandori, R. Needleman, J. K. Lanyi, and A. Maeda, 1996, *Biochemistry*, 35, 4063–4068.
[83] M. Tanio, S. Inoue, K. Yokota, T. Seki, S. Tuzi, R. Needleman, J. K. Lanyi, A. Naito, and H. Saitô, 1999, *Biophys. J.*, 77, 431–442.
[84] M. Tanio, S. Tuzi, S. Yamaguchi, R. Kawaminami, A. Naito, R. Needleman, J. K. Lanyi, and H. Saitô, 1999, *Biophys. J.*, 77, 1577–1584.
[85] H. Saitô, R. Kawaminami, M. Tanio, T. Arakawa, S. Yamaguchi, and S. Tuzi, 2002, *Spectroscopy*, 16, 107–120.

[86] J. K. Lanyi, 1998, *J. Struct. Biol.*, 124, 164–178.
[87] W. D. Hoff, K. -W. Jung, and J. L. Spudich, 1997, *Annu. Rev. Biophys. Biomol. Struct.*, 26, 223–258.
[88] R. Seidel, B. Scharf, M. Gautel, K. Kleine, D. Oesterhelt, and M. Engelhard, 1995, *Proc. Natl. Acad. Sci. USA*, 92, 3036–3040.
[89] X.-N. Zhang, J. Zhu, and J. L. Spudich, 1999, *Proc. Natl. Acad. Sci. USA*, 96, 857–862.
[90] S. Kimura, A. Naito, S. Tuzi, and H. Saitô, 2001, *Biopolymers*, 58, 78–88.
[91] H. Saitô, 2004, *Chem. Phys. Lipids*, 132, 101–112.
[92] V. I. Gordeliy, J. Labahn, R. Moukhametzianov, R. Efremov, J. Granzin, R. Schlesinger, G. Buldt, T. Savopol, A. J. Scheidig, J. P. Klare, and M. Engelhard, 2002, *Nature*, 419, 484–487.
[93] O. Vinogradova, P. Badola, L. Czerski, F. D. Sonnichsen, and C. R. Sanders II, 1997, *Biophys. J.*, 72, 2688–2701.
[94] B. M. Gorzelle, J. K. Nagy, K. Oxenoid, W. L. Lonzer, D. S. Cafiso, and C. R. Sanders, 1999, *Biochemistry*, 38, 16373–16382.
[95] V. A. Lightner, R. M. Bell, and P. Modrich, 1983, *J. Biol. Chem.*, 258, 10856–10861.
[96] C. R. Loomis, J. P. Walsh, and R. M. Bell, 1985, *J. Biol. Chem.*, 260, 4091–4097.

Chapter 14

MEMBRANE PROTEINS II: 3D STRUCTURE

The general approach for obtaining NMR constraints essential for determination of 3D structure of proteins as well as its theoretical background were already discussed in Section 6.1.2. In particular, 3D structures of membrane proteins can be determined by means of orientational or distance constraints available from mechanically or spontaneously, magnetically oriented or unoriented rigid samples, respectively, based on the methods discussed in Chapter 6, as far as the portion under consideration is static. Fully hydrated membrane proteins, however, are not always suitable for determination of such 3D structure by these approaches, because they are in many instances flexible, undergoing a variety of fluctuation motions depending upon portions such as N- or C-terminus, loops, and amphiphilic and transmembrane α-helices, even though their 3D structures are still available from fully hydrated MOVS, as discussed in Section 6.1.2.2. NMR measurements on mechanically oriented system are usually carried out under the conditions either at lower temperature or lower hydration up to 70% relative humidity, in order to minimize the effects of such fluctuation motions especially at the transmembrane (and/or amphiphilic) α-helices, if any, as much as possible. Such attempts, however, are not always successful for the N- or C-terminus or loops, which are fully exposed to aqueous phase.

14.1. 3D Structure of Mechanically Oriented Membrane Proteins[1–7]

As described in Section 6.1.2, mechanically oriented membrane proteins are prepared by dissolving the protein–lipid mixtures in an organic solvent such as 2,2,2-trifluoroethanol (TFE), chloroform, or methanol, depositing them on the glass slide to form a thin film after evaporation and hydration and annealing under atmosphere of controlled humidity. Several thousands of mechanically oriented bilayers are available by stacking about 10–40 slides. For 3D structure determination, it is crucial to prepare completely oriented samples in order to obtain orientational constraints. Unfortunately, generation of structural and orientational information about membrane proteins has been limited mainly to peptides, since no general methods have been available for aligning native and reconstituted membranes containing larger integral membrane proteins in sufficient quality and quantity without affecting their structural integrity or function. Gröbner *et al.*[8] proposed a general procedure to align fully hydrated functional biological membranes containing large membrane proteins based on isopotential spin-dry ultracentrifugation technique, relying on the centrifugation of membrane fragments onto a support with simultaneous, or subsequent, partial evaporation of the solvent which aids alignment.[8] The quality of orientation, as shown by the mosaic spread of the samples, was monitored by static solid state ^{31}P NMR for the phospholipids and by ^{2}H NMR for a deuterated retinal in bovine rhodopsin.[8,9] The generality of this method is demonstrated with three different membranes containing bovine rhodopsin in reconstituted bilayers, natural membranes with the red cell anion-exchange transport protein in erythrocytes, band 3, and the nicotinic acetylcholine receptor (nAcChoR).

The conformation of phosphocholine headgroup in lipid bilayers is extremely sensitive to the level of hydration, at water concentrations below 10 water molecules per lipid ($n_{\mathrm{w}} \approx 10$), but is not affected by additional water above this number.[10] Instead, hydration-optimized planar lipid bilayer samples were prepared by equilibrating the oriented POPC sample at 5% RH for 96 h,[11] in order to reduce the water concentration from $n_{\mathrm{w}} \approx 10$ to $n_{\mathrm{w}} \approx 2$, to improve the low mechanical stability and high conductance properties of the sample and to alleviate its susceptibility to dehydration during the course of NMR measurements. The phase transition temperature for DOPC bilayer increases from -18 °C at $n_w \approx 10$ to 4 °C at $n_w \approx 1$,[12] because the phase-transition temperature is hydration-dependent. As a result, these samples have greater stability over the course of multidimensional NMR experiments and have lower sample conductance for greater rf power efficiency and enable greater filling factors within rf coil to be obtained for improved experimental sensitivity.

So far, six 3D structures of the membrane proteins, including gA, nAchR M2 channel, the transmembrane region of influenza M2 channel, HIV Vpu channel, α-factor receptor M6, and fd coat protein in membrane, have been solved by solid state NMR utilizing completely oriented samples and deposited in the PDB, as illustrated in Figure 14.1.[6] Naturally, revealed structures are confined to either the transmembrane or amphipathic α-helices, reflecting the orientational constraints available from the mechanically

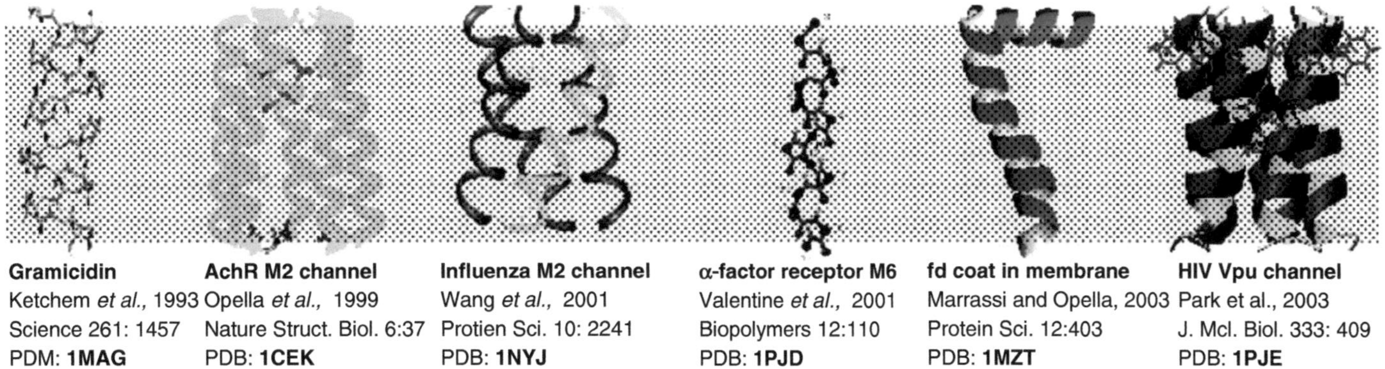

Figure 14.1. Structures determined by solid state NMR in oriented bilayers. The PDB file numbers are in bold face (from Opella and Marassi[6]).

oriented samples. In addition, no structural information is available from unoriented N- or C-terminal residues as well as interhelical loops, however.

14.1.1. Influenza M2 Channel[13]

On endocytosis of the influenza A virus, the low pH environment of the endosome activates the M2 protein H^+ channel, presumably through protonation of the histidine side chain in the transmembrane section of the protein. The acidification of the viral interior disrupts the ribonucleoprotein core, leading to fusion of the viral envelope with the endosomal wall and continuation of the infectious life cycle.[14] The M2 protein has 97 amino acids, with a putative single 19-residue transmembrane helix, a 24-residue extraviral segment, and 54 intraviral residues. As a tetramer, the protein forms the H^+ channel. Wang *et al.*[13] synthesized uniformly or selectively ^{15}N-labeled 25-residue transmembrane peptide of the M2 protein (M2-TMP), spanning the putative transmembrane segment with a few hydrophilic residue on either end and exhibiting channel activity similar to M2 protein. The PISEMA[15] spectra, as described in Section 6.1, has proved to be a very powerful tool for studying membrane protein structure by solid state NMR. Figure 14.2, as an extension of Figure 6.8, illustrates the ^{15}N PISEMA spectra of single- and multiple-site-labeled M2-TMP peptide from hydrated DMPC bilayers (in a 1:16 mole ratio) uniformly aligned with respect to the applied magnetic field axis: they are superimposed in three sequential groups: (a) sites 26–31, (b) sites 32–37, and (c) sites 38–43. Directly, by observation of this resonance pattern, it is now clear that residues 26–43 are entirely helical. These spectra correlate the anisotropic ^{15}N–H dipolar coupling with the anisotropic ^{15}N chemical shift. A mirror image pair of PISA wheel[16,17] is available for M2-TMP from the PISEMA spectra shown above, based on the orientation of the principal axis frames, which are fixed in the molecular frame. Then, a high-resolution structure of the monomer backbone and a detailed description of the helix orientation tilted by 38° with respect to the bilayer normal were achieved by using the orientational restraints from the solid state NMR data. With this unique information, the tetrameric structure of this H^+ channel is constrained substantially, by MD simulation using the CHARMM empirical function, as illustrated in Figure 14.1.

Subsequently, Tian *et al.*[18] expressed, purified, and reconstituted the full-length M2 protein from influenza A virus into DMPC/DMPG liposomes in which a stable tetrameric preparation is present as revealed by SDS-PAGE

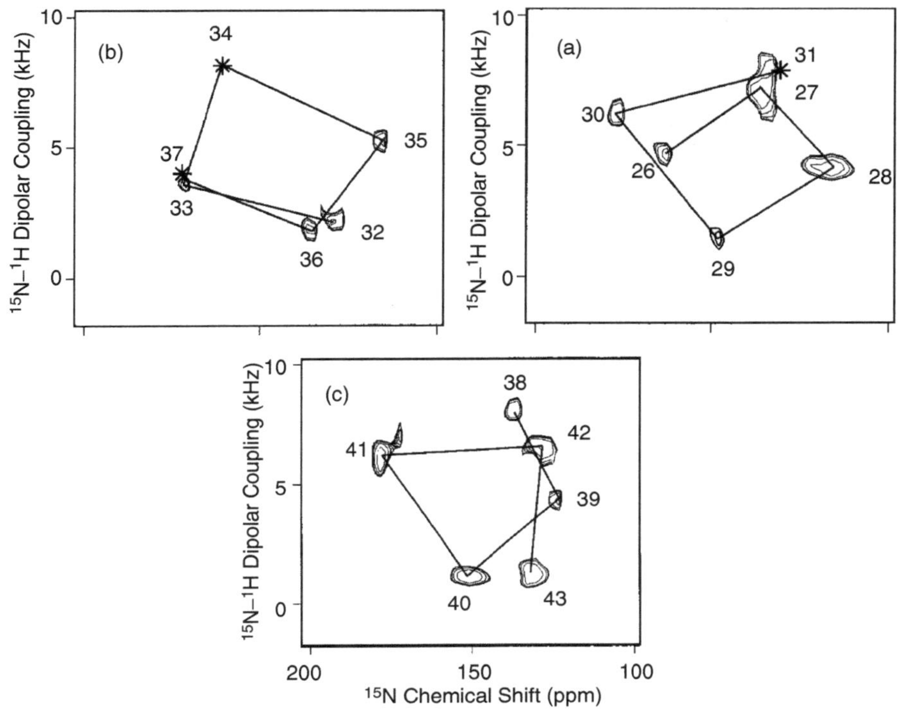

Figure 14.2. PISEMA spectra of ^{15}N-multiple-site-labeled M2-TMP. (a) Sites 26–31, (b) sites 32–37, and (c) sites 38–43. Hydration is ≈50% (from Wang *et al.*[13]).

analysis. The PISA wheel data based on the PISEMA spectra of ^{15}N-uniformly labeled and ^{15}N-Val or Leu-labeled M2 reconstituted in DMPC/DMPG bilayer and uniformly aligned preparation suggested the existence of a transmembrane helix having a tilt angle of approximately 25°, quantitatively similar to results obtained on the above-mentioned M2-TMP reconstituted in DMPC bilayer (38°).

14.1.2. Channel-Forming Domain of Virus Protein "u" (Vpu) From HIV–1

Vpu is a relatively small, 81-residue, helical membrane protein ($Vpu_{2\text{–}81}$) that can be obtained by expression in bacteria.[19] It has one long hydrophobic membrane spanning helix and two shorter amphipathic helices that reside in the plane of the membrane as components of its cytoplasmic domain. It has been suggested that this function is related to its ability to act as an ion channel on the basis of its similarity to the influenza virus M2 protein, as described above. As illustrated in Figure 14.3, the 1D solid state ^{15}N NMR signals of uniformly ^{15}N-labeled full-length VPu are clearly segregated into two distinct bands at chemical shift frequencies associated with NH bonds in the transmembrane helix perpendicular to the membrane surface (200 ppm) and with NH bonds in both cytoplasmic helices parallel to the membrane surface (70 ppm). The spectra of truncated preparations, Vpu_{2-51} (residues 2–51), which contains the transmembrane α-helix and the first amphipathic helix of the cytoplasmic domain, and of Vpu_{28-81} (residues 28–81), which contains only the cytoplasmic domains, support the models as shown in Figure 14.3. Indeed, truncated Vpu_{2-51}, which has the transmembrane helix, gives rise to signals characteristic of discrete channels in lipid bilayers, whereas the cytoplasmic domain Vpu_{28-81}, which lacks the transmembrane helix, does not. Further, 2D PISEMA (^{1}H–^{15}N dipolar coupling/^{15}N chemical shift) spectra of uniformly ^{15}N-labeled full-length Vpu in an oriented bilayer showed that approximately 16 resonances can be discerned in the

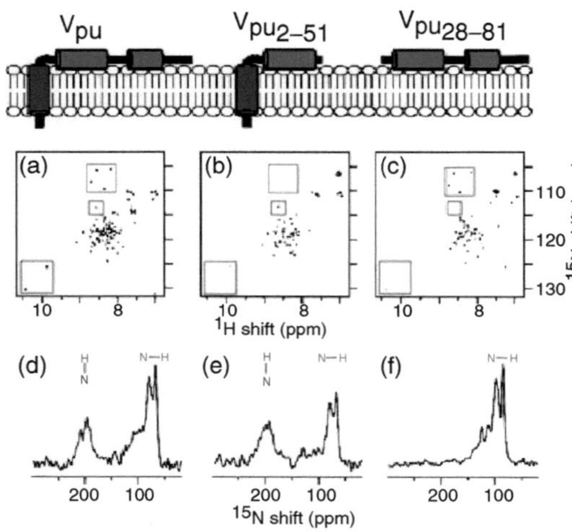

Figure 14.3. NMR spectra of three uniformly ^{15}N-labeled recombinant Vpu constructs. The overall architecture of the three constructs (top). 2D heteronuclear correlation spectra in dihexanoyl phosphatidylcholine micelles (middle). Representative assigned resonances are highlighted in the boxes: amide resonance from Gly 53, 58, 67, and 71 (red boxes), Ser 23 (blue boxes), and indole resonance from Trp 22 and 76 (red/blue boxes). 1D solid state ^{15}N NMR spectra of the three Vpu constructs obtained at 0 °C in oriented lipid bilayers (bottom). The orientations of transmembrane (blue) and in-plane (red) amide NH bonds are indicated above the spectra (from Marassi *et al.*[19]).

spectral region (8 kHz dipolar coupling and 200 ppm chemical shift) associated with a transmembrane helix. The "wheel-like" pattern of resonances observed in this region of the spectrum is characteristic of a tilted transmembrane α-helix and provides an index of the slant and polarity of the helix.

To focus on the principal structural features of the transmembrane domain, 2D $^{1}H/^{15}N$ PISEMA spectrum of ^{15}N-labeled Vpu_{2-30+} was examined in lipid bilayers aligned between glass plates.[20] As illustrated in Figure 14.4, the sequence in the PISEMA spectrum is divided into two segments: residues 8–16 (blue) and residues 17–25 (red). The second column (Figure 14.4(d) and (e)) represents the ideal PISA wheels that correspond to the experimental data, showing that the two segments of the helix have different tilt angles. The magnitudes of the ^{1}H–^{15}N dipolar couplings obtained from the completely aligned bilayer and weakly aligned micelles are plotted as a function of the residue number in Figure 14.4 (f) and (g), respectively, as described by dipolar waves.[21] Namely, a sinusoid with a period of 3.6 residues was fit to the experimental ^{1}H–^{15}N dipolar couplings as a function of residue number to characterize the length and orientation of each helical segment relative to the

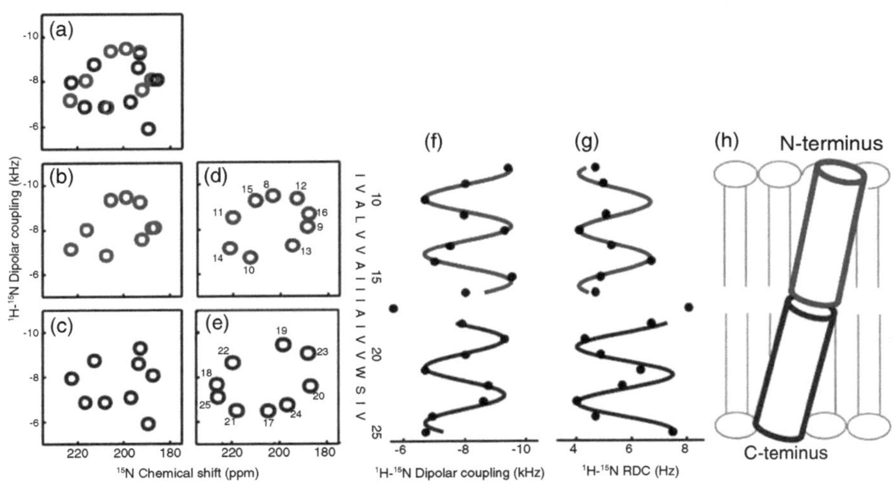

Figure 14.4. (a)–(c) Representations of the experimental PISEMA spectrum of Vpu_{2-30+} in completely aligned bilayers. (a) Residues 8–25, (b) residues 8–16, (c) residues 17–25. (d)–(e) Ideal PISA wheels with uniform dihedral angles ($\phi = -57°$; $\psi = -47°$) corresponding to the experimental data. (d) Residues 8–16 with the tilt angle of 12°. (e) Residues 17–25 with the tilt angle of 15°, (f) Dipolar waves of ^{1}H–^{15}N unaveraged dipolar couplings obtained from completely aligned bilayers. (g) Dipolar waves of ^{1}H–^{15}N residual dipolar couplings obtained from weakly aligned micelles. (h) A tube representation of the transmembrane helix of Vpu_{2-3+} in lipid bilayers (from Park *et al.*[20]).

direction of the applied magnetic field.[21] It is therefore noteworthy that the results for the transmembrane helix domain of Vpu are consistent in showing a kink at residue 17 in both micelle and bilayer. A model of pentamer responsible for the ion-channel activity was obtained by means of MD simulation based on the ^{15}N restraints thus obtained, as illustrated in Figure 14.1.

14.1.3. Bacteriorhodopsin

^2H NMR spectra on selectively deuterated retinal of hydrated bR from PM which are uniaxially oriented on stack of glass plates were utilized to reveal their orientation relative to the bilayer normal both in the ground state[22–25] and in the M-photointermediate.[26,27] The specific orientations of the three labeled methyl groups ([2,4,4,16,16,16,17,17,17,18,18-$^2H_{11}$]retinal) on the cyclohexene ring were determined by analysis of the deuterium quadrupole splittings. The two adjacent methyl groups at C_{16} and C_{17} (on C_1) of the retinal were found to make respective angles of 94° ± 2° and 75° ± 3° with the membrane normal (see the angle β_{PM} in Figure 6.9).[22] The third group (on C_5) points toward the cytoplasmic side with an angle of 46° ± 3° with the membrane normal, indicating that it has a 6-*s-trans* conformation.[27] This conformation was also confirmed by 18-CD_3 methyl group dynamics in bR.[28] Further, the angles for the C_{18}, C_{19}, and C_{20} groups are found to be 37° ± 1°, 40° ± 1°, and 32° ± 1°, respectively.[24] The angle between the C_1–(1*R*)–1-CD_3 bond and the bilayer normal was determined with high accuracy from the simultaneous analysis of a series of ^2H NMR spectra recorded at different inclinations of the uniaxially oriented sample in the magnetic field at 20° and −50 °C.[26] The M-photointermediate of bR trapped using guanidium hydrochloride at −60 °C showed that the angle between the C–CD_3 and the magnetic field at the C_{19} methyl group changed to 44° ± 2°, which is consistent with a slight upward tilting of the polyene chain. Additional line broadening compared with that of the ground state suggested some two-state heterogeneity.[25] Further, it was shown that the C_9–CD_3 bond shows the largest orientational change of 7° from the ground state to the M-photointermediate.[27]

As already discussed in Section 6.1.2.3, the orientation of the deuterated methyl group in [18-C^2H_3]-retinal in oriented bR was studied in both the ground state and M_{412}-photointermediate by means of ^2H MAOSS measurements utilizing MAS with improved spectral sensitivity and resolution as compared to static NMR on oriented samples as discussed above (see Figures 6.9 and 6.10).[29] Indeed, the β_{PM} angle for bR_{568} is 36° with a mosaic spread $\Delta\beta$ between ±0° and ±8°, while the angle for M_{412}-photo intermediate state is 22° with $\Delta\beta$ between ±10° and ±18°.[29] ^{15}N MAOSS spectrum of [^{15}N-Met]bR which consists of five resolved peaks among nine Met residues was recorded, in order to reveal the orientation of the principal axes of the ^{15}N CSA tensors with respect to the membrane normal.[30] For an oriented sample, it was shown that ^{15}N PISEMA provides good signal separation for transmembrane helices with modest tilt angles relative to the bilayer normal, while HETCOR provides good signal separations for tilt angles in the 60–90° regime being typical for in-plane structures or loop regions.[31,32] Therefore, ^{15}N PISEMA and HETCOR spectra of oriented [^{15}N-Met]bR on the surface of 30 thin glass plates were recorded to reveal tilt angles for individual transmembrane α-helices in bR.[33] By deconvolution of the helix signals using SIMMOL[34] and SIMPSON[35] simulation software based on crystal structures, constraints for some helix tilt angles were established. It was estimated that the extracellular section of helix B has a tilt of less than 5° from the membrane normal, while the tilt of helix A was estimated to be 18–22°, both of which are in agreement with most crystal structures.

14.1.4. Rhodopsin (Rho)

G-protein-coupled receptors (GPCRs) form the largest known family among integral membrane proteins, up to 5% of all genes encoded in the genomes of higher eukaryotes.[36] Their highly conserved topology is made up of seven transmembrane helices like bR and functions as a transducer of extracellular signals like hormone, pheromones, odeorants, or light into the activation of intracellular G-protein complex. Rho is a typical GPCR active as a mammalian photoreceptor protein of 40 kDa covalently linked to 11-*cis* retinal through Lys 296. Absorption of a photon by the 11-*cis* retinal causes its isomerization to all-*trans* retinal, leading to a conformational change of the protein moiety, including the cytoplasmic surface. Its 3D structure of 2.8 Å has been resolved by X-ray diffraction[37] and served as an important molecular basis of understanding the vision as well as signal transductions for a variety of GPCRs.

Static ^{2}H NMR spectra of uniaxially oriented bovine Rho containing 11-*cis* retinal in which methyl groups are specifically deuterated at C_{19} or C_{20} positions were recorded.[38] Analysis of the obtained ^{2}H NMR spectra provided angles for the individually labeled chemical bond vectors, as a similar manner to that obtained for bR as described above: the orientation angles for the C_{19} and C_{20} methyl groups with respect to the bilayer normal are 42° ± 5° and 30° ± 5°, respectively. ^{2}H MAS NMR (MAOSS) spectra of oriented regenerated bovine rhodopsin with specifically ^{2}H-labeled retinal stacked on glass plates both at the dark and M_1 (metarhodopsin I, meta I) were studied to

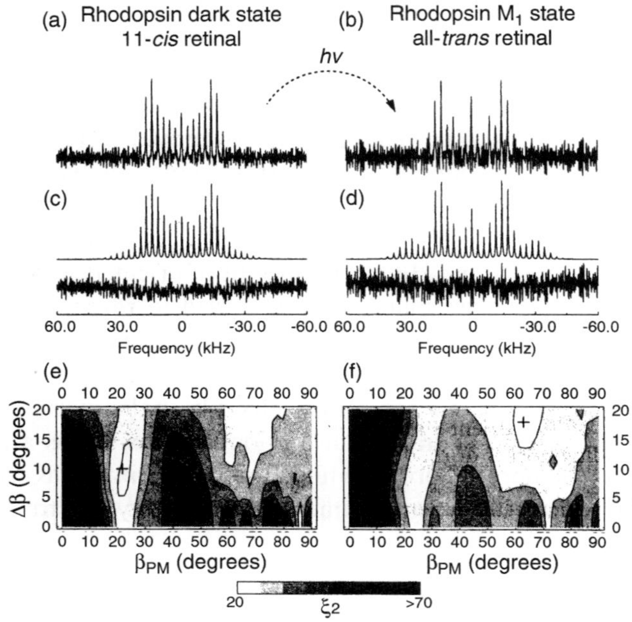

Figure 14.5. ^{2}H-MAOSS NMR spectra of [18-C^{2}H$_3$]-retinal in dark-adapted rhodopsin (a) and upon photo-activation in the M$_1$ state (illuminated below 273K) (b), at $\omega_r = 2680$ Hz and $T = 213$ K. The spectra were analyzed by minimizing the rms deviation of the MAS sideband intensities in the [β_{PM}, $\Delta\beta$]-parameter space, as shown by ζ^2-contour plots (e) and (f). The C–C^{2}H$_3$ tilt angle is determined to be $\beta_{PM} = 21° \pm 5°$ in the dark-adapted state and $62° \pm 7°$ in the M$_1$ state. It can be seen that $\Delta\beta$ changes slightly from 10° in the ground state to 18° in M$_1$. The difference between best fit and experimental spectra are shown in (c) and (d) (from Gröbner *et al.*[39]).

determine the orientation of the methyl group vectors C_5–C_{18} and C_9–C_{19}, and C_{13}–C_{20} carrying C–^{2}H$_3$ relative to the membrane normal (Figure 14.5).[39] In particular, the C–^{2}H$_3$ tilt angle is determined to be $\beta_{PM} = 21° \pm 5°$ in the dark-adapted state and $62° \pm 7°$ in the M$_1$ state from the ^{2}H NMR MAOSS spectra of [18-C^{2}H$_3$]retinal-labeled Rho. It can be seen that $\Delta\beta$ changes slightly from 10° in the ground state to 18° in M$_1$. The retinal, however, has a 6-*s-trans* conformation of the β-ionone ring as previously found for retinal in bR, opposite to the crystal structure of 11-*cis* retinal[40] and Rho[37,41,42] and solid state ^{13}C NMR data.[43–45] Indeed, it was shown that β-ionone ring takes a twisted 6-*s-cis* conformation based on chemical shift tensor of bovine Rho regenerated with selectively ^{13}C-enriched retinal at C$_5$[43,44] and distance constraints, C$_8$ to C$_{16}$ and C$_{17}$ and from C$_8$ to C$_{18}$, available from RR studies of Rho regenerated with 11-Z-[8,18, $^{13}C_2$]retinal or 11-Z-[18, 16/17 $^{13}C_2$]retinal.[45] To clarify this contradiction in more detail, static ^{2}H NMR spectra of oriented bovine Rho regenerated with selectively ^{2}H-labeled 11-*cis* retinal at the C_5, C_9, or C_{13} methyl positions in POPC bilayers were examined at temperature below the gel to liquid crystalline phase transition, where rotational and translational diffusion of Rho is effectively quenched.[46] The experimental tilt series of ^{2}H NMR spectra were fit to a theoretical line shape analysis giving the retinylidene bond orientations with respect to the membrane normal in the dark state. Moreover, the relative orientations of pairs of methyl groups were used to calculate effective torsion angles between different planes of unsaturation of the retinal chromophore. Their results are consistent with significant conformational distortion of retinal, and they have important implications for quantum mechanical calculations of its electronic spectral properties. In particular, the β-ionone ring was shown to have a twisted 6-*s-cis* conformation, whereas the polyene chain is twisted 12-*s-trans* as illustrated in Figure 14.6.[46]

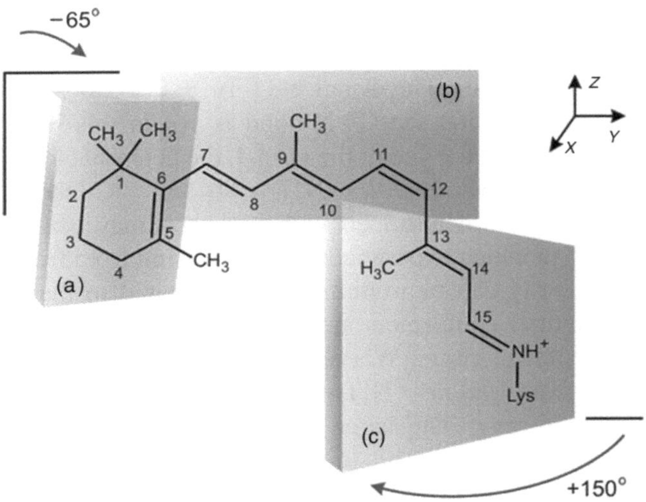

Figure 14.6. Calculated structure of retinal within the binding pocket of rhodopsin at −15 °C based on ^{2}H NMR angular restraints. The rotational degrees of freedom are approximated in terms of three planes, (a), (b), and (c); i.e., additional twisting of the polyene chain is not considered. Planes (b) and (c) have a relative orientation specified by the torsion angle of the C_{12}–C_{13} bond, which is +150°. The relative orientation of the (a) and (b) planes, i.e., the conformation of the β-ionone ring relative to the polyene chain, is due to the C_6–C_7 torsion angle with a value of −65° (from Salgado *et al.*[46]).

14.2. Secondary Structure Based on Distance Constraints

Local secondary structures are readily available also from distance constraints obtained from unoriented samples in which plausible conformational fluctuations could be frozen either at low temperatures or dehydrated preparations such as crystalline or lyophilized state.

14.2.1. Bacteriorhodopsin and Rhodopsin

The interatomic distances of ^{13}C-labeled positions in PM werfe determined by RR measurements.[47] The distances for the selectively ^{13}C-labeled retinal are 4.1 Å $\pm$ 0.4 Å for C_8–C_{18} pair and 3.3–3.5 Å $\pm$ 0.4 Å for the average C_8–C_{16}/C_8–C_{17} pairs, indicating that retinal is in a 6-*s*-*trans* conformation. The same approach was also used to determine the distances from [14-^{13}C]retinal to [ε-^{13}C]Lys216 in dark-adapted bR in order to examine the structure of the retinal–protein linkage and its role in coupling the isomerization of retinal to unidirectional proton transfer.[48,49] The 4.1 Å $\pm$ 0.3 Å distance is found for bR_{568} and 3.9 Å $\pm$ 0.1 Å distance is in the thermally trapped M_{412} state. This finding demonstrates that the C=N bond is *anti* in these states. The distance 3.0 Å $\pm$ 0.2 Å observed in bR_{555}, however, demonstrates that the C=N bond is *syn* in this state. To reveal more structural details at the site of the retinal, the heteronuclear distances of dark-adapted bR between [*indole*-^{15}N]Trp 86 and ^{13}C-labeled retinal at position 14 or 15 were determined with high accuracy using ^{13}C SFAM REDOR NMR,[50] in which experimental errors arising from frequency offset and rf inhomogeneity can be minimized as compared to the conventional REDOR. Two retinal conformers are distinguished by their different isotropic 14-^{13}C chemical shifts. Whereas the C_{14} position of 13-*cis*–15-*syn*-retinal is 4.2 Å from [*indole*-^{15}N]Trp 86, this distance is 3.9 Å in the all-*trans*–15-*anti* conformer. The latter distance allows one to check on the details of the active center of bR in the various published models derived from X-ray and electron diffraction data. ^{15}N REDOR difference measurement of [20-^{13}C]retinal, [*indole*-^{15}N]Trp-bR showed that the distance between the C_{20} of retinal and the indole nitrogen of Trp 182, which is the closest residue to retinal, changes only slightly from the light-adapted state (3.36 Å $\pm$ 0.2 Å) to the early M state (3.16 Å $\pm$ 0.4 Å).[51] All Xaa-Pro peptide bonds are in the *trans* configuration on the basis of ^{13}C NMR spectra of [3-^{13}C]Pro- and [4-^{13}C]Pro-labeled bR.[52] ^{13}C and ^{15}N chemical shifts from X-Pro peptide bonds in bR were assigned from REDOR difference spectra of pairwise labeled samples, and correlations of chemical shifts with structure are explored in a series of X-Pro model compounds.[53] Results for the three-membrane embedded X-Pro bonds of bR indicate only slight changes in the transition from the resting state of the protein to either the early or late M state of the proton motive photocycle.

The frequencies of maximum visible absorbance and the ^{15}N chemical shifts of the 13-*cis* and all *trans* compounds are found to be linearly related to the strength of the protonated Schiff base (*p*SB) counterion (CI) interaction as measured by $(1/d^2)$,[54–56] where d is the center-to-center distance between the *p*SB and the CI charge, as illustrated in Figure 14.7. With these calibrations, $d = 4.0, 3.9, 3.7, 3.6$, and 3.8 Å (± 0.3 Å) was estimated for the J_{625}, K_{590}, L_{550}, N_{520}, and bR_{555} states for 13-*cis*-compounds of bR, respectively. These distances are compared with similarly determined values of about 4.16 Å ± 0.03 Å and 4.66 Å ± 0.04 Å for the all-*trans* bR_{568} and O_{640} states, respectively. ^{13}C NMR signals of [4-^{13}C]Asp-bR yielded changes between ground state and M intermediate upon protonation and deprotonation as viewed from their peak positions.[57] The distances of the carboxyl carbons of the Asp 85 and Asp 212 side chains to the C_{14} of retinal were measured by 2D RFDR and rf-driven spin-diffusion experiments:[58] 4.4 Å ± 0.6 Å and 4.8 Å ± 1.0 Å for the [4-^{13}C]Asp 212 to

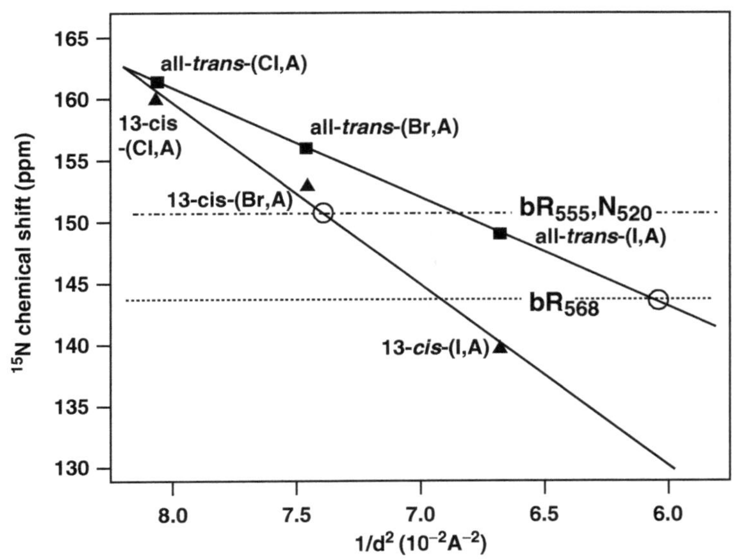

Figure 14.7. ^{15}N isotropic chemical shift of the indicated salts of retinal *p*SB with aniline vs. $1/d^2$, where d is the sum of the crystallographic radius of N^{3-} and the halide counterion. Squares represent all-*trans* retinal compounds, and triangles represent 13-*cis* retinal compounds. The specific *p*SBs are further identified by the notation (X, A), where X indicates the halide (Cl, Br, or I) and A identifies aniline (from Hu *et al.*[55]).

[14-^{13}C]retinal distances in bR_{568} and M_{412}, respectively. The spin-diffusion data are consistent with these results and indicate that the Asp 212 to C_{14} retinal distance increases by 16 $\pm$ 10% upon conversion to the M-state. The absence of cross-peaks from [14-^{13}C]retinal to [4-^{13}C]Asp 85 in all states and between any [4-^{13}C]Asp residue and [14-^{13}C]retinal in bR_{555} indicates that these distances exceed 6.0 Å. Further, distances between the Schiff base and Asp 85 and Asp 212 of uniformly ^{13}C, ^{15}N-labeled bR were measured by FSR at -80 °C.[59] The distances for Asp 85 and Asp 212 were found to be 4.7 $\pm$ 0.3 and 4.9 Å $\pm$ 0.5 Å, respectively. They are in good agreement with the data from diffraction studies. One arginine, probably Arg 82, is proposed to be perturbed in the M intermediate trapped at -44 °C in the presence of 0.3 M guanidine chloride and D85N, respectively[60] on the basis of ^{15}N NMR study on [$\eta_{1,2}$-$^{15}N_2$]-labeled bR and D85N mutant. It should be expected that these conformational changes cause transient pK_a changes in the protein, which drives the proton transfer.

The C_{10}–C_{20} and C_{11}–C_{20} distances of selectively ^{13}C-labeled retinylidene chromophore incorporated in bovine Rho were measured by RR experiment,[61] to examine spatial structure of the $C_{10} - C_{11} = C_{12} - C_{13} - C_{20}$ motif in the native chromophore, its 10-methyl analogue, and predischarge photoproduct meta I. The observed distances, $r_{10,20} = 0.304\,\text{nm} \pm 0.015\,\text{nm}$ and $r_{11,20} = 0.293\,\text{nm} \pm 0.015\,\text{nm}$ confirm that the retinylidene is 11-*Z* and the C_{10}–C_{13} unit is conformationally twisted. ^{13}C NMR signals of regenerated Rho with 11-Z-[8,9,10,11,12,13,14,15,19, 20-$^{13}C_{10}$] retinal were recorded by using 2D-correlation spectroscopy of dipolar recoupling experiment to clarify interaction between the retinylidene ligand and GPCR target.[62] The obtained chemical shift data were used to measure the extent of delocalization of positive charge into the polyene. Further, ^{13}C and ^{1}H chemical shifts of regenerated Rho with uniformly ^{13}C-labeled 11-*cis* retinal and its 9-*cis* analogue isorhodopsin were recorded by means of 2D homonuclear (^{13}C–^{13}C) and heteronuclear (^{1}H–^{13}C) dipolar correlation experiments.[63,64] The ligation shifts, $\Delta\sigma_{\text{lig}} = \sigma_{\text{lig}} - \sigma_{p\text{SB}}$ for ^{1}H and ^{13}C nuclei were defined to reflect the spatial and electronic structure of the chromophore in the active site of rhodopsin (σ_{lig}) relative to the *p*SB model dissolved in $CDCl_3$ solution ($\sigma_{p\text{SB}}$). Pronounced shifts $\Delta\sigma^{\text{H}}_{\text{lig}}$ are observed for the methyl protons of the ring moiety arising from interactions between the chromophore and the protein-binding pocket involving the aromatic amino acid residues Phe 208, 212, and Trp 265 which are in close contact with H_{16}/H_{17} and H_{18}, respectively.[63] The relatively large downfield shifts $\Delta\sigma^{\text{C}}_{\text{lig}}$ observed for C_{20} methyl group are consistent with a nonplanar conformation of 12-s bond. The 9-*cis* substrate conforms to the opsin-binding pocket in isorhodopsin in a manner very similar to that of the 11-*cis* form in rhodopsin, but the NMR data revealed an improper fit of the 9-*cis* chromophore in this binding site.[64]

As described in Section 6.3.1, a H–C_{10}–C_{11}–H molecular torsion angle of the chromophore in bovine rhodopsin (as illustrated in Figure 6.25)[65] and its meta I photointermediate[66] was determined by double quantum heteronuclear local field spectroscopy. The torsion angles were estimated to be $|\phi| = 160° \pm 10°$ for rhodopsin and $\phi = 180° \pm 25°$ for meta I. The result is consistent with current models of the photoinduced conformational transitions in the chromophores, in which the 11-*Z* retinal ground state is twisted, while the latter photointermediate has a planar all-*E* conformation. The conformation of the retinal chromophore in the late photointermediate meta I was determined by observation of ^{13}C nuclei introduced into the β-ionone ring (at the C_{16}, C_{17}, and C_{18} methyl groups) and into the adjoining segment of the polyene chain (at C_8).[67] Conformation of meta I as viewed from the C_8 chemical shift and RR experiment is unchanged from the ground state of the protein as 6-*s-cis* forms.

Apart from the above-mentioned studies using ^{13}C-labeled chromophore, ^{15}N and ^{13}C NMR study was also performed on regenerated bovine Rho from the apoprotein expressed by using suspension cultures of HEK293S cells in defined media containing 6-^{15}N-lysine and 2-^{13}C glycine, with the yield of protein 1.5–1.8 mg/L.[68] After generation of the rhodopsin pigment by the addition of 11-*cis* retinal, the whole cell were pelleted and solubilized with octyl-β-glucoside. For MAS NMR experiments, the labeled rhodopsin after purification by Sepharose column was concentrated and reconstituted into DOPC vesicles by dialysis. The recombinant baculovirus Sf9 cell line (ATTC CRL–1711) was used to prepare 10 mg of [α,ε-^{15}N]Lys-labeled Rho from two 5 L culture batches and the protein was reconstituted into bovine retina lipids.[69] The effective SB counterion distance in Rho was estimated using the empirical relationship as demonstrated in Figure 14.7.[54,55] The resulting effective Schiff base counterion distance of greater than 4 Å is consistent with structural water in the binding site hydrogen bonded with Schiff base nitrogen and Glu 113 counterion.

^{13}C DARR spectra of uniformly ^{13}C-labeled Rho were recorded in order to establish its ^{13}C–^{13}C correlation in the ground state.[70] In Rho containing [4′-^{13}C]Tyr and [8,9-^{13}C]retinal, two distinct tyrosine-to-retinal correlations were observed. The most intense cross-peak arises from a correlation between Tyr 268 and the [19-$^{13}CH_3$]retinal, which are 4.8 Å apart in the Rho crystal structure. A second cross-peak arises from a correlation between Tyr 191 and the [19-$^{13}CH_3$]retinal, which is 5.5 Å apart in the crystal structure. ^{13}C DARR spectra of uniformly ^{13}C-labeled Rho in metarhodopsin II (meta II) intermediate, the active intermediate of Rho, were recorded in order to determine how retinal isomerization is coupled to receptor activation of GPCRs.[71] In view of the retinal-binding pocket in Rho from crystalline structure, ^{13}C labels were incorporated into tyrosine, serine, glycine, and threonine of the apoprotein opsin and Rho was regenerated with the pigment 11-*cis* retinal that have been ^{13}C-labeled at the C_{12}, C_{14}, C_{15}, C_{19}, and C_{20} carbons. Figure 14.8 shows regions of the 2D NMR spectra of Rho (black) and meta II (red), showing specific retinal–protein interaction contacts.[71] A single tyrosine to C_{20} cross-peak in Rho (black) is observed at 155.2 ppm, ascribable to the C_{20}–Tyr 268 (at a distance of 4.4 Å). Importantly, upon conversion to meta II, the 155.2 ppm cross-peak disappears and a new C_{20}–Tyr cross-peak appears at 156.1 ppm. In Figure 14.8(b), two tyrosine-retinal cross-peaks are observed. Upon conversion to meta II, no C_{19}–Tyr peaks are observed despite

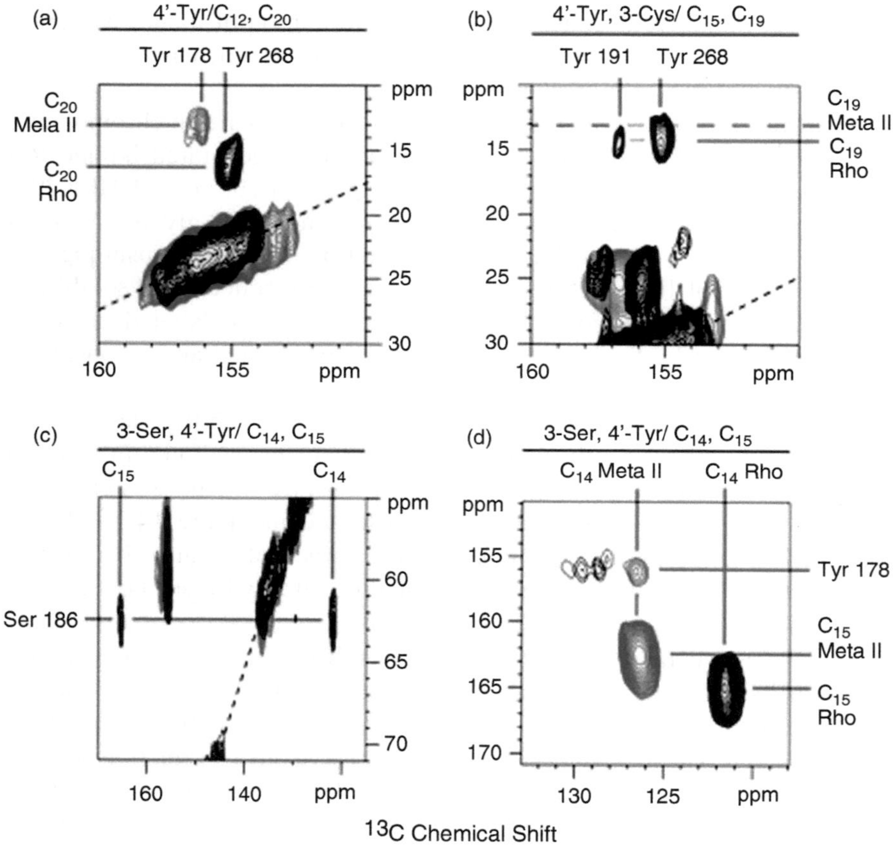

Figure 14.8. The 2D DARR NMR of rhodopsin and meta II. The expanded region of the full 2D NMR spectra of rhodopsin (black) and meta II (red) shown contain cross-peaks between the retinal and protein ^{13}C-labels. The protein and retinal labels are as follows: [4′-^{13}C]Tyr and [12,20-^{13}C]retinal (a); [4′-^{13}C]Tyr, [3-^{13}C]Cys, and [15,19-^{13}C]retinal (b); and [3-^{13}C]Ser, [4′-^{13}C]Tyr, [14,15-^{13}C]retinal (c and d) (from Patel *et al.*[71]).

longer data accumulations, indicating that C_{19} methyl groups has moved away from all of the tyrosines in the retinal-binding site. In the dark, the [14,15-^{13}C]retinal exhibit cross-peaks to Ser 186 (Figure 14.8(c)) but not to any of the tyrosines in the binding pocket (Figure 14.8(d)). In meta II, the C_{14}–Ser 186 and C_{15}–Ser 186 cross-peaks disappear and new cross-peaks are observed between C_{14} and Tyr 178 (Figure 14.8(d)). The essential aspects of the isomerization trajectory are large rotation of the C_{20} methyl group toward extracellular loops 2 and a 4- to 5-Å translation of the retinal chromophore toward transmembrane helix 5. ^{15}N NMR spectra of [U-^{13}C, U-^{15}N]Trp-labeled Rho were examined in order to probe the changes in hydrogen bonding upon Rho activation.[72] Mapping of binding site contacts in Rho and retinal was also examined by HETCOR spectra obtained with selective interface detection spectroscopy (SIDY) using REDOR dephasing and LG decoupling, to resolve $^{1}H_{GPCR}$ (Rho)–$^{13}C_{lig}$ (uniformly ^{13}C-labeled ligand, 11-*cis* retinal) signals:[72a] Thr 118–C_{19}, Phe 261–C_4, and Thr 265–C_8. In such studies, prior knowledge about 3D structure by X-ray diffraction is essential to locate such hidden pairs, however.

14.2.2. Photosynthetic Reaction Center

The photosynthetic reaction centers (RCs) from purple bacteria such as *Rps. viridis* and *Rb. spaeroides* are the first membrane proteins described at atomic resolution by X-ray diffraction.[73–75] The RC from *Rb. spaeroides* is a transmembrane protein complex consisting of three polypeptide chains (L, M, and H) and nine cofactors (two bacteriochlorophylls $(BChl)_2$ forming the so-called special pair (P), two accessory bacteriochlorophylls, two bacteriopheophytins (P_A and P_B), two quinones (Q_A and Q_B), and one Fe^{2+} ion arranged in an almost C_2 symmetry. Solid-state NMR studies[76–82] for RC from *Rb. spaeroides* have been performed to locate ^{13}C NMR signals of [4-^{13}C]Tyr-enriched preparations,[76] to characterize the functionally asymmetric Q_A binding using ^{13}C-enriched ubiquinone–10[77] and pheophytin cofactors,[78] and to study RC by 1D or 2D CP-MAS techniques under the condition of frozen detergent-solubilized state. Light-induced strongly polarized ^{15}N or ^{13}C NMR signals were observed from the cofactors of natural abundance or ^{15}N-labeled one[79–82] when forward electron transfer from the special pair P was blocked either by removal or prereduction of the quinone Q_A as electron acceptor from samples of dialysis precipitates against detergent-free buffer and then water or frozen detergent-solubilized preparation. This photo-CIDNP[83] proved to be very valuable means to examine information about the electron density in P in the radical pair, spin-density distribution of tetrapyrrole cofactors, and asymmetric electronic structure of $P^{\cdot +}$ in primary donor. Further, almost complete set of chemical shifts of the aromatic ring carbons of a single Chl *a* molecule was assigned to the P_2-cofactor of the primary electron donor P_{700}, as detected by photo-CIDNP in photosystem I of plants.[84]

The LH2 complex is a peripheral photosynthetic antenna pigment proteins utilized to absorb light and to transfer the excited-state energy to the light-harvesting LH1 complex surrounding the RC.[85] The crystal structure of the LH2 of *Rps. acidophilia*[86] shows a ring structure of nine identical units, each containing an α- and β-polypeptide of 53 and 41, respectively, both of which span the membrane once as α-helices. The two polypeptides bind a total of three chlorophyll molecules and two carotenoids. The nine hetero-dimeric units form a hollow cylinder with the α-chains forming the inner wall and the β-chains forming the outer walls.[85] Instead of whole complex, Alia *et al.*[87] recorded ^{15}N and ^{13}C NMR spectra of one protomer of [$^{13}C_6$, $^{15}N_3$]-, [π-^{15}N]-, and [τ-^{15}N]-histidine-labeled LH2 complex solubilized in detergent at 225 K, to gain insight into charged state of histidines and hydrogen-bonding status in this complex. Specific ^{15}N labeling confirmed that it is the τ-nitrogen of histidines, which is ligated to Mg^{2+} ion of B50 BChl molecules. Heteronuclear 2D correlation spectra of uniformly [^{13}C, ^{15}N]labeled LH2 complex were recorded at frozen state of detergent-solubilized preparation.[88] Instead, narrowed peaks of the intrinsic trans-membrane LH2 complex were achieved by extensive and selective biosyn-thesis prepared from [1,4-^{13}C]succinic acid or [2,3-^{13}C]succinic acid-labeled media and sequence-specific assignment of peak based on 2D dipolar correlation experiments.[89,90] Further, selective chemical shift assign-ment of B800 and B850 bacteriochlorophylls (BChl) in uniformly [^{13}C,^{15}N]-labeled LH2 complexes was performed by 2D RFDR correlation spectros-copy.[91] By correction of the ring current shifts by DFT calculations, the ^{13}C shift effects due to the interactions with the protein matrix with reference to those of monomeric BChl dissolved in acetone were attributed to the dielectrics of the protein environment, in contrast with local effects due to interaction with specific amino acid residues. Considerable shifts of $-6.2 < \Delta\sigma < +8.5$ ppm are detected for ^{13}C nuclei for both B800 and B850 bacteriochlorin rings.

14.2.3. Bacterial Chemoreceptors

Bacterial chemoreceptors mediate chemotaxis by recognizing specific chemicals and regulating a noncovalently associated histidine kinase. Ligand binding to the external domain of the membrane-spanning receptor generates a transmembrane signal that modulates kinase activity inside the cell.[92] Solid state REDOR NMR distances of serine bacterial chemotaxis receptor were measured to characterize specific structural feature of the ligand-binding site interactions in the intact, membrane-bound receptor, by preparing [^{15}N]Ser bound to a [^{13}C]Phe-receptor of lyophilized homodimer of 60 kDa.[93] The results indicate two 4.0 Å $\pm$ 0.2 Å distances, in excellent agreement with the X-ray crystal structure of a soluble fragment of the homologous aspartate receptor.[94] To clarify a plausible ligand-induced change in the transmembrane α-helices, ^{13}C $\cdots$ ^{19}F distances were measured by REDOR for [1-^{13}C]cysteine (α_1-helix) and [*ring*–4-^{19}F]phenylalanine (α_4-helix)-labeled serine bacterial chemoreceptor in the periplasmic domain, to result in the change of 1.0 Å $\pm$ 0.3 Å in the α_1 and α_4 chains.[95] Indeed this result is consistent with a 1.6 Å ''swinging piston'' motion of α_4/TM2 relative to α_1/TM1 within a subunit[96] rather than a ligand-induced 4° ''scissoring'' motion across the dimer interface,[94] proposed based on comparison of the crystal structures of the aspartate receptor ligand-binding domain fragment in the presence and absence of ligand molecule. To clarify this problem further, specifically ^{13}C-labeled serine bacterial chemoreceptor was prepared[97] to determine the distance between a unique Cys residue and a nonunique low abundance residue (Tyr or Phe). A ^{13}C–^{13}C internuclear distance measurement by RR method from ^{13}CO (i) to $^{13}C_\beta$ ($i+3$) at the periplasmic edge of the second membrane-spanning helix (TM2) of 5.1 Å $\pm$ 0.2 Å is consistent with the predicted α-helical structure. A second ^{13}C–^{13}C distance between the transmembrane helices 5.0 and 5.3 Å in the presence and absence of ligand is consistent with the structural model of this protein.[98]

14.2.4. Helical Structures and Helix–Helix Interaction of Membrane Proteins in Lipid Bilayers

Helix–helix interactions are important in the folding and function of membrane proteins with multiple membrane-spanning helices.[99] Dimerization of human glycophorin A in erythrocyte membrane is mediated by specific interactions within the helical transmembrane domain of the protein. Magnetization exchange rates of the transmembrane peptide (29-residue long) by RR measurements were measured between [^{13}C]methyl labels in the hydrophobic sequence –G79-V80-M81-A82-G83-V84– located in the middle of the transmembrane domain and specific [^{13}C]carbonyl labels along the peptide backbone across the dimer interface.[100] Significant magnetization exchange was observed only between V80 ($^{13}CH_3$) and G79 ($^{13}C{=}O$) and between V84 ($^{13}CH_3$) and G83 ($^{13}C{=}O$), indicating that these residues are packed in the dimer interface in a ''ridges-in-grooves'' arrangement. Further, it was shown that direct packing contacts occur between glycine residues at positions 79 and 83, between Ile 76 and Gly 79 and between Val 80 and Gly 83.[101] One of the unusual features of the seven-residue motif for dimerization, LIxxGVxxGVxxT, is the large number of β-branched amino acids that may limit the entropic cost of dimerization by restricting side-chain motion in the monomeric transmembrane helix.[102] RR measurement of a 2.9-Å distance between the γ-methyl and backbone carbonyl carbons of Thr 87 is consistent with a *gauche*-conformation for the χ_1 torsion angle. REDOR measurements demonstrate close packing (4.0 Å $\pm$ 0.2 Å) of the Thr 87 γ-methyl group with the backbone nitrogen of Ile 88 across the dimer interface. The short interhelical distance places the β-hydroxyl of Thr 87 within hydrogen-bonding range of the backbone carbonyl of Val 84 on the opposing helix. Deuterium NMR spectroscopy was used to characterize the dynamics of fully deuterated Val 80 and Val 84, two essential amino acids of the dimerization motif.[103] In fact, the deuterium line shapes in both the monomer or dimer are characterized by the fast methyl group rotation with virtually no motion about the C_α–C_β bond. This is consistent with restriction of the side chain in both the monomer and dimer due to intrahelical packing interactions involving the β-methyl groups.

Phospholamban (PLB) is a 52-residue integral membrane protein that is involved in reversibly inhibiting the Ca^{2+} pump and regulating the flow of Ca ions across the sarcoplasmic reticulum membrane during muscle contraction and relaxation. The internuclear distances of the synthetic protein (full-length) between [1-^{13}C]Leu 7 and [3-^{13}C]Ala 11 in the cytoplasmic region, between [1-^{13}C]Pro 21 and [3-^{13}C]Ala 24 in the juxtamembrane region, and between [1-^{13}C]Leu 42 and [3-^{13}C]Cys 46 in the transmembrane domain of PLB determined by RR measurements are consistent with α-helical secondary structure.[104] Additional REDOR measurements confirmed that the secondary structure is helical in the region of Pro 21 and that there are no large conformational changes upon phosphorylation. These results support the model of the PLB pentamer as a bundle of five long α-helices. In addition, ^{2}H NMR spectra of [C^2H_3]Leu-labeled TM-PLB (22 amino acid residues) at 28, 39, and 51 positions exhibited line shapes characteristic of either methyl group reorientation about the C_γ–C_δ bond axis or by additional librational motion about the C_α–C_β and C_β–C_γ bond axes.[105] It was shown that all of the residues are located inside the lipid bilayer and Leu 51 side chain has less motion than Leu 39 or Leu 28, which is attributed to its incorporation in the pentameric PLB leucine zipper motif. The ^{15}N powder spectra of Leu 39 and Leu 42 residues indicated no backbone motion, while

Leu 28 exhibited slight backbone motion. The ^{15}N chemical shift value from the mechanically aligned spectrum of ^{15}N-labeled Leu 39 PLB in DOPC/DOPE bilayers was 220 ppm and is characteristic of a TM peptide that is nearly parallel with the bilayer normal. ^{13}C NMR spectra of a double ^{13}C-labeled 24-residue synthetic peptide ([$^{13}C_2$]CAPLB$_{29-52}$), corresponding to the membrane-spanning sequence of PLB, showed that peptide motion became constrained in the presence of the SERCA1 (sarco(endo)plasmic reticulum) isoform of Ca^{2+}-ATPase.[106] ^{13}C–^{13}C distance determined by RR method confirmed that the sequence spanning Phe 32 and Ala 36 was α-helical and that this structure was not disrupted by interaction with Ca^{2+}-ATPase. Further, it was proposed on the basis of REDOR experiments that the sequence Ala 24-Gln 26 switches from an α-helix in pure lipid membrane to a more extended structure in the presence of SERCA1, which may reflect local structural distortions which change the orientations of the transmembrane and cytoplasmic domains.[106a] These results suggest that Ca^{2+}-ATPase has a long-range effect on the structure of PLB around residue–25, which promotes the functional association of the two proteins.

Receptor tyrosine kinases (RTK) are integral membrane proteins containing a large extracellular ligand-binding, a single transmembrane helix, and a large intracellular kinase domain. The Neu RTK provides the most direct evidence that specific interactions between transmembrane helices can mediate dimerization and receptor activation.[107] The local secondary structure and interhelical contacts in the region of position Val 664 in peptide models of the activated Neu receptor were analyzed by RR and REDOR NMR methods. Intrahelical ^{13}C RR distance measurements were made between [1-^{13}C]Thr 662 and [2-^{13}C]Gly 665 on peptides corresponding to the wild-type Neu and activated Neu transmembrane sequences containing valine and glutamate at position 664, respectively. Similar internuclear distances (4.5 Å $\pm$ 0.2 Å) in both Neu and Neu*, indicating that the region near residue–664 is helical and is not influenced by mutation. Interhelical $^{15}N \cdots {}^{13}C$ REDOR measurements between Gln 664 side chains on opposing helices were not consistent with hydrogen bonding between the side-chain functional groups. However, interhelical RR measurements between [1-^{13}C]Glu 664 and [2-^{13}C]Gly 665 and between [1-^{13}C]Gly 665 and [2-^{13}C]Gly 665 demonstrated close contacts (4.3–4.5 Å) consistent with the packing of Gly 665 in the Neu* dimer interface.

The orientational restraints for mechanically aligned transmembrane peptides are very precise structural restraints as described in Section 14.1, but they are not able to position domains or potentially even secondary structural elements relative to each other from experimental restraints alone.[108] As an additional restraint, for this reason, (interhelical) distance between $^{15}N_{\pi}$-labeled His 37 and $^{13}C_{\gamma}$-labeled Trp 41 of influenza viral coat protein

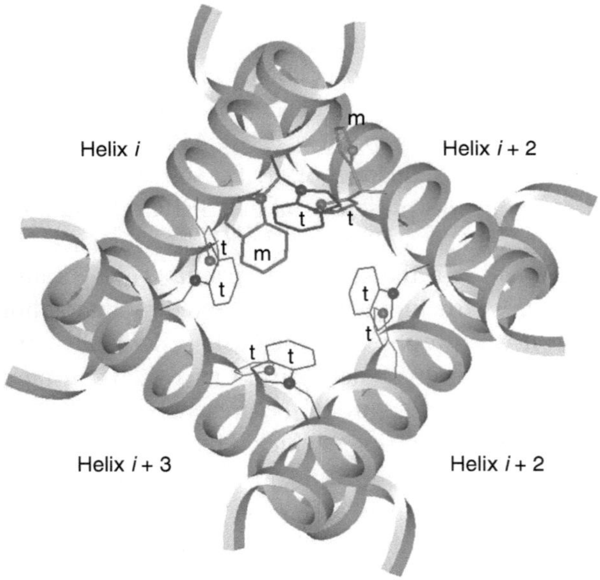

Figure 14.9. For an intermolecular interaction in this tetrameric structure, Trp 41 of helix i must be interacting with His 37 of helix $i+1$, and it is readily seen that both χ_1 rotamers must be in the t ($-177°$) state. Both t and m ($-65°$) states of Trp 41 are only shown in helix i, and t states of Trp 41 are shown in other helices. Both t and m states of His 37 are shown in helix $i+1$ and t states of His 37 are shown in other helices (from Nishimura *et al.*[108]).

in hydrated lipid bilayers was determined to be less than 3.9 Å by REDOR experiment. The distance restraints were utilized to search the side-chain orientations in the four helix bundle structural characterization by molecular dynamics calculation as illustrated in Figure 14.9. The channel appears to be closed by the proximity of the four indoles consistent with electrophysiology and mutagenesis studies of the intact protein at pH 7.0. The observation of a 2-kHz coupling in the PISEMA spectrum of $^{15}N_{\pi}$His 37 validates the orientation of the His 37 side chain based on the observed REDOR distance.

14.2.5. Ligand–Receptor Interactions and Ion-Channel Blocking

Solid state NMR is a suitable means to reveal conformation and dynamics of small ligand molecules bound to membrane-bound receptor molecules or to ion channel as an excellent means for discovery of potentially useful drug. Excellent review articles for this purpose are available.[109,110]

The nAcChoR is a ligand-gated ion channel (≈280 kDa) consisting of five glycosylated subunits ($\alpha_2\beta\gamma\delta$) for which acetylcholine is the agonist. The receptor is a major component of the electric organ of electric fish such as *Torpedeo marmorata*, from which enriched membranes containing ≈30–35% of the total protein is the receptor.[110] These membranes are used extensively for pharmacological characterizations and the receptor from this source is a model for the mammarian receptor.[110–114] As direct probes of the agonist-binding site, ^{13}C- or ^{2}H-labeled $N^+(CH_3)_3$-bromoacetylcholine (BAC) was utilized to probe directly the binding site.[111–113] The constrained agonist was observed by delayed dephasing, CP to detect only bound agonist in this freely exchanging ligand–receptor system by ^{13}C NMR.[111] ^{2}H NMR spectra of oriented nAcChoR membranes labeled with $N^+(CD_3)_3$-BAC on glass slides, prepared by using an isopotential spinning technique, were obtained at 0° and 90° to allow the orientational dependence of the ligand to be determined with respect to the membrane normal, as illustrated in Figure 14.10. Numerical simulations of this series of spectra permitted the orientation of the quaternary ammonium ion group of BAC to be deduced with respect to the membrane normal, revealing an angle of 42° (Figure 14.9(c)).

The M2 transmembrane domain of the δ-subunit of nAChoR from *Torpedo marmorata* in oriented system was analyzed in detail by ^{15}N PISEMA spectra to yield its secondary structure, as already illustrated in Figure 14.1.[114] Conformational feature of a synthetic 35-residue peptide representing the extended αM1 domain (206–240) of *Torpedo californica* nAChoR in lipid bilayer was explored by means of ^{13}C NMR.[115] The membrane-embedded αM1 segment forms an unstable α-helix, particularly near residue Leu 18 (αLeu 223 in the entire nAchoR).

Highly resolved ^{13}C NMR spectra were obtained from a system composed of a receptor–toxin complex.[116] The NMR sample consists of membrane preparation of the nAChoR of *T. californica* which was incubated with uniformly ^{13}C, ^{15}N-labeled neurotoxin II (61-residues, NTT II), found in the venom of the Asian cobra *Naja oxiana*. The comparison with so-called NMR data of the free toxin indicates that its overall structure is very similar when bound to the receptor, but significant changes were observed for one isoleucine.

Several biologically active derivatives of the cardiotonic steroid ouabain have been made by using NMR isotopes (^{13}C, ^{2}H, and ^{19}F) in the rhamnose sugar and steroid moieties, and examined at the digitalis receptor site of renal Na^+/K^+ATPase by a combination of solid state NMR methods.[117] Deuterium NMR spectra of ^{2}H-labeled inhibitors revealed that the sugar

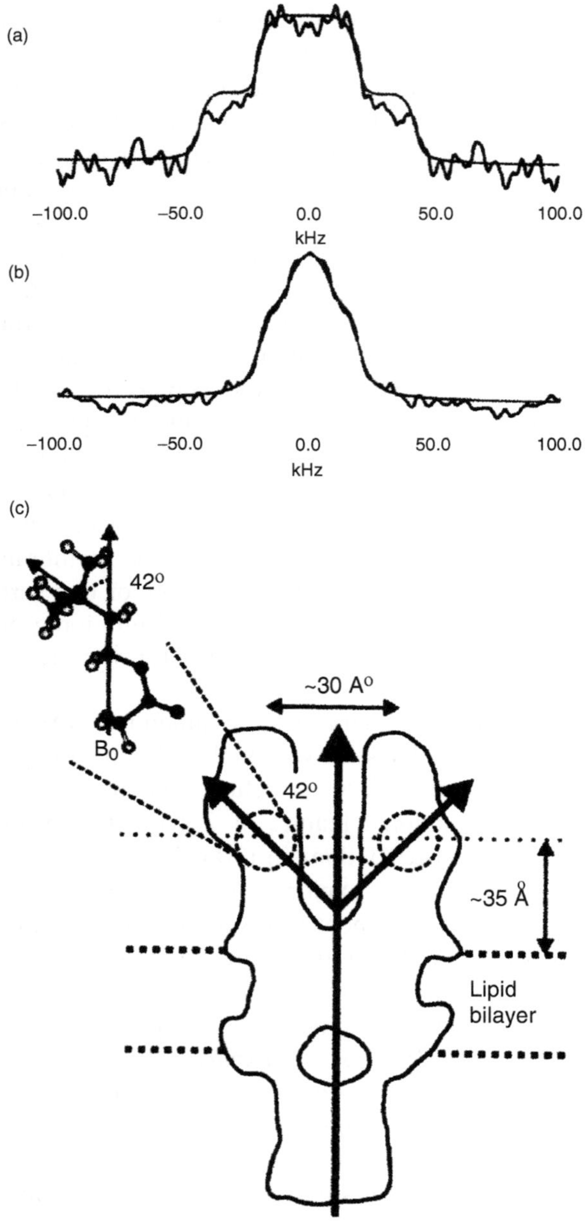

Figure 14.10. ^{2}H NMR spectra of the deuterium-labeled bromoacetylcholine bound to oriented nAcChoR oriented at 0° (a) and 90° (b) with respect to the magnetic field, together with simulated spectra. (c) Diagram showing the orientation of the ligand with respect to the overall morphology of the receptor membrane. The quaternary ammonium group is oriented at 42° ± 5° with respect to the membrane normal (from Williamson *et al.*[113]).

group was only loosely associated with the binding site, whereas the steroid group was more constrained. A ^{19}F, ^{13}C-REDOR was used to determine the structure of a selectively ^{19}F, ^{13}C-labeled inhibitor (Figure 14.11) in the digitalis receptor site of Na^+/K^+ATPase membranes, and it showed that the ouabain derivatives adopt a conformation in which the sugar extends out of the plane of the steroid ring system. The combined structural and dynamic information favors a model for inhibition in which the ouabain analogues lie across the surface of the Na^+/K^+-ATPase α-subunit with the sugar group facing away from the surface of the membrane but free to move into contact with one or more aromatic residues. A range of reversible-substituted imidazole [1,2-*a*]pyridine and irreversible inhibitors are available as a result of the need to design both types of inhibitors for potential treatment of peptic ulcers.[100] RR or ^{13}C REDOR measurements were performed to measure a precise distance of [10, 14-$^{13}C_2$]-labeled TMPIP (1,2,3-trimethyl–8-(phenylomethoxyl)-imidazo[1,2-*a*]pyridinium cation)[118] (Figure 14.12) or pentafluorphenylmethoxy analogue of TMPIP[119] bound to H^+/K^+-ATPase, respectively. The $\phi_1 = 165° \pm 15°$ determined for TMPIP by RR measurement was combined to deduce ϕ_2 and ϕ_3 by REDOR distances.

NMR methods have been adopted to observe directly the characteristics of specific substrate binding to natural membranes or liposomes containing sugar H^+ symport proteins Galp or FucP in the presence or absence of inhibitor, in which CP technique was utilized to discriminate bound substrate as little as 250 nmol.[120–123] Substrate affinities for membrane transport proteins, a nucleoside transporter NupC and glucuronide transporter

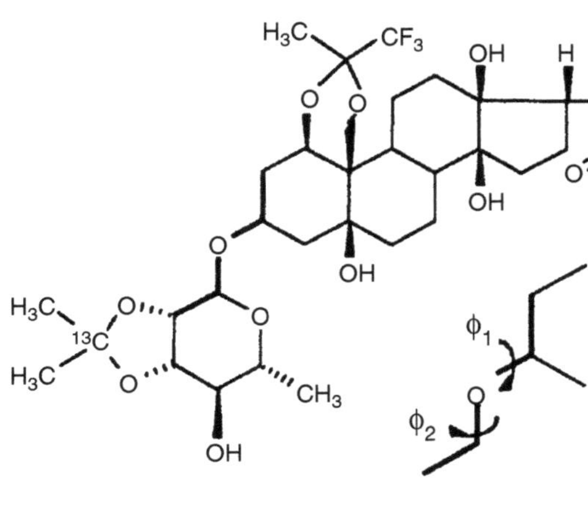

Figure 14.11. Ouabain diacetonide chemically modified to incorporate ^{19}F and ^{13}C at sites between which the distance has been determined using (^{19}F dephase, ^{13}C observe) REDOR experiment (from Middleton *et al.*[117]).

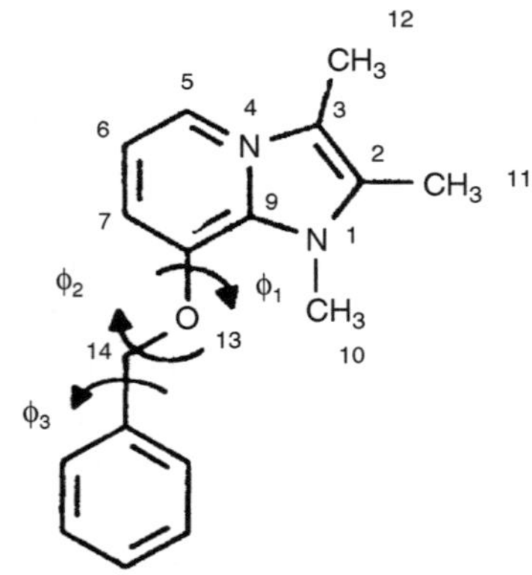

Figure 14.12. [10,14-$^{13}C_2$]labeled TMPIP (from Middleton *et al.*[118]).

GusB were also determined to yield binding affinity by ^{13}C CP-MAS NMR.[124] The binding of tetraphenylphosphonium (TPP$^+$) to EmrE, a membrane-bound, 110-residue *E. coli* multidrug transport protein was observed by ^{31}P CP-MAS NMR.[125] A population of bound ligand appears shifted 4 ppm to lower frequency compared to free ligand in solution, which suggests a rather direct and specific type of interaction of the ligand with the protein.

Functionally active analogues of neurotensin, NT (8–13) and NT (9–13), were observed by NMR while bound to the agonist-binding site of detergent-solubilized rat neurotensin receptor expressed as a recombinant protein in *E. coli.*[126,127] Significant shifts of NT (8–13) were observed under slow MAS and high-power proton decoupling in both the carboxyl terminus and tyrosine side chain of NT (8–13), suggesting that these sites are important in the interaction of the neurotensin with the agonist-binding site on the receptor.[126] Upon receptor binding, the peptides undergo conformational change to a β-stranded conformation, as viewed from the conformation-dependent chemical shifts determined by 2D ^{13}C–^{13}C correlation spectra of frozen uniformly ^{13}C, ^{15}N-labeled peptides/receptor preparation in detergent solution.[127]

References

[1] S. J. Opella and P. L. Stewart, 1989, *Methods Enzymol.*, 176, 242–275.
[2] T. Cross, 1997, *Methods Enzymol.*, 289, 672–697.
[3] S. J. Opella, 1997, *Nat. Struct. Biol.*, 4 (Suppl.), 845–848.

[4] S. J. Opella, C. Ma, and F. M. Marassi, 2001, *Methods Enzymol.*, 339, 285–313.
[5] F. M. Marassi, 2001, 80, *Biophys. J.*, 994–1003.
[6] S. J. Opella and F. M. Marassi, 2004, *Chem. Rev.*, 104, 3587–3606.
[7] A. Watts, S. K. Straus, S. L. Grage, M. Kamihira, Y. H. Lam, and X. Zhao, 2004, in *Protein NMR Techniques*, Second Edition, A. K. Downing, Ed., Methods in Molecular Biology, vol. 278, Humana Press, New Jersey, pp. 403–473.
[8] G. Gröbner, A. Taylor, P. T. F. Williamson, G. Choi, C. Glaubitz, J. A. Watts, W. J. de Grip, and A. Watts, 1997, *Anal. Biochem.*, 254, 132–138.
[9] G. Gröbner, G. Choi, I. J. Burnett, C. Glaubitz, P. J. E. Verdegem, J. Lugtenburg, and A. Watts, 1998, *FEBS Lett.*, 422, 201–204.
[10] R. G. Griffin, 1976, *J. Am. Chem. Soc.*, 98, 851–853.
[11] F. M. Marassi and K. J. Crowell, 2003, *J. Magn. Reson.*, 161, 64–69.
[12] A. S. Ulrich, M. Sami, and A. Watts, 1994, *Biochim. Biophys. Acta*, 1191, 225–230.
[13] J. Wang, S. Kim, F. Kovacs, and T. A. Cross, 2001, *Protein Sci.*, 10, 2241–2250.
[14] R. A. Lamb, L. J. Holsinger, and L. H. Pinto, 1994, in *Receptor Mediated Virus Entry into Cell*, E. Wimmer, Ed., Cold Spring Harbor Laboratory Press, New York, pp. 303–321.
[15] C. H. Wu, A. Ramamoorthy, and S. J. Opella, 1994, *J. Magn. Reson.*, 109, 270–272.
[16] F. M. Marassi and S. J. Opella, 2000, *J. Magn. Reson.*, 144, 150–155.
[17] J. Wang, J. Denny, C. Tian, S. Kim, Y. Mo, F. Kovacs, Z. Song, K. Nishimura, Z. Gan, R. Fu, J. R. Quine, and T. A. Cross, 2000, *J. Magn. Reson.*, 144, 162–167.
[18] C. Tian, K. Tobler, R. A. Lamb, L. H. Pinto, and T. A. Cross, 2002, *Biochemistry*, 41, 11294–11300.
[19] F. M. Marassi, C. Ma, H. Gratkowski, S. K. Straus, K. Strebel, M. Oblatt-Montal, M. Montal, and S. J. Opella, 1999, *Proc. Natl. Acad. Sci. USA*, 96, 14336–14341.
[20] S. H. Park, A. A. Mrse, A. A. Nevzorov, M. F. Mesleh, M. Oblatt-Montal, M. Montal, and S. J. Opella, 2003, *J. Mol. Biol.*, 333, 409–424.
[21] M. F. Mesleh, G. Veglia, T. M. DeSilva, F. M. Marassi, and S. J. Opella, 2002, *J. Am. Chem. Soc.*, 124, 4206–4207.
[22] A. S. Ulrich, M. P. Heyn, and A. Watts, 1992, *Biochemistry*, 31, 10390–10399.
[23] A. S. Ulrich and A. Watts, 1994, *Biophys. J.*, 66, 1441–1449.
[24] A. S. Ulrich, A. Watts, I. Wallat, and M. P. Heyn, 1994, *Biochemistry*, 33, 5370–5375.
[25] A. S. Ulrich, I. Wallat, M. P. Heyn, and A. Watts, 1995, *Nat. Struct. Biol.*, 2, 190–192.
[26] S. Moltke, A. A. Nevzorov, N. Sakai, I. Wallat, C. Job, K. Nakanishi, M. P. Heyn, and M. F. Brown, 1998, *Biochemistry*, 37, 11821–11835.
[27] S. Moltke, I. Wallat, N. Sakai, K. Nakanishi, M. F. Brown, and M. P. Heyn, 1999, *Biochemistry*, 38, 11762–11772.
[28] V. Copie, A. E. McDermott, K. Beshah, J. C. Williams, M. Spijker-Assink, R. Gebhard, J. Lugtneburg, J. Herzfeld, and R. G. Griffin, 1994, *Biochemistry*, 33, 3280–3286.
[29] C. Glaubitz, I. J. Burnett, G. Gröbner, A. J. Mason, and A. Watts, 1999, *J. Am. Chem. Soc.*, 121, 5787–5794.
[30] A. J. Mason, S. L. Grage, S. K. Straus, C. Glaubitz, and A. Watts, 2004, *Biophys. J.*, 86, 1610–1617.
[31] F. M. Marassi, 2001, *Biophys. J.*, 80, 994–1003.

[32] T. Vosegaard and N. C. Nielsen, 2002, *J. Biomol. NMR*, 22, 225–247.

[33] M. Kamihira, T. Vosegaard, A. J. Mason, S. K. Straus, N. C. Nielsen, and A. Watts, 2005, *J. Struct. Biol.*, 149, 7–16.

[34] M. Bak, R. Schultz, T. Vosegaard, and N. C. Nielsen, 2002, *J. Magn. Reson.*, 154, 28–45.

[35] M. Bak, J. T. Rasmussen, and N. C. Nielsen, 2000, *J. Magn. Reson.*, 147, 296–330.

[36] L.-O. Essen, 2001, *Chembiochem*, 2, 513–516.

[37] K. Palczewski, T. Kumasaka, T. Hori, C. A. Behnke, H. Motoshima, B. A. Fox, I. Le Trong, D. C. Teller, T. Okada, R. E. Stenkamp, M. Yamamoto, and M. Miyano, 2000, *Science*, 289, 739–745.

[38] G. Gröbner, G. Choi, I. J. Burnett, C. Glaubitz, P. J. E. Verdegem, J. Lugtenburg, and A. Watts, 1998, *FEBS Lett.*, 422, 201–204.

[39] G. Gröbner, I. J. Burnett, C. Galubitz, G. Choi, A. J. Mason, and A. Watts, *Nature*, 2000, 405, 810–813.

[40] R. D. Gilardi, I. L. Karle, and J. Karle, 1972, *Acta Crystallogr.*, B28, 2605–2612.

[41] D. C. Teller, T. Okada, C. A. Behnke, K. Palczewski, and Stenkamp, 2001, *Biochemistry*, 40, 7761–7772.

[42] T. Okada, Y. Fujiyoshi, M. Silow, J. Navarro, E. M. Landau, and Y. Shichida, 2002, *Proc. Natl. Acad. Sci. USA*, 99, 5982–5987.

[43] L. C. P. J. Mollevanger, A. P. M. Kentgens, J. A. Pardoen, J. M. L. Courtin, W. S. Veeman, J. Lugtenburg, and W. J. de Grip, 1987, *Eur. J. Biochem.*, 163, 9–14.

[44] S. O. Smith, I. Palings, V. Copie, D. P. Raleigh, J. Courtin, J. A. Pardoen, J. Lugtenburg, R. A. Mathies, and R. G. Griffin, 1987, *Biochemistry*, 26, 1606–1611.

[45] P. J. R. Spooner, J. M. Sharples, M. A. Verhoeven, J. Lugtenburg, C. Glaubitz, and A. Watts, 2002, *Biochemistry*, 41, 7549–7555.

[46] G. F. J. Salgado, A. V. Struts, K. Tanaka, N. Fujioka, K. Nakanishi, and M. F. Brown, 2004, *Biochemistry*, 43, 12819–12828.

[47] A. E. McDermott, F. Creuzet, R. Gebhard, K. van der Hoef, M. H. Levitt, J. Herzfeld, J. Lugtenburg, and R. G. Griffin, 1994, *Biochemistry*, 33, 6129–6136.

[48] L. K. Thompson, A. E. McDermott, J. Raap, C. M. van der Wielen, J. Lugtenburg, J. Herzfeld, and R. G. Griffin, 1992, *Biochemistry*, 31, 7931–7938.

[49] K. V. Lakshmi, M. Auger, J. Raap, J. Lugtenburg, R. G. Griffin, and J. Herzfeld, 1993, *J. Am. Chem. Soc.*, 115, 8515–8516.

[50] M. Helmle, H. Patzelt, A. Ockenfels, W. Gärtner, D. Oesterhelt, and B. Bechinger, 2000, *Biochemistry*, 39, 10066–10071.

[51] A. T. Petkova, M. Hatanaka, C. P. Jaroniec, J. G. Hu, M. Belenky, M. Verhoeven, J. Lugtenburg, R. G. Griffin, and J. Herzfeld, 2002, *Biochemistry*, 41, 2429–2437.

[52] M. Engelhard, S. Finkler, G. Metz, and F. Siebert, 1996, *Eur. J. Biochem.*, 235, 526–533.

[53] J. C. Lansing, J. G. Hu, M. Belenky, R. G. Griffin, and J. Herzfeld, 2003, *Biochemistry*, 42, 3586–3593.

[54] J. Hu, R. G. Griffin, and J. Herzfeld, 1994, *Proc. Natl. Acad. Sci. USA*, 91, 8880–8884.

[55] J. G. Hu, R. G. Griffin, and J. Herzfeld, 1997, *J. Am. Chem. Soc.*, 119, 9495–9498.

[56] J. G. Hu, B. Q. Sun, A. T. Petkova, R. G. Griffin, and J. Herzfeld. 1997, *Biochemistry*, 36, 9316–9322.

[57] G. Metz, F. Siebert, and M. Engelhard, 1992, *FEBS Lett.*, 303, 237–241.

[58] J. M. Griffiths, A. E. Bennett, M. Engelhard, F. Siebert, J. Raap, J. Lugtenburg, J. Herzfeld, and R. G. Griffin, 2000, *Biochemistry*, 39, 362–371.

[59] C. P. Jaroniec, J. C. Lansing, B. A. Tounge, M. Belenky, J. Herzfeld, and R. G. Griffin, 2001, *J. Am. Chem. Soc.*, 123, 12929–12930.

[60] A. T. Petkova, J. G. Hu, M. Bizounok, M. Simpson, R. G. Griffin, and J. Herzfeld, 1999, *Biochemistry*, 38, 1562–1572.

[61] P. J. E. Verdegem, P. H. M. Bovee-Geurts, W. J. de Grip, J. Lugtenburg, and H. J. M. de Groot, 1999, *Biochemistry*, 38, 11316–11324.

[62] M. A. L. Verhoven, A. F. L. Creemers, P. H. M. Bovee-Geurts, W. J. de Grip, J. Lugtenburg, and H. J. M. de Groot, 2001, *Biochemistry*, 40, 3282–3288.

[63] A. F. Creemers, S. Kiihne, P. H. Bovee-Geurts, W. J. de Grip, J. Lugtenburg, and H. J. M. de Groot, 2002, *Proc. Natl. Acad. Sci. USA*, 99, 9101–9106.

[64] A. F. L. Creemers, P. H. Bovee-Geurts, W. J. de Grip, J. Lugtenburg, and H. J. M. de Groot, 2004, *Biochemistry*, 43, 16011–16018.

[65] X. Feng, P. J. E. Verdegem, Y. K. Lee, D. Sandström, M. Edén, P. Bovee-Geurts, W. J. de Grip, J. Lugtenburg, H. J. M. de Groot, and M. H. Levitt., 1997, *J. Am. Chem. Soc.*, 119, 6853–6857.

[66] X. Feng, P. J. E. Verdegem, M. Eden, D. Sandstrom, Y. K. Lee, P. H. M. Bovee-Geurts, W. J. de Grip, J. Lugtenburg, H. J. M. de Groot, and M. H. Levitt, 2000, *J. Biomol. NMR*, 16, 1–8.

[67] P. J. R. Spooner, J. M. Sharples, S. C. Goodall, H. Seedorf, M. A. Verhoeven, J. Lugtenburg, P. H. M. Bovee-Geurts, W. J. de Grip, and A. Watts, 2003, *Biochemistry*, 42, 13371–13378.

[68] M. Eilers, P. J. Reeves, W. Ying, H. G. Khorana, and S. O. Smith, 1999, *Proc. Natl. Acad. Sci. USA*, 96, 487–492.

[69] A. F. L. Creemers, C. H. W. Klaassen, P. H. M. Bovee-Geurts, R. Kelle, U. Kragl, J. Raap, W. J. de Grip, J. Lugtneburg, and H. J. M. de Groot, 1999, *Biochemistry*, 38, 7195–7199.

[70] E. Crocker, A. B. Patel, M. Eilers, S. Jayaraman, E. Getmanova, P. J. Reeves, M. Ziliox, H. G. Khorana, M. Sheves, and S. O. Smith, 2004, *J. Biomol. NMR*, 29, 11–20.

[71] A. B. Patel, E. Crocker, M. Eilers, A. Hirshfeld, M. Sheves, and S. O. Smith, 2004, *Proc. Natl. Acad. Sci. USA*, 101, 10048–10053.

[72] A. B. Patel, E. Crocker, P. J. Reeves, E. V. Getmanova, M. Eilers, H. G. Khorana, and S. O. Smith, 2005, *J. Mol. Biol.*, 347, 803–812.

[72a] S. R. Kiihne, A. F. L. Creemers, W. J. de Grip, P. H. M. Bovee-Geurts, J. Lugtenburg, and H. J. M. de Groot, 2005, *J. Am. Chem. Soc.*, 127, 5734–5735.

[73] J. Deisenhofer and H. Michel, 1989, *EMBO J.*, 8, 2149–2170.

[74] T. O. Yeates, H. Komiya, A. Chirino, D. C. Rees, J. P. Allen, and G. Feher, 1988, *Proc. Natl. Acad. Sci. USA*, 85, 7993–7997.

[75] D. C. Rees, H. Komiya, T. O. Yeates, J. P. Allen, and G. Feher, 1989, *Annu. Rev. Biochem.*, 58, 607–633.

[76] M. R. Fischer, H. J. M. de Groot, J. Raap, C. Winkel, A. J. Hoff, and J. Lugtenburg, 1992, *Biochemistry*, 31, 11038–11049.

[77] W. B. S. van Liemt, G. J. Boender, P. Gast, A. J. Hoff, J. Lugtenburg, and H. J. M. de Groot, 1995, *Biochemistry*, 34, 10229–10236.

[78] T. A. Egorova-Zachernyuk, B. van Rossum, G. J. Boender, E. Franken, J. Ashurst, J. Raap, P. Gast, A. J. Hoff, H. Oschkinat, and H. J. M. de Groot, 1997, *Biochemistry*, 36, 7513–7519.

[79] M. G. Zysmilich and A. McDermott, 1994, *J. Am. Chem. Soc.*, 116, 8362–8363.

[80] M. G. Zysmilich and A. McDermott, 1996, *J. Am. Chem. Soc.*, 118, 5867–5873.

[81] M. G. Zysmilich and A. McDermott, 1996, *Proc. Natl. Acad. Sci. USA*, 93, 6857–6860.

[82] J. Matysik, Alia, P. Gast, H. J. van Gorkom, A. J. Hoff, and H. J. M. de Groot, 2000, *Proc. Natl. Acad. Sci. USA*, 97, 9865–9870.

[83] R. Kaptein, 1977, in *Introduction to Chemically Induced Magnetic Polarization*, L. T. Muss, Ed., D. Reidel, Dordrecht, The Netherlands.

[84] Alia, E. Roy, P. Gast, H. J. van Gorkom, H. J. M. de Groot, G. Jeschke, and J. Matysik, 2004, *J. Am. Chem. Soc.*, 126, 12819–12826.

[85] C. Branden and E. J. Tooze, 1999, *Introduction to Protein Structure*, Second Edition, Chapter 12, Garland Publishing, New York.

[86] G. McDermott, S. M. Prince, A. A. Freer, A. M. Hawthornthwaite-Lawless, M. Z. Papiz, R. J. Cogdell, and N. W. Isaacs, 1995, *Nature*, 374, 517–521.

[87] Alia, J. Matysik, C. Soede-Huijbregts, M. Baldus, J. Raap, J. Lugtenburg, P. Gast, H. J. van Gorkom, A. J. Hoff, and H. J. M. de Groot, *J. Am. Chem. Soc.*, 2001, 123, 4803–4809.

[88] T. A. Egorova-Zachernyuk, J. Hollander, N. Fraser, P. Gast, A. J. Hoff, R. Cogdell, H. J. M. de Groot, and M. Baldus, 2001, *J. Biomol. NMR*, 19, 243–253.

[89] A. J. van Gammeren, F. B. Hulsbergen, J. G. Hollander, and H..J. M de Groot, 2004, *J. Biomol. NMR*, 30, 267–274.

[90] A. J. van Gammeren, F. B. Hulsbergen, J. G. Hollander, and H. J. M. de Groot, 2005, *J. Biomol. NMR*, 31, 279–293.

[91] A. J. van Gammeren, F. Buda, F. B. Hulsbergen, S. Kiihne, J. G. Hollander, T. A. Egorova-Zachernyuk, N. J. Fraser, R. J. Cogdell, and H. J. M. de Groot, 2005, *J. Am. Chem. Soc.*, 127, 3213–3219.

[92] J. J. Falke and G. L. Hazelbauer, 2001, *Trends Biochem. Sci.*, 26, 257–265.

[93] J. Wang, Y. S. Balazs, and L. K. Thompson, 1997, *Biochemistry*, 36, 1699–1703.

[94] M. V. Milburn, G. G. Prive, D. L. Milligan, W. G. Scott, J. Yeh, J. Jancarik, D. E. Koshland, Jr., and S. H. Kim, 1991, *Science*, 254, 1342–1347.

[95] O. J. Murphy III, F. A. Kovacs, E. L. Sicard, and L. K. Thompson, 2001, *Biochemistry*, 40, 1358–1366.

[96] S. A. Chervitz and J. J. Falke, 1996, *Proc. Natl. Acad. Sci. USA*, 93, 2545–2550.

[97] B. Isaac, G. J. Gallagher, Y. S. Balazs, and L. K. Thompson, 2002, *Biochemistry*, 41, 3025–3036.

[98] K. K. Kim, H. Yokota, and S. H. Kim, 1999, *Nature*, 400, 787–792.

[99] J. L. Popot and D. M. Engelman, 2000, *Annu. Rev. Biochem.*, 69, 881–922.

[100] S. O. Smith and B. J. Bormann, 1995, *Proc. Natl. Acad. Sci. USA*, 92, 488–491.

[101] S. O. Smith, D. Song, S. Shekar, M. Groesbeek, M. Ziliox, and S. Aimoto, 2001, *Biochemistry*, 40, 6553–6558.

[102] S. O. Smith, M. Eilers, D. Song, E. Crocker, W. Ying, M. Groesbeek, G. Metz, M. Ziliox, and S. Aimoto, 2002, *Biophys. J.*, 82, 2476–2486.

[103] W. Liu, E. Crocker, D. J. Siminovitch, and S. O. Smith, 2003, *Biophys. J.*, 84, 1263–1271.

[104] S. O. Smith, T. Kawakami, W. Liu, M. Zilix, and S. Aimoto, 2001, *J. Mol. Biol.*, 313, 1139–1148.

[105] E. K. Tiburu, E. S. Karp, P. C. Dave, K. Damodaran, and G. A. Lorigan, 2004, *Biochemistry*, 43, 13899–13909.

[106] Z. Ahmed, D. G. Reid, A. Watts, and D. A. Middleton, 2000, *Biochim. Biophys. Acta*, 1468, 187–198.

[106a] E. Hughes and D. A. Middleton, 2002, *J. Biol. Chem.*, 278, 20835–20842.
[107] N. E. Hynes and D. F. Stern, 1994, *Biochim. Biophys. Acta*, 1198, 165–184.
[108] K. Nishimura, S. Kim, L. Zhang, and T. A. Cross, 2002, *Biochemistry*, 41, 13170–13177.
[109] A. Watts, 1999, *Curr. Opin. Biotechnol.*, 10, 48–53.
[110] A. Watts, 2002, *Mol. Membr. Biol.*, 19, 267–275.
[111] P. T. F. Williamson, G. Grobner, P. J. Spooner, K. W. Miller, and A. Watts, 1998, *Biochemistry*, 37, 10854–10859.
[112] P. T. F. Williamson, J. A. Watts, G. H. Addona, K. W. Miller, and A. Watts, 2001. *Proc. Natl. Acad. Sci. USA*, 98, 2346–2351.
[113] P. T. F. Williamson, B. H. Meier, and A. Watts, 2004, *Eur. Biophys. J.*, 33, 247–254.
[114] S. J. Opella, F. M. Marassi, J. J. Gesell, A. P. Valente, Y. Kim, M. Oblatt-Montal, and M. Montal, 1999, *Nat. Struct. Biol.*, 6, 374–379.
[115] M. R. R. de Planque, D. T. S. Rijkers, J. I. Fletcher, R. M. J. Liskamp, and F. Separovic, 2004, *Biochim. Biophys. Acta*, 1665, 40–47.
[116] L. Krabben, B.-J. van Rossum, F. Castellani, E. Bocharov, A. A. Schulga, A. S. Arseniev, C. Weise, F. Hucho, and H. Oschkinat, 2004, *FEBS Lett.*, 564, 319–324.
[117] D. A. Middleton, S. Rankin, M. Esmann, and A. Watts, 2000, *Proc. Natl. Acad. Sci. USA*, 97, 13602–13607.
[118] D. A. Middleton, R. Robins, X. Feng, M. H. Levitt, I. D. Spiers, C. H. Schwalbe, D. G. Reid, and A. Watts, 1997, *FEBS Lett.*, 410, 269–274.
[119] J. A. Watts, A. Watts, and D. A. Middleton, 2001, *J. Biol. Chem.*, 276, 43197–43204.
[120] P. J. R. Spooner, N. G. Rutherford, A. Watts, and P. J. F. Henderson, 1994, *Proc. Natl. Acad. Sci. USA*, 91, 3877–3881.
[121] A. N. Appleyard, R. B. Herbert, P. J. F. Henderson, A. Watts, and P. J. R. Spooner, 2000, *Biochim. Biophys. Acta*, 1509, 55–64.
[122] P. J. R. Spooner, W. J. O'Reilly, S. W. Homans, N. G. Rutherford, P. J. F. Henderson, and A. Watts, 1998, *Biophys. J.*, 75, 2794–2800.
[123] S. G. Patching, A. R. Brough, R. B. Herbert, J. A. Rajakarier, P. J. F. Henderson, and D. A. Middleton, 2004, *J. Am. Chem. Soc.*, 126, 3072–3080.
[124] C. Glaubitz, A. Gröger, K. Gottschalk, P. Spooner, A. Watts, S. Schuldiner, and H. Kessler, 2000, *FEBS Lett.*, 480, 127–131.
[125] P. T. F. Williamson, S. Bains, C. Chung, R. Cooke, and A. Watts, 2002, *FEBS Lett.*, 518, 111–115.
[126] S. Luca, J. F. White, A. K. Sohal, D. V. Filippov, J. H. van Boom, R. Grisshammer, and M. Baldus, 2003, *Proc. Natl. Acad. Sci. USA*, 100, 10706–10711.
[127] H. Heise, S. Luca, B. L. de Groot, H. Grubmüller, and M. Baldus, 1994, *Biophys.* J., 89, 2113–2120.

Chapter 15

BIOLOGICALLY ACTIVE MEMBRANE-ASSOCIATED PEPTIDES

A variety of biologically active membrane-associated peptides, such as channel-forming, antimicrobial or fusion peptides, and ligand molecules to receptors such as GPCRs, play important role for their respective biological activities. A general approach toward determination of secondary structure and dynamics for such peptides by solid state NMR is the same as that of membrane proteins as discussed already in Chapters 13 and 14. As a result, a large number of conformational constraints are available from membrane-bound peptides either at frozen or dried condition, based on measured distances or dihedral angles using the recoupling techniques (Sections 6.2 and 6.3). Further, the tilt angles of helices with respect to the bilayer normal are accessible by using either mechanically or spontaneously, magnetically aligned system, as already discussed in Section 6.1.2. Nevertheless, it should be always taken into account that such peptides are very flexible in lipid bilayers under physiological condition either at ambient temperature or fully hydrated state.

15.1. Channel-Forming Peptides

15.1.1. Gramicidin A

Gramicidin A (gA), a major synthetic product of *Bacillus brevis*, is a polypeptide of 15 amino acid residues whose primary sequence was described in Section 6.1.2, and forms a monovalent cation selective channel that is dimeric, but single stranded. The high-resolution structure of the channel monomer has been defined with 120 precise orientation constraints from solid state NMR of uniformly aligned sample in bilayers,[1–3] as illustrated in Figure 6.4. It forms a single-stranded helix with a right-handed sense, 6.5 residues per turn, and β-strand torsion angles. The monomer–monomer geometry (amino-terminal-to-amino-terminal) has been characterized by solution NMR in SDS micelles[4] in which the monomer fold proved to be the same as in lipid bilayers. Naturally, the refined solid state NMR constraints for gA in a lipid bilayer[4] are not consistent with an X-ray crystallographic structure for gramicidin having a double-stranded, right-handed helix with 7.2 residues per turn in the absence of lipids.[5]

The intermolecular distance measurements generated here provide a straightforward approach for characterizing the dimeric structure of gA in lipid bilayers.[6] There are two choices for specific $^{13}C/^{15}N$ isotropic labeling of gA. The ^{13}C and ^{15}N labels are incorporated into each monomer, but in different sites. Based on the high-resolution monomer structure,[2] the intra-monomer distance is 4.21 Å and the orientation of the internuclear vector with respect to the motional axis is 35.3°, yielding a scaling factor of 0.5 for the dipolar interaction. The intermonomer distance between these $^{13}C/^{15}N$ labels is 9.47 Å, too long to yield a detectable dipolar coupling, based on the model of the amino-terminal-to-amino-terminal hydrogen-bonded single-stranded dimer. In the $^{13}C_1$-Val 1, ^{15}N-Ala 5 gA sample, the intramonomer distance is 8.19 Å with a scaling factor of 0.36, resulting from an orientation of 40.9° with respect to the global motion axis. Such a dipolar coupling is too weak to be detectable. On the other hand, the intermonomer distance between the labels is 4.03 Å with an angle of 2.2° between the internuclear vector and the motional axis. The scaling factor resulting from this small angle is negligible, 0.997. Such a distance constraint could be observable.

In the case of $^{13}C_1$-Val 7, ^{15}N-Gly 2 gA in unoriented hydrated DMPC bilayers, ^{13}C–^{15}N interatomic distance was determined to be 4.2 ± 0.2 Å between the labeled sites. This result is in good agreement with the one calculated from the high-resolution structure of the monomer.[2] Similarly, interatomic distance of $^{13}C_1$-Val 1, ^{15}N-Ala 5 gA in unoriented hydrated DMPC bilayers was determined to be 4.3 ± 0.1 Å. This result agrees well with the gA dimers structure in SDS micelles.

15.1.2. Melittin

Melittin is a hexacosapeptide with a primary structure of Gly-Ile-Gly-Ala-Val-Leu-Lys-Val-Leu-Thr-Thr-Gly-Leu-Pro-Ala-Leu-Ile-Ser-Trp-Ile-Lys-Arg-Lys-Arg-Gln-Gln-NH_2, and is a main component of bee venom.[7] Melittin has a powerful hemolytic activity[8] in addition to voltage-dependent ion-conductance across planar lipid bilayers at low concentration.[9] It also causes selective micellization of bilayers as well as membrane fusion at high concentration.[10] As the temperature is lowered to the gel phase, the membranes break down into small particles. Upon raising the temperature above the gel-to-liquid crystalline phase transition temerature (T_c), the small particles re-form to unilamellar vesicles. It is proposed that the bilayer discs surrounded by a belt of melittin molecules are formed at a temperature below the T_c.[11]

15.1.2.1. Morphological changes of lipid bilayers

Giant vesicles with diameters of ~20 μm were observed by optical microscopy for melittin–DMPC bilayers at 27.9 °C.[12] When the temperature was lowered to 24.9 °C ($T_c = 23$ °C for the neat DMPC bilayers), the surface of vesicles became blurred and dynamic pore formation was visible in the microscopic picture taken at different exposure times.[13] These vesicles disappeared completely at 22.9 °C. It was thus found that the melittin–lecithin bilayers reversibly undergo their fusion and disruption near the respective T_c. The fluctuation of lipids is responsible for the membrane fusion above the T_c and association of melittin molecules causes membrane lysis below the T_c.

Figure 15.1 shows the static ^{31}P DD-NMR spectra of melittin–DMPC bilayers hydrated with Tris buffer recorded at various temperatures.[13] Immediately after the sample was placed in the magnetic field, the ^{31}P NMR spectrum of an axially symmetric powder pattern characteristic of the liquid crystalline phase was initially recorded at 40 °C. The upper field edge ($\delta_\perp$) is more intense than the lower field edge ($\delta_\parallel$) as compared with a normally axially symmetric powder pattern. This finding indicates that the DMPC bilayer is partially aligned to the applied magnetic field with the bilayer plane being parallel by forming elongated vesicles. When the temperature was lowered to 30 °C, the intensity of the upper field edge of the powder pattern was further increased, leading to the spectrum of the almost complete alignment to the magnetic field. At 25 °C, the axially symmetric powder pattern appeared again. This spectrum changed to a broad envelope of the powder pattern with round edges at 20 °C due to the presence of a large amplitude motion in addition to a rotational motion about the molecular axis by lateral diffusion of the lipid molecule. At 10 °C, the isotropic ^{31}P NMR signal is dominated near 0 ppm, because of the isotropic rapid tumbling motion of small particles caused by melittin-induced lysis of larger vesicles. The same axially symmetric powder patterns appeared again when the temperature was raised from 10 °C to 25 °C, as a result of fusion to form larger spherical vesicles. At a temperature above 30 °C, the single

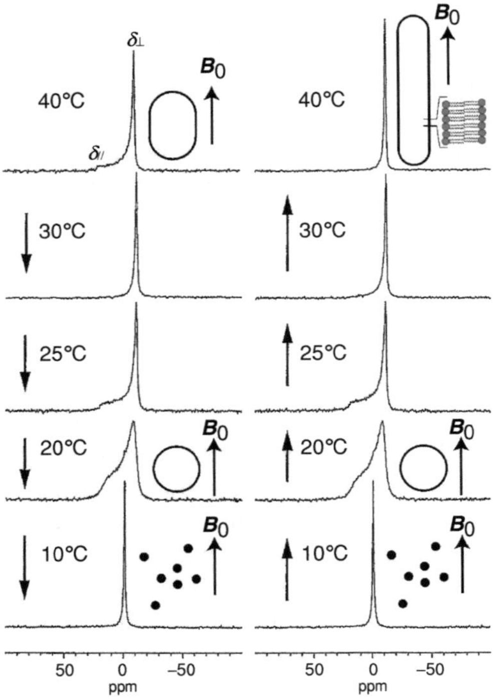

Figure 15.1. Temperature variation of the ^{31}P NMR spectra of the melittin–DMPC bilayer systems. The arrows indicate the direction of the temperature variation. $\boldsymbol{B}_0$ indicates the direction of the static magnetic field. The shapes of the vesicles are also depicted.[13]

perpendicular component appeared at -12 ppm, arising from the anisotropic ^{31}P chemical shift tensor of liquid crystalline bilayers. This result indicates that the lipid bilayer surface is oriented parallel to the magnetic field with a higher order of alignment by forming longer elongated vesicles referred to the magnetically oriented vesicle systems (MOVS).[14]

Morphology of lipid bilayer induced by melittin is shown in Figure 15.2. In the microscopic observation, it is evident that melittin is strongly bound to the vesicles and distributed homogeneously at a temperature above the T_c. When a temperature is close to the T_c, melittin molecules associated each other to cause phase separation as observed in fluorescent microscopy. Consequently, a large amplitude fluctuation of lipid molecules occurs near the T_c as shown in Figure 15.2. It was shown that melittin forms the pseudo-transmembrane α-helix with amphiphilic nature.[14] Nevertheless, melittin can stay homogeneously as a monomeric form in the hydrophobic environments when the lipid bilayer takes liquid crystalline phase above the T_c. At a temperature close

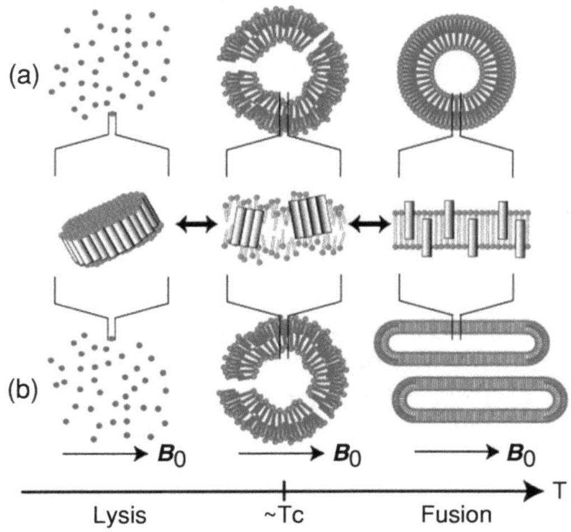

Figure 15.2. Schematic representation of the process of morphological changes in the melittin–lecithin bilayer systems in the absence (a) and presence (b) of applied magnetic field.

to the T_c, melittin molecules associate with each other by facing hydrophilic side together and facing hydrophobic side to the lipid to cause a larger phase separation and partial disorder of lipid. At a temperature below the T_c, a large number of melittin molecules associate each other. Consequently, small lipid bilayer particles are surrounded by the belt of melittin to be released from the vesicle, resulting in the membrane disruption. Subsequently, entire vesicles are dissolved in the buffer solution. On the other hand, when a temperature is slightly above the T_c, a small number of associated melittin molecules still cause a large amplitude motion of lipids in addition to the motion about the molecular axis. This large amplitude motion of lipids may fluctuate the surface of vesicles to cause the mixing of lipids between the two vesicles and consequently cause vesicle fusion.

15.1.2.2. Mechanically or spontaneous aligned lipid bilayer

Static ^{13}C NMR spectra of ^{13}C-labeled melittin incorporated into bilayers of the diether lipid, ditetradecylphosphatidylcholine, which is mechanically aligned between stacked glass plates were recorded[15] as a function of the angle between the bilayer planes and the magnetic field at temperatures above and below the lipid gel-to-liquid crystalline phase transition temperature T_c. For bilayers aligned with the normal along the applied magnetic field, there was no shift in the carbonyl resonance of residues Ile 2, Ala 4, Leu 9, Leu 13, or Ala 15, with minor changes for residues Val 18 and Ile 20, and small changes at Val 5, Val 6, and Ile 17 on immobilization of the peptide below T_c. In contrast, the spectra for bilayers aligned at right angles to the field showed greatly increased anisotropy below T_c for all analogues. From these experiments it was evident that the peptide was well-aligned in the bilayers and reoriented about the bilayer normal. The reduced CSA and the chemical shifts were consistent with melittin adopting a helical conformation with a transbilayer orientation in the lipid membranes. With the exception of Ile 17, there was no apparent difference between the behavior of residues in the two segments that form separate helices in the water-soluble form of the peptide, suggesting that in membranes the angle between the helices is greater than the 120° observed in the crystal form.[16,17]

As described in Section 6.1.3.2., fully hydrated melittin–DMPC lipid bilayer is also spontaneously aligned along the static magnetic field by forming elongated vesicles with the long axis parallel to the magnetic field as viewed from slow DD-MAS ^{13}C NMR spectra (Figure 6.12). In such case, no ^{13}C CP-MAS NMR signal is visible from incorporated peptide. Using this magnetically orienting property of the membrane, structure, orientation, and dynamics of melittin have been extensively studied.[12] Static ^{13}C DD-NMR spectra of [1-^{13}C]Ile 20-melittin bound to the DMPC bilayer hydrated with Tris buffer recorded at −60 °C show a broad asymmetrical powder pattern characterized by $\delta_{11} = 241$, $\delta_{22} = 189$, and $\delta_{33} = 96$ ppm (Figure 15.3(a)).[12] The presence of this broad signal indicates that any motion of melittin bound to the DMPC bilayer is completely frozen at −60°C. A narrowed ^{13}C NMR signal was observed at 174.8 ppm for Ile^{20} C=O by fast DD-MAS experiment at 40 °C, and its position was displaced upfield by 4.6 ppm in the oriented bilayer at 40 °C, as observed in the magnetically oriented state. An axially symmetrical powder pattern with an anisotropy of 14.9 ppm was recorded at 40 °C in the slow DD-MAS experiment as shown in Figure 15.3 (b). Because the line width due to the anisotropy at 40 °C is not as broad as that at −60 °C, it is expected that the α-helical segment undergoes rapid reorientation about the helical axis at 40 °C. Secondary structure of melittin bound to DMPC bilayers can be determined by the conformation-dependent ^{13}C chemical shifts with reference to those of model systems as discussed in Section 6.4 (see Table 6.1.). Because the isotropic ^{13}C chemical shifts of [1-^{13}C]Gly 3, [1-^{13}C]Val 5, [1-^{13}C]Gly 12, [1-^{13}C]Leu 16, and [1-^{13}C]Ile 20 residues in melittin are found to be 172.7,175.2, 171.6, 175.6, and 174.8 ppm, respectively, all of the residues mentioned above are involved in the α-helix.[12]

Orientation of peptides bound to the magnetically aligned lipid bilayer can be determined by examination of the CSA of the backbone carbonyl

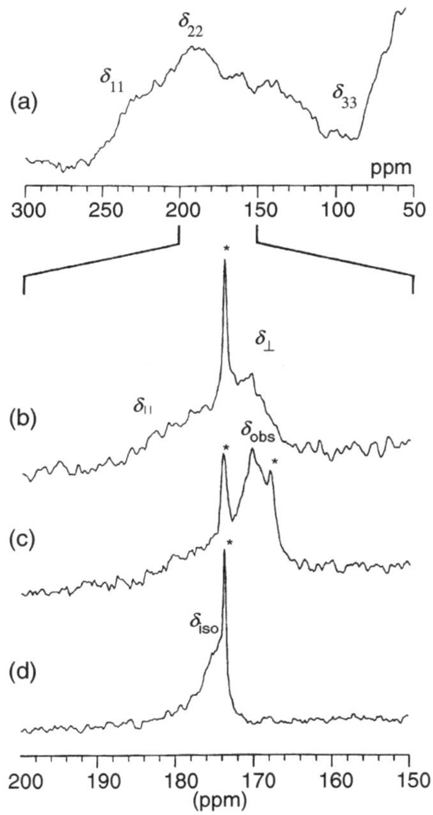

Figure 15.3. Temperature variation of ^{13}C NMR spectra of a DMPC bilayer in the presence of [1-^{13}C]Ile20-melittin in the static condition at −60 °C (a) and slow MAS (b), static (c), and fast MAS (d) conditions at 40 °C. The signals (marked by asterisks) appearing at 173 ppm in (b), 173 and 168 ppm in (c), and 173 ppm in (d) are assigned to the C=O groups of DMPC.[12]

carbon in the peptide chain. Chemical shift tensors of the carbonyl carbons are well characterized as shown in Figure 6.13 (Section 6.1.3.2). Indeed, it is expected that 150 ppm of CSA should be observed at low temperature, when the peptide is completely static. In contrast, the molecules exhibit large amplitude motion about the bilayer normal, when the temperature is raised until T_c, resulting in the axially symmetric powder pattern as shown in Figure 15.3(b). In fact, slow DD-MAS (≈100 Hz) experiment yields the axially symmetric powder pattern, since the bilayers tend to be aligned with the magnetic field.[12] It turned out that the molecule actually

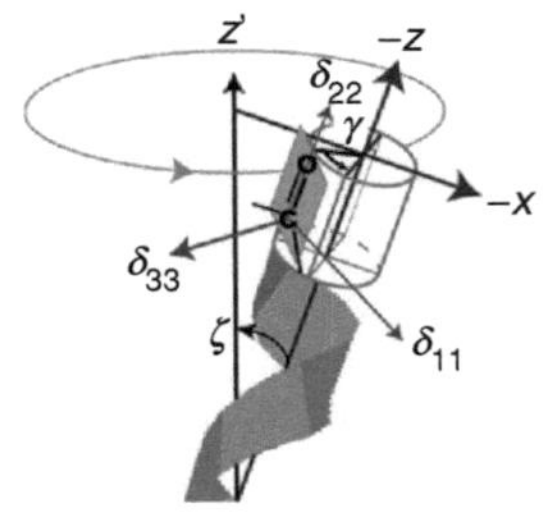

Figure15.4. Direction of the principal axis of the ^{13}C chemical shift tensor of the C=O group, the helical axis, and the static magnetic field ($\boldsymbol{B}_0$).

undergoes reorientational motion about the bilayer normal, and the axially symmetric pattern was seen in the DMPC–melittin bilayer systems[12]. When the spinning was stopped, lipid bilayer will be again spontaneously oriented to the magnetic field, and the membrane bound molecule also aligns to the magnetic field (Figure 15.3(c)). In most case, membrane bound biomolecule rotates about the membrane normal because of the lateral diffusion of the lipid molecule in the liquid crystalline phase. Entire molecule forms a α-helix in the membrane bound state and the helix axis is rotated about the bilayer normal with the tilt angle of ζ and the phase angle γ (Figure 15.4). Under this dynamic state, the CSA, $\Delta\delta = \delta_{\|} - \delta_{\perp}$, can be expressed as

$$\Delta\delta = \frac{3}{2}\sin^2\varsigma(\delta_{11}\cos^2\gamma + \delta_{33}\sin^2\gamma - \delta_{22}) + \left(\delta_{22} + \frac{\delta_{11}+\delta_{33}}{2}\right). \quad (15.1)$$

In this equation, $\Delta\delta$ values oscillate as a function of γ with the oscillation amplitude of (3/2)sin2ζ, which is referred to as the chemical shift oscillation. When the α-helical peptide has a large tilt angle, $\Delta\delta$ value changes to large extent. Using this property of $\Delta\delta$, the tilt angle of the α-helix with respect to the bilayer normal can be determined by comparing the anisotropic patterns of carbonyl carbons of consecutive amino acid residues, which form α-helix, with the chemical shift values of the corresponding magnetically aligned state. When the peptide participates in an ideal α-helix structure, the interpeptide plane angle for consecutive peptide planes can be assumed to be 100°. This tilt angle can be accurately obtained by taking RMSD values between the observed and calculated CSAs as given by

$$\mathrm{RMSD} = \left[\sum_{i=1}^{N}\left\{(\Delta\delta_{\mathrm{obs}})_i - (\Delta\delta_{\mathrm{cal}})_i\right\}^2/\mathrm{N}\right]^{\frac{1}{2}}. \quad (15.2)$$

Because the lipid bilayer is oriented, with respect to the magnetic field, with the bilayer surface parallel to the magnetic field and the α-helical axis of melittin precessed about the averaged helical axis, θ reflects the direction of averaged α-helix with respect to the surface of the lipid bilayer. Actually, the static ^{13}C chemical shift δ_{obs} of Ile 20 C=O in the magnetically oriented state was displaced upfield by 4.6 ppm from the isotropic value (δ_{obs}). This value allows one to determine the averaged α-helical axis inclined nearly 90° to the bilayer plane. On the other hand, that of Gly 3 C=O was displaced downfield by 6.8 ppm (Figure 6.13), whereas the axially symmetrical powder pattern was reversed in shape as compared to that of Ile^{20} C=O, and hence the $\delta_{\parallel} - \delta_{\perp}$ value is negative in this case. This result leads to the conclusion that the average axis of the α-helix is inclined again 90° to the bilayer plane. Therefore, the transmembrane α-helices of melittin are embedded in the lipid bilayer systems, and both N- and C-terminal helices are reoriented about the average helical axis, which is parallel to the lipid bilayer normal. It is emphasized that the charged amino acid residues such as Lys 7 in the N-terminus and Lys 21, Arg 22, Lys 23, and Arg 24 in the C-terminus may be closely located at the opposite sides of the polar head-groups of lipid bilayers, although melittin forms the amphiphilic helix within the lipid bilayers.

Slow DD-MAS NMR results indicate that melittin forms the transmembrane α-helix in the lipid bilayer, whose average axis is parallel to the bilayer normal. It was also shown that the transmembrane α-helix is not static, but undergoes motion, namely, the N- and C-terminal α-helical rods rotate or reorient rapidly about the average helical axis. Although the average direction of the α-helical axis is parallel to the bilayer normal, the local helical axis may precess about the bilayer normal by making an angle of 30° and 10° for the N- and C-terminal helical rods, respectively.[12]

^{13}C static, slow and fast DD-MAS NMR spectra were recorded in melittin–lecithin vesicles composed of 1,2-dilauroyl-*sn*-glycero-3-phosphocholine (DLPC) or DPPC.[14] Highly ordered magnetic alignments were achieved with the membrane surface parallel to the magnetic field above the gel-to-liquid crystalline phase transition temperature (T_c). Using these magnetically oriented vesicle systems, dynamic structures of melittin bound to the vesicles were investigated by analyzing the ^{13}C anisotropic and isotropic chemical shifts of selectively ^{13}C-labeled carbonyl carbons of melittin in static and MAS conditions. These results indicate that melittin molecules adopt an α-helical structure and laterally diffuse to rotate rapidly around the membrane normal with tilt angles of the N-terminal helix being −33° and −36° and those of the C-terminal helix being 21° and 25° for DLPC and DPPC vesicles, respectively (Figure 15.5).[17a] REDOR was used to measure the interatomic distance between [1-^{13}C]Val 8 and [^{15}N]Leu 13 to further

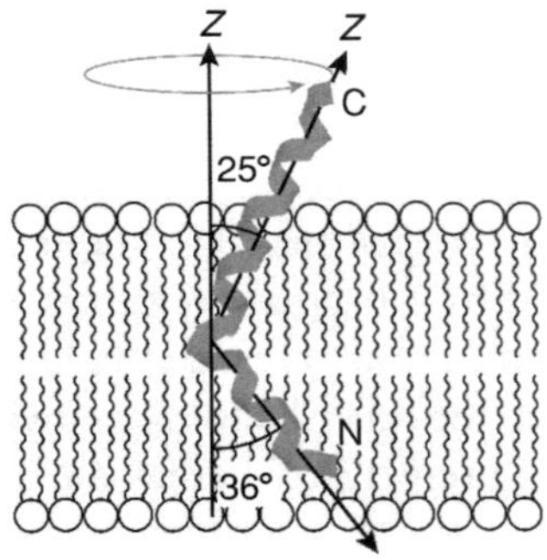

Figure 15.5. Schematic representations of the structure of melittin bound to the DPPC bilayers determined from the ^{13}C CSAs. The tilt angles of N- and C-terminals to the bilayer normal are 36° and 25°, respectively.[14]

identify the bending α-helical structure of melittin to possess the interhelical angles of 126° and 119° in DLPC and DPPC membranes, respectively. These analyses further lead to the conclusion that the α-helices of melittin molecules penetrate the hydrophobic core of the bilayers incompletely as a pseudo-transmembrane structure, and induce fusion and disruption of vesicles.

15.1.3. Alamethicin

Alamethicin is a 20 amino acid antibiotic peptide with the amino acid sequence of Ac-Aib-Pro-Aib-Ala-Aib-Ala-Gln-Aib-Val-Aib-Gly-Leu-Aib-Pro-Val-Aib-Aib-Glu-Gln-Phol, where the N-terminal residue is acetylated and the C-terminal residue is L-phenylalaninol.[18] Because of its tractable size and its significant voltage-dependent conductance in lipid bilayer system, alamethicin has been regarded an ideal model for studying voltage-gated conformational changes in α-helical antibacterial peptides, transmembrane ion transport, and helix–membrane interactions for membrane proteins in more general terms.[19] Due to the presence of a central proline residue within the potentially membrane-buried part of the amino acid sequence, alamethicin resembles a transmembrane domain in a typical transport/channel protein with several bilayer-spanning helices.[20]

Conformation and orientation of membrane-bound alamethicin were studied using magnetically oriented lipid bilayer systems by means of solid state NMR spectroscopy.[21] ^{13}C chemical shifts of isotopically labeled alamethicin indicated that alamethicin forms α-helical structure in lipid bilayer, and is oriented along the bilayer normal. It was found that the CSA was substantially reduced by the rotation of alamethicin helix. Therefore, alamethicin exists as a monomeric form in the absence of membrane potential.

The conformation of antibiotic ionophore alamethicin in mechanically oriented phospholipid bilayer has been studied using ^{15}N solid state NMR in combination with molecular modeling and MD simulations.[22,23] ^{15}N-labeled variants at different positions of alamethicin along with three of Aib (α-amino isobutyric acid) residues replaced by Ala were examined to establish experimental structural constraints and determine the orientation of alamethicin in hydrated phospholipid bilayers, and to investigate the potential for a major kink in the region of the central Pro 14 residue. From the anisotropic ^{15}N chemical shifts and ^{1}H–^{15}N dipolar couplings determined for alamethicin with ^{15}N-labeling on the Ala 6, Val 19, and Val 15 residues and incorporated into phospholipid bilayer with a peptide-to-lipid molar ratio of 1:8, it was shown that alamethicin has a largely linear α-helical structure spanning the membrane with the molecular axis tilting by 10°–20° relative to the bilayer normal. In particular, it was found that the compatibility with a straight α-helix was tilted by 17° and a slightly kinked molecular dynamics structure was tilted by 11° relative to the bilayer normal. In contrast, the structural constraints derived by solid state NMR appear not to be compatible with any of several model structure crossing the membrane with vanishing tilt angle or the earlier reported X-ray diffraction structure. The solid state NMR-compatible structure may support the formation of a left-handed and parallel multimeric ion channel.

15.2. Antimicrobial Peptides

It is important to characterize the peptide–membrane interaction to understand the activity of peptide in membranes such as antimicrobial activity. All antimicrobial peptides interact with membranes and tend to be divided into two mechanistic classes: membrane disruptive and nonmembrane disruptive. Cationic antimicrobial peptides have multiple actions on cells ranging from membrane permeabilization to cell wall and division effects to macromolecular synthesis inhibition, and that the action responsible for killing bacteria at the minimal effective concentration varies from peptide to peptide and from bacterium to bacterium for a given peptide.[24,25]

One mechanism of interaction of cationic antimicrobial peptides with the cell envelope of Gram-negative bacteria is discussed as shown in Figure 15.6.[24] Passage across the outer membrane is proposed to occur by

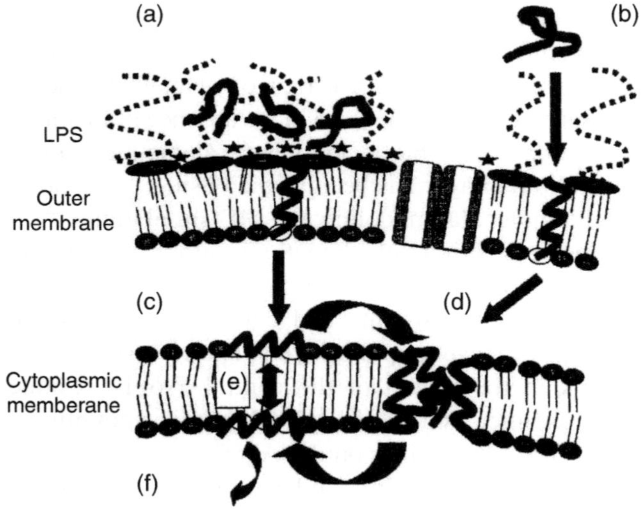

Figure 15.6. Proposed mechanism of interaction of cationic antimicrobial peptides with the cell envelope of Gram-negative bacteria. Passage across the outer membrane by creating cracks (a) or by disrupting the membrane (b). The amphipathic peptide will insert into the membrane interface (c) or a micelle-like complex spans the membrane (d) or flip–flop across the membrane (e). Some monomer will be translocated into the cytoplasm and can dissociate from the membrane (f).[24]

self-promoted uptake. Unfolded cationic peptides are proposed to associate with the negatively charged surface of the outer membrane and either neutralize the charge over a patch of the outer membrane, creating cracks through which the peptide can cross the outer membrane (a), or actually bind to the divalent cation binding sites on LPS and disrupt the membrane (b). Once the peptide has transited the outer membrane, it will bind to the negatively charged surface of the cytoplasmic membrane, created by the head groups of phosphatidylglycerol and cardiolipin, and the amphipathic peptide will insert into the membrane interface (c). It is not known at which point in this process the peptide actually folds into its amphipathic structure. Many peptide molecules will insert into the membrane interface and are proposed to then either aggregate into a micelle-like complex which spans the membrane (d) or flip–flop across the membrane under the influence of the large transmembrane electrical potential gradient (~140 mV) (e). The micelle-like aggregates (d) are proposed to have water associated with them, and this provides channels for the movement of ions across the membrane and possibly leakage of larger water-soluble molecules. These aggregates would be variable in size and lifetime and will dissociate into monomers that may be disposed at either side of the membrane. The net effect of (d) and (e) is that some monomers will be translocated into the cytoplasm, and can dissociate from the membrane and bind to cellular polyanions such as DNA and RNA. It is of equal importance to determine the structure and orientation of peptide bound to membrane if one wants to understand the action of peptides on membrane. To date, the structures of a number of membrane associated peptides have been studied using solid state NMR techniques as described in the following section.

15.2.1. Protegrin-1

Protegrin-1 (PG-1, 2154 Da) is an 18-residue broad-spectrum antimicrobial peptide found in porcine leukocytes.[26] It kills Gram-positive and Gram-negative bacteria and fungi, and has modest antiviral activities against HIV-1. PG-1 forms an antiparallel β-strand in solution,[27] where the two strands are stabilized by two disulfide bonds among the four Cys residues. This disulfide-stabilized β-hairpin motif is common to a number of other anti-microbial peptides such as human defensins and tachyplesin.

Evidence of PG-1 aggregation in lipid bilayer is so far sparse in the literature. In aqueous and DMSO solution, PG-1 is monomeric.[27] In DPC micelles, ^{1}H NOE data suggested dimer formation.[28] Neutron diffraction patterns of fully hydrated DMPC membranes containing 1:30 (P:L) PG-1 suggest that the peptide aggregates to form stable pores.[29] Lowering the temperature or hydration level of the bilayers resulted in crystallization of these pores, implying cooperative action of several PG-1 molecules in disrupting the bilayer.

The dynamics and aggregation of a β-sheet antimicrobial peptide, PG-1, are investigated using solid state NMR spectroscopy.[30] The CSAs of F12 and V16 carbonyl carbons are uniaxially averaged in DLPC bilayers but approach to rigid limit values results in the thicker POPC bilayers. The C_α–H_α dipolar coupling of L5 is scaled by a factor of 0.16 in DLPC bilayers, but has a near-unity order parameter of 0.96 in POPC bilayers. The larger couplings of PG-1 in POPC bilayers indicate immobilization of the peptide, suggesting that PG-1 forms oligomeric aggregates at the biologically relevant bilayer thickness. Exchange NMR experiments on [1-^{13}CO]Phe12-labeled PG-1 show that the peptide undergoes slow reorientation with a correlation time of 0.7 ± 0.2 s in POPC bilayers. This long correlation time suggests that in addition to aggregation, geometric constraints in the membrane may also contribute to PG-1 immobilization. The PG-1 aggregates contact both the surface and the hydrophobic center of the POPC bilayer, as determined by

^{1}H spin diffusion measurements. Thus, solid state NMR provides a wide range of information about the molecular details of membrane peptide immobilization and aggregation in lipid bilayers.

15.2.2. PGLa

PGLa (peptide between Glycine and Leucine amide (GMASKAGAIAG-KIAKVALKAL-NH2) is an antibacterial peptide of the magainin family, isolated from the skin of the African clawed frog *Xenopus laevis*.[31] It is nonhemolytic but exhibits a broad spectrum of antimicrobial activity against Gram-positive and Gram-negative bacteria, fungi, and protozoa by enhancing the permeability of the biological membrane.[32] The low toxicity of the magainins against eukaryotic cells makes them a potential candidate for the development of new antibacterial drugs. Even though the sequence homology of PGLa with magainin 1 or 2 is low, the peptides share a number of spectral features. Like magainins 1 and 2, the 21-residue peptide PGLa is positively charged and adopts a random-coil conformation in water.[33] Upon membrane binding, the peptides fold into amphipathic α-helical conformations.[33] Solid state ^{15}N NMR spectroscopy has shown that the PGLa helix is aligned essentially parallel to the membrane surface at low peptide-to-lipid ratios.[34] Membrane permeabilization, however, was suggested to be induced by the formation of a peptide–lipid pore, where the PGLa helices are oriented perpendicular to the membrane surface.[35] The selectivity of the cationic PGLa for prokaryotic systems is caused, at least partially, by a preferred interaction with negatively charged membranes. CD experiments showed that the helicity of PGLa is larger with negatively charged vesicles than with electrically neutral vesicles.[36] Furthermore, the thermotropic phase behavior of negatively charged phosphatidylglycerol membranes was distinctly changed upon addition of PGLa, whereas that of neutral membrane was not affected. These results are relevant for biological membranes since the outer leaflet of most bacterial membranes contains indeed a high amount of negatively charged phospholipids, while that of eukaryotic membrane is generally composed of neutral lipids.

The membrane-disruptive antimicrobial peptide PGLa is found to change its orientation in a DMPC bilayer when its concentration is increased to biologically active levels. The alignment of the α-helix was determined by highly sensitive solid state NMR measurement of ^{19}F dipolar couplings on CF_3-labeled side chains, and supported by a nonperturbing ^{15}N label.[37] At low peptide:lipid ratio of 1:200, the amphiphilic peptide resides on the membrane surface in the so-called S-state, as expected. However, at high peptide concentration (>1:50 molar ratio) the helix axis changes its tilt angle from about 95° to approximately 125°, with the C-terminus pointing towards the bilayer interior. This tilted ''T-state'' represents a novel feature of antimicrobial peptides, which is distinct from a membrane inserted I-state. At intermediate concentration, PGLa is in exchange between the S- and T-state in the timescale of the NMR experiment. In both states the peptide molecules undergo fast rotation around the membrane normal in liquid crystalline bilayers, hence large peptide aggregates do not form. Very likely the obliquely tilted T-state represents an antiparallel dimer of PGLa that is formed in the membrane at increasing concentration.

15.2.3. LL-37

LL-37 is a 4.5-kDa cationic, amphipathic α-helical antimicrobial peptide with the sequence LLGDFFRKSKEKIGKEFKRIVQRIKDFLRNLVPRTES.[38] LL-37 has antimicrobial activity *in vivo*[39] and induces vesicle leakage and permeabilization of the inner and outer membranes of *E. coli* cells in a dose-dependent manner at concentrations comparable to its minimum inhibitory concentration.[40] ^{15}N PISEMA spectroscopy of site-specifically labeled LL-37 in oriented lipid bilayer indicates that the amphipathic helix is oriented parallel to the surface of the bilayer. ^{31}P NMR and differential scanning calorimetry (DSC) experiments revealed that LL-37 induces positive curvature strain but does not break the membrane into small fragments or micelles.[41] The role of curvature strain has been discussed with regard to lipid–lipid modulation of membrane function. These results support a toroidal pore model of bilayer disruption or formation of less organized transient defects in the membrane. The DSC results demonstrate that LL-37 penetrates into the hydrophobic interior of the bilayer and disrupts acyl chain packing and cooperativity. Site-specific resolution and quantitative information on the influence of LL-37 on acyl chain motion is obtained from ^{2}H NMR spectra of acyl chain perdeuterated lipids.[42] The ^{2}H NMR experiments show that LL-37 disorders the acyl chains, and are used to characterize the extent of the perturbation at different depths in the hydrophobic core. The results are consistent with the amphipathic helix axis aligned parallel to the membrane surface at the hydrophobic–hydrophilic interface of the bilayer with an estimated 5–6 Å depth of penetration of the hydrophobic face of the amphipathic helix into the hydrophobic interior of the bilayer. In addition, the NMR data are analyzed to determine the effect of LL-37 on the properties of the lipid bilayers, such as hydrophobic thickness, area per lipid, and coefficient of thermal expansion, since the effect of LL-37 varies greatly with its environment. These results indicate that bilayer order influences the depth of insertion of LL-37 into the hydrophobic–hydrophilic interface of the bilayer, altering the balance of electrostatic and hydrophobic interaction between the peptide and lipids.

15.3. Opioid Peptides

15.3.1. Dynorphin

Dynorphin is an endogeneous opioid heptadecapeptide, Tyr-Gly-Gly-Phe-Leu-Arg-Arg-Ile-Arg-Pro-Lys-Leu-Lys-Leu-Lys-Trp-Asp-Asn-Gln-OH, which was originally isolated from porcine pituitary.[43] Dynorphin A(1–17) binds to κ-opioid receptor with high affinity. The functional positions of dynorphin A(1–13) have been examined by means of enzymatic digestion.[44] It was found that Lys 13, Lys 11, and Arg 7 residues play an essential role in the affinity of receptor binding.

Lipid bilayers of DMPC containing opioid peptide dynorphin A(1–17) are found to be spontaneously aligned to the applied magnetic field near the phase transition temperature between the gel and liquid crystalline state (T_c = 23 °C), as examined by ^{31}P NMR spectroscopy.[45] The specific interaction between the peptide and lipid bilayer leading to this property was also examined by optical microscopy, light scattering, and potassium ion-selective electrode, together with a comparative study on dynorphin A(1–13). A substantial change in the light scattering intensity was noted for DMPC containing dynorphin A(1–17) near the T_c but not for the system containing A(1–13). Besides, reversible change in morphology of bilayer, from small lipid particles to large vesicles, was observed by optical microscopy at T_c. These results indicate that lysis and fusion of the lipid bilayers are induced by the presence of dynorphin A(1–17). It turned out that the bilayers are spontaneously aligned to the magnetic field above T_c in parallel with the bilayer surface, because a single ^{31}P NMR signal appeared at the perpendicular position of the ^{31}P chemical shift tensor. In contrast, no such magnetic ordering was noted for DMPC bilayers containing dynorphin A(1–13).[45] It was proved that DMPC bilayer in the presence of dynorphin A(1–17) forms vesicle above T_c, because leakage of potassium ion from the lipid bilayers was observed by potassium ion-selective electrode after adding Triton X-100. It is concluded that DMPC bilayer consists of elongated vesicle with the long axis parallel to the magnetic field, together with the data of microscopic observation of cylindrical shape of the vesicles.

Further, the long axis is found to be at least five times longer than the short axis of the elongated vesicles in view of simulated ^{31}P NMR line shape.[45]

Infrared-Attenuated Total Reflection (IR-ATR) spectroscopy and capacitance minimization (CM) were used to study the secondary structure, orientation, and accumulation of dynorphin A(1–13) molecules on the surface of planar membranes prepared from DLPC.[44] CM studies showed adsorption localized on the side of the lipid bilayer to which the dynorphin A(1–13) had been added. The binding was reversible and a preliminary analysis yielded an apparent dissociation constant of $K_d = 11$ mM and a maximum surface density of 1 molecule/110 nm^2 in the presence of 10 mM KCl in the aqueous phase. Shielding studies at higher ionic strength suggested that a charged part of the molecule extended into the aqueous phase. It is reported that dynorphin A(1–17) interacts strongly with DMPC, causing lysis and fusion of membrane. This membrane system also shows magnetic ordering. It is further found that the interaction is much smaller for the case of dynorphin A(1–13) as compared to dynorphin A(1–17).[45]

It is assumed α-helical structure extends from Tyr 1 to Pro 10 in the presence of lipid bilayers. The helix is oriented perpendicularly to the membrane surface. The extended conformation of the C-terminal Lys-Leu-Lys-OH segment with peptide bonds (H–N–C=O dipoles) perpendicular to the membrane surface was chosen. In this structural model, the message segment, dynorphin (1–4), contacts hydrophobic membrane layers and the address segment, dynorphin (5–13), the aqueous phase. The results with IR-ATR spectroscopy indicate a strong increase of helicity on contact with neutral lecithin membranes. Moreover, CD spectra showed an increased helicity in dynorphin A(1–13) in 2,2,2-trifluoro-ethanol (TFE)–water mixtures, similar to that observed in SDS solutions. The formation of helix in dynorphin A(1–17) has been examined by molecular dynamic simulation[47] and solid state NMR.[48] The results indicate that N-terminal helix inserted into the membrane is tilted to membrane normal.[44] Secondary structure and orientation of dynorphin bound to DMPC bilayer were investigated by solid state ^{13}C NMR spectroscopy using MOVS.[48] It was found that dynorphin adopts an α-helical structure in the N-terminus from Gly 2 to Leu 5 with reference to the conformation-dependent ^{13}C chemical shifts. In contrast, disordered conformations are evident from the center to the C-terminus and is located on the membrane surface, and the N-terminal α-helix is inserted into the membrane with a tilt angle of 21° to the bilayer normal, as shown in Figure 15.7.[48] Since a blue shift was induced in tryptophan emission spectrum, it is proposed that the C-terminal region of dynorphin A(1–17) is proposed to bend back onto the bilayer–water interface, and the Trp 14 residue is stabilized in a hydrophobic region near the interface by interaction with polar headgroup of

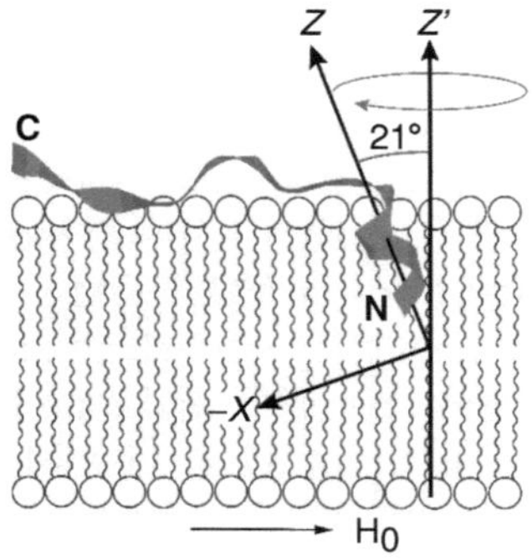

Figure 15.7. Schematic representation of the structure of dynorphin bound to lipid bilayers. The N-terminal helix is inserted into the lipid bilayers with the tilt angle of 21°. The central to the C-terminal regions shows disordered and lies on the surface of the lipid bilayers.[48]

phospholipids. This structure suggests a possibility that dynorphin interacts with the extracellular loop II of the κ-receptor through a helix–helix interaction.

15.3.2. β-Endorphin

β-Endorphin was isolated from camel pituitary glands, and its amino acid sequence, identical to the C-terminal 31 amino acid residues of sheep β-endorphin, Tyr-Gly-Gly-Phe-Met-Thr-Ser-Glu-Lys-Ser-Glu-Lys-Ser-Gln-Thr-Pro-Leu-Val-Thr-Leu-Phe-Lys-Asn-Ala-Ile-Ile-Lys-Asn-Ala-Tyr-Lys-Lys-Gly-Gly, was determined.[49] Its structure–activity relationship indicated that the peptide binds to two distinct sites on the receptor, namely one interacts with the N-terminus and the other with the C-terminus of the peptide.[50] In addition, the binding of the C-terminal fragment appears to be necessary for the binding of the N-terminus and for biological activity. CD studies indicate that the interaction of β-endorphin with lipids with negative charges induces a conformational change to a partial helical conformation in the peptide.[51]

Solution ^{1}H NMR spectroscopy has been used for structure determination of β-endorphin in solution.[47] Photo-CIDNP was utilized to characterize Tyr 1, Tyr 27 of human β-endorphin by comparing to Tyr 1 of Met-enkephalin and His 27 of camel β-endorphin with and without the presence of SDS and *n*-dodecylphosphorylcholine micelles.[53] Overall secondary structural characterization has been determined using ^{1}H NMR spectroscopy with COSY and NOESY experiments.[54] The regions between Tyr 1 and Thr 12 and Leu 14 and Lys 28 of β-endorphin turned out to take α-helical structures and that between Lys 28 and Glu 31 form a turn structure, in methanol and water-mixed solvent.[54] Similarly, β-endorphin takes helical structure in SDS micelle.[55] In contrast, random-coil structure was found in an aqueous solution.[54,55]

15.3.3. κ-opioid receptor bound to membrane

The κ-receptor, as well as the μ- and δ-receptors, belongs to the superfamily of G-protein-coupled receptors, which are integral membrane proteins presumed to have the common structural motif of seven transmembrane helices, connecting loops, and long N- and C-terminal tail at the extracellular and intracellular domains. X-ray structure of the κ-receptor is not available because of difficulty in crystallization. Investigations with κ-opioid receptor chimeras have shown that the putative second extracellular loop (ECL II) contributes substantially to the κ-receptor's selectivity in dynorphin ligand recognition.[56] In addition, the ECL II domain contains eight acidic amino acids. Dynorphin A(1–13), on the other hand, includes five basic amino acids (from Arg 6 to Lys 13). Their removal or substitution by Ala causes a marked decrease in κ selectivity, especially in the case of Arg 7.[39] While the residues 12–17 in dynorphin A(1–17) can be truncated without a significant loss in bioactivity, it is reasonable to speculate that dynorphin A binds to the κ-receptor through Columbic interactions. However other investigations performed docking studies based on a homology model of the κ-receptor and have proposed that α helical domain of the ECL II interacts with dynorphin A(1–10) through hydrophobic interactions.[57]

To explore the structural features contributing to the specific binding of ECL II of the κ-receptor to its selective analogues, a 33-amino acid peptide composed of the sequence (residue 196–228) based on the ECL II sequence of human κ-opioid receptor was synthesized.[52] In addition to the amino acid residues believed to constitute the ECL II of the receptor, this peptide included N-terminal and C-terminal amino acids believed to reside, respectively, in the transmembrane IV and transmembrane V domains of the κ-receptor. C210 in the ECL II was replaced with A210, to increase the stability of the synthetic peptide. In addition, the N-terminus was acetylated and the C-terminus amidated. The sequence is Ac-Leu 196-Gly-Gly-Thr-Lys-Val-Arg-Glu-Asp-Val-Asp-Val-Ile-Glu-Ala-Ser-Leu-Gln-Phe-Pro-Asp-Asp-Asp-Tyr-Ser-Trp-Trp-Asp-Leu-Phe-Met-Lys-Ile 228-NH_2.

The structures deduced from distant geometry and simulated annealing present a family of structures that display α-helical array from V6 to A15 with an RMSD of 0.48 Å for the backbone atoms. The peptide is relatively undefined in the middle hinge region, because the S16-P20 sequence yield few nuclear NOEs. Following D21 is a β-turn spanning three consecutive Asp residues (D21–D23) and Y24. The C-terminus of the peptide displays α-helical tendency. Additionally, the residues following V6 form a turn of approximately 90° from the N-terminus to accommodate the strong NOE between the NH group of R7 and methyl group of T4.[58]

The binding profile of the κ-receptor is relatively unique among the opioid receptors, while those of the μ- and δ-receptors are similar to each other. The NMR data show that the ECL II of the κ-receptor displays an amphiphilic helical region from V201 to A210 (C210 in the original sequence), which is not present in the μ- and δ-receptors.[58] It supports the idea that an amphiphilic helical domain of the ECL II interacts with dynorphin A(1–10), based on sequence analysis and homology modeling of the κ-opioid receptor. In light of the existence of an amphiphilic helical region in dynorphin A(1–17) from residue Gly 3 to Arg 10, which was previously studied in a DPC micelle environment, the complementary helix–helix binding mode is a reasonable proposal for the binding of endogeneous dynorphin. Moreover, the helix–helix interaction is proposed to be hydrophobic in character because the binding of dynorphin A(1–13) is essentially not affected by charge neutralization of the ECL II. A β-turn around the D22 (D218) and D23 (D218) residues represents another feature of the ECL II. The side chains of these two aspartic acids are exposed to the aqueous medium. The significance of this β-turn in the ECL II of the κ-opioid receptor has not been determined yet.

An understanding of the molecular mechanism by which these dynorphin ligands interact with and activate the κ-receptor is an important step toward understanding the development of addiction associated with κ-opioid receptor agonist without causing the respiratory depression .

15.4. Fusion Peptides

Membrane fusion between enveloped viruses and host cell membranes is an obligatory process of viral infection mediated by viral glycoproteins, e.g. by influenza virus hemagglutinin (HA)[59] and membrane-spanning subunit (gp41) of human immunodeficiency virus type 1 (HIV-1).[60] To increase the fusion rate, such viruses employ a "fusion peptide" which represents an ≈20-residue apolar domain at the N-terminus of the viral envelope fusion protein.[61] During fusion, this domain is believed to interact with target cells and possibly viral membranes.

The influenza HA organizes as a homotrimer and each monomer is comprised of two disulfide-linked polypeptide chains, HA1 and HA2.[62] The HA1 subunit contains the sialic acid receptor binding domain, while the HA2 subunit is the fusogenic forming component and contains the fusion peptide at the N-terminus: GLFGAIAGFIENGWEGMIDG-amide. A common property of a number of wild and mutated fusion peptides is that they lower the bilayer to hexagonal phase transition temperature of phosphatidylethanolamine, indicating that they promote negative curvature.[63] Many viral fusion peptides have been found to be α-helical when inserted into a membrane. There is also some evidence for viral fusion peptides having β-structure, although it is not easy to understand how the fusion peptide of this type interacts with membranes. In relation to this question, Han and Tamm[64] showed on the basis of CD and FTIR measurements that monomeric peptides insert into the bilayers in a predominantly α-helical conformation, whereas self-associated fusion peptides adopt predominantly antiparallel β-sheet structures at the membrane surface. The two forms are readily interconvertible and the equilibrium between them is determined by the pH and ionic strength of surrounding solution. Lowering the pH favors the monomeric α-helical conformations, whereas increasing the ionic strength shifts the equilibrium toward the membrane-associated aggregates. REDOR-filtered technique was utilized[65,66] to contrast the signals of the labeled ^{13}C–^{15}N pairs in fusion peptides: the influenza peptide IFP-L2CF3N contained a ^{13}C carbonyl label at Leu-2 and an ^{15}N label at Phe 3 while the HIV-1 peptide HFP-UF8L9G10 was uniformly ^{13}C- and ^{15}N-labeled at Phe 8, Leu 9, and Gly 10. Secondary shift analysis was consistent with a β-strand structure over these three residues in the latter.

In fact, conformation of the membrane-bound ^{13}C carbonyl-labeled HIV-1 fusion peptide turned out to take oligomeric β-stranded form which is an equilibrium rather than a kinetically trapped structure, as viewed from ^{13}C chemical shift data as well as REDOR experiments.[67–69]

15.5. Membrane Model System

Model membrane systems consisting of phospholipids and synthetic transmembrane peptides have been used extensively to address the relationship between hydrophobic matching and membrane organization.[70] It was shown that polyleucine–alanine peptides that are flanked on both sides by tryptophan residues can significantly influence the organization of phosphatidylcholine model membranes when the peptides are relatively long or short with respect to bilayer hydrophobic thickness. However, because the Trp residues of the studied peptides are positioned at or near the termini of the peptides, hydrophobic mismatch also implies that these residues become exposed to a different interfacial environment. As a result, the observed membrane response to the incorporation of relatively short or long peptides could be driven not only by a tendency to avoid hydrophobic mismatch situations but also by a tendency of the flanking residues to avoid displacement from specific interfacial sites.

Membrane model systems consisting of phosphatidylcholines and hydrophobic α-helical peptides with tryptophan flanking residues, a characteristic motif for transmembrane protein segments, were used to investigate the contribution of tryptophans to peptide–lipid interaction.[71] Peptides of different lengths and with the flanking tryptophans at different positions in the sequence were incorporated in relatively thick or thin lipid bilayers. The organization of the systems was assessed by NMR methods and by hydrogen/deuterium exchange in combination with mass spectrometry. It was found that relatively short peptides induce nonlamellar phases and that relatively long analogues order the lipid acyl chains in response to peptide–bilayer mismatch. These effects do not correlate with the total hydrophobic peptide length, but instead with the length of the stretch between the flanking tryptophan residues. The tryptophan indole ring was consistently found to be positioned near the lipid carbonyl moieties, regardless of the peptide–lipid combination, as indicated by MAS-NMR measurements. These observations suggest that lipid adaptations are not primarily directed to avoid a peptide–lipid hydrophobic mismatch, but instead to prevent displacement of the tryptophan side chains from the polar–apolar interface. In contrast, long lysine-flanked analogues fully associate with a bilayer without significant lipid adaptations, and hydrogen/deuterium exchange experiments indicate that this is achieved by simply exposing more (hydrophobic) residues to the lipid head group region. The results highlight the specific properties that are imposed on transmembrane protein segments by flanking tryptophan residues.

References

[1] R. R. Ketchem, W. Hu, and T. A. Cross, 1993, *Science*, 261, 1457–1460.

[2] R. R. Ketchem, B. Roux, and T. A. Cross, 1997, *Structure*, 5, 1655–1669.

[3] F. Kovacs, J. Quine, and T. A. Cross, 1999, *Proc. Natl. Acad. Sci. USA*, 96, 7910–7915.

[4] A. L. Lomize, V. Y. Orekhov, and A. S. Arseniev, 1992, *Bioorg. Khim.*, 18, 182–200.

[5] B. M. Burkhart, N. Li, D. A. Langs, W. A. Pangborn, and W. L. Duax, 1998, *Proc. Natl. Acad. Sci. USA*, 95, 12950–12955.
[6] R. Fu, M. Cotten, and T. A. Cross, 2000, *J. Biomol. NMR*, 16, 261–268.
[7] E. Habermann and J. Jentsch, 1967, *Hoppe-Seyler's Z. Physiol. Chem.*, 348, 37–50.
[8] G. Sessa, J. H. Free, G. Colacicco, and G. Weissmann, 1969, *J. Biol. Chem.*, 244, 3575–3582.
[9] M. T. Tosteson and D. C. Tosteson, 1981, *Biophys. J.*, 36, 109–116.
[10] C. E. Dempsey, 1990, *Biochim. Biophys. Acta*, 1031, 143–161.
[11] J. Dufourcq, J.-F. Faucon, G. Fourche, J. L. Dasseux, M. Le Maire, and T. Gulik-Krzywicki, 1986, *Biochim. Biophys. Acta*, 859, 33–48.
[12] A. Naito, T. Nagao, K. Norisada, T. Mizuno, S. Tuzi, and H. Saitô, 2000, *Biophys. J.*, 78, 2405–2417.
[13] S. Toraya, T. Nagao, K. Norisada, S. Tuzi, H. Saitô, S. Izumi, and A. Naito, 2005, *Biophys. J.*, 89, 3214–3222.
[14] S. Toraya, K. Nishimura, and A. Naito, 2004, *Biophys. J.*, 87, 3323–3335.
[15] R. Smith, F. Separovic, T. J. Milne, A. Whittaker, F. M. Bennett, B. A. Cornell, and A. Makriyannis, 1994, *J. Mol. Biol.*, 241, 456–466.
[16] T. C. Terwilliger and D. Eisenberg, 1982, *J. Biol. Chem.*, 257, 6010–6015.
[17] T. C. Terwilliger, L. Weissman, and D. Eisenberg, 1992, *Biophys. J.*, 37, 353–361.
[17a] A. Naito, S. Toraya, and K. Nishimura, 2006, Modern Magnetic Resonance, G. A. Webb Ed. Springer, in press.
[18] R. C. Pandey, J. C. Cook, and K. L. Rinehart, Jr., 1977, *J. Am. Chem. Soc.*, 99, 8469–8483.
[19] D. S. Cafiso, 1994, *Annu. Rev. Biophys. Biomol. Struct.*, 23, 141–165.
[20] C. J. Brandl and C. M. Deber, 1986, *Proc. Natl. Acad. Sci. USA*, 83, 917–921.
[21] T. Nagao, A. Naito, S. Tuzi, and H. Saitô, 1999, *Peptide Sci.*, 341–344.
[22] C. L. North, M. Barranger-Mathys, and D. S. Cafiso, 1995, *Biophys. J.*, 69, 2392–2397.
[23] M. Bak, R. P. Bywater, M Hohwy, J. K. Thomsen, K. Adelhorst, H. J. Jakobsen, O. W. Sorensen, and N. C. Nielsen, 2001, *Biophys. J.*, 81, 1684–1698.
[24] J.-P. S. Powers and R. E. W. Hancock, 2003, *Peptides*, 24, 1681–1691.
[25] B. Bechinger, 1999, *Biochim. Biophys. Acta*, 1462, 157–183.
[26] V. N. Kokyakov, S. S. L. Harwig, E. A. Panyutich, A. A. Shevchenko, G. M. Aleshina, O. V. Shamova, H. A. Korneva, and R. I. Lehrer, 1993, *FEBS Lett.*, 327, 231–236.
[27] A. Aumelas, M. Mangoni, C. Roumestand, L. Chiche, E. Despaux, G. Grassy, B. Calas, and A. Chavanieu, 1996, *Eur. J. Biochem.*, 237, 575–583.
[28] C. Roumestand, V. Louis, A. Aumelas, G. Grassy, B. Calas, and A. Chavanieu, 1998, *FEBS Lett.*, 421, 263–267.
[29] L. Yang, T. M. Weiss, R. I. Lehrer, and H. W. Huang, 2000, *Biophys. J.*, 79, 2002–2009.
[30] J. J. Buffy, A. J. Waring, R. I. Lehrer, and M. Hong, 2003, *Biochemistry*, 42, 13725–13734.
[31] W. Hoffmann, K. Richter, and G. Kreil, 1983, *EMBO J.*, 2, 711–714.
[32] E. Soravia, G. Martini, and M. Zasloff, 1988, *FEBS Lett.*, 228, 337–340.
[33] M. Jackson, H. H. Hantsch, and J. H. Spencer, 1992, *Biochemistry*, 31, 7289–7293.
[34] B. Bechinger, M. Zasloff, and S. J. Opella, 1998, *Biophys. J.*, 74, 981–987.
[35] K. Matsuzaki, Y. Mitani, K. Y. Kada, O. Murase, S. Yoneyama, M. Zasloff, and K. Miyajima, 1998, *Biochemistry*, 37, 15144–15153.

[36] A. Latal, G. Degovics, R. F. Epand, R. M. Epand, and K. Lohner, 1997, *Eur. J. Biochem.*, 248, 938–946.
[37] R. W. Glaser, C. Sachse, U. H. N. Durr, P. Wadhwani, S. Afonin, E. Strandberg, and A. S. Ulrich, 2005, *Biophys. J.*, 88, 3392–3397.
[38] G. H. Gudmundsson, B. Agerberth, J. Odeberg, T. Bergman, B. Olsson, and R. Salcedo, 1996, *Eur. J. Biochem.*, 238, 325–332.
[39] A. Di Nardo, A. Vitiello, and R. L. Gallo, 2003, *J. Immunol.*, 170, 2274–2278.
[40] J. Turner, Y. Cho, N. N. Dinh, A. J. Waring, R. I. Lehrer, 1998, *Antimicrob. Agents Chemother.*, 42, 2206–2214.
[41] K. A. Henzler Wildman, D. K. Lee, and A. Ramamoorthy, 2003, *Biochemistry*, 42, 6545–6558.
[42] K. A. Henzler Wildman, G. A. Martinez, M. F. Brown, and A. Ramamoorthy, 2004, *Biochemistry*, 43, 8459–8469.
[43] A. Goldstein, S. Tachibana, L. I. Lowney, M. Hunkapiller, and L. Hood, 1979, *Proc. Natl. Acad. Sci. USA*, 76, 6666–6670.
[44.] C. Chavikin and A. Goldstein, 1981, *Proc. Natl. Acad. Sci. USA*, 78, 6543–6547.
[45] A. Naito, T. Nagao, M. Obata, Y. Shindo, M. Okamoto, S. Yokoyama, S. Tuzi, and H. Saitô, 2002, *Biochim. Biophys. Acta*, 1558, 34–44.
[46] D. Erne, D. F. Sargent, and R. Schwyzer, 1985, *Biochemistry*, 24, 4261–4263.
[47] R. Sankararamakrishnan and M. Weistein, 2000, *Biophys. J.*, 79, 2331–2344.
[48] T. Uezono, S. Toraya, M. Obata, K. Nishimura, S. Tuzi, H. Saitô, and A. Naito, 2005, *J. Mol. Struct.*, 749, 13–19.
[49] C. H. Li and D. Chung, 1976, *Proc. Natl. Acad. Sci. USA*, 73, 1145–1148.
[50] C. H. Li, 1982, *Cell*, 31, 504–505.
[51] C.-S. C. Wu, N. M. Lee, H. H. Loh, J. T. Yang, and C. H. Li, 1979, *Proc. Natl. Acad. Sci. USA*, 76, 3656–3659.
[52] F. Cabassi and L. Zetta, 1982, *Int. J. Pept. Protein Res.*, 20, 154–158.
[53] L. Zetta, P. J. Hore, and R. Kaptein, 1983, *Eur. J. Biochem.*, 134, 371–376.
[54] O. Lichtarge, O. Jardetzky, and C. H. Li, 1987, *Biochemistry*, 26, 5916–5925.
[55] L. Zetta, R. Consonni, A. Demargo, G. Vecchio, G. Gazzela, and R. Longhi, 1990, *Biochemistry*, 30, 899–909.
[56] F. Meng, M. T. Hoversten, R. C. Thompson, L. Taylor, S. J. Watson, and H. Akil, 1995, *J. Biol. Chem.*, 270, 12730–12736.
[57] G. Paterlini, P. S. Portoghese, and D. M. Ferguson, 1997, *J. Med. Chem.*, 40, 3254–3262.
[58] L. Zhang, R. N. DeHaren, and M. Goodman, 2002, *Biochemistry*, 41, 61–68.
[59] I. A. Wilson, J. J. Skehel, and D. C. Wiley, 1981, *Nature*, 289, 366–373.
[60] D. C. Chan, D. Fass, J. M. Berger, and P. S. Kim, 1977, *Cell*, 89, 263–273.
[61] D. E. Eckert and P. S. Kim, 2001, *Annu. Rev. Biochem.*, 70, 777–810.
[62] L. D. Hernandez, L. R. Hoffman, T. G. Wolfsberg, and J. M. White, 1996, *Annu. Rev. Cell Dev. Biol.*, 12, 627–661.
[63] R. M. Epand, 2003, *Biochim. Biophys. Acta*, 1614, 116–121.
[64] X. Han and L. K. Tamm, 2000, *J. Mol. Biol.*, 304, 953–965.
[65] J. Yang, P. D. Parkanzky, M. L. Bodner, C. A. Duskin, and D. P. Weliky, 2002, *J. Magn. Reson.*, 159, 101–110.
[66] M. L. Bodner, C. M. Gabrys, P. D. Parkanzky, J. Yang, C. A. Duskin, and D. P. Weliky, 2004, *Magn. Reson. Chem.*, 42, 187–194.
[67] J. Yang, C. M. Gabrys, and D. P. Weliky, 2001, *Biochemistry*, 40, 8126–8137.
[68] J. Yang and D. P. Weliky, 2003, *Biochemistry*, 42, 11879–11890.

[69] J. Yang, M. Prorok, F. J. Castellino, and D. P. Weliky, 2004, *Biophys. J.*, 87, 1951–1963.

[70] J. A. Killian, 1998, *Biochim. Biophys. Acta*, 1376, 401–415.

[71] M. R. R. de Planque, B. B. Bonev, J. A. A. Demmers, D. V. Greathouse, R. E. Koeppe II, F. Separovic, A. Watts, and J. A. Killian, 2003, *Biochemistry*, 42, 5341–5348.

Chapter 16
AMYLOID AND RELATED BIOMOLECULES

16.1. Amyloid β-Peptide (Aβ)

Formation of amyloid fibril, from normally innocuous, soluble proteins or peptides to polymerize to form insoluble fibrils, is now seen in biochemically diverse conditions. Although a number of nonfibrillar proteins or peptides have been identified to form the amyloid fibrils, all of which exhibit similar morphology as viewed from electron micrographs.[1] Some of the phenomena are related to the misfolding of proteins, leading to severe diseases such as the fibril deposit in the brain in the case of Alzheimer's disease, in the pancreas in the case of type II diabetes, etc. Therefore, elucidation of the molecular structure of amyloid fibrils is important for understanding the mechanism of self-aggregation. However, it is difficult to determine high-resolution molecular structures by using ordinary spectroscopic methods, because the fibrils are heterogeneous solids. In the last decade, solid-state NMR spectroscopy has demonstrated its advantage for the conformational determination of Alzheimer's amyloid β-peptides (Aβ), which are comprised of 39–43 amino acids residues and are the main component of the amyloid plaques of Alzheimer's disease. Not only the intrachain conformation of the Aβ molecule in the fibrils but also the intermolecular alignment has been analyzed to explore the mechanism of molecular association to form fibrils.

RR method has been used to characterize the structures of fragments of amyloid.[2–5] Griffin and coworkers synthesized the β-amyloid fragment Aβ(34–42) (H_2N-Leu-Met-Val-Gly-Gly-Val-Val-Ile-Ala-CO_2H), which is the C-terminus of the β-amyloid protein. The fragment was chosen because the region is implicated in the initiation of amyloid formation. The structure of this molecule was determined by the ^{13}C–^{13}C interatomic distances and the ^{13}C chemical shifts using 10–20% of the doubly ^{13}C-labeled Aβ(34–42) diluted with the unlabeled peptide. The α-carbon of the *i*th residue and the carbonyl carbon of the *i*+1th residue were doubly labeled, and the A[α*i*,*i*+1] interatomic distance was observed by RR method. Similarly, B[i,α(i+2)] and C[i,α(i+3)] interatomic distances were determined. Since RR signal of A does not show a dilution effect, an intermolecular contribution does not exist. On the other hand, B and C show strong intermolecular contributions from B* and C*. Therefore, it turns out that the fragment forms an antiparallel β-sheet. Furthermore, the intermolecular contribution indicates that β-strands consist of antiparallel β-sheets forming hydrogen bonds with the position that is offset from the N-terminus position. It is interesting that the information on the intermolecular contribution made it possible to reveal the assembly of the amyloid molecules.

On the other hand, DRAWS solid-state NMR technique was applied to characterize the peptide conformation of the Aβ(10–35) fragment and the supramolecular organization of fibrils formed by Aβ(10–35), which contains both hydrophobic and nonhydrophobic segments of Aβ. The peptide backbones in the fibrillated precipitation were confirmed to form an extended conformation in view of the ^{13}C chemical shifts of carbonyl carbon and the interatomic distance between i- and i+1-labeled carbonyl carbons obtained by DRAWS method. Interpeptide distances between the ^{13}C nuclei measured by the DRAWS method using singly ^{13}C-labeled peptides were shown to be 4.9–5.8 ($\pm$0.4) Å, respectively, throughout the entire length of the peptide. These results indicated that an in-register parallel organization of β-sheets is formed in the fibril of Aβ(10–35).[6–8]

Fibril structure of another fragment of Aβ, which comprises residues 16–22 of the Alzheimer's β-amyloid peptide [Aβ(16–22)] and is the shortest fibril forming one, was determined.[9,10] 2D MAS exchange and constant time double quantum-filtered dipolar recoupling (CTDQFD) techniques revealed that the torsion angles of the peptide backbone showed the extended conformation. 1D and 2D spectra of selectively and uniformly labeled samples exhibit ^{13}C NMR line width of <2 ppm, showing that the peptide, including amino acid side chains, has a well-ordered conformation in the fibril. The ^{13}C chemical shifts determined from 2D chemical shift correlation spectra also indicated that the entire hydrophobic segment forms a β-strand conformation. ^{13}C MQ NMR and ^{13}C–^{15}N REDOR data indicate an antiparallel organization of β-sheets with 17+k $\leftrightarrow$ 21–k registry in Aβ(16–22) fibrils.[11,12] On the other hand, the NMR data for Aβ(11–25) fibrils grown at pH 7.4 indicate an antiparallel β-sheet structure with a 17+k $\leftrightarrow$ 20–k registry. In contrast, the NMR data for Aβ(11–25) fibril grown at pH 2.4 indicate an antiparallel β-sheet structure with a 17+k $\leftrightarrow$ 22–k registry.[12]

Full-length β-amyloid fibrils of Aβ(1–40) were examined by solid-state NMR to generate sufficient structural constraints to permit the development of a structural model by constraint energy minimization.[13] The model is consistent with a large body of data that includes measurements of intermolecular distances by fpRFDR-CT,[12] MQ ^{13}C NMR spectra,[15] ^{13}C and ^{15}N chemical shift and line width measurements from 2D spectroscopy,[13] torsion angle constraints from tensor correlation,[13] X-ray diffraction data,[16] and measurements of the mass per length (MPL) of Aβ fibrils by scanning transmission electron microscope (STEM).[17] In the first stage, the β-sheet segments are identified from ^{13}C chemical shifts and the alignment and registry of β-strands within a β-sheet are determined from dipolar recoupling and MQ NMR data. Remaining few degrees of freedom are further restricted by the fibril dimensions and MPL data.

Features of the structural model shown in Figure 16.1[18] that are supported by the experimental data are as follows:

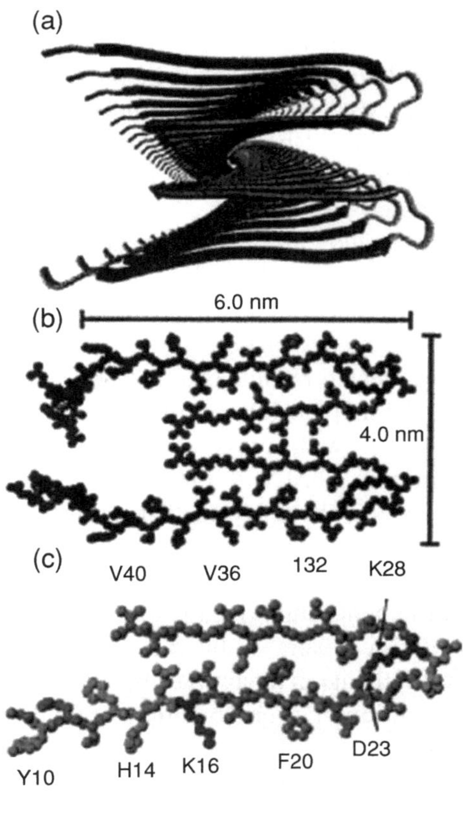

Figure 16.1. Structural model for Aβ(1–40) fibrils, considered with solid-state NMR constraints on the molecular conformation from isotropic chemical shifts, intermolecular distances, and incorporating the cross-β-motif (from Tycko[18]).

1. Approximately the first 10 residues of Aβ(1–40) are structurally disordered in the fibrils, as indicated by relatively large ^{13}C MAS NMR line width[13] (>3 ppm for CO, C_{α}, and C_{β} site in residues 2–9 vs. <2.5 ppm in residues 12–39) and by significantly weaker intermolecular ^{13}C–^{13}C dipole–dipole coupling than in the rest of the sequence.[14,15]
2. Residues 12–24 and 30–40 form two β-strand segments that are separated by a ''bend'' segment with non-β-strand conformation at Gly 25, Ser 26, and Gly 29. These secondary structure elements are indicated by ^{13}C chemical shifts in 2D spectra and by tensor correlation data.[13]
3. The two β-strand segments form two separate parallel β-sheets. In-register, parallel alignment of the peptide chains from Gly 9 to Val 39 is indicated by intermolecular ^{13}C–^{13}C dipolar couplings detected in dipolar coupling and MQ NMR measurements on a series of singly ^{13}C-labeled sample.[12,13] The net bend angle of 180° in residues 24–29 in Figure 16.1, which brings the two β-sheets in contact through side chain–side chain interactions is inconsistent with the dimension of the narrowest Aβ(1–40) fibrils observed in electron microscope (EM) images, called ''protofilaments,'' which have widths of 5 ± 1 nm. Without a large net bend between the two β-strands, the structurally ordered part of each Aβ(1–40) molecule would have a length of approximately 10 nm, at variance with the observed protofilament width. Thus, a single molecular layer in the cross-β-motif is a double-layered β-sheet in this model.
4. In Figure 16.1(b), the two cross-β-units make corrected at the hydrophobic surfaces created by side chains of residues 30–40. This mode of association seems plausible on physical grounds and results in a four-layered model with cross-sectional dimensions that closely match the minimum fibril width observed experimentally and with distances between β-sheet layers that agree well with the 8.9-nm spacing suggested by equatorial scattering peaks in fiber diffraction data.[16] Fibrils with large widths may be formed by lateral association of these protofilaments and typically exhibit morphology, suggesting that they are twisted pairs or bundles of finer filaments. Contacts between protofilaments in a paired or bundled fibril may be along β-sheet surfaces, through a combination of hydrophobic interaction (possibly involving Val 18 and Phe 20) and electrostatic interaction (possibly involving Lys 16 and Glu 22).[13]

The structure in Figure 16.1 resolves an important fundamental feature regarding the intermolecular interaction that stabilize Aβ(1–40) fibrils. The experimental observation that the β-sheets in Aβ(1–40) fibrils have an in-register, parallel structure maximizes contacts among hydrophobic side chains. However, an in-register, parallel β-sheet structure also places all charged side chains in rows with spacings on the order of 0.5 nm. In the low dielectric environment that may exist in the interior of an amyloid fibril, repulsions between like charges could destabilize the fibril structures by roughly 70 kCal/mol, overwhelming the stabilizing effect of the hydrophobic contact (estimated to be of order 2 kCal/mol). In Figure 16.1, the only charged side chains in the interior are Asp 23 and Lys 28, which form salt bridges that may actually stabilize the structure. Apart from Asp 23 and Lys 28, the protofilament structure has a purely hydrophobic core. Other charged and polar side chains are on the exterior of the structure, in a high-dielectric environments the fibrils grow in aqueous solution. ^{13}C and ^{15}N chemical shifts indicate that side chains of Asp 23 and Lys 28 are indeed charged in Aβ(1–40) fibrils grown at pH 7.4. Dipolar couplings between the side-chain carboxylate carbon of Asp 23 and the side-chain amino nitrogen of Lys 28 indicate an interatomic distance of roughly 0.4 nm, consistent with salt bridge formation.[13] Similar cross-β-motif is also studied using molecular dynamic simulation for Aβ(10–35) peptide.[17]

Although the 40-residue form of full-length Aβ is present at highest concentrations in the human body, it has been shown that elevated levels of the 42-residue form [Aβ(1–42), with additional I31 and A42 residues at the C-terminus] are associated with familiar form of Alzheimer's disease and that Aβ(1–42) peptides are the major component of immature senile plaques and cerebrovascular amyloid deposits. *In vitro*, Aβ(1–42) forms fibrils more rapidly and at lower concentration than Aβ(1–40). Solid-state NMR measurements indicate the same in-register, parallel alignment of peptide chains in β-sheet in Aβ(1–42) fibrils as demonstrated for Aβ(1–40) fibrils.[19]

16.2. Calcitonin

Calcitonin (CT) is a peptide hormone consisting of 32 amino acid residues that contains an intrachain disulfide bridge between Cys 1 and Cys 7 and a proline amide at the C-terminus. CT has been a useful drug for various bone disorders such as Paget's disease and osteoporosis. However, human calcitonin (hCT) has a tendency to associate to form fibril precipitate in aqueous solution. This fibril is known to be the same type as amyloid fibril, and hence the fibril formation has been studied as a model of amyloid fibril formation.

Conformations of hCT in several solvents were studied by solution NMR. In TFE–H_2O, hCT forms a helical structure between the residues 9 and 21.[20]

A short double-stranded antiparallel β-sheet form, however, was observed in the central region made by residues 16–21 in DMSO–H_2O.[21] It was suggested that hCT exhibits an amphiphilic nature when it forms an α-helix. However, the secondary structure of hCT in H_2O was shown to be a totally random coil as determined from the ^1H chemical shift data. Subsequent studies by 2D NOESY and CD measurements indicated that it adopts an extended conformation with high flexibility in aqueous solution.[21] The α-helical conformation was present in the central region of hCT in aqueous acidic solution, although it is shorter than in TFE–H_2O.[22] Similarly, NMR studies showed that salmon calcitonin (sCT) also forms an amphiphilic α-helical structure in the central region in TFE–H_2O, methanol–H_2O, and SDS micelles. However, sCT has a high activity and stability as a drug and its fibrillation process is much slower than in hCT.

2D solution NMR measurements of the time course of the fibrillation process showed that peaks from residues in the N-terminal (Cys 1–Cys 7) and the central (Met 8–Pro 23) regions are broadened and disappear earlier than those in the C-terminal region.[23] These findings indicate that the α-helices are bundled together in the first homogeneous nucleation process. Subsequently, it appears that larger fibrils grow in a second heterogeneous process. However, local structures and characteristics of hCT in the fibril were not directly obtained by solution NMR, because the intrinsically broadened signals from the fibril components cannot be observed.

Conformational transition of hCT during fibril formation in the acidic and neutral conditions were investigated by DD- and CP-MAS ^{13}C NMR using site specifically ^{13}C-labeled hCTs.[24] In aqueous acetic acid solution (pH 3.3), a local α-helical structure is present around Gly,[10] whereas a random coil is dominant as viewed from Phe 22, Ala 26, and Ala 31 in the monomer using DD-MAS NMR on the basis of the conformation-dependent ^{13}C chemical shifts. On the other hand, the CP-MAS spectra are visible from the fibril formed. The ^{13}C resonances of fibrils were shifted to upper field by 1.9 and 1.3 ppm for [1-^{13}C]Gly 10 and [1-^{13}C]Phe 22, respectively, and a local β-sheet form was detected in the fibril at pH 3.3 as viewed from Gly 10 and Phe 22. The results indicate that conformational transitions from the α-helix to β-sheet, and from random coil to β-sheet forms occurred around Gly 10 and Phe 22, respectively, during the fibril formation. Whereas two or three signals that are assigned to be the β-sheet and random coils were observed for Ala 26 and Ala 31. It is noticeable that the random-coil components still remain in the fibril around Ala 26 and Ala 31 residues in addition to the converted β-sheet.[24]

In contrast to the fibril at pH 3.3, the fibril at pH 7.5 formed a local β-sheet conformation at the central region and exhibited a random coil at the C-terminus region. In particular, the signals of [1-^{13}C]Gly 10 and

[1-^{13}C]Phe 22 at pH 7.5 in the CP-MAS spectra are displaced upfield by 0.8 and 1.3 ppm, respectively, as compared with those at pH 3.3. The extent of the conformational heterogeneity of the fibril prepared at pH 7.5 is less than that at pH 3.3 as viewed from their line width, except for Phe 22 C=O. These results indicate that the fibril structure is substantially changed under a particular pH condition. This difference is caused by the changes in molecular interactions among the charged side chains such as Asp 15, Lys 18, and His 20. The homogeneous nucleation process can be attenuated by the unfavorable electrostatic interactions between the positively charged side chains of Lys 18 and His 20, and the amino group of Cys 1 at pH 3.3. On the other hand, at pH 7.5 the side chain of Lys 18 is protonated with a positive charge and that of Asp 15 is deprotonated with a negative charge leading to the fast fibrillation. The electrostatic interaction of the positively and negatively charged side chains between the hCT molecules may assemble the hCT in an antiparallel way at pH 7.5, where the positively charged side chains associate in both parallel and antiparallel ways at pH 3.3 (Figure 16.2). At pH 4.1, antiparallel β-sheet is formed in the central core region, while random coils are persistent, although the random coil region at pH 4.1 was larger than that at pH 7.5.[25] The local conformations and structure of D15N–hCT fibrils at pH 7.5 and 3.2 were found to be similar to each other, and those of hCT at pH 3.3 were interpreted as a mixture of antiparallel and parallel β-sheets, whereas such structures were different from the hCT fibrils at pH 7.5.[26] It is concluded that not only a hydrophobic interaction among the amphiphilic α-helices, but also an electrostatic interaction between charged side chains can play an important role in the fibril formation at pH 7.5, 4.1, and 3.3 acting as electrostatically favorable and unfavorable interaction, respectively, as summarized in Figure 16.2.

The increased or decreased ^{13}C resonance intensities from the fibrils and monomers were recorded by CP- and DD-MAS NMR, respectively, after a certain delay time was observed. In such case, recording both CP- and DD-MAS NMR is essential to record entire ^{13}C NMR signals. The plots suggest that the fibrillation can be explained by a two-step reaction mechanism, in which the first step is a homogeneous association to form a nucleus and the second step is an autocatalytic heterogeneous fibrillation. The most striking feature in the kinetic analysis at pH 3.3 is that the rate constants for the first step, k_1 values, are three to five and one to three orders of magnitude smaller than those of the second step, k_2 and ak_2 (effective rate constant) values, respectively, for the two labeled samples. These results suggest that the first homogeneous nucleation process is much slower than the second heterogeneous fibrillation process. It is also noted that the k_2 values are similar for the samples, whereas the k_1 values are quite different and sensitive to salt concentration or to a small amount of impurity, etc. A fibrillation process

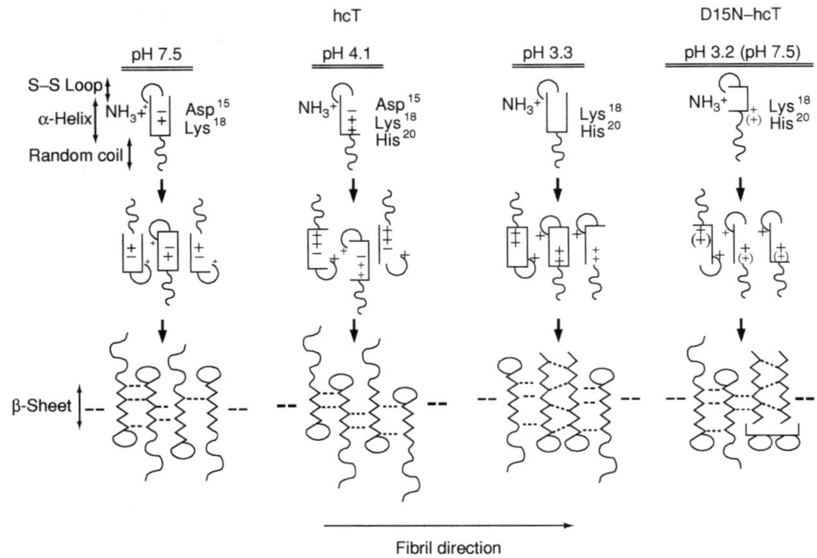

Figure 16.2. Schematic representation for fibril formation of hCT at pH 7.5, 4.1, and 3.3, and D15N-hCT at pH 3.2 and 7.5.

derived by the kinetic analysis as well as the conformational transition is illustrated in Figure 16.2. Using this model, an α-helical bundle (micelle) has to change its conformation from α-helix to a β-sheet simultaneously in the first nucleation process, while one α-helix can be converted to a β-sheet in the second heterogeneous fibrillation process. These findings in ^{13}C NMR experiments imply that it is sufficient to consider the two-step reactions for the fibrillation kinetics.

Time course of the fibril formation at pH 7.5 was also examined by CD measurements, because it is not possible to determine the fast fibrillation rate at pH 7.5 by ^{13}C NMR measurements. Monomeric component assigned to random coil (205 nm) in the CD spectra decreased gradually as a result of fibril formation in the much lower concentration than those for NMR measurements. Consequently, the kinetic parameters are obtained by the same way, and it is found that the fibril formation after the nucleation at pH 7.5 occurred much faster than that at pH 3.3 because k_2 values at pH 7.5 were three order of magnitude larger than that at pH 3.3, although whole k_1 values were not different among them. The k_1 values are not correlated with pH and concentration from the experimental observation.[24]

Molecular packing in the hCT fibril grown at pH 7.5 can be discussed in terms of the accurately measured interatomic distances in the different molecules other than the dipolar pair in the same molecule.[24] ^{13}C and ^{15}N

nuclei in [1-^{13}C]Phe 16 and [^{15}N]Phe 19 of the fibril of the NH_2-DFNKFCOOH fragment of calcitonin are located one residue away from the actual hydrogen bonding location. DFNKF fragment is chosen because this short aromatic charged peptide fragment of calcitonin is known to form amyloid fibril.[27] Therefore, the fibril structure may arise from the manner of molecular packing as shown in the top part of Figure 16.3. This packing scheme clearly shows that most fragments form exactly a head-to-tail anti-parallel β-sheet structure. It is of interest that the phenyl rings of Phe 16 residues are facing each other on the same side of the β-sheet, to be able to contact with π−π interactions as illustrated in Figure 16.3 (top). Actually, the distance between the two phenyl rings was estimated to be 4.6 Å using a precision molecular structure model for proteins, which is a distance that will allow the π−π interactions as illustrated in Figure 16.3. This interaction may be considered as a crucial factor for further stabilization of the hCT fibril in the core region. By assuming that the DFNKF pentapeptide forms the core part

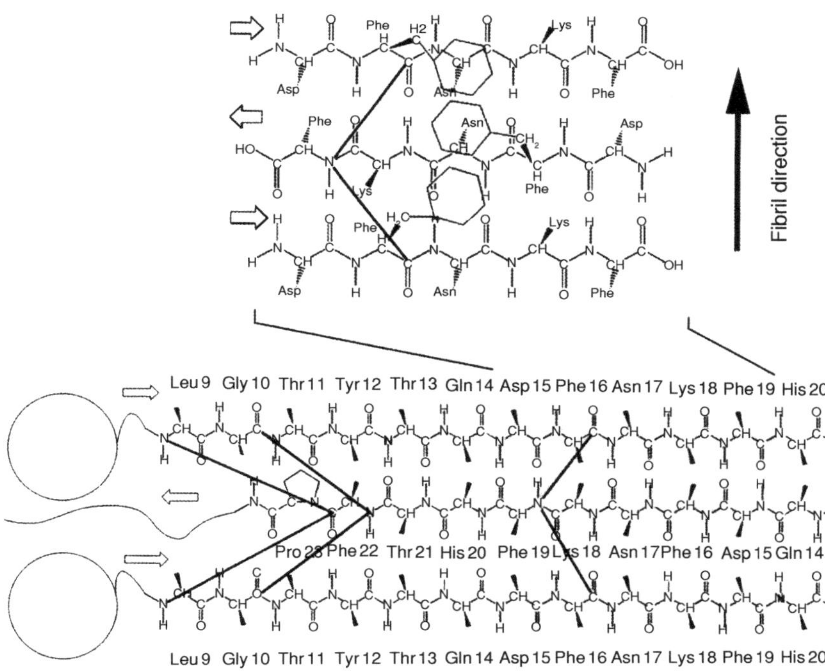

Figure 16.3. Schematic representation of fibril structures of DFNKF (top) and hCT (bottom). The solid lines show the ^{13}C…^{15}N interatomic distances measured in the ^{13}C REDOR experiments, phenyl rings of Phe 16 indicate the possible π−π interactions among molecules.[25]

of hCT fibrils, the whole packing scheme of hCT fibril can be drawn as in the bottom part of Figure 16.3 as an antiparallel β-sheet configuration. It is reasonable to explain that Phe 22 actually forms a hydrogen bond with Tyr 12 and therefore Gly 10 is two residues away and Leu 9, three residues away, showing the weak REDOR effects for intact [1-^{13}C]Gly 10, [^{15}N]Phe 22-hCT, and [^{15}N]Leu 9, [1-^{13}C]Phe 22-hCT.

It is interesting to compare the structure of β-amyloid fibrils with that of hCT fibrils. In the case of amyloid β-peptide [Aβ1–40)] protofilaments, two β-strands form separate parallel β-sheets in a double-layered cross-β-motif.[13] On the other hand, an antiparallel organization of β-sheets formed in Aβ(16–21) fibrils.[9,10] In the case of hCT fibrils, an antiparallel β-sheet structure is formed instead of a parallel β-sheet, as evidenced by REDOR NMR. Because of the existence of a bulky loop at the N-terminus in hCT, the molecular assembly of hCT may be different from that of Aβ peptides.

References

[1] J. D. Sipe, 1992, *Annu. Rev. Biochem.*, 61, 947–975.

[2] R. G. S. Spencer, K. J. Halverson, M. Auger, A. E. McDermott, R. G. Griffin, and P. T. Lansbury, Jr., 1991, *Biochemistry*, 30, 10382–10387.

[3] P. T. Lansbury, Jr., P. R. Costa, J. M. Griffiths, E. J. Simon, M. Auger, K. J. Halverson, D. A. Kocisko, Z. S. Hendsch, T. T. Ashburn, R. G. S. Spencer, B. Tidor, and R. G. Griffin, 1995, *Nat. Struct. Biol.*, 2, 990–998.

[4] J. M. Griffiths, T. T. Ashburn, M. Auger, P. R. Costa, R. G. Griffin, and P. T. Lansbury, Jr., 1995, *J. Am. Chem. Soc.*, 117, 3539–3546.

[5] P. R. Costa, D. A. Kocisko, B. Q. Sun, P. T. Lansbury, Jr., and R. G. Griffin, 1997, *J. Am. Chem. Soc.*, 119, 10487–10493.

[6] T. L. S. Benzinger, D. M. Gregory, T. S. Burkoth, H. Miller-Auer, D. G. Lynn, R. E. Botto, and S. C. Meredith, 1998, *Proc. Natl. Acad. Sci. USA*, 95, 13407–13412.

[7] T. L. S. Benzinger, D. M. Gregory, T. S. Burkoth, H. Miller-Auer, D. G. Lynn, R. E. Botto, and S. C. Meredith, 2000, *Biochemistry*, 39, 3491–3499.

[8] T. S. Burkoth, T. L. S. Benzinger, V. Urban, D. M. Morgan, D. M. Gregory, P. Thiyagarajan, R. E. Botto, S. C. Meredith, and D. G. Lynn, 2000, *J. Am. Chem. Soc.*, 122, 7883–7889.

[9] J. J. Balbach, Y. Ishii, O. N. Antzutkin, R. D. Leapman, N. W. Rizzo, F. Dyda, J. Reed, and R. Tycko, 2000, *Biochemistry*, 39, 13748–13759.

[10] O. N. Antzutkin, J. J. Balbach, R. D. Leapman, N. W. Rizzo, J. Reed, and R. Tycko, 2000, *Proc. Natl. Acad. Sci. USA*, 97, 13045–13050.

[11] R. Tycko and Y. Ishii, 2003, *J. Am. Chem. Soc.*, 125, 6606–6607.

[12] A. T. Petkova, G. Buntkowsky, F. Dyda, R. D. Leapman, W. M. Yau, and R. Tycko, 2004, *J. Mol. Biol.*, 335, 247–260.

[13] A. T. Petkova, Y. Ishii, J. J. Balbach, O. N. Antzutkin, R. D. Leapman, F. Delaglio, and R. Tycko, 2002, *Pro. Natl. Acad. Sci. USA*, 99, 16742–16747.

[14] J. J. Balbach, A. T. Petkova, N. A. Oyler, O. N. Antzutkin, D. J. Gordon, S. C. Meredith, and R. Tycko, 2002, *Biophys. J.*, 83, 1205–1216.
[15] O. N. Antzutkin, J. J. Balbach, R. D. Leapman, N. W. Rizzo, J. Reed, and R. Tycko, 2000, *Proc. Natl. Acad. Sci. USA*, 97, 13045–13050.
[16] S. B. Malinchik, H. Inouye, K. E. Szumowski, and D. A. Kirschner, 1998, *Biophys. J.*, 74, 537–545.
[17] B. Ma and R. Nussinov, 2002, *Proc. Natl. Acad. Sci. USA*, 99, 14126–14131.
[18] R. Tycko, 2003, *Biochemistry*, 42, 3151–3159.
[19] O. N. Antzutkin, R. D. Leapman, J. J. Balbach, and R. Tycko, 2002, *Biochemistry*, 41, 15436–15450.
[20] M. Doi, Y. Kobayashi, Y. Kyogoku, M. Takimito, and K. Gota, 1989, in *Peptides: Chemistry, Structure & Biology*, 11th Proc. Symp. July 9–14, La Jolla, California, pp. 165–167.
[21] A. Motta, P. A. Temussi, E. Wunsch, and G. Bovermann, 1991, *Biochemistry*, 30, 2364–2371.
[22] Y. H. Jeon, K. Kanaori, H. Takashima, T. Koshiba, and Y. A. Nosaka, 1998, *Proc. ICMRBS XVIII*, Tokyo, p. 61.
[23] K. Kanaori and A. Y. Nosaka, 1995, *Biochemistry*, 34, 12138–12143.
[24] M. Kamihira, A. Naito, S. Tuzi, A. Y. Nosaka, and H. Saitô, 2000, *Protein Sci.*, 9, 867–877.
[25] A. Naito, M. Kamihira, R. Inoue, and H. Saitô, 2004, *Magn. Reson. Chem.*, 42, 247–257.
[26] M. Kamihira, Y. Oshiro, S. Tuzi, A. Y. Nosaka, H. Saitô, and A. Naito, 2003, *J. Biol. Chem.*, 278, 2859–2865.
[27] M. Reches, Y. Porat, and E. Gazit, 2002, *J. Biol. Chem.*, 277, 35475–35480.

GLOSSARY

Aβ, *a*myloid *β*-peptide

Ach(o)R, *a*cetyl*cho*line *r*eceptor

Aib, α-*A*mino*i*so*b*utyric acid

$(Ala)_n$, poly(L-alanine)

AMCP, *a*mplitude *m*odulated *CP*

Bchl, *b*acterio*chl*orophyll

bicelles, *bi*layered mi*celles*

bO, *b*acterio*o*psin

bR, *b*acterio*r*hodopsin

AHH CP, *a*diabatic passages through the *H*artmann–*H*ahn *c*ross *p*olarization

BLYP, *B*eck and *L*ee–*Y*ang–*P*arr gradient-corrected exchange correlation functionals

CM, *c*apacitance *m*inimization

CHAPSO, 3-[(3-cholamidopropyl)-dimethylammonio]–1-propanesulfonate

CD, *c*ircular *d*ichroism

CM, *c*apacitance *m*inimization

COSY, *co*rrelated *s*pectroscop*y*

CP, *c*ross *p*olarization

CP-MAS, *c*ross *p*olarization-*m*agic *a*ngle *s*pinning

CPMG, *C*arr–*P*urcell–*M*eiboom–*G*ill

CRAMPS, *c*ombined *r*otational *a*nd *m*ulti*p*ulse *s*pectroscopy

CSA, *c*hemical *s*hift *a*nisotropy

C7, seven-fold symmetric irradiation

CT, *c*alci*t*onin

CTDQFD, *c*onstant-*t*ime *d*ouble-quantum-*f*iltered *d*ipolar recoupling

CW, *c*ontinuous *w*ave

DANTE, *d*elays *a*lternating with *n*utation for *t*ailored *e*xcitation

DARR, *d*ipolar *a*ssisted *r*otational *r*esonance

DAS, *d*ynamic *a*ngle *s*pinning

DD, *d*ipolar *d*ecoupled

DDph, *d*ipolar *d*e*ph*asing

DD-MAS, *d*ipolar *d*ecoupled-*m*agic *a*ngle *s*pinning

DECORDER, *d*irection *e*xchange with *c*orrelation for *o*rientation *d*istribution *e*valuation and *r*econstruction

DFT, *d*ensity *f*unctional *t*heory

DGK, *d*iacyl*g*lycerol *k*inase

DIPSHIFT, *dip*olar chemical *shift*

DM, *n*-*d*ecyl-β-*m*altoside

DMSO, *dim*ethyl*s*ulf*o*xide

DNA, *d*eoxyribo*n*ucleic *a*cid

DMEM, *D*ulbecco's *m*odified *E*agle's *m*edium

DMPC, 1,2-*dim*yristoyl-sn-glycero–3-*p*hospho*c*holine

DMPG, 1,2-*dim*yristoyl-sn-glycero–3-[*p*hospho-*rac*-(1-*g*lycerol)]

DMPS, 1,2-*d*imyristoyl-sn-glycero–3-[*p*hosphor-L-*s*erine]

DPPC, 1,2- *dip*almitoyristoyl-sn-glycero–3-*p*hospho*c*holine

DHPC, 1,2-*dih*exanoyl-sn-glycero–3-*p*hospho*c*holine

2D NMR, two-dimensional NMR

DOQSY, *d*ouble-*q*uantum *s*pectroscop*y*

DOR, *do*uble *r*otation

DP or DP_n, *d*egree of *p*olymerization or number average degree of polymerization

DQCSA, *d*ouble-*q*uantum *c*hemical *s*hift *a*nisotropy

DQSY, *d*ouble-*q*uantum *s*pectroscop*y*

DQDRAWAS, *d*ouble-*q*uantum *d*ipolar *r*ecoupling with *a* windowles*s* (multiple irradiation)

DRAMA, *d*ipolar *r*ecovery *a*t the *m*agic *a*ngle

DRAWS, *d*ipolar *re*coupling with *a* *w*indowles*s* (multiple irradiation)

DREAM, *d*ipolar *r*ecoupling *e*nhancement through *a*mplitude *m*odulation

EFG, *e*lectric *f*ield *g*radient

EM, *e*lectron *m*icroscope

EmrE, E*scherichia coli m*ultidrug *r*esistance protein *E*

FPT INDO, *f*inite *p*erturbation *t*heory *i*ntermediate *n*eglect of *d*ifferential *o*verlap

fpRFDR-CT, *c*onstant-*t*ime *f*inite-*p*ulse *r*adio*f*requency-*d*riven *r*ecoupling

FSLG, *f*requency-*s*witched *L*ee–*G*oldburg

FSR, *f*requency *s*elective *R*EDOR

FWHM, *f*ull *w*idth at *h*alf *m*aximum

gA, *g*ramicidin *A*

GIAO, *g*auge *i*ndependent *a*tomic *o*rbital

GIAO-MP2, *g*auge *i*ndependent *a*tomic *o*rbital-second-order perturbation expansion by *M*oller–*P*lesset

$(Gly)_n$, poly(glycine)

GPCR, *G*-*p*rotein *c*oupled *r*eceptor

HA, *h*em*a*gglutinin

HB, *h*ydrogen *b*ond

HETCOR, *het*eronuclear *cor*relation

HGT, *h*igh-*G*ly-*T*yr protein

HHCP, *H*artmann–*H*ahn *c*ross *p*olarization

HIV–1, *h*uman *i*mmunodeficiency *v*irus type 1

HMQ, *h*eteronuclear *m*ultiple *q*uantum

hCT, *h*uman *c*alci*t*onin

INADEQUATE, *i*ncredible *n*atural *a*bundance doubl*e* *qua*ntum *t*ransfer *e*xperiment

LC, *l*iquid *c*rystal

LCAO, *l*inear *c*ombination of *a*tomic *o*rbital

LG decoupling, *L*ee–*G*oldburg decoupling

LH2 complex, *l*ight *h*arvesting complex

MAS, *m*agic *a*ngle *s*pinning

MELODRAMA, *mel*ding *o*f spin-locking and *DRAMA*

MO, *m*olecular *o*rbital

MPL, *m*ass *p*er *l*ength

MREV, *M*ansfield–*R*him–*E*lleman–*V*aughan

MQ, *m*ultiple *q*uantum

MQMAS, *m*ultiple *q*uantum *m*agic *a*ngle *s*pinning

NMR, *n*uclear *m*agnetic *r*esonance

NOE, *n*uclear *O*verhauser *e*nhancement

NOESY, *n*uclear *O*verhauser and *e*xchange spectroscop*y*

ORD, optical *r*otatory *d*ispersion

PAS, *p*rincipal *a*xis *s*ystem

PBLG, *p*oly(γ-*b*enzyl L-*g*lutamate)

PDSD, *p*roton-*d*riven *s*pin *d*iffusion

Photo-CIDNP, photo*c*hemically *i*nduced *d*ynamic *n*uclear *p*olarization

pHtrII, *pharaonis* cognate transducer

PISA, *p*olarity *i*ndex *s*lant *a*ngle

PISEMA, *p*olarization *i*nversion *s*pin *e*xchange at the *m*agic *a*ngle

PLB, *p*hospho*l*am*b*in

PM, *p*urple *m*embrane

PMLG, *p*hase-*m*odulated *L*ee–*G*oldburg

POLG, *p*oly(γ-*o*ctadecyl L-*g*lutamate)

POLLG, *p*oly(γ-*o*leyl L-*g*lutamate)

POPC, 1-*p*almitoyl–2-*o*leyl-sn-glycero–3-*p*hospho*c*holine

PDA, *p*oly(D-*a*lanine)

PLA, *p*oly(L-*a*lanine)

PG, *p*oly*g*lycine

PLIL, *p*oly(L-*i*so*l*eucine)

PLV, *p*oly(L-*v*aline)

*p*pR, p*haraonis* *p*hobo*r*hodopsin

pSB, *p*rotonated *S*chiff *b*ase

p-ZQ, *p*seudo *z*ero *q*uantum

RACO, *r*elayed *a*nisotropy *co*rrelation

RC, (photosynthetic) *r*eaction *c*enter

RDX, REDOR of $^{13}C_x$

REDOR, *r*otationary *e*cho *do*uble *r*esonance

rf, *r*adio *f*requency

RFDR, *r*adio *f*requency *d*riven *r*esonance

Rho, *rho*dopsin

RMSD, *r*oot-*m*ean-*s*quare-*d*eivation

RNA, *r*ibo*n*ucleic *a*cid

ROCSA, *r*ecoupling *o*f *c*hemical *s*hift *a*nisotropy

RR or R^2, *r*otational *r*esonance

RTK, *r*eceptor *t*yrosine *k*inase

SB, *S*chiff *b*ase

SCMK, *S*-(*c*arboxyl-*m*ethyl)*k*eratin

SCMKA, low-sulfur protein of SCMK

SCMKB, high-sulfur protein of SCMK

SCMKM, protecting the thiol groups with iodoacetic acid to form SCMK

SEDRA, *s*imple *e*xcitation for the *d*ephasing of *r*otational echo *a*mplitude

SEMA, *s*pin *e*xchange at *m*agic *a*ngle

SERCA, *s*arco(*e*ndo)plasmic *r*eticulum

SIDY, *s*elective *i*nterface *d*etection spectroscop*y*

SIMPSON, *s*imulation *p*rogram for *so*lid state *N*MR spectroscopy

SLF, *s*eparated-*l*ocal-*f*ield

S/N, *s*ignal to *n*oise

SR, *s*ynchrotron *r*adiation

SOS, *s*um-*o*ver-*s*tates

SPECIFIC CP, *spec*trally *i*nduced *f*iltering in *c*ombination with *c*ross *p*olarization

SPICP, *s*ynchronous *p*hase *i*nversion *CP*

SFAM REDOR, *s*imultaneous *f*requency *a*mplitude *m*odulation *r*otational *e*cho *d*ouble *r*esonance

STEM, *s*canning *t*ransmission *e*lectron *m*icroscope

STMAS, *s*atellite *t*ransitions and *m*agic *a*ngle *s*pinning

TEDOR, *t*ransferred *e*cho *do*uble *r*esonance

TBMO, *t*ight-*b*inding *m*olecular *o*rbital

TMV, *t*obacco *m*osaic *v*irus

TM, *t*rans*m*embrane

TFA, *tri*f*luoro*a*cetic acid

TFE, 2,2,2-*trif*luoro*e*thanol

TMP, *t*ransme*m*brane *p*eptide

TMS, *t*etra*m*ethyl*s*ilane

TOBSY, *to*tal through-*b*ond correlation *s*pectroscop*y*

TOCSY, *to*tal *c*orrelated *s*pectroscop*y*

TOSS, *to*tal *s*uppression of *s*pinning sidebands

TPPM, *t*wo-*p*ulse *p*hase-*m*odulation

U2QF COSY, *u*niform-sign cross-peak double-*q*uantum-*f*iltered *co*rrelation *s*pectroscop*y*

VACP, *v*ariable-*a*mplitude *CP*

$(\mathrm{Val})_n$, poly(L-valine)

WAXD, *w*ide-*a*ngle *X*-ray *d*iffraction

WISE, *wi*de-line *se*paration

XiX, *X*-*i*nverse-*X*

ZQ, *z*ero *q*uantum

One-letter description of amino acids

A	Ala
C	Cys
D	Asp
E	Glu
F	Phe
G	Gly
H	His
I	Ile
K	Lys
L	Leu
M	Met
N	Asn
P	Pro
Q	Gln
R	Arg
S	Ser
T	Thr
W	Trp
Y	Tyr
V	Val

INDEX

MIX
Papier aus verantwortungsvollen Quellen
Paper from responsible sources
FSC® C105338

If you have any concerns about our products,
you can contact us on
ProductSafety@springernature.com

In case Publisher is established outside the EU,
the EU authorized representative is:
Springer Nature Customer Service Center GmbH
Europaplatz 3, 69115 Heidelberg, Germany

Printed by Libri Plureos GmbH
in Hamburg, Germany